Vesicle Trafficking in Cancer

Yosef Yarden • Gabi Tarcic
Editors

Vesicle Trafficking in Cancer

Springer

Editors
Yosef Yarden
Department of Biological Regulation
Weizmann Institute of Science
Rehovot, Israel

Gabi Tarcic
Department of Biological Regulation
Weizmann Institute of Science
Rehovot, Israel

ISBN 978-1-4614-6527-0 ISBN 978-1-4614-6528-7 (eBook)
DOI 10.1007/978-1-4614-6528-7
Springer New York Heidelberg Dordrecht London

Library of Congress Control Number: 2013933446

Printed on acid-free paper

Springer is part of Springer Science+Business Media (www.springer.com)

Preface

The engulfment of small portions of the plasma membrane, along with an assortment of specific surface proteins, and their packaging in vesicles, which subsequently travel to various organellar destinations, is a vital process tuned by multiple lipids, nucleotides, and proteins, which undergo ubiquitination, phosphorylation, and other reversible covalent modifications. The highly complex endocytic process is both ubiquitous and robust. The reason why endocytosis and intracellular trafficking are essential for life is dictated by the nature of their diverse surface-bound cargo; nutrient receptors and transporters, intercellular and matrix adhesion molecules, as well as a plethora of receptors for growth factors, chemokines, and cytokines are all transported to and from the plasma membrane by means of vesicular trafficking. Hence, the endocytic process critically regulates metabolism, signal transduction, and cell polarity/migration. The remarkable robustness of intracellular trafficking is the outcome of a unique hub-centric design: distinct modular hubs (e.g., AP2-EPS15 and HRS-LST2) comprising a phosphoinositol-binding, membrane-anchoring component, an ubiquitin-binding module, and a machinery enabling homo-assembly are distributed along the pathway. Their bistable regulation entails a set of small GTP binders of the RAB family. Scheduled hub transitions define points of commitment to vesicle docking, fusion, scission, and and subtly manipulates conferring the characteristic unidirectional nature of intracellular trafficking.

While viruses and other cellular invaders utilize the endocytic machinery as their port of entry, according to observations reported over the last decade, cancer multiplies and subtly manipulates vesicular trafficking to imbalance energy and metabolism, signal transduction, and cellular invasion. This volume reviews the plethora of molecular mechanisms that manipulate vesicular trafficking in tumors. The notion of derailed endocytosis in cancer first emerged from studies of receptors for growth factors, such as the epidermal growth factor receptor (EGFR) and c-MET, the receptor for the hepatocyte growth factor (HGF). However, later studies extended the notion to additional families of surface molecules, such as integrins and cadherins, and G-protein-coupled receptors. At this rather initial phase of research, it is still difficult generalizing the strategies enabling deliberate manipulation of the endocytic process by malignant transformation. Nevertheless, it seems safe arguing that malignancies target the fundamental attributes ensuring unidirectional cargo progress, such as phosphoinositol metabolism, specific RAB proteins, certain E3 ubiquitin ligases or deubiquitinating enzymes (DUBs), as well as the actin/tubulin meshwork.

Importantly, mammalian cells maintain several endocytic pathways and portals, such as the caveolae-mediated pathway and macropinocytosis, but the best understood and apparently most relevant to cancer is the clathrin-mediated pathway. In the opening chapter of this book, **Alexander Sorkin** and **Manojkumar Puthenveedu** review the sequence of events taking place within clathrin-coated areas of the plasma membrane when cargo is actively recruited to the coated area. They also describe how such receptors influence nucleation, stabilization, and size of the clathrin-coated pits, which are considered the bottleneck of receptor endocytosis. Accordingly, overexpression of c-MET and other growth factor receptors, a frequent aberration in carcinomas, appears to saturate the pit and thereby slows down the rate of receptor inactivation. In another chapter, **Tal Hirschhorn** and **Marcelo Ehrlich** describe one of the major regulators of endocytosis, namely, actin dynamics, which control not only invagination and scission but also vesicle movement along actin cables. One critical regulator, the huntingtin-interacting protein-1 related (Hip1R), binds both clathrin and actin, hence acting as a coordinator of actin remodeling and vesicle dynamics. Remarkably, overexpression of HIP1R has been observed in brain and prostate tumors. In a subsequent chapter, **Eli Zamir**, **Nachiket Vartak**, and **Philippe I. H. Bastiaens** highlight the importance of the concentration parameter of membrane proteins. This is determined by the spatial distribution of proteins, their translocation to membrane surfaces, and the interactions between mutant and wild-type versions.

Giorgio Scita and colleagues explain how endocytosis and recycling ensure the asymmetric distribution of membrane proteins, which is crucial for proper polarized cellular functions, including directed cell migration. Viewing the topic from a different perspective, **Shreya Mitra** and **Gordon Mills** propose that abnormal vesicular trafficking disturbs cell polarity by eliminating tight junctions and diminishing apical-basal polarity, which regulate a myriad of cellular functions, including metabolism and asymmetric division of stem cells. Concentrating on cadherins and integrins, especially on their roles in epithelial to mesenchymal transition, a process thought to precede epithelial cell migration and metastasis, **Crislyn D'Souza-Schorey** and **Guangpu Li** argue that sustained signaling from endosomes leads to the formation of invasive structures reminiscent of tumorigenic phenotypes. In line with this view, **Iwona Pilecka** and **Marta Miaczynska** consider endosomes as platforms that can sustain signals generated by internalized G-protein-coupled receptors and receptor tyrosine kinases, thereby enhancing downstream biological outcomes, such as cell migration. They also highlight the presence of endosomal proteins in the nucleus, where they might regulate transcription or chromatin remodeling. **Ying-Nai Wang, Jennifer Hsu, and Mien-Chie Hung** extend this to evidence favoring shuttling of internalized receptors, for example, EGFR and HER2, a cancer-promoting kin of EGFR, to the nucleus and to other subcellular compartments, where they act as transcriptional regulators. Furthermore, they review evidence linking the translocation of EGFR into the nucleus with poor clinical prognosis, as well as with the outcome of anticancer treatments (e.g., specific kinase inhibitors and monoclonal antibodies).

Cargo ubiquitination, along with ubiquitination of the endocytic machinery, has emerged in the last decade as drivers or by-products of malignant transformation. This is the reason why four chapters of this volume are devoted to ubiquitin-mediated regulation of the endocytic process. **Daniela Hoeller** and **Ivan Dikic** provide an introduction to the versatility of the ubiquitin system and the crosstalk to other posttranslational modification. In another chapter, **Elena Maspero**, **Hans-Peter Wollscheid,** and **Simona Polo** describe ubiquitination of a vast array of mammalian signaling receptors, such as growth factor receptors, G-protein-coupled receptors, NOTCH, various channels, and interferon receptors. They highlight putative roles of monoubiquitination of a set of endocytic adaptors, which bind ubiquitin. In addition, they review implications for cancer, such as the ability of HER2, an internalization-defective receptor, to shunt internalized EGFRs to the recycling route. Several E3 ubiquitin ligases play critical roles in the regulation of endocytosis and malignancy. The list includes Hakai, Nedd4, Deltex, and the CBL family. **Stanley Lipkowitz** and colleagues devote a chapter to the three CBL proteins, emphasizing, on the one hand, their regulation by phosphorylation and more than 50 interacting proteins, and, on the other hand, the variety of oncogenic mutations that inactivate CBL. Remarkably, over the past five years, CBL mutations have been found in ~5% of a wide variety of myeloid neoplasms, including the myelodysplastic syndrome, myelofibrosis, refractory anemia, acute myeloid leukemia, atypical chronic myelogenous leukemia (aCML), and juvenile myelomonocytic leukemia (JMML; up to 15%).The action of CBL and other E3 ubiquitin ligases is reversed by a large set of deubiquitinating enzymes (DUBs). **Han Liu**, **Sylvie Urbé**, and **Michael Clague** describe in depth the DUBs engaged in the regulation of vesicular trafficking. For example, two endosome localized DUBs, AMSH and USP8, accelerate recycling of receptors by reducing active sorting to lysosomal degradation. Predictably, such DUBs might act as tumor suppressors, but their actions appear more complicated than expected, as clarified by Clague and colleagues.

To deepen the description of aberrant endocytosis, several chapters of the book concentrate on specific cargos: c-MET, EGFR, and integrins. **Stéphanie Kermorgant** and colleagues underscore the importance of c-MET's signaling from endosomes, as well as the ability of certain oncogenic mutants of the receptor to enhance downstream signals, by means of defying normal endocytosis. Similarly, **Sergio Anastasi**, **Stefano Alemà**, and **Oreste Segatto** present endocytosis as an element of spatial receptor regulation. They focus on the diverse mechanisms through which receptors escape from downregulation in cancer cells. Specifically, they highlight the intrinsic refractoriness of brain and lung mutants of EGFR to endocytosis-mediated downregulation. In contrast to the well-understood behavior of ligand-activated receptors, ligand-independent activation mechanisms of EGFR remain poorly understood. **Tzipora Goldkorn**, **Simone Filosto**, and **Samuel Chung** highlight stress-dependent activation, internalization, and trafficking of EGFR. Accordingly, under cellular oxidative stress, p38-MAPK, c-SRC, caveolin-1, and ceramides, membrane sphingolipids generated during oxidative stress undertake driver roles in vesicular sorting. In a subsequent chapter, **Elena Rainero**, **Peter V.E. van den Berghe**, **and Jim Norman** argue that endocytosis and recycling of integrins is important during tumor progression and clarify how mutations of p53 drive invasion and metastasis by altering integrin and EGFR recycling.

Finally, because of the pivotal roles played by derailed endocytosis in cancer, future studies will likely translate this new understanding to improved cancer therapy, as well as to efforts that link aberrant trafficking to mechanisms promoting patient resistance to specific drugs. An exemplification is provided in Jim Norman's chapter: the overall lack of efficacy of anti-integrin drugs (i.e., $\alpha v\beta 3$ inhibitors) in tumor angiogenesis is attributable to enhanced recycling of both $\alpha 5\beta 1$ integrin and receptor tyrosine kinases. Along this vein, the closing chapter, written by **Gabi Tarcic** and **Yosef Yarden**, discusses the possibility that antireceptor monoclonal antibodies commonly used to treat various types of cancer (e.g., trastuzumab, an anti-HER2 antibody) actually force endocytosis of their oncogenic antigens and subsequently target them to degradation in lysosomes. Thus, future studies will not only shed new, endocytosis-relevant light on molecular modes of oncogenic processes; they also might open new avenues in cancer therapy.

Contents

Contributors

Stefano Alemà Institute of Cell Biology and Neurobiology, CNR, Monterotondo, Italy

Sergio Anastasi Department of Experimental Oncology, Regina Elena Cancer Institute, Rome, Italy

Department of Physiology and Pharmacology, University of Rome La Sapienza, Rome, Italy

Rachel Barrow Spatial Signalling Team, Centre for Tumour Biology, Barts Cancer Institute – a Cancer Research UK Centre of Excellence, Queen Mary University of London, John Vane Science Centre, London UK

Philippe I.H. Bastiaens Department of Systemic Cell Biology, Max Planck Institute of Molecular Physiology, Dortmund, Germany

Technische Universität Dortmund, Dortmund, Germany

Peter V.E. van den Berghe Beatson Institute for Cancer Research, Glasgow, UK

Samuel Chung Center for Comparative Respiratory Biology and Medicine (CCRBM), University of California School of Medicine, Davis, CA, USA

Michael J. Clague Cellular and Molecular Physiology, Institute of Translational Medicine, The University of Liverpool, Liverpool, UK

Crislyn D'Souza-Schorey Department of Biological Sciences, University of Notre Dame, Notre Dame, IN, USA

Ivan Dikic Buchmann Institute for Molecular Life Sciences and Institute of Biochemistry II, Goethe University School of Medicine, Frankfurt, Germany

Andrea Disanza IFOM Foundation, Institute FIRC of Molecular Oncology, Milan, Italy

Marcelo Ehrlich Department of Cell Research and Immunology, Tel Aviv University, Tel Aviv, Israel

Simone Filosto Center for Comparative Respiratory Biology and Medicine (CCRBM), University of California School of Medicine, Davis, CA, USA

Emanuela Frittoli IFOM Foundation, Institute FIRC of Molecular Oncology, Milan, Italy

Tzipora Goldkorn Center for Comparative Respiratory Biology and Medicine (CCRBM), University of California School of Medicine, Davis, CA, USA

Signal Transduction Internal Medicine, University of California School of Medicine, Davis, CA, USA

Tal Hirschhorn Department of Cell Research and Immunology, Tel Aviv University, Tel Aviv, Israel

Daniela Hoeller Buchmann Institute for Molecular Life Sciences and Institute of Biochemistry II, Goethe University School of Medicine, Frankfurt, Germany

Jennifer L. Hsu Department of Molecular and Cellular Oncology, The University of Texas MD Anderson Cancer Center, Houston, TX, USA

Center for Molecular Medicine and Graduate Institute of Cancer Biology, China Medical University, Taichung, Taiwan, R.O.C

Mien-Chie Hung Department of Molecular and Cellular Oncology, The University of Texas MD Anderson Cancer Center, Houston, TX, USA

Center for Molecular Medicine and Graduate Institute of Cancer Biology, China Medical University, Taichung, Taiwan, R.O.C

Carine Joffre Spatial Signalling Team, Centre for Tumour Biology, Barts Cancer Institute – a Cancer Research UK Centre of Excellence, Queen Mary University of London, John Vane Science Centre, London, UK

Institut Curie, INSERM U-830, Paris, France

Stephen C. Kales Laboratory of Cellular and Molecular Biology, Center for Cancer Research, National Cancer Institute, National Institutes of Health, Bethesda, MD, USA

Stéphanie Kermorgant Spatial Signalling Team, Centre for Tumour Biology, Barts Cancer Institute – a Cancer Research UK Centre of Excellence, Queen Mary University of London, John Vane Science Centre, London, UK

Guangpu Li Department of Biochemistry and Molecular Biology, University of Oklahoma Health Sciences Center, Oklahoma City, OK, USA

Stanley Lipkowitz Laboratory of Cellular and Molecular Biology, Center for Cancer Research, National Cancer Institute, National Institutes of Health, Bethesda, MD, USA

Han Liu Cellular and Molecular Physiology, Institute of Translational Medicine, The University of Liverpool, Liverpool, UK

Ke Ma Laboratory of Cellular and Molecular Biology, Center for Cancer Research, National Cancer Institute, National Institutes of Health, Bethesda, MD, USA

Elena Maspero IFOM, Fondazione Istituto FIRC di Oncologia Molecolare, Milan, Italy

Ludovic Ménard Spatial Signalling Team, Centre for Tumour Biology, Barts Cancer Institute – a Cancer Research UK Centre of Excellence, Queen Mary University of London, John Vane Science Centre, London, UK

Marta Miaczynska International Institute of Molecular and Cell Biology, Laboratory of Cell Biology, Warsaw, Poland

Gordon B. Mills The Department of Systems Biology, University of Texas MD Anderson Cancer Center, Houston, TX, USA

Shreya Mitra The Department of Systems Biology, University of Texas MD Anderson Cancer Center, Houston, TX, USA

Marion M. Nau Laboratory of Cellular and Molecular Biology, Center for Cancer Research, National Cancer Institute, National Institutes of Health, Bethesda, MD, USA

Jim C. Norman Beatson Institute for Cancer Research, Glasgow, UK

Andrea Palamidessi IFOM Foundation, Institute FIRC of Molecular Oncology, Milan, Italy

Iwona Pilecka International Institute of Molecular and Cell Biology, Laboratory of Cell Biology, Warsaw, Poland

Simona Polo IFOM, Fondazione Istituto FIRC di Oncologia Molecolare, Milan, Italy

Dipartimento di Medicina, Chirurgia ed Odontoiatria, Universita' degli Studi di Milano, Milan, Italy

Manojkumar A. Puthenveedu Department of Biological Sciences, Carnegie Mellon University, Pittsburgh, PA, USA

Elena Rainero Beatson Institute for Cancer Research, Glasgow, UK

Giorgio Scita IFOM Foundation, Institute FIRC of Molecular Oncology, Milan, Italy

Dipartimento di Medicina, Chirurgia ed Odontoiatria, Università degli Studi di Milano, Milan, Italy

Oreste Segatto, MD Department of Experimental Oncology, Regina Elena Cancer Institute, Rome, Italy

Sara Sigismund IFOM Foundation, Institute FIRC of Molecular Oncology, Milan, Italy

Alexander Sorkin Department of Cell Biology, University of Pittsburgh School of Medicine, Pittsburgh, PA, USA

Gabi Tarcic Department of Biological Regulation, Weizmann Institute of Science, Rehovot, Israel

Sylvie Urbé Cellular and Molecular Physiology, Institute of Translational Medicine, The University of Liverpool, Liverpool, UK

Nachiket Vartak Department of Systemic Cell Biology, Max Planck Institute of Molecular Physiology, Dortmund, Germany

Ying-Nai Wang Department of Molecular and Cellular Oncology, The University of Texas MD Anderson Cancer Center, Houston, TX, USA

Center for Molecular Medicine and Graduate Institute of Cancer Biology, China Medical University, Taichung, Taiwan, R.O.C

Hans-Peter Wollscheid Dipartimento di Medicina, Chirurgia ed Odontoiatria, Universita' degli Studi di Milano, 20122 Milan, Italy

Yosef Yarden Department of Biological Regulation, Weizmann Institute of Science, Rehovot, Israel

Eli Zamir Department of Systemic Cell Biology, Max Planck Institute of Molecular Physiology, Dortmund, Germany

Chapter 1
Clathrin-Mediated Endocytosis

Alexander Sorkin and Manojkumar A. Puthenveedu

Abstract Clathrin-mediated endocytosis is the main portal of entry into the cell for many soluble and membrane molecules. Clathrin-coated vesicles are formed from the plasma membrane in a sequence of coordinated protein-lipid and protein-protein interactions, starting with adaptor-mediated recruitment of clathrin to the membrane, proceeding to clathrin polymerization and assembly into deeply curved coated buds, and ending with the dynamin-dependent scission of a coated vesicle. Clathrin coats trap and concentrate endocytic cargo by using a multitude of adaptor proteins that recognize specific sequence motifs in the cytosolic domains of receptors and other transmembrane cargo molecules. Endocytic cargo that is concentrated in this manner, such as signaling receptors, may regulate the stability, size, and dynamics of individual clathrin coats and thereby influence endocytosis.

1.1 Introduction

Endocytosis is a process by which cells internalize extracellular and cell-surface materials. These materials include membrane proteins, which comprise of a third of the genome. Many of these proteins have critical functions at the plasma membrane or need to pass through the cell surface on the way to their intracellular sites of function. Therefore, the localization and function of all these proteins (also referred as endocytic "cargo") are regulated by endocytosis. Endocytic cargo is captured in small areas of the plasma membrane that invaginate to form buds, which eventually

A. Sorkin (✉)
Department of Cell Biology, University of Pittsburgh School of Medicine, Pittsburgh, PA, USA
e-mail: sorkin@pitt.edu

M.A. Puthenveedu (✉)
Department of Biological Sciences, Carnegie Mellon University, Pittsburgh, PA, USA
e-mail: map3@andrew.cmu.edu

Y. Yarden and G. Tarcic (eds.), *Vesicle Trafficking in Cancer*,
DOI 10.1007/978-1-4614-6528-7_1, © Springer Science+Business Media New York 2013

pinch off to form vesicles that move into the cytoplasm. Newly formed endocytic vesicles typically fuse with early endosomes, and the endocytosed cargo is then sorted into recycling, lysosomal, or other trafficking pathways.

Endocytic vesicles are formed by several mechanisms. Clathrin-mediated endocytosis (CME) is an evolutionary conserved pathway that is the main and best characterized pathway of endocytosis, although several clathrin-independent endocytosis pathways have also been described [1, 2]. CME is essential for many fundamental cellular processes, such as synaptic transmission, signal transduction, nutrient uptake into the cell, and membrane homeostasis. It is the fastest and highly regulated pathway of endocytosis. The specific internalization rate constants measured for the CME cargo, such as transferrin and epidermal growth factor (EGF) receptors, can be as high as 0.3–0.4 min^{-1} (30–40% of surface receptors are endocytosed in 1 min) [3].

The process of formation of an endocytic clathrin-coated vesicle (CCV) has been studied intensively for about 50 years using various methodologies. CCV is formed by the assembly of clathrin-coated pits (CCPs)—small areas of the cytosolic surface of the plasma membrane that are covered by a protein coat consisting mainly of clathrin—followed by their scission from the membrane generating free vesicles. In recent years, remarkable progress has been made in understanding the molecular mechanisms of this process through the development of two main experimental approaches: One, resolving the atomic structures of clathrin and other components of clathrin assembly has ultimately led to the high-resolution structure of an entire CCV formed in vitro. Two, total interference reflection fluorescence (TIRF) and other live-cell optical microscopy methods have allowed detailed real-time imaging of the CME process and have unveiled a sequence of events and protein recruitment to the clathrin coat during endocytosis with very high temporal resolution. Many comprehensive review articles have been dedicated to the mechanisms of CME (for example [2, 4, 5]). In this chapter, we will discuss recent advances in understanding the molecular machineries involved in CME in mammalian cells. The endocytosis of "signaling" receptors, e.g., receptors that are capable of triggering cellular signal transduction processes, will be specifically described, and examples of the regulatory mechanisms by which these receptors can modulate CME will also be illustrated.

1.2 Structure of Clathrin and Clathrin Vesicles In Vitro

Clathrin is the main component of coated pits and vesicles. It is a soluble cytosolic protein that cannot bind lipids directly and is therefore recruited to the membrane by a specialized set of proteins called adaptors (discussed below) to assemble into coated pits and buds. The ability of clathrin to polymerize into a lattice and assemble into vesicles in vitro under mildly acidic pH allowed comprehensive biochemical analysis of the assembled state of clathrin. Elucidation of the high-resolution structure of the clathrin molecule and the clathrin lattice has led to a remarkable level of understanding of the mechanisms of lattice assembly and clathrin interactions

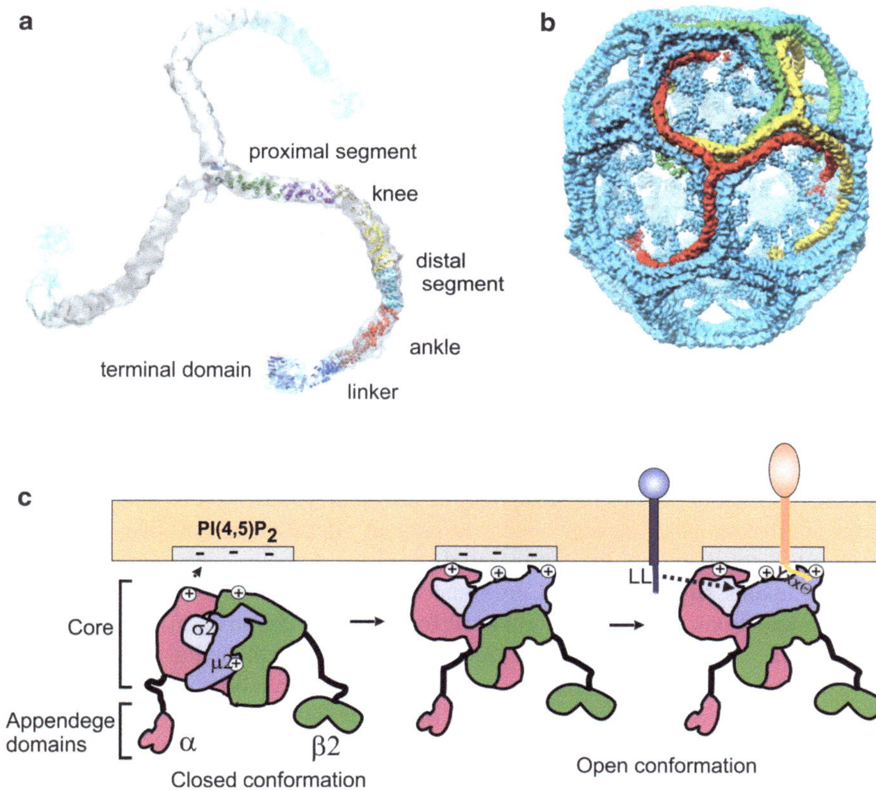

Fig. 1.1 Schematic structures of clathrin triskelion, polymerized clathrin, and AP-2. (**a**) The clathrin triskelion consists of three CHC and three CLC (not shown) molecules. Names of segments of one heavy chain are indicated with the terminal domain at the amino-terminus and the vertex at the carboxyl-terminus (adapted from [6, 11] with permission from *Nature*). (**b**) Image reconstruction at 7.9 Å resolution of a clathrin coat (in a shape of "hexagonal barrel") assembled in vitro from the bovine-brain clathrin (heavy chains only) and AP-2 using cryo-electron microscopy images. There are 36 clathrin triskelions in the structure. The colored triskelions show three independent triskelions. Noisy central density, from spatially disordered and substoichiometric AP-2 complexes, has been flattened (adapted from [6, 11] with permission from *Nature*). (**c**) Clathrin adaptor complex AP-2 consists of the core and appendage domains connected to the core by flexible hinge domains. AP2 binds initially to negatively charged PI (4,5)P2 headgroups in the membrane mainly through the positively charged interfaces in the α subunit and, additionally, in the β2 subunit. The large conformational change in AP-2 is then triggered by the electrostatic attraction of C-terminal lipid-binding patches of μ2 to the negatively charged membrane, which results in an "open" conformation form of AP-2. AP-2 in open conformation can bind tyrosine- and leucine-based internalization motifs present in the cytosolic domains of the membrane cargo

involved in this process [6]. Clathrin functions as a heterohexamer, adopting a three-legged triskelion structure of three heavy chains and three light chains [7, 8] (Fig. 1.1a). Approximately 100 clathrin triskelion units are present in the assembled coated vesicle.

Clathrin heavy chains (CHC) serve as structural element of the clathrin lattice, whereas clathrin light chains have mainly regulatory functions.

The CHC consists of eight CHC repeat (CHCR) motifs (CHCR0-7) forming a right-handed super-helix coil of α-helices [6, 9], arranged into proximal, distal, and terminal domains [10]. The proximal domains (CHCR6 and 7) of three CHC molecules interact to form the vertex of the triskelion. CHCR5 makes the "knee," CHCR3 and 4 constitute the distal domain, and CHCR1 and 2 with the carboxyl (C)-terminal part of CHCR0 make the ankle domain (Fig. 1.1a). The clathrin vertex and three proximal domains form the clathrin hub. The hub provides stability to the triskelion and allows assembly into the characteristic polyhedral lattice. High-resolution cryo-electron microscopy structure analysis revealed a helical tripod, composed of the C-terminal domains of CHCs and positioned beneath the vertex towards the center of the lattice. This tripod contacts the ankles from neighboring triskelia, thus stabilizing the lattice (Fig. 1.1b) [6, 11].

The clathrin knee is flexible, and this allows the lattice to adapt to varying vesicle sizes in vitro and varying membrane curvatures in vivo. Each triskelion vertex is centered at a lattice vertex, and the heavy chain legs form two adjusted edges of a planar (consisting mostly of hexagons) or polyhedral lattice (consisting of pentagons and hexagons). The legs appear to interact via proximal and distal domains, each edge consisting of two antiparallel proximal domains with two antiparallel distal domains situated beneath proximal domains. Because the knee is flexible, these interactions can generate either hexagons or pentagons. An increase in the number of pentagons leads to increased curvature of the polyhedral lattice.

The amino (N)-terminal domain is attached to the distal domain by an α-zigzag linker that positions the terminal domain inside the lattice. The N-terminal domain is structured as a seven-bladed β-propeller [12], in which the β-sheets are organized regularly around the central axis. This β-propeller has at least three distinct binding sites for many proteins, including adaptors that recruit clathrin to the membrane. Proteins containing a "clathrin box" motif Lϕxϕ[D/E], where ϕ is a bulky hydrophobic amino acid and x is any residue, have been shown to bind the site between blades 1 and 2 [12]. Peptides with W-based motif (PWxxW) are thought to fit into a pocket in the center of the domain, formed by the upper regions of blades 1, 4, 6, and 7 [13]. A third binding site, between blades 4 and 5, binds specific sequences via hydrophobic interactions [12].

Clathrin light chains (CLCa and CLCb) bind residues 1438-51 in the proximal domains of CHC through their carboxyl-terminal domains [14–16]. CLCs have a helical rodlike shape and face the outside of the lattice when they contact the proximal domain [6]. Interestingly, in contrast to RNA interference (RNAi) knockdown of CHC, which results in complete disappearance of coated pits at the plasma membrane and strong inhibition of the endocytosis of many types of cargo, knockdown of CLC does not lead to inhibition of endocytosis of classical CME cargo such as the transferrin receptor [17]. Thus, although CLC has been proposed to stabilize the lattice [18], this function of CLCs is apparently not essential for general endocytosis. Recently, CLCs have been implicated in clathrin-mediated endocytic processes that require actin [19, 20]. In this situation, CLCs bind huntingtin-interacting protein related protein (HipR1) that in turn binds cortactin and F-actin, thus linking the clathrin coat to actin filaments [21, 22]. Further, knockdown of CLC or overexpression of its non-phosphorylatable version has been shown to impair the endocytosis of G protein-coupled receptors (GPCRs) [23]. Thus, CLCs might play a role in specialized endocytic scenarios.

1.3 Steps of Coated Vesicle Formation

1.3.1 Coat Nucleation/Initiation

Formation of an endocytic coated pit at the membrane is initiated by concentrating several clathrin triskelions in a small area of the inner leaflet of the plasma membrane leading to triskelion interactions, clathrin polymerization, and lattice assembly. Clathrin triskelions are recruited from the cytosol to the membrane by adaptor proteins that are capable of simultaneous interaction with triskelions and lipids containing negatively charged head groups. Clathrin lattices are also formed on membranes of endosomes and the Golgi apparatus. The specificity of the formation of CCPs at the plasma membrane is achieved by adaptors with preferential binding to phosphatidylinositol-4,5-biphosphate [PI (4,5)P2], a lipid that is enriched in the plasma membrane (Fig. 1.2).

Historically, adaptor protein complex-2 (AP-2) has been considered to be the main adaptor that recruits clathrin to the plasma membrane (Fig. 1.1c). AP-2, present with the highest stoichiometry to clathrin among all other CCV components, is a heterotetramer consisting of tightly associated α, $\beta2$, $\mu2$, and $\sigma2$ subunits [24, 25]. The 200-kDa core domain of AP-2, that contacts the membrane, consists of the trunk domains of large α and $\beta2$ subunits, assembled together with the $\mu2$ and $\sigma2$ subunits [26, 27]. α and $\beta2$ subunits have 30 kDa bilobal C-terminal appendages connected to the trunks with long flexible linkers [28]. Appendage domains are capable of binding to different sequence motifs found on many accessory/regulatory proteins (reviewed in [29]). The $\beta2$ hinge domain contains two clathrin-binding motifs and strongly interacts with the terminal domain of CHC. Structures of the AP-2 core bound to a phosphoinositol phosphate headgroup analog (inositol hexakisphosphate) were solved [26]. Four positively charged interfaces, that can bind PI (4,5)P2, on α, $\beta2$, and $\mu2$ subunits were identified [26, 30–33]. The α subunit site appears to play a key role in the initial docking of AP-2 onto PI (4,5)P2 [30].

Several other proteins, such as epsin, clathrin assembly lymphoid myeloid leukemia (CALM), and its neuronal homolog AP180, satisfy the criteria that define clathrin adaptors: an ability to bind lipids and the terminal domain of CHC [34–37] (Fig. 1.2). The formation of the clathrin lattice in vitro on liposomes and the entire process of coated vesicle formation was reconstituted using a membrane-bound fusion protein of epsin-1 and, with lesser efficiency, AP180 [38]. In these

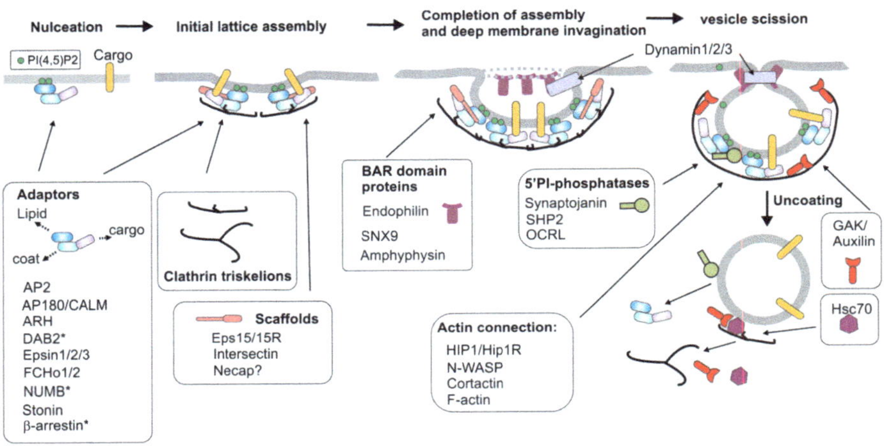

Fig. 1.2 Formation of clathrin-coated vesicle at the plasma membrane. Schematic representation of consecutive stages of the cycle of CCV assembly and disassembly in mammalian cells, based on the heroic systematic analysis by Merrifield and coworkers [61] and numerous other studies. Various adaptors, scaffolds, and other accessory proteins are recruited to the site of forming or assembled clathrin structures at different stages of the process (indicated by *arrows*). Adaptor proteins are shown in generic modular shape consisting of cargo, clathrin coat, and lipid-binding interfaces. *Asterisk* points out on adaptors that require AP-2 binding for linking their cargo to a CCP. Scaffolds do not bind lipids directly but interact with multiple membrane-associated adaptors and may interact with cargo (Eps15/Eps15R)

experiments, the truncated mutant of epsin-1, containing clathrin-binding motifs and lacking ENTH (lipid-binding) domain, but fused to an artificial membrane attachment moiety, was incorporated into liposomes. Recruitment of soluble clathrin to the membrane-bound epsin mutant resulted in the assembly of a slightly invaginated lattice adopting the concave shape of individual clathrin triskelions and ultimately formation and scission of CCVs. These experiments supported the previously proposed hypothesis that the membrane-bending ability of polymerized clathrin is sufficient for generating an initial curvature of coated pits [6, 39]. The question still remains—what is the relative contribution of individual adaptors in the coated pit formation in an intact cell? Is AP-2 sufficient for the initiation of the lattice assembly, or does an assembly process initiated by monomeric adaptors take place in parallel to the AP-2-mediated assembly? Single-molecule imaging analysis demonstrated that two consecutive events of sequential recruitment of two molecules of AP-2 and one clathrin triskelion to a future endocytic site on the plasma membrane typically precede the assembly of a clathrin-coated structure [40]. This recent study, therefore, suggested that membrane AP-2 recruitment is sufficient and essential for coat initiation.

How initial clathrin assembly at the membrane and the initial membrane invagination are coordinated and the mechanisms of this invagination are currently under debate. Recent studies by McMahon and coworkers implicated proteins containing the membrane shaping F-BAR domain (see below), FCH domain only (FCHo1 and 2) in the nucleation step [41]. FCHo proteins bind Eps15, highly homologous Eps15-related protein, Eps15R, and intersectin, all three proteins with scaffold properties, capable of interacting with multiple clathrin coat accessory/regulatory proteins. It was proposed that binding of the FCHo1/2 “module” to the future endocytic site on the membrane takes place prior to AP-2 recruitment and that F-BAR domain recognizes low membrane curvatures and generates initial curvature of the coated pit. siRNA knockdown of FCHo1/2 blocked formation of coated pits and buds [41]. Subsequent studies confirmed the role of FCHo proteins in coat nucleation [42]. However, translational silencing of FCHo1/2 did not eliminate clathrin- and AP-2 positive structures in the plasma membrane or inhibit endocytosis of the transferrin receptor [43], and FCHo siRNA had only partial effects on endocytosis [44]. Likewise, RNAi experiments demonstrated that depletion of Eps15 and Eps15R causes no significant inhibition of CME [17] or at best a partial inhibition [45]. Furthermore, at least in some cells, all cellular Eps15 could be co-immunoprecipitated with AP-2, suggesting that an AP-2 independent function of Eps15 is unlikely [46]. Therefore, further research is necessary to clarify the precise role of FCHo1/2 and associated proteins and reconcile contrasting models of coordinated coat nucleation and membrane bending.

A number of other adaptor proteins found in CCPs, such as Disabled-2, NUMB, β-arrestins, and autosomal recessive hypercholesterolaemia (ARH), are capable of binding PI (4,5)P2, the terminal domain of clathrin (directly or through AP-2), and transmembrane cargo (Fig. 1.1) but are not directly demonstrated to be sufficient to promote coat assembly [47]. For instance, the clathrin box on β-arrestins is exposed only after they interact with their transmembrane cargo, GPCRs [48, 49]. The common theme is that all clathrin adaptors including AP-2 and monomeric adaptors are capable of binding to the transmembrane endocytic cargo, suggesting that cargo recruitment into a forming coat may have a regulatory function during the CCV formation [47]. In fact, FCHo1 has been proposed to act also as the cargo adaptor, as the μ-homologous domain in FCHo can interact with the BMP receptor [43]. The possibility that cargo alone is sufficient to initiate coat assembly is difficult to envision, because coated pits are not formed on other intracellular membranes where the same cargo molecules are present. It is more plausible to propose that cargo binding to the adaptor proteins occurs after the nucleation step and plays a regulatory role in the kinetics of the CCV cycle. Indeed, the hypothesis was proposed whereby cargo recruitment into forming clathrin coat is necessary for the completion of the CCV formation process and that coats not bearing cargo undergo disassembly before full assembly of the coated bud (see below) [50].

1.3.2 Formation of the Clathrin-Coated Bud

Following initiation of clathrin lattice assembly on the membrane that is slightly bent by assembled clathrin and/or F-BAR proteins, the coat expands simultaneously with the increase in the membrane curvature, thus forming the coated "bud" connected to the membrane by the "neck" (Fig. 1.2). The intrinsic curvature of assembled clathrin due to pentagonal faces of the polyhedral lattice makes it capable of deforming membranes into stable highly curved buds. During the growth of the coat and formation of the bud, proteins such as Eps15 that are involved in the initial nucleation step and, possibly, cargo recruitment are "pushed" to the edge of the coat [46] (Fig. 1.2). In the same time, epsin, which interacts with lipids and is directly bound to clathrin, remains distributed throughout the clathrin coat [51].

The mechanisms of coat assembly on highly curved membranes are not fully understood. According to one model, the shape of assembled clathrin provides sufficient energy to promote high curvature of the membrane ([38] and references therein). Another model suggests that the energy provided by assembled clathrin coat would not be sufficient to support highly curved membranes, because clathrin triskelions bind to flexible regions of adaptors. This model proposes that additional proteins must provide energy to bend the membrane and form coated buds [2]. Several proteins might bind to slightly curved areas of the membrane and further increase the membrane curvature, either by inserting an amphipathic helix into the inner layer of the membrane (in the case of the ENTH domain of epsin) or through membrane shaping by BAR (Bin/amphiphysin/Rvs) and N-BAR domain scaffolds [52, 53] (Fig. 1.2). Dimerization of BAR domains produces a crescent-shaped structure with a positively charged concave interface, thus enabling generation of the membrane curvature by electrostatic interactions of this surface with negatively charged head groups of lipids [53]. Proteins containing N-BAR and BAR domains also have capacity to bind accessory proteins located in coated pits (Fig. 1.2). This second model proposes that clathrin plays a role in stabilizing the curved membrane and the entire coat structure. The ability to reconstitute the process of clathrin-coated bud formation and vesicle scission from liposomes in the absence of BAR or other lipid-binding domains supports the first model [38]. Furthermore, lack of effects of siRNA knockdown of epsins and BAR domain proteins on endocytosis rates of the conventional CME cargo like the transferrin receptor [17, 51, 54] argues that clathrin is sufficient to carry out all stages of membrane remodeling during CME until vesicle scission. However, it is possible that membrane-bending proteins are required for more efficient membrane bending in the situation when cargo molecules with larger extracellular domains are being packaged into the CCV. Furthermore, there is significant redundancy among membrane-bending proteins, and the absence of the effects of single and double RNAi knockdowns on endocytosis should be interpreted with caution.

1.3.3 Vesicle Scission

As the CCP nears completion, the large GTPase dynamin concentrates at the narrow neck that attaches the forming vesicle to the membrane and promotes vesicle scission (reviewed by Ferguson and De Camilli [55] and Schmid and Frolov [56]).

There are three mammalian dynamin genes, which encode proteins that are 80% homologous. Dynamin 1 is expressed at high levels exclusively in neurons, although low levels of dynamin 1 are detected in many non-neuronal cultured cell lines. Dynamin 2 is expressed ubiquitously. Dynamin 3 is found predominantly in the testis and, at low levels, in brain and other tissues. All three have the same domain organization: an amino-terminal G domain, a "middle" or "stalk" region, a pleckstrin homology (PH) domain, a GTPase effector domain (GED), and a Pro-rich carboxy-terminal region (Pro-rich domain, PRD). Although dynamin can bind to the membrane through its PH domain, its concentration at the neck appears to require the interaction of the PRD with the SH3 domains of amphiphysin, endophilin, and/or SNX9, proteins that contain BAR and N-BAR domains. At the neck, dynamin assembles into dimers and higher oligomers, and this assembly stimulates GTP hydrolysis and results in a conformational change in the dynamin oligomeric complex. This conformational change provides energy necessary for constricting the neck of the nascent vesicle, fission of the membrane at the constricted neck, and pinching off a CCV [55, 56]. The mechanisms of dynamin-dependent fission are not fully understood. When the activity of dynamin is inhibited by addition of non-hydrolyzable analogs of GTP or mutations, long membrane necks covered by the dynamin collar are observed, indicative of the function of long dynamin oligomers at the late stage of membrane fission [57]. However, long necks are not observed under normal conditions, and it is likely that fission is mediated by short dynamin scaffolds. Analysis of dynamin's interaction with membrane nanotubes suggested that dynamin catalyzes membrane remodeling by generating regulated curvature constraints and bringing membranes to the point of spontaneous fission [58].

In living cells, the fission process is very rapid, and it is likely that the dynamin activity is coordinated with the activity of other accessory proteins. Recently, the ENTH domain of epsin was proposed to be involved in membrane fission by inserting the amphipathic helix into the membrane [59]. Under certain experimental conditions, epsin was shown to rescue the inhibition of fission in the dynamin-depleted cells. On the other hand, accumulation of N-BAR domain containing proteins at the forming neck of coated buds was found to inhibit fission, suggesting that release of these proteins from the nascent vesicle is an important checkpoint during the vesicle scission process. Furthermore, in another recent study, a distinct mechanism of coated pit closure was proposed [60]. In this study, coated pits located on the top surface of substrate-adherent cultured cells were often seen in the close proximity to microvilli. Actin-based structures in the microvilli were proposed to participate in the closure of the clathrin-coated buds, presumably by projecting over an open neck from one side of the narrow and fusion of the end of a microvilli with the opposite side of the neck [60]. In mammalian cells, F-actin and regulators of actin branch assembly such as Arp2/3 (actin-related protein 2/3), N-WASP (Wiscott-Aldrich syndrome-like), and cortactin are transiently recruited to clathrin structures at the time of or immediately before vesicle scission [61, 62]. Actin cytoskeleton has been implicated in CME, although, in mammalian cultured cells, actin polymerization is not essential for coated pit assembly and endocytosis of various cargos [19]. The role of actin in endocytosis is discussed in detail in the Chap. 2.

1.3.4 Uncoating

After pinching off from the plasma membrane, CCVs are rapidly uncoated by the ATPase heat shock cognate 70 (Hsc70) and its co-chaperone, the J domain-containing protein auxilin [63, 64]. Auxilin 1 is expressed in neurons, whereas auxilin 2/GAK (cyclin G-associated kinase) is ubiquitously expressed. The LLGLE motif of auxilin binds to the terminal domain of CHC [65]. High-resolution structure of the clathrin vesicle demonstrated that auxilin also interacts with an ankle region of CHC [11]. Binding of auxilin to the clathrin lattice formed in vitro causes change in CHC interactions and moves the clathrin terminal domains slightly outwards. This repositioning causes substantial structural alterations in the lattice, increasing the diameter of a barrel-shaped CCV. The J domain of clathrin-bound auxilin recruits Hsc70, thus positioning Hsc70 in the proximity to several critical interactions of assembled clathrin. In vitro, one auxilin and three or less Hsc70 molecules per triskelia are necessary for uncoating. Hsc70 is a chaperone protein involved in many folding, degradation, and translation processes, and therefore it is difficult to conduct functional experiments to analyze the uncoating function of Hsc70 in intact cells. Hsc70 binding causes a global distortion in the lattice, presumably by interfering with interactions of proximal and distal domains of CHC. In cells, it is likely that budded vesicles have a coat that is interrupted at the site of membrane fission. Therefore, it is possible that the process of uncoating starts from the exposed edge of the coat [2].

The lipid phosphatase, synaptojanin, is recruited to nascent vesicle at the time of scission with kinetics similar to that of dynamin and several BAR domain proteins (Fig. 1.2) [61, 66]. Binding of synaptojanin to the SH3 domain of endophilin is proposed to mediate the association of synaptojanin with forming vesicles [67]. Synaptojanin dephosphorylates the head group of PI (4,5)P2 at position 5′ [68]. Decrease in PI (4,5)P2 concentrations releases AP-2, AP180, CALM, and other PI (4,5)P2-binding adaptors to the cytosol. Phosphoinositol-3-kinases that use PI (4,5) P2 as substrate and that are capable of binding to clathrin may also contribute to the reduction of PI (4,5)P2 concentration [30]. Another lipid phosphatase, SHIP2, is shown to be recruited to CCPs earlier than synaptojanin and may also participate in PI (4,5)P2 dephosphorylation [66]. Additionally, oculocerebrorenal syndrome of Lowe (OCRL) lipid phosphatase is proposed to participate in lipid remodeling at the late stages of endocytosis, e.g., immediately after vesicle scission [61, 66, 69]. Finally, released clathrin triskelions, adaptors, and accessory proteins recycle back to the plasma membrane to form new clathrin-coated structures.

1.4 Cargo Recruitment and Endocytosis

A key feature of endocytosis of membrane proteins is that it is highly selective. That is, some proteins are concentrated in CCPs compared to the surrounding membrane, while many others are not. Endocytic proteins are concentrated by a

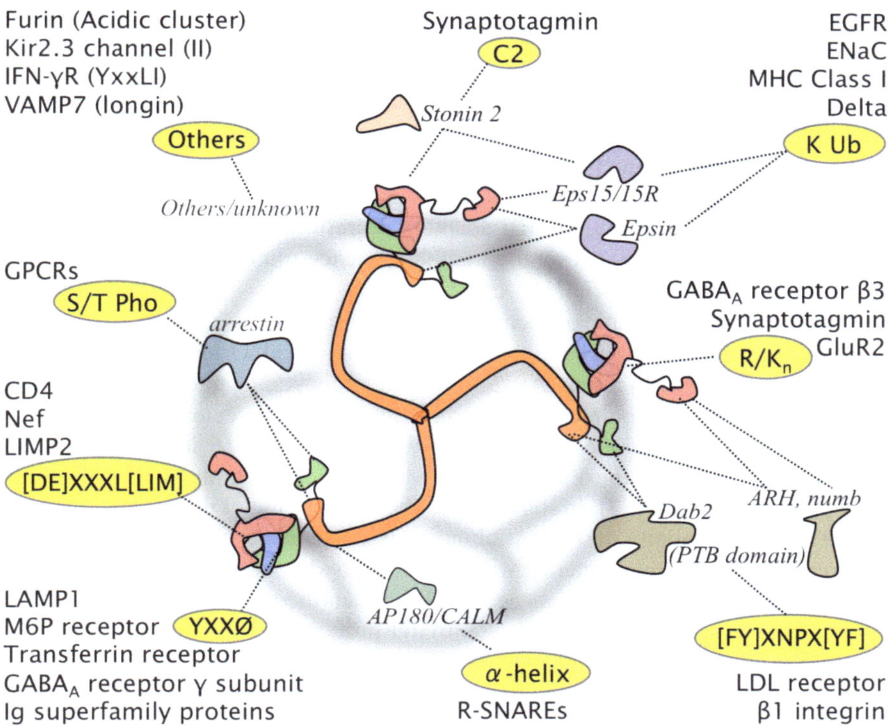

Fig. 1.3 Endocytic sorting signals and adaptors. Schematic of the main endocytic sorting motifs, example cargo molecules that contain these motifs, and the adaptors that mediate their interactions with clathrin. *Dotted lines* indicate the proposed interactions. Single letter amino acid notations are used for the sequence motifs, and *square brackets* indicate alternate residues at the same position. X indicates any residue, and ϕ indicates a bulky hydrophobic residue. Pho denotes phosphorylation, and Ub denotes ubiquitination

simple affinity principle, whereby they are physically linked to the clathrin coat machinery by "adaptor" proteins that recognize specific sorting sequences on cargo (Fig. 1.3). The sorting sequences of some proteins bind adaptors irrespective of whether they are bound to extracellular ligands, leading to "constitutive" internalization of these receptors [47]. For many other proteins, such as signaling receptors, adaptor-cargo binding requires posttranslational modifications and conformational changes of the cargo protein, often induced by ligand binding. Therefore, the internalization of these cargo molecules is regulated [70]. The molecular mechanisms involved in cargo sorting into CCPs in both these modes of endocytosis are discussed below.

1.4.1 Constitutive Endocytosis

Many surface proteins, such as nutrient receptors like the transferrin receptor (TfR) and the low-density lipoprotein (LDL) receptor, are continuously internalized from the plasma membrane and recycled back to the cell surface from the endosome. Most of these receptors use specific sorting sequences on their cytoplasmic surfaces to either directly or indirectly bind adaptors. Careful analysis of many examples over the years has yielded several common sequence motifs that mediate adaptor binding and endocytosis [47, 71].

The best-known example of such endocytosis signals is the Yxxϕ motif, in which x is any amino acid and ϕ a bulky hydrophobic amino acid. Originally identified on TfR, this motif directly binds AP-2, the main endocytic adaptor [72]. Structural studies have since pinpointed this binding interface. The extended Yxxϕ motif forms a transient antiparallel β-strand with the C-terminal β-16 strand of μ2, stabilized by interactions of the Y and the ϕ residues with compatible pockets on either side of the β-16 strand [73]. In some cargo, like the γ subunit of the $GABA_A$ receptor (YGYECL), the affinity of Yxxϕ-μ2 binding is substantially increased by residues upstream of Yxxϕ, which interact with additional hydrophobic pockets on μ2 [74]. A highly related YDYCRV sequence has been identified in BST-2/tetherin [75], a clinically relevant protein due to its antiviral properties. However, this sequence appears to bind the α-appendage domain outside the AP-2 core [76].

Interestingly, while the binding of Yxxϕ to AP-2 is typically independent of the presence of ligand, the accessibility of the binding domain on μ2 is regulated to prevent uncontrolled internalization of cargo. The μ2 subunit is normally in a closed conformation, with the Yxxϕ-binding region masked by the β2 subunit [26, 31]. During CCP initiation and assembly, binding of AP-2—in particular, positively charged patches in the C-lobe of μ2—to PI (4,5)P2 frees the μ2 from the β2 subunit and switches it into an open conformation that allows Yxxϕ binding [31, 77] (Fig. 1.1c). It has also been proposed that phosphorylation of a specific threonine 156 residue on μ2 by adaptor-associated kinase-1 (AAK-1), stimulated by clathrin assembly, plays an additional stabilizing role in this switch in conformation [31, 78–83]. Further, the affinity of Yxxϕ with μ2 may be inhibited by phosphorylation of this motif, such as with the $GABA_A$ receptor and CTLA-4 [74, 84]. In the case of the $GABA_A$ receptor, phosphorylation of either tyrosine in the YGYECL sequence (where the last four residues form the Yxxϕ motif) by Src family kinases inhibits AP-2 binding [74, 85].

The acidic di-leucine ([DE]xxxL[LIM]) motif is another well-studied internalization motif. These motifs bind AP-2 on the σ2 subunit adjoining the PI (4,5) P2-binding region of the α subunit in the AP-2 core [32]. Much like with the tyrosine motif, diLeu binding to AP-2 is also regulated at multiple levels. The binding pocket on σ2 is masked by the N-terminal domain of the β2 subunit and must be made accessible before cargo binding. Lipid binding of multiple subunits of AP-2 [31] might contribute to the large-scale movements required for moving the β2 segment away (Fig. 1.1c). Evidence also suggests that phosphorylation of a specific tyrosine residue on β2 that packs against the binding domain [86–88] may stabilize a conformation of AP-2 that is accessible to diLeu motifs [32]. Interestingly, structural studies show that the μ2 subunit can remain in the closed conformation even under conditions where diLeu can bind σ2 [32], suggesting that AP-2 might differentially bind these two motifs. However, evidence also suggests that different sorting motifs can influence the binding of each other to adaptors [89, 90]. A more straightforward way of regulating diLeu binding to AP-2 is seen in the regulation of E-cadherin turnover by p120 catenin [91]. p120 contains a diLeu-binding motif that competes with AP-2 for E-cadherin binding and prevents its internalization. However, p120 itself has an internalization motif that binds Numb, an alternate adapter discussed below, and evidence suggests that the complex itself may be internalized in a Numb-dependent manner [92–94]. The molecular details of how all these diverse cargo can influence adaptor binding and endocytosis of each other are still not fully understood.

In addition to these well-studied sequence motifs, a cluster of basic residues has been shown to act as internalization signals for a set of cargo proteins, including the GluR2 subunit of AMPA receptors, the β3 subunit of the $GABA_A$ receptor, and the synaptic Ca^{2+} sensor Synaptotagmin 1. While these are thought to bind the μ2 subunit of AP-2, the exact binding interface and regulation are not clear.

A fourth signal, the [FY]xNPx[YF] motif, was in fact the first internalization signal identified on any cargo protein, in the classic experiments by Brown and Goldstein on LDL receptor (LDLR) internalization [95, 96]. This tyrosine-based motif has since been identified on several proteins, including β-integrins and amyloid precursor protein (APP). Surprisingly, despite the similarity to the Yxxϕ motif, this motif does not directly bind AP-2. Further, the internalization of LDLR proceeds efficiently even when most of AP-2 is depleted from cells, even though TfR internalization is inhibited [97, 98]. This led to the idea that alternate adaptors, not AP-2, drive LDLR internalization. Several alternate adaptors that show remarkable specificity to [FY]xNPx[YF], including ARH, Dab2, and Numb, have been now identified [99–102]. Overexpression of exogenous Dab2 selectively increases the internalization of β-integrins without affecting TfR [103]. Further, depletion of both Dab2 and ARH, but not either alone, inhibits the internalization of LDLR, but not TfR [104–106]. This suggests that ARH and Dab2 are largely functionally redundant in LDLR internalization, although ARH might depend on AP-2 for CCP localization [105, 106].

A characteristic of these adaptors is that they contain "phosphotyrosine-binding" (PTB) domains, which are 100–150 amino acid modules that were originally identified as protein folds that bind phosphorylated tyrosines in an NPxY motif [107]. PTB folds are characterized by an orthogonal β-sandwich, on which the NPxY peptide forms a temporary antiparallel β-strand with the NP residues stabilizing a tight β-turn to present the Y into its binding pocket [108]. In the case of these adaptors, however, it seems that the PTB domain is a misnomer, as they mostly prefer non-phosphorylated Y or F residues [101, 109]. A recent crystal structure suggests that ARH recognizes a longer sequence of the LDLR motif and that it uses an atypical hydrophobic pocket to bind the critical tyrosine [110]. This variation might explain the flexibility of ARH in recognizing pY/Y/F on that position. At the other end, ARH uses a consensus helical motif that is shared by several adaptors such as epsin and arrestin (see below) to bind the β2 appendage of AP-2 and link cargo to the coat [111].

Much like AP-2, these alternate adapters are also regulated. A well-studied example is Numb, which mediates the internalization of APP, Notch receptor, and integrins, although it requires AP-2 for its localization in CCPs [102, 112, 113]. Phosphorylation of Thr102 by AAK1 induces Numb redistribution from the plasma membrane into endosomes, while a T102A mutant is constitutively localized to surface puncta [114]. Further, phosphorylation of multiple serines by calcium/calmodulindependent protein kinase or by atypical protein kinase C also modulate Numb activity by inhibiting AP-2 binding [94, 115, 116]. Similarly, ARH function could be potentially regulated by phosphorylation of Tyr888 on the ARH-binding domain on the β2 appendage of AP-2 [111, 117]. Such selective phosphorylation of different adaptors by spatially restricted kinases provides a mechanism for localized endocytosis of cargo from defined areas of cells, as is required for cell migration [115].

In addition to these general mechanisms, several cargo proteins use relatively distinct alternate signals and adapters. Arginine (R)-soluble NSF attachment protein receptor (SNARE) proteins—key mediators of membrane fusion—are interesting examples [118]. These need to be included in forming CCVs, as they mediate fusion steps including that of endocytic vesicles and endosomes. However, SNAREs do not have the linear internalization motifs discussed above. Instead, small R-SNAREs, such as synaptobrevin and VAMP8, 3, and 2, are internalized by specific and direct interactions between the N-terminal halves of their SNARE motifs and the ANTH (AP180 N-terminal homology) domain of the endocytic clathrin adaptor CALM [119, 120]. Other SNAREs such as the R-SNAREs VAMP7, and Vti1b use a folded N-terminal domain, called the longin domain, to interact with the ArfGAP Hrb and EpsinR, which might serve as the respective clathrin adapters for these SNAREs [121, 122]. Other adaptor proteins implicated for specific cargo include stonin 2, which binds a set of basic residues on the C2A domain of Synaptotagmin 1 [123, 124], and endophilin, which, in addition to its role in membrane bending, might moonlight as a sorting adapter for the transporter VGLUT-1 [125].

1.4.2 Regulated Endocytosis

In contrast to many of the proteins above, various signaling receptors, channels, and transporters are internalized in response to specific triggers. In the case of signaling receptors, the most common trigger is the binding of extracellular ligands. These triggers typically initiate one of two covalent and reversible modifications on the cargo—phosphorylation and ubiquitination.

Phosphorylation on defined serine (Ser) and threonine (Thr) residues form the internalization signal for many signaling receptors. This has been best established for members of the GPCR family of signaling receptors [126]. Agonist-binding and activation of GPCRs on the cell surface causes hyperphosphorylation of multiple Ser/Thr residues mainly on the third intracellular loop and C-terminal tail of the receptors [127, 128]. This recruits the specific adaptor β-arrestin (or nonvisual arrestin) to the GPCRs [129, 130]. β-arrestins consist of an N- and a C-terminal globular domain linked by a flexible region and an extended C-terminal tail that contain one or two regions with consensus LϕxϕD/E clathrin-binding sequences [131–133] and an IVFxxFxRxR domain that binds the β2 appendage domain of AP-2 [134, 135]. Before GPCR activation, β-arrestins are kept in an inactive or "closed" conformation by intramolecular interactions of the IV residues in the AP-2 binding domain with a hydrophobic pocket on the N-terminal domain [133, 136]. GPCR activation and binding induce a conformational shift that releases the C-terminal tail and exposes the clathrin- and AP-2-binding motifs [137]. Interestingly, the AP-2-binding motif exists as a part of a β-strand in the closed conformation of β-arrestin. Upon activation, the released sequence undergoes a strand-to-helix transition to adopt a structure highly similar to the AP-2 binding motif on ARH [111]. Arrestin activity is also regulated by PI (4,5)P2 binding [138], phosphorylation [139], and ubiquitination [140, 141], but the precise roles of these regulations appear to be complex and not fully understood. Emerging data suggest that arrestins and related proteins might serve as adaptors for non-GPCR cargo molecules such as the transforming growth factor beta receptors [142] and surface transporters [143], while some GPCRs might use alternate/additional adapters such as disheveled 2 [144].

Ubiquitination is widely used as an internalization signal by many endocytic cargo including growth factor receptors [70], GPCRs [145, 146], and various channels and transporters [147]. Ubiquitination is a posttranslational modification, where ubiquitin (Ub), a conserved 76 amino acid peptide, is covalently conjugated onto typically lysine residues by sequential reactions involving Ub-activating (E1), Ub-conjugating (E2), and Ub-ligase (E3) enzymes [148]. These Ub signals are proposed to be recognized primarily by epsins, Eps15 and Eps15R, clathrin- and AP-2-associated proteins (Figs. 1.2 and 1.3) which contain tandem arrays of ubiquitin-interacting motifs (UIMs), and which are capable of binding simultaneously to multiple Ub moieties conjugated to cargo [149]. The distance between individual UIMs in this array might define the specificity of epsins and Eps15 to different Ub-linked cargo [150, 151]. Epsins and Eps15/Eps15R use a helical motif similar to ARH and arrestins to bind the α- and β-appendages of AP-2 [47]. Epsins also bear a clathrin box allowing their direct binding to the clathrin terminal domain [152], and a colocalization study suggested that clathrin binding might negatively regulate epsin's ability to bind Ub [21] (for further reading, please see Chap. 9).

The EGF-receptor (EGFR), a receptor tyrosine kinase (RTK) frequently overactive in cancer [153], has been extensively studied as an example of Ub-dependent endocytic cargo. Upon ligand binding, EGFR is dimerized, which leads to activation of its intrinsic tyrosine kinase and cross-phosphorylation of several tyrosines on the C-terminus [154, 155]. Among the many effectors that these phosphotyrosines recruit is the adapter Grb2, which in turn recruits the E3 Ub-ligase Cbl, which also binds EGFR directly on Tyr1045 [156–159]. Cbls (c-Cbl, Cbl-b, and Cbl-c) are RING domain E3 ligases that recruit an E2 enzyme that catalyzes ubiquitination of the kinase domain of the receptor [160, 161] (for further reading, see Chap. 11).

However, there is considerable controversy regarding the role of Ub in EGFR endocytosis. A version of EGFR with 15 Lys mutated, which was ubiquitinated less than 1%, internalized efficiently, suggesting that Ub is not required for EGFR internalization [162]. However, adding two Ub sites onto a kinase- and internalization-deficient EGFR, or an in-frame fusion of a single Ub polypeptide to an EGFR without its C-terminus, was sufficient to promote receptor internalization [162, 163]. Additional lysines in the distal C-terminus of EGFR were found to be acetylated and, when mutated together with the kinase domain lysines, resulted in a mutant of EGFR with slow internalization [86]. Moreover, classical linear Yxxϕ [88] and diLeu [87] internalization motifs have also been identified on EGFR. Elimination of these signals together with multiple lysine mutations results in dramatic inhibition of clathrin-mediated EGFR internalization [86]. Further, ligand-dependent recruitment of adaptors to activated RTKs can directly mediate their recruitment to CCPs, such as in the case of Grb2-Tom1L bridging of EGFR to clathrin and RALT/MIG-6 linking EGFR to AP-2 [164, 165]. The current model of EGFR endocytosis assumes that multiple redundant, and arguably cell-type-dependent, mechanisms mediate EGFR endocytosis [166, 167] (for further reading, see Chap. 14).

Cbl proteins bind to and mediate ubiquitination of numerous RTKs, such as fibroblast growth factor, platelet-derived growth factor, vascular endothelial growth factor, colony stimulating factor-1, hepatocyte growth factor (c-Met), macrophage-stimulating protein (Ron), and ephrin receptors, as well as other RTKs [168–174]. In addition, HECT domain Ub E3 ligases of NEDD4 family are implicated in ubiquitination of insulin, insulin growth factor 1, fibroblast growth factor, ErbB4, and TrkA receptors [175–179]. As in case of Cbl, NEDD4 family E3s bind either directly to these receptors or indirectly through the SH2 adaptor Grb10 (see references above). The overall theme in the regulation of RTK endocytosis is the substantial redundancy of the mechanisms involving ubiquitination, AP-2 binding motifs, and other mechanisms. Moreover, multiple E3 ligases are often implicated in ubiquitination of a single RTK. It is important to note that in many instances, the involvement of ubiquitination in downregulation/degradation of RTKs is well established but the direct role of ubiquitination in the internalization step is not. One of the main reasons for such an uncertainty is technical difficulties in demonstrating and analyzing ligand-induced and ubiquitination-dependent recruitment of RTKs into CCPs and their binding to ubiquitin adaptors in CCP, such as epsins and Eps15/R.

Ubiquitin is emerging as the major endocytic sorting signal in many channel and transport proteins [180]. The best-studied example is the epithelial Na^+ channel ENaC [180, 181]. PPxY motifs on the α and β subunits of ENaC recruit the HECT family E3 Ub-ligase Nedd4-2, which binds the PPxY motif using its WW domain and ubiquitinates the Lys residues in several subunits of ENaC. Requirement of Ub in ENaC endocytosis has been demonstrated in cells, where mutating the Lys to Arg resulted in an increased number of ENaC on the cell surface [182]. In addition,

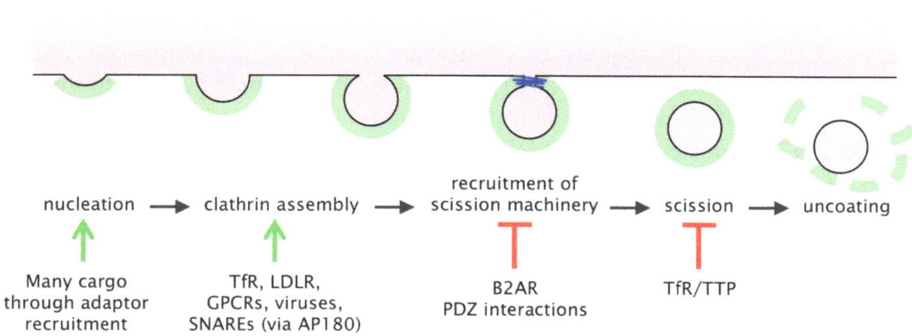

Fig. 1.4 Cargo-mediated control of endocytosis. This figure shows a representation of example cargo and the steps they control. While the mechanistic details of cargo-mediated control, however, still need to be fully investigated, evidence suggests that it is based on local interactions of cargo with adaptors or accessory proteins involved in CME

several transporters, including those for dopamine [183], glutamate [184], glycine [185, 186], and cationic amino acids [187], are also internalized via NEDD4/4-2-mediated ubiquitination that is triggered by activation of protein kinase C.

1.4.3 Cargo-Mediated Control of Endocytosis

In the recent years, it has become increasingly evident that cargo molecules play a far more active role in dictating the behavior of CCPs than was previously appreciated. As mentioned above, early observations, based on monitoring clathrin and dynamin in live cells, suggested that many CCPs were aborted early on after nucleation without continued clathrin assembly or scission. Most of these "abortive" events were observed in CCPs before they recruited detectable fluorescent cargo like TfR. Based on this, it was proposed that the presence of cargo stabilized randomly nucleated CCPs and allowed them to proceed to completion of coat formation and scission [50] (Fig. 1.4). This was further confirmed in a large-scale automated analysis study, which showed that TfR overexpression, in an AP-2 dependent manner, increased the fraction of CCPs that went on to completion [188]. In the case of LDLR, however, overexpression resulted in more CCP nucleation events, but most of them were abortive [189]. Signaling receptor cargo like the EGFR have also been proposed to initiate the formation of CCPs [190, 191], maybe through the efficient recruitment of adaptors [40, 47]. However, TIRF microscopy analysis demonstrated that active EGFR are recruited into preexisting coated pits and that EGF stimulation does not result in formation of new CCPs [192]. Similarly, activation of beta-2 adrenergic receptor (B2AR) did not result in the increase in the number of CCPs [193]. Therefore, the question of cargo-induced de novo CCPs warrants further investigation using quantitative microscopic methods.

In addition to nucleation and stabilization, it is now clear that cargo molecules can actively control the features of formed CCPs (Fig. 1.4). One such feature is the size of CCPs. Typically, coated pits are ~100 nm in diameter in cultured cells and in tissues in vivo. Recruitment of a large cargo, such as the stomatitis virus, to a clathrin structure results in a larger coat (up to 200 nm) [194]. Moreover, recruitment of GPCRs [193] or clustering of TfR [195] can also increase the size and clustering of CCPs. In addition to cargo, ANTH domain containing proteins AP180 and CALM are proposed to regulate size of coated pits; however, the mechanisms of this regulation are not well understood [196–198]. Further, large, flat clathrin lattices are often found on the substrate-bound membrane of many widely used lines of cultured cells [199]. It is unclear why these large planar lattices are formed in some but not in other cell types [200].

Further, cargo molecules can also modify the dynamics of CCPs (Fig. 1.4). This was first observed for the B2AR, a prototypical GPCR, which delayed the lifetimes of the specific CCPs in which they were localized after activation. This delay was mediated by a PDZ ligand sequence present in the C-terminal cytoplasmic tail of the receptor, which slowed the recruitment of dynamin and therefore scission [201]. Interestingly, binding of this tail to actin was both required and sufficient for CCP delay. This suggested that such "cargo-mediated control" of CCPs is a general mechanism, as many signaling receptors have PDZ or related actin-binding sequences on their cytosolic domains [202]. How this interplays with the proposed role for actin in promoting vesicle scission is not clear. Recent data also suggest that even though LDLR increases the fraction of abortive events, it also increases the lifetimes of productive CCPs by approximately 10% [189]. Clustering of TfR can also induce increases in CCP sizes and lifetimes, but the mechanisms of either of these are not understood [195], though it has been proposed that TfR might use a specific adapter that changes dynamin activity [203]. Further, the significance of these relatively small but robust increases in lifetimes is not clear at present. In the case of signaling receptors, considering that small changes in the rates of internalization can generate larger changes in surface levels of receptors over time, the threefold increase in lifetimes observed [201] might have significant effects on signaling in vivo.

1.4.4 Selectivity in Cargo Endocytosis

Having just one common mechanism—CCPs—to internalize hundreds of different cargo molecules causes a potential problem for the cell. Considering that these coated pits have limited capacity, one would expect cargos to compete with each other and interfere in each others' endocytosis. But, surprisingly, they do not. For example, TfR endocytosis can be saturated by overexpression, as expected with limited binding sites for these sequences [204]. But, even under these conditions, EGFR endocytosis continues at near-normal rates. This noninterference (noncompetitive nature) is important for physiologically sensitive cargo like signaling receptors, as their endocytosis has delicate consequences in their downstream signaling.

How do cells reduce interference between cargo? This is a fundamental question in trafficking not limited to endocytosis, as most steps in trafficking rely on one or two core machineries, such as coat complexes, to transport hundreds of different cargo molecules. One straightforward way to achieve this is to use separate adaptors for separate cargo. While this is certainly part of the answer [204, 205], there are far fewer adaptors than there are cargo molecules, and even cargos that bind the same adaptor do not compete with each other [206]. A potential explanation for this is that different signals bind different domains on the same adapter [47]. As described above, the Yxxϕ and diLeu motifs bind distinct regions on adapters, and data suggest that the binding of these signals might be independent of each other [32]. Elucidation of the exact binding domains for various other signals and cargo will provide a better understanding of how multiple cargos can be endocytosed without too much interference.

An additional question that is gaining increasing interest is whether there are biochemically distinct subsets of CCPs, as defined by distinct cargo molecules. Live-cell imaging of endocytosis by TIRF microscopy still remains the best tool to address this question, although it is subject to the limitations of detection limits and the effects of exogenous expression of tagged proteins. Within these limitations, it has been demonstrated that cargo, including GPCRs [201, 207, 208], TfR [195, 203], LDLR, and EGFR [209, 210], localizes to a subset of CCPs present on the surface. In the case of GPCRs and TfR, as mentioned above, this results in a change in behavior of those CCPs, suggesting that functionally different subsets of CCPs may be generated on demand by such cargo segregation. One possible explanation for these subsets could be that adaptor proteins are typically limiting in many cells. Consistent with this, when arrestins or Dab2/ARH were also co-expressed, the fraction of CCPs occupied by GPCRs [201] or LDLR [105], respectively, increased. The existence of cargo-selective CCPs, the mechanisms generating them, and their physiological relevance warrant further careful investigation.

1.5 Concluding Remarks

In this chapter, we briefly described multiple biochemical reactions, macromolecular conformational changes, molecular interactions, and membrane-remodeling events that occur during the cycle of CCV assembly and disassembly. CME is probably the best-studied membrane-trafficking process. The number of accessory and other regulatory proteins identified as being involved in CME continues to grow, but at the same time, examples of experiments demonstrating the essentiality of individual proteins during CME in mammalian cells are strikingly rare. Several explanations can account for this disparity. First, many accessory components are functionally redundant at least in some cell types, and elimination of one of the redundant components often has a very minimal effect on the efficiency of the CME process. Second, some clathrin-associated components, such as cargo adaptors, are specific to a particular class of cargo, and their functional significance in CME can be revealed only if endocytosis of this particular cargo is examined. In this respect, understanding what determines the heterogeneity of coated structures and specificity of populations of coated pits to individual types of cargo will be important. This is especially true for physiologically specialized proteins like signaling receptors, which have different roles in distinct cell types in our body. Third, while TIRF microscopy and in vitro protein-protein interaction experiments provided the most critical data about the involvement and dynamics of many participating molecules, these data may need to be revised when the same measurements are performed by monitoring endogenous proteins under physiological conditions, such as by using cells expressing their tagged versions prepared using emerging genome-editing techniques [211]. It will also be critical to standardize the analytic methods that are used to define and monitor an endocytic event. Such attempts towards objective classification of these events are in progress in several laboratories [212–215]. Another outstanding technical challenge is to develop methodologies that would allow measurements of endocytosis kinetics over the whole surface of the cell with the same time and space resolution that is achieved on the bottom membrane of cultured cells using TIRF microscopy. Finally, there is clearly a lack of information about the mechanisms of CME in vivo in mammals. Further development of high-resolution and quantitative intravital imaging will be critical to elucidation and comparative analysis of these mechanisms in various normal tissues and tumors [216].

Acknowledgements Supported by NIH grants CA089151 (AS), DA014204 (AS) and DA024698 (MAP).

References

1. Howes MT, Mayor S, Parton RG (2010) Molecules, mechanisms, and cellular roles of clathrin-independent endocytosis. Curr Opin Cell Biol 22(4):519–527
2. McMahon HT, Boucrot E (2011) Molecular mechanism and physiological functions of clathrin-mediated endocytosis. Nat Rev Mol Cell Biol 12(8):517–533
3. Lund KA, Opresko LK, Strarbuck C, Walsh BJ, Wiley HS (1990) Quantitative analysis of the endocytic system involved in hormone-induced receptor internalization. J Biol Chem 265:15713–15723
4. Doherty GJ, McMahon HT (2009) Mechanisms of endocytosis. Annu Rev Biochem 78: 857–902. doi:10.1146/annurev.biochem.78.081307.110540
5. Young A (2007) Structural insights into the clathrin coat. Semin Cell Dev Biol 18(4):448–458
6. Fotin A, Cheng Y, Sliz P, Grigorieff N, Harrison SC, Kirchhausen T, Walz T (2004) Molecular model for a complete clathrin lattice from electron cryomicroscopy. Nature 432(7017): 573–579
7. Kirchhausen T, Harrison SC (1981) Protein organization in clathrin trimers. Cell 23(3): 755–761
8. Undewickell E, Branton D (1981) Assembly units of clathrin coats. Nature 289:420–422
9. Ybe JA, Brodsky FM, Hofmann K, Lin K, Liu SH, Chen L, Earnest TN, Fletterick RJ, Hwang PK (1999) Clathrin self-assembly is mediated by a tandemly repeated superhelix. Nature 399(6734):371–375

10. Kirchhausen T, Harrison SC (1984) Structural domains of clathrin heavy chains. J Cell Biol 718 99(5):1725–1734
11. Fotin A, Cheng Y, Grigorieff N, Walz T, Harrison SC, Kirchhausen T (2004) Structure of an auxilin-bound clathrin coat and its implications for the mechanism of uncoating. Nature 432(7017):649–653
12. ter Haar E, Musacchio A, Harrison SC, Kirchhausen T (1998) Atomic structure of clathrin: a β propeller terminal domain joins an α zigzag linker. Cell 95:563–575
13. Miele AE, Watson PJ, Evans PR, Traub LM, Owen DJ (2004) Two distinct interaction motifs in amphiphysin bind two independent sites on the clathrin terminal domain beta-propeller. Nat Struct Mol Biol 11(3):242–248
14. Chen CY, Reese ML, Hwang PK, Ota N, Agard D, Brodsky FM (2002) Clathrin light and heavy chain interface: alpha-helix binding superhelix loops via critical tryptophans. EMBO J 21(22):6072–6082
15. Nathke IS, Heuser J, Lupas A, Stock J, Turck CW, Brodsky FM (1992) Folding and trimerization of clathrin subunits at the triskelion hub. Cell 68(5):899–910
16. Ungewickell E (1983) Biochemical and immunological studies on clathrin light chains and their binding sites on clathrin triskelions. EMBO J 2(8):1401–1408
17. Huang F, Khvorova A, Marshall W, Sorkin A (2004) Analysis of clathrin-mediated endocytosis of epidermal growth factor receptor by RNA interference. J Biol Chem 279(16):16657–16661
18. Pishvaee B, Munn A, Payne GS (1997) A novel structural model for regulation of clathrin function. EMBO J 16(9):2227–2239
19. Boulant S, Kural C, Zeeh J-C, Ubelmann F, Kirchhausen T (2011) Actin dynamics counteract membrane tension during clathrin-mediated endocytosis. Nat Cell Biol 13(9):1124–1131
20. Wilbur JD, Chen CY, Manalo V, Hwang PK, Fletterick RJ, Brodsky FM (2008) Actin binding by Hip1 (huntingtin-interacting protein 1) and Hip1R (hip1-related protein) is regulated by clathrin light chain. J Biol Chem 283(47):32870–32879
21. Chen H, De Camilli P (2005) The association of epsin with ubiquitinated cargo along the endocytic pathway is negatively regulated by its interaction with clathrin. Proc Natl Acad Sci U S A 102(8):2766–2771
22. Engqvist-Goldstein AE, Warren RA, Kessels MM, Keen JH, Heuser J, Drubin DG (2001) The actin-binding protein Hip1R associates with clathrin during early stages of endocytosis and promotes clathrin assembly in vitro. J Cell Biol 154(6):1209–1223
23. Ferreira F, Foley M, Cooke A, Cunningham M, Smith G, Woolley R, Henderson G, Kelly E, Mundell S, Smythe E (2012) Endocytosis of g protein-coupled receptors is regulated by clathrin light chain phosphorylation. Curr Biol 22(15):1361–1370
24. Keen JH, Willingham MC, Pastan I (1981) Clathrin and coated vesicle proteins immunological characterization. J Biol Chem 256(5):2538–2544
25. Pearse BM, Robinson MS (1984) Purification and properties of 100-kd proteins from coated vesicles and their reconstitution with clathrin. EMBO J 3(9):1951–1957
26. Collins BM, McCoy AJ, Kent HM, Evans PR, Owen DJ (2002) Molecular architecture and functional model of the endocytic AP2 complex. Cell 109(4):523–535
27. Heuser JE, Keen J (1988) Deep-etch visualization of proteins involved in clathrin assembly. J Cell Biol 107(3):877–886
28. Zaremba S, Keen JH (1983) Assembly polypeptides from coated vesicles mediate reassembly of unique clathrin coats. J Cell Biol 97(5 pt 1):1339–1347
29. Owen DJ, Collins BM, Evans PR (2004) Adaptors for clathrin coats: structure and function. Annu Rev Cell Dev Biol 20:153–191
30. Gaidarov I, Chen Q, Falck JR, Reddy KK, Keen JH (1996) A functional phosphatidylinositol 3,4,5-trisphosphate/phosphoinositide binding domain in the clathrin adaptor AP-2 alpha subunit. Implications for the endocytic pathway. J Biol Chem 271(34):20922–20929
31. Jackson LP, Kelly BT, McCoy AJ, Gaffry T, James LC, Collins BM, Höning S, Evans PR, Owen DJ (2010) A large-scale conformational change couples membrane recruitment to cargo binding in the AP2 clathrin adaptor complex. Cell 141(7):1220–1229

32. Kelly BT, McCoy AJ, Spate K, Miller SE, Evans PR, Honing S, Owen DJ (2008) A structural explanation for the binding of endocytic dileucine motifs by the AP2 complex. Nature 456(7224):976–979
33. Rohde G, Wenzel D, Haucke V (2002) A phosphatidylinositol (4,5)-bisphosphate binding site within mu2-adaptin regulates clathrin-mediated endocytosis. J Cell Biol 158(2): 209–214
34. Ahle S, Ungewickell E (1986) Purification and properties of a new clathrin assembly protein. EMBO J 5(12):3143–3149
35. Chen H, Fre S, Slepnev VI, Capua MR, Takei K, Butler MH, Di Fiore PP, De Camilli P (1998) Epsin is an EH-domain-binding protein implicated in clathrin-mediated endocytosis. Nature 394(6695):793–797
36. Ford MG, Pearse BM, Higgins MK, Vallis Y, Owen DJ, Gibson A, Hopkins CR, Evans PR, McMahon HT (2001) Simultaneous binding of Ptdins (4,5)P2 and clathrin by AP180 in the nucleation of clathrin lattices on membranes. Science 291(5506):1051–1055
37. Tebar F, Bohlander S, Sorkin A (1999) Interactions of clathrin assembly lymphoid myeloid (CALM) protein with clathrin and its impact on endocytosis. Mol Biol Cell 10:2687–2702
38. Dannhauser PN, Ungewickell EJ (2012) Reconstitution of clathrin-coated bud and vesicle formation with minimal components. Nat Cell Biol 14(6):634–639
39. Hinrichsen L, Meyerholz A, Groos S, Ungewickell EJ (2006) Bending a membrane: how clathrin affects budding. Proc Natl Acad Sci U S A 103(23):8715–8720
40. Cocucci E, Aguet F, Boulant S, Kirchhausen T (2012) The first five seconds in the life of a clathrin-coated pit. Cell 150(3):495–507
41. Henne WM, Boucrot E, Meinecke M, Evergren E, Vallis Y, Mittal R, McMahon HT (2010) FCHo proteins are nucleators of clathrin-mediated endocytosis. Science 328(5983): 1281–1284
42. Nunez D, Antonescu C, Mettlen M, Liu A, Schmid SL, Loerke D, Danuser G (2011) Hotspots organize clathrin-mediated endocytosis by efficient recruitment and retention of nucleating resources. Traffic 12(12):1868–1878
43. Umasankar PK, Sanker S, Thieman JR, Chakraborty S, Wendland B, Tsang M, Traub LM (2012) Distinct and separable activities of the endocytic clathrin-coat components fcho1/2 and AP-2 in developmental patterning. Nat Cell Biol 14(5):488–501
44. Mulkearns EE, Cooper JA (2012) FCH domain only-2 organizes clathrin-coated structures and interacts with disabled-2 for low-density lipoprotein receptor endocytosis. Mol Biol Cell 23(7):1330–1342
45. Teckchandani A, Mulkearns EE, Randolph TW, Toida N, Cooper JA (2012) The clathrin adaptor dab2 recruits EH domain scaffold proteins to regulate integrin beta1 endocytosis. Mol Biol Cell 23(15):2905–2916
46. Tebar F, Sorkina T, Sorkin A, Ericsson M, Kirchhausen T (1996) Eps15 is a component of clathrin-coated pits and vesicles and is located at the rim of coated pits. J Biol Chem 271(46):28727–28730
47. Traub LM (2009) Tickets to ride: selecting cargo for clathrin-regulated internalization. Nat Rev Mol Cell Biol 10(9):583–596
48. Claing A, Laporte SA, Caron MG, Lefkowitz RJ (2002) Endocytosis of G protein-coupled receptors: roles of G protein-coupled receptor kinases and beta-arrestin proteins. Prog Neurobiol 66(2):61–79
49. Kim Y-M, Benovic JL (2002) Differential roles of arrestin-2 interaction with clathrin and adaptor protein 2 in G protein-coupled receptor trafficking. J Biol Chem 277(34): 30760–30768
50. Ehrlich M, Boll W, Van Oijen A, Hariharan R, Chandran K, Nibert ML, Kirchhausen T (2004) Endocytosis by random initiation and stabilization of clathrin-coated pits. Cell 118(5):591–605
51. Kazazic M, Bertelsen V, Pedersen KW, Vuong TT, Grandal MV, Rodland MS, Traub LM, Stang E, Madshus IH (2009) Epsin 1 is involved in recruitment of ubiquitinated EGF receptors into clathrin-coated pits. Traffic 10(2):235–245

52. Ford MG, Mills IG, Peter BJ, Vallis Y, Praefcke GJ, Evans PR, McMahon HT (2002) Curvature of clathrin-coated pits driven by epsin. Nature 419(6905):361–366
53. Peter BJ, Kent HM, Mills IG, Vallis Y, Butler PJ, Evans PR, McMahon HT (2004) BAR domains as sensors of membrane curvature: the amphiphysin BAR structure. Science 303(5657):495–499
54. Sigismund S, Woelk T, Puri C, Maspero E, Tacchetti C, Transidico P, Di Fiore PP, Polo S (2005) From the cover: clathrin-independent endocytosis of ubiquitinated cargos. Proc Natl Acad Sci U S A 102(8):2760–2765
55. Ferguson SM, De Camilli P (2012) Dynamin, a membrane-remodelling gtpase. Nat Rev Mol Cell Biol 13(2):75–88
56. Schmid SL, Frolov VA (2011) Dynamin: functional design of a membrane fission catalyst. Annu Rev Cell Dev Biol 27(3):1–27
57. Takei K, McPherson PS, Schmid SL, De Camilli P (1995) Tubular membrane invaginations coated by dynamin rings are induced by GTP-gamma S in nerve terminals [see comments]. Nature 374(6518):186–190
58. Bashkirov PV, Akimov SA, Evseev AI, Schmid SL, Zimmerberg J, Frolov VA (2008) GTPase cycle of dynamin is coupled to membrane squeeze and release, leading to spontaneous fission. Cell 135(7):1276–1286
59. Boucrot E, Pick A, Camdere G, Liska N, Evergren E, McMahon HT, Kozlov MM (2012) Membrane fission is promoted by insertion of amphipathic helices and is restricted by crescent BAR domains. Cell 149(1):124–136
60. Shevchuk AI, Novak P, Taylor M, Diakonov IA, Ziyadeh-Isleem A, Bitoun M, Guicheney P et al (2012) An alternative mechanism of clathrin-coated pit closure revealed by ion conductance microscopy. J Cell Biol 197(4):499–508
61. Taylor MJ, Perrais D, Merrifield CJ (2011) A high precision survey of the molecular dynamics of mammalian clathrin-mediated endocytosis. PLoS Biol 9(3):e1000604
62. Taylor MJ, Lampe M, Merrifield CJ (2012) A feedback loop between dynamin and actin recruitment during clathrin-mediated endocytosis. PLoS Biol 10(4):e1001302
63. Schlossman DM, Schmid SL, Braell WA, Rothman JE (1984) An enzyme that removes clathrin coats: purification of an uncoating atpase. J Cell Biol 99(2):723–733
64. Ungewickell E, Ungewickell H, Holstein SE, Lindner R, Prasad K, Barouch W, Martin B, Greene LE, Eisenberg E (1995) Role of auxilin in uncoating clathrin-coated vesicles. Nature 378(6557):632–635
65. Scheele U, Kalthoff C, Ungewickell E (2001) Multiple interactions of auxilin 1 with clathrin and the AP-2 adaptor complex. J Biol Chem 276(39):36131–36138
66. Nakatsu F, Perera RM, Lucast L, Zoncu R, Domin J, Gertler FB, Toomre D, De Camilli P (2010) The inositol 5-phosphatase SHIP2 regulates endocytic clathrin-coated pit dynamics. J Cell Biol 190(3):307–315
67. Schuske KR, Richmond JE, Matthies DS, Davis WS, Runz S, Rube DA, van der Bliek AM, Jorgensen EM (2003) Endophilin is required for synaptic vesicle endocytosis by localizing synaptojanin. Neuron 40(4):749–762
68. McPherson PS, Garcia EP, Slepnev VI, David C, Zhang X, Grabs D, Sossin WS, Bauerfeind R, Nemoto Y, De Camilli P (1996) A presynaptic inositol-5-phosphatase. Nature 379(6563):353–357
69. Erdmann KS, Mao Y, McCrea HJ, Zoncu R, Lee S, Paradise S, Modregger J, Biemesderfer D, Toomre D, De Camilli P (2007) A role of the lowe syndrome protein OCRL in early steps of the endocytic pathway. Dev Cell 13(3):377–390
70. Sorkin A, von Zastrow M (2009) Endocytosis and signalling: intertwining molecular networks. Nat Rev Mol Cell Biol 10(9):609–622
71. Rao Y, Rückert C, Saenger W, Haucke V (2012) The early steps of endocytosis: from cargo selection to membrane deformation. Eur J Cell Biol 91(4):226–233
72. Ohno H, Stewart J, Fournier MC, Bosshart H, Rhee I, Miyatake S, Saito T, Gallusser A, Kirchhausen T, Bonifacino JS (1995) Interaction of tyrosine-based sorting signals with clathrin-associated proteins. Science 269(5232):1872–1875

73. Owen DJ, Evans PR (1998) A structural explanation for the recognition of tyrosine-based endocytotic signals. Science 282(5392):1327–1332
74. Kittler JT, Chen G, Kukhtina V, Vahedi-Faridi A, Zhenglin G, Tretter V, Smith KR et al (2008) Regulation of synaptic inhibition by phospho-dependent binding of the AP2 complex to a YECL motif in the GABAA receptor gamma2 subunit. Proc Natl Acad Sci U S A 105(9):3616–3621
75. Rollason R, Korolchuk V, Hamilton C, Schu P, Banting G (2007) Clathrin-mediated endocytosis of a lipid-raft-associated protein is mediated through a dual tyrosine motif. J Cell Sci 120(pt 21):3850–3858
76. Masuyama N, Kuronita T, Tanaka R, Muto T, Hirota Y, Takigawa A, Fujita H, Aso Y, Amano J, Tanaka Y (2009) HM1.24 is internalized from lipid rafts by clathrin-mediated endocytosis through interaction with alpha-adaptin. J Biol Chem 284(23):15927–15941
77. Höning S, Ricotta D, Krauss M, Späte K, Spolaore B, Motley A, Robinson M, Robinson C, Haucke V, Owen DJ (2005) Phosphatidylinositol- (4,5)-bisphosphate regulates sorting signal recognition by the clathrin-associated adaptor complex AP2. Mol Cell 18(5):519–531
78. Conner SD, Schmid SL (2002) Identification of an adaptor-associated kinase, AAK1, as a regulator of clathrin-mediated endocytosis. J Cell Biol 156(5):921–929
79. Conner SD, Schröter T, Schmid SL (2003) AAK1-mediated micro2 phosphorylation is stimulated by assembled clathrin. Traffic 4(12):885–890
80. Henderson DM, Conner SD (2007) A novel AAK1 splice variant functions at multiple steps of the endocytic pathway. Mol Biol Cell 18(7):2698–2706
81. Jackson AP, Flett A, Smythe C, Hufton L, Wettey FR, Smythe E (2003) Clathrin promotes incorporation of cargo into coated pits by activation of the AP2 adaptor micro2 kinase. J Cell Biol 163(2):231–236
82. Olusanya O, Andrews PD, Swedlow JR, Smythe E (2001) Phosphorylation of threonine 156 of the mu2 subunit of the AP2 complex is essential for endocytosis in vitro and in vivo. Curr Biol 11(11):896–900
83. Ricotta D, Conner SD, Schmid SL, von Figura K, Honing S (2002) Phosphorylation of the AP2 mu subunit by AAK1 mediates high affinity binding to membrane protein sorting signals. J Cell Biol 156(5):791–795
84. Shiratori T, Miyatake S, Ohno H, Nakaseko C, Isono K, Bonifacino JS, Saito T (1997) Tyrosine phosphorylation controls internalization of CTLA-4 by regulating its interaction with clathrin-associated adaptor complex AP-2. Immunity 6(5):583–589
85. Jurd R, Tretter V, Walker J, Brandon NJ, Moss SJ (2010) Fyn kinase contributes to tyrosine phosphorylation of the GABA (A) receptor gamma2 subunit. Mol Cell Neurosci 44(2):129–134, http://www.sciencedirect.com/science/article/pii/S1044743110000461
86. Goh LK, Huang F, Kim W, Gygi S, Sorkin A (2010) Multiple mechanisms collectively regulate clathrin-mediated endocytosis of the epidermal growth factor receptor. J Cell Biol 189(5):871–883
87. Huang F, Jiang X, Sorkin A (2003) Tyrosine phosphorylation of the beta2 subunit of clathrin adaptor complex AP-2 reveals the role of a di-leucine motif in the epidermal growth factor receptor trafficking. J Biol Chem 278(44):43411–43417
88. Sorkin A, Mazzotti M, Sorkina T, Scotto L, Beguinot L (1996) Epidermal growth factor receptor interaction with clathrin adaptors is mediated by the tyr974-containing internalization motif. J Biol Chem 271(23):13377–13384
89. Haucke V, De Camilli P (1999) AP-2 recruitment to synaptotagmin stimulated by tyrosine-based endocytic motifs. Science 285(5431):1268–1271
90. Lee I, Doray B, Govero J, Kornfeld S (2008) Binding of cargo sorting signals to AP-1 enhances its association with ADP ribosylation factor 1-GTP. J Cell Biol 180(3):467–472
91. Ishiyama N, Lee S-H, Liu S, Li G-Y, Smith MJ, Reichardt LF, Ikura M (2010) Dynamic and static interactions between p120 catenin and e-cadherin regulate the stability of cell-cell adhesion. Cell 141(1):117–128
92. Davis MA, Ireton RC, Reynolds AB (2003) A core function for p120-catenin in cadherin turnover. J Cell Biol 163(3):525–534

93. Miyashita Y, Ozawa M (2007) Increased internalization of p120-uncoupled e-cadherin and a requirement for a dileucine motif in the cytoplasmic domain for endocytosis of the protein. J Biol Chem 282(15):11540–11548
94. Sato K, Watanabe T, Wang S, Kakeno M, Matsuzawa K, Matsui T, Yokoi K et al (2011) Numb controls e-cadherin endocytosis through p120 catenin with apkc. Mol Biol Cell 22(17):3103–3119
95. Chen WJ, Goldstein JL, Brown MS (1990) NPXY, a sequence often found in cytoplasmic tails, is required for coated pit-mediated internalization of the low density lipoprotein receptor. J Biol Chem 265(6):3116–3123
96. Davis CG, Lehrman MA, Russell DW, Anderson RG, Brown MS, Goldstein JL (1986) The J.D. Mutation in familial hypercholesterolemia: amino acid substitution in cytoplasmic domain impedes internalization of LDL receptors. Cell 45(1):15–24
97. Hinrichsen L, Harborth J, Andrees L, Weber K, Ungewickell EJ (2003) Effect of clathrin heavy chain- and alpha-adaptin-specific small inhibitory RNAs on endocytic accessory proteins and receptor trafficking in HeLa cells. J Biol Chem 278(46):45160–45170
98. Motley A, Bright NA, Seaman MNJ, Robinson MS (2003) Clathrin-mediated endocytosis in ap-2-depleted cells. J Cell Biol 162(5):909–918
99. Garcia CK, Wilund K, Arca M, Zuliani G, Fellin R, Maioli M, Calandra S et al (2001) Autosomal recessive hypercholesterolemia caused by mutations in a putative LDL receptor adaptor protein. Science 292(5520):1394–1398
100. Mishra SK, Keyel PA, Hawryluk MJ, Agostinelli NR, Watkins SC, Traub LM (2002) Disabled-2 exhibits the properties of a cargo-selective endocytic clathrin adaptor. EMBO J 21(18):4915–4926
101. Morris SM, Cooper JA (2001) Disabled-2 colocalizes with the LDLR in clathrin-coated pits and interacts with AP-2. Traffic 2(2):111–123
102. Santolini E, Puri C, Salcini AE, Gagliani MC, Pelicci PG, Tacchetti C, Di Fiore PP (2000) Numb is an endocytic protein. J Cell Biol 151(6):1345–1352
103. Chetrit D, Ziv N, Ehrlich M (2009) Dab2 regulates clathrin assembly and cell spreading. Biochem J 418(3):701–715
104. Eden ER, Sun X-M, Patel DD, Soutar AK (2007) Adaptor protein disabled-2 modulates low density lipoprotein receptor synthesis in fibroblasts from patients with autosomal recessive hypercholesterolaemia. Hum Mol Genet 16(22):2751–2759
105. Keyel PA, Mishra SK, Roth R, Heuser JE, Watkins SC, Traub LM (2006) A single common portal for clathrin-mediated endocytosis of distinct cargo governed by cargo-selective adaptors. Mol Biol Cell 17(10):4300
106. Maurer ME, Cooper JA (2006) The adaptor protein dab2 sorts LDL receptors into coated pits independently of AP-2 and ARH. J Cell Sci 119(pt 20):4235–4246
107. Uhlik MT, Temple B, Bencharit S, Kimple AJ, Siderovski DP, Johnson GL (2005) Structural and evolutionary division of phosphotyrosine binding (PTB) domains. J Mol Biol 345(1):1–20
108. Stolt PC, Bock HH (2006) Modulation of lipoprotein receptor functions by intracellular adaptor proteins. Cell Signal 18(10):1560–1571
109. Howell BW, Lanier LM, Frank R, Gertler FB, Cooper JA (1999) The disabled 1 phosphotyrosine-binding domain binds to the internalization signals of transmembrane glycoproteins and to phospholipids. Mol Cell Biol 19(7):5179–5188
110. Dvir H, Shah M, Girardi E, Guo L, Farquhar MG, Zajonc DM (2012) Atomic structure of the autosomal recessive hypercholesterolemia phosphotyrosine-binding domain in complex with the LDL-receptor tail. Proc Natl Acad Sci U S A 109(18):6916–6921
111. Edeling MA, Mishra SK, Keyel PA, Steinhauser AL, Collins BM, Roth R, Heuser JE, Owen DJ, Traub LM (2006) Molecular switches involving the AP-2 beta2 appendage regulate endocytic cargo selection and clathrin coat assembly. Dev Cell 10(3):329–342
112. Berdnik D, Török T, González-Gaitán M, Knoblich JA (2002) The endocytic protein alpha-adaptin is required for numb-mediated asymmetric cell division in drosophila. Dev Cell 3(2):221–231

113. Kyriazis GA, Wei Z, Vandermey M, Jo D-G, Xin O, Mattson MP, Chan SL (2008) Numb endocytic adapter proteins regulate the transport and processing of the amyloid precursor protein in an isoform-dependent manner: implications for Alzheimer disease pathogenesis. J Biol Chem 283(37):25492–25502
114. Sorensen EB, Conner SD (2008) AAK1 regulates numb function at an early step in clathrin-mediated endocytosis. Traffic 9(10):1791–1800
115. Nishimura T, Kaibuchi K (2007) Numb controls integrin endocytosis for directional cell migration with apkc and PAR-3. Dev Cell 13(1):15–28
116. Tokumitsu H, Hatano N, Yokokura S, Sueyoshi Y, Nozaki N, Kobayashi R (2006) Phosphorylation of numb regulates its interaction with the clathrin-associated adaptor AP-2. FEBS Lett 580(24):5797–5801
117. Schmid EM, Ford MG, Burtey A, Praefcke GJ, Peak-Chew SY, Mills IG, Benmerah A, McMahon HT (2006) Role of the AP2 beta-appendage hub in recruiting partners for clathrin-coated vesicle assembly. PLoS Biol 4(9):e262, http://dx.doi.org/10.1371%2Fjournal.pbio.0040262
118. Südhof TC, Rothman JE (2009) Membrane fusion: grappling with SNARE and SM proteins. Science 323(5913):474–477
119. Koo SJ, Markovic S, Puchkov D, Mahrenholz CC, Beceren-Braun F, Maritzen T, Dernedde J, Volkmer R, Oschkinat H, Haucke V (2011) SNARE motif-mediated sorting of synaptobrevin by the endocytic adaptors clathrin assembly lymphoid myeloid leukemia (CALM) and AP180 at synapses. Proc Natl Acad Sci U S A 108(33):13540–13545
120. Miller SE, Sahlender DA, Graham SC, Honing S, Robinson MS, Peden AA, Owen DJ (2011) The molecular basis for the endocytosis of small r-snares by the clathrin adaptor CALM. Cell 147(5):1118–1131
121. Miller SE, Collins BM, McCoy AJ, Robinson MS, Owen DJ (2007) A snare-adaptor interaction is a new mode of cargo recognition in clathrin-coated vesicles. Nature 450(7169): 570–574
122. Pryor PR, Jackson L, Gray SR, Edeling MA, Thompson A, Sanderson CM, Evans PR, Owen DJ, Paul Luzio J (2008) Molecular basis for the sorting of the SNARE VAMP7 into endocytic clathrin-coated vesicles by the arfgap hrb. Cell 134(5):817–827
123. Diril MK, Wienisch M, Jung N, Klingauf J, Haucke V (2006) Stonin 2 is an AP-2-dependent endocytic sorting adaptor for synaptotagmin internalization and recycling. Dev Cell 10(2):233–244
124. Martina JA, Bonangelino CJ, Aguilar RC, Bonifacino JS (2001) Stonin 2: an adaptor-like protein that interacts with components of the endocytic machinery. J Cell Biol 153(5): 1111–1120
125. Voglmaier SM, Kam K, Yang H, Fortin DL, Hua Z, Nicoll RA, Edwards RH (2006) Distinct endocytic pathways control the rate and extent of synaptic vesicle protein recycling. Neuron 51(1):71–84
126. Magalhaes AC, Dunn H, Ferguson SSG (2012) Regulation of GPCR activity, trafficking and localization by gpcr-interacting proteins. Br J Pharmacol 165(6):1717–1736
127. Benovic JL, Strasser RH, Caron MG, Lefkowitz RJ (1986) Beta-adrenergic receptor kinase: identification of a novel protein kinase that phosphorylates the agonist-occupied form of the receptor. Proc Natl Acad Sci U S A 83(9):2797–2801
128. Fredericks ZL, Pitcher JA, Lefkowitz RJ (1996) Identification of the G protein-coupled receptor kinase phosphorylation sites in the human beta2-adrenergic receptor. J Biol Chem 271(23):13796–13803
129. Ferguson SS, Downey WE III, Colapietro AM, Barak LS, Ménard L, Caron MG (1996) Role of beta-arrestin in mediating agonist-promoted G protein-coupled receptor internalization. Science 271(5247):363–366
130. Goodman OB Jr, Krupnick JG, Santini F, Gurevich VV, Penn RB, Gagnon AW, Keen JH, Benovic JL (1996) Beta-arrestin acts as a clathrin adaptor in endocytosis of the beta2-adrenergic receptor. Nature 383(6599):447–450

131. Kang DS, Kern RC, Puthenveedu MA, von Zastrow M, Williams JC, Benovic JL (2009) Structure of an arrestin2-clathrin complex reveals a novel clathrin binding domain that modulates receptor trafficking. J Biol Chem 284(43):29860–29872
132. Krupnick JG, Goodman OB Jr, Keen JH, Benovic JL (1997) Arrestin/clathrin interaction. Localization of the clathrin binding domain of nonvisual arrestins to the carboxy terminus. J Biol Chem 272(23):15011–15016
133. Milano SK, Pace HC, Kim Y-M, Brenner C, Benovic JL (2002) Scaffolding functions of arrestin-2 revealed by crystal structure and mutagenesis. Biochemistry 41(10):3321–3328
134. Laporte SA, Oakley RH, Holt JA, Barak LS, Caron MG (2000) The interaction of beta-arrestin with the AP-2 adaptor is required for the clustering of beta 2-adrenergic receptor into clathrin-coated pits. J Biol Chem 275(30):23120–23126
135. Laporte SA, Oakley RH, Zhang J, Holt JA, Ferguson SS, Caron MG, Barak LS (1999) The beta2-adrenergic receptor/betaarrestin complex recruits the clathrin adaptor AP-2 during endocytosis. Proc Natl Acad Sci U S A 96(7):3712–3717
136. Hirsch JA, Schubert C, Gurevich VV, Sigler PB (1999) The 2.8 A crystal structure of visual arrestin: a model for arrestin's regulation. Cell 97(2):257–269
137. Gurevich VV, Gurevich EV (2004) The molecular acrobatics of arrestin activation. Trends Pharmacol Sci 25(2):105–111
138. Gaidarov I, Krupnick JG, Falck JR, Benovic JL, Keen JH (1999) Arrestin function in G protein-coupled receptor endocytosis requires phosphoinositide binding. EMBO J 18(4):871–881
139. Lin FT, Krueger KM, Kendall HE, Daaka Y, Fredericks ZL, Pitcher JA, Lefkowitz RJ (1997) Clathrin-mediated endocytosis of the beta-adrenergic receptor is regulated by phosphorylation/dephosphorylation of beta-arrestin1. J Biol Chem 272(49):31051–31057
140. Ahmed MR, Zhan X, Song X, Kook S, Gurevich VV, Gurevich EV (2011) Ubiquitin ligase parkin promotes Mdm2-arrestin interaction but inhibits arrestin ubiquitination. Biochemistry 50(18):3749–3763
141. Shenoy SK, McDonald PH, Kohout TA, Lefkowitz RJ (2001) Regulation of receptor fate by ubiquitination of activated beta 2- adrenergic receptor and beta-arrestin. Science 294(5545):1307–1313
142. Chen W, Kirkbride KC, How T, Nelson CD, Mo J, Frederick JP, Wang XF, Lefkowitz RJ, Blobe GC (2003) Beta-arrestin 2 mediates endocytosis of type III TGF-beta receptor and down-regulation of its signaling. Science 301(5638):1394–1397
143. Lin CH, MacGurn JA, Chu T, Stefan CJ, Emr SD (2008) Arrestin-related ubiquitin-ligase adaptors regulate endocytosis and protein turnover at the cell surface—supplemental info. Cell 135(4):714–725
144. Yu A, Rual J-F, Tamai K, Harada Y, Vidal M, He X, Kirchhausen T (2007) Association of dishevelled with the clathrin AP-2 adaptor is required for frizzled endocytosis and planar cell polarity signaling. Dev Cell 12(1):129–141
145. Canals M, Scholten DJ, de Munnik S, Han MK, Smit MJ, Leurs R (2012) Ubiquitination of CXCR7 controls receptor trafficking. PLoS One 7(3):e34192
146. Wolfe BL, Marchese A, Trejo JA (2007) Ubiquitination differentially regulates clathrin-dependent internalization of protease-activated receptor-1. J Cell Biol 177(5):905–916
147. Miranda M, Sorkin A (2007) Regulation of receptors and transporters by ubiquitination: new insights into surprisingly similar mechanisms. Mol Interv 7(3):157–167
148. Komander D, Rape M (2012) The ubiquitin code. Annu Rev Biochem 81:203–229
149. Polo S, Sigismund S, Faretta M, Guidi M, Capua MR, Bossi G, Chen H, De Camilli P, Di Fiore PP (2002) A single motif responsible for ubiquitin recognition and monoubiquitination in endocytic proteins. Nature 416(6879):451–455
150. Sato Y, Yoshikawa A, Mimura H, Yamashita M, Yamagata A, Fukai S (2009) Structural basis for specific recognition of lys 63-linked polyubiquitin chains by tandem UIMs of RAP80. EMBO J 28(16):2461–2468
151. Sims JJ, Cohen RE (2009) Linkage-specific avidity defines the lysine 63-linked polyubiquitin-binding preference of rap80. Mol Cell 33(6):775–783

152. Drake MT, Downs MA, Traub LM (2000) Epsin binds to clathrin by associating directly with the clathrin-terminal domain. Evidence for cooperative binding through two discrete sites. J Biol Chem 275(9):6479–6489
153. Yarden Y, Pines G (2012) The ERBB network: at last, cancer therapy meets systems biology. Nat Rev Cancer 12(8):553–563
154. Ferguson KM, Berger MB, Mendrola JM, Cho HS, Leahy DJ, Lemmon MA (2003) EGF activates its receptor by removing interactions that autoinhibit ectodomain dimerization. Mol Cell 11(2):507–517
155. Lemmon MA, Schlessinger J (2010) Cell signaling by receptor tyrosine kinases. Cell 141(7):1117–1134
156. Jiang X, Huang F, Marusyk A, Sorkin A (2003) Grb2 regulates internalization of EGF receptors through clathrin-coated pits. Mol Biol Cell 14(3):858–870
157. Jiang X, Sorkin A (2003) Epidermal growth factor receptor internalization through clathrin-coated pits requires cbl RING finger and proline-rich domains but not receptor polyubiquitylation. Traffic 4(8):529–543
158. Levkowitz G, Waterman H, Zamir E, Kam Z, Oved S, Langdon WY, Beguinot L, Geiger B, Yarden Y (1998) C-Cbl/Sli-1 regulates endocytic sorting and ubiquitination of the epidermal growth factor receptor. Genes Dev 12(23):3663–3674
159. Waterman H, Katz M, Rubin C, Shtiegman K, Lavi S, Elson A, Jovin T, Yarden Y (2002) A mutant EGF-receptor defective in ubiquitylation and endocytosis unveils a role for Grb2 in negative signaling. EMBO J 21(3):303–313
160. Huang F, Kirkpatrick D, Jiang X, Gygi S, Sorkin A (2006) Differential regulation of EGF receptor internalization and degradation by multiubiquitination within the kinase domain. Mol Cell 21(6):737–748
161. Umebayashi K, Stenmark H, Yoshimori T (2008) Ubc4/5 and C-cbl continue to ubiquitinate EGF receptor after internalization to facilitate polyubiquitination and degradation. Mol Biol Cell 19(8):3454–3462
162. Huang F, Goh LK, Sorkin A (2007) EGF receptor ubiquitination is not necessary for its internalization. Proc Natl Acad Sci U S A 104(43):16904–16909
163. Haglund K, Sigismund S, Polo S, Szymkiewicz I, Di Fiore PP, Dikic I (2003) Multiple monoubiquitination of RTKs is sufficient for their endocytosis and degradation. Nat Cell Biol 5(5):461–466
164. Frosi Y, Anastasi S, Ballarò C, Varsano G, Castellani L, Maspero E, Polo S, Alemà S, Segatto O (2010) A two-tiered mechanism of EGFR inhibition by RALT/MIG6 via kinase suppression and receptor degradation. J Cell Biol 189(3):557–571
165. Liu NS, Loo LS, Loh E, Seet L-F, Hong W (2009) Participation of Tom1L1 in EGF-stimulated endocytosis of EGF receptor. EMBO J 28(22):3485–3499
166. Madshus IH, Stang E (2009) Internalization and intracellular sorting of the EGF receptor: a model for understanding the mechanisms of receptor trafficking. J Cell Sci 122 (pt 19):3433–3439
167. Sorkin A, Goh LK (2009) Endocytosis and intracellular trafficking of erbbs. Exp Cell Res 315(4):683–696
168. Duval M, Bédard-Goulet S, Delisle C, Gratton J-P (2003) Vascular endothelial growth factor-dependent down-regulation of flk-1/KDR involves cbl-mediated ubiquitination. Consequences on nitric oxide production from endothelial cells. J Biol Chem 278(22):20091–20097
169. Fasen K, Cerretti DP, Huynh-Do U (2008) Ligand binding induces Cbl-dependent EphB1 receptor degradation through the lysosomal pathway. Traffic 9(2):251–266
170. Levkowitz G, Waterman H, Ettenberg SA, Katz M, Tsygankov AY, Alroy I, Lavi S et al (1999) Ubiquitin ligase activity and tyrosine phosphorylation underlie suppression of growth factor signaling by c-Cbl/Sli-1. Mol Cell 4(6):1029–1040
171. Marmor MD, Yarden Y (2004) Role of protein ubiquitylation in regulating endocytosis of receptor tyrosine kinases. Oncogene 23(11):2057–2070

172. Miyake S, Mullane-Robinson KP, Lill NL, Douillard P, Band H (1999) Cbl-mediated negative regulation of platelet-derived growth factor receptor-dependent cell proliferation. A critical role for cbl tyrosine kinase-binding domain. J Biol Chem 274(23):16619–16628
173. Penengo L, Rubin C, Yarden Y, Gaudino G (2003) C-Cbl is a critical modulator of the Ron tyrosine kinase receptor. Oncogene 22(24):3669–3679
174. Wilhelmsen K, Burkhalter S, van der Geer P (2002) C-Cbl binds the CSF-1 receptor at tyrosine 973, a novel phosphorylation site in the receptor's carboxy-terminus. Oncogene 21(7):1079–1089
175. Arévalo JC, Waite J, Rajagopal R, Beyna M, Chen Z-Y, Lee FS, Chao MV (2006) Cell survival through trk neurotrophin receptors is differentially regulated by ubiquitination. Neuron 50(4):549–559
176. Monami G, Emiliozzi V, Morrione A (2008) Grb10/Nedd4-mediated multiubiquitination of the insulin-like growth factor receptor regulates receptor internalization. J Cell Physiol 216(2):426–437
177. Persaud A, Alberts P, Hayes M, Guettler S, Clarke I, Sicheri F, Dirks P, Ciruna B, Rotin D (2011) Nedd4-1 binds and ubiquitylates activated FGFR1 to control its endocytosis and function. EMBO J 30(16):3259–3273
178. Sundvall M, Korhonen A, Paatero I, Gaudio E, Melino G, Croce CM, Aqeilan RI, Elenius K (2008) Isoform-specific monoubiquitination, endocytosis, and degradation of alternatively spliced ErbB4 isoforms. Proc Natl Acad Sci U S A 105(11):4162–4167
179. Vecchione A, Marchese A, Henry P, Rotin D, Morrione A (2003) The Grb10/Nedd4 complex regulates ligand-induced ubiquitination and stability of the insulin-like growth factor I receptor. Mol Cell Biol 23(9):3363–3372
180. Rotin D, Staub O (2012) Nedd4-2 and the regulation of epithelial sodium transport. Front Physiol 3:212
181. Butterworth MB (2010) Regulation of the epithelial sodium channel (ENaC) by membrane trafficking. Biochim Biophys Acta 1802(12):1166–1177
182. Staub O, Gautschi I, Ishikawa T, Breitschopf K, Ciechanover A, Schild L, Rotin D (1997) Regulation of stability and function of the epithelial Na+ channel (ENaC) by ubiquitination. EMBO J 16(21):6325–6336
183. Miranda M, Wu CC, Sorkina T, Korstjens DR, Sorkin A (2005) Enhanced ubiquitylation and accelerated degradation of the dopamine transporter mediated by protein kinase C. J Biol Chem 280(42):35617–35624
184. García-Tardón N, González-González IM, Martínez-Villarreal J, Fernández-Sánchez E, Giménez C, Zafra F (2012) Protein kinase C (PKC)-promoted endocytosis of glutamate transporter GLT-1 requires ubiquitin ligase nedd4-2-dependent ubiquitination but not phosphorylation. J Biol Chem 287(23):19177–19187
185. Fernández-Sánchez E, Martínez-Villarreal J, Giménez C, Zafra F (2009) Constitutive and regulated endocytosis of the glycine transporter GLYT1b is controlled by ubiquitination. J Biol Chem 284(29):19482–19492
186. de Juan-Sanz J, Zafra F, López-Corcuera B, Aragón C (2011) Endocytosis of the neuronal glycine transporter GLYT2: role of membrane rafts and protein kinase c-dependent ubiquitination. Traffic 12(12):1850–1867
187. Vina-Vilaseca A, Bender-Sigel J, Sorkina T, Closs EI, Sorkin A (2011) Protein kinase c-dependent ubiquitination and clathrin-mediated endocytosis of the cationic amino acid transporter CAT-1. J Biol Chem 286(10):8697–8706
188. Loerke D, Mettlen M, Yarar D, Jaqaman K, Jaqaman H, Danuser G, Schmid SL (2009) Cargo and dynamin regulate clathrin-coated pit maturation. PLoS Biol 7(3):e57
189. Mettlen M, Loerke D, Yarar D, Danuser G, Schmid SL (2010) Cargo- and adaptor-specific mechanisms regulate clathrin-mediated endocytosis. J Cell Biol 188(6):919–933
190. Johannessen LE, Pedersen NM, Pedersen KW, Madshus IH, Stang E (2006) Activation of the epidermal growth factor (EGF) receptor induces formation of EGF receptor- and grb2-containing clathrin-coated pits. Mol Cell Biol 26(2):389–401

191. Puri C, Tosoni D, Comai R, Rabellino A, Segat D, Caneva F, Luzzi P, Di Fiore PP, Tacchetti C (2005) Relationships between EGFR signaling-competent and endocytosis-competent membrane microdomains. Mol Biol Cell 16(6):2704–2718
192. Rappoport JZ, Simon SM (2009) Endocytic trafficking of activated EGFR is AP-2 dependent and occurs through preformed clathrin spots. J Cell Sci 122(pt 9):1301–1305
193. Santini F, Gaidarov I, Keen JH (2002) G protein-coupled receptor/arrestin3 modulation of the endocytic machinery. J Cell Biol 156(4):665–676
194. Cureton DK, Massol RH, Saffarian S, Kirchhausen TL, Whelan SP (2009) Vesicular stomatitis virus enters cells through vesicles incompletely coated with clathrin that depend upon actin for internalization. PLoS Pathog 5(4):e1000394
195. Liu AP, Aguet F, Danuser G, Schmid SL (2010) Local clustering of transferrin receptors promotes clathrin-coated pit initiation. J Cell Biol 191(7):1381–1393
196. Meyerholz A, Hinrichsen L, Groos S, Esk PC, Brandes G, Ungewickell EJ (2005) Effect of clathrin assembly lymphoid myeloid leukemia protein depletion on clathrin coat formation. Traffic 6(12):1225–1234
197. Morgan JR, Zhao X, Womack M, Prasad K, Augustine GJ, Lafer EM (1999) A role for the clathrin assembly domain of AP180 in synaptic vesicle endocytosis. J Neurosci 19(23):10201–10212
198. Nonet ML, Holgado AM, Brewer F, Serpe CJ, Norbeck BA, Holleran J, Wei L, Hartwieg E, Jorgensen EM, Alfonso A (1999) UNC-11, a Caenorhabditis elegans AP180 homologue, regulates the size and protein composition of synaptic vesicles. Mol Biol Cell 10(7): 2343–2360
199. Heuser J, Kirchhausen T (1985) Deep-etch views of clathrin assemblies. J Ultrastruct Res 92(1–2):1–27
200. Saffarian S, Cocucci E, Kirchhausen T (2009) Distinct dynamics of endocytic clathrin-coated pits and coated plaques. PLoS Biol 7(9):e1000191
201. Puthenveedu MA, von Zastrow M (2006) Cargo regulates clathrin-coated pit dynamics. Cell 127(1):113–124
202. Ritter SL, Hall RA (2009) Fine-tuning of GPCR activity by receptor-interacting proteins. Nat Rev Mol Cell Biol 10(12):819–830
203. Tosoni D, Puri C, Confalonieri S, Salcini AE, De Camilli P, Tacchetti C, Di Fiore PP (2005) TTP specifically regulates the internalization of the transferrin receptor. Cell 123(5):875–888
204. Warren RA, Green FA, Enns CA (1997) Saturation of the endocytic pathway for the transferrin receptor does not affect the endocytosis of the epidermal growth factor receptor. J Biol Chem 272(4):2116–2121
205. Warren RA, Green FA, Stenberg PE, Enns CA (1998) Distinct saturable pathways for the endocytosis of different tyrosine motifs. J Biol Chem 273(27):17056–17063
206. Marks MS, Woodruff L, Ohno H, Bonifacino JS (1996) Protein targeting by tyrosine- and di-leucine-based signals: evidence for distinct saturable components. J Cell Biol 135(2):341–354
207. Cao TT, Mays RW, von Zastrow M (1998) Regulated endocytosis of G protein-coupled receptors by a biochemically and functionally distinct subpopulation of clathrin-coated pits. J Biol Chem 273(38):24592–24602
208. Mundell SJ, Luo J, Benovic JL, Conley PB, Poole AW (2006) Distinct clathrin-coated pits sort different G protein-coupled receptor cargo. Traffic 7(10):1420–1431
209. Lakadamyali M, Rust MJ, Zhuang X (2006) Ligands for clathrin-mediated endocytosis are differentially sorted into distinct populations of early endosomes. Cell 124(5):997–1009
210. Leonard D, Hayakawa A, Lawe D, Lambright D, Bellve KD, Standley C, Lifshitz LM, Fogarty KE, Corvera S (2008) Sorting of EGF and transferrin at the plasma membrane and by cargo-specific signaling to eea1-enriched endosomes. J Cell Sci 121(pt 20):3445–3458
211. Doyon JB, Zeitler B, Cheng J, Cheng AT, Cherone JM, Santiago Y, Lee AH et al (2011) Rapid and efficient clathrin-mediated endocytosis revealed in genome-edited mammalian cells. Nat Cell Biol 13(3):331–337

212. Eliceiri KW, Berthold MR, Goldberg IG, Ibáñez L, Manjunath BS, Martone ME, Murphy RF et al (2012) Biological imaging software tools. Nat Methods 9(7):697–710
213. Jaqaman K, Loerke D, Mettlen M, Kuwata H, Grinstein S, Schmid SL, Danuser G (2008) Robust single-particle tracking in live-cell time-lapse sequences. Nat Methods 5(8): 695–702
214. Liang L, Shen H, De Camilli P, Duncan JS (2010) Tracking clathrin coated pits with a multiple hypothesis based method. Med Image Comput Comput Assist Interv 13(pt 2):315–322
215. Meijering E, Dzyubachyk O, Smal I (2012) Methods for cell and particle tracking. Methods Enzymol 504:183–200
216. Weigert R, Sramkova M, Parente L, Amornphimoltham P, Masedunskas A (2010) Intravital microscopy: a novel tool to study cell biology in living animals. Histochem Cell Biol 133(5):481–491

Chapter 2
Intimate and Facultative? Regulation of Clathrin-Mediated Endocytosis by the Actin Cytoskeleton

Tal Hirschhorn and Marcelo Ehrlich

Abstract Clathrin-mediated endocytosis (CME) involves the timely coordination of plasma membrane deformation, clathrin coat assembly, and cargo inclusion. CME culminates with vesicle release through membrane scission, followed by internalization and uncoating en route to the endosome. The biochemical and biophysical requirements of CME are supplied by a broad number of regulators, many of which bind actin and/or modify actin dynamics. The multidomain structure of these regulators enables their integration into signal-based cellular programs. The architecture and dynamic nature of the actin cytoskeleton establish its influence on distinct aspects of CME, ranging from membrane compartmentalization and acquisition of membrane curvature to coated pit invagination and vesicle movement. However, in mammalian cells, the regulation of CME by the actin cytoskeleton may be facultative, as CME still occurs, in certain cellular contexts, in the absence of actin polymerization. This chapter addresses the complexity of CME regulation by actin and expands on the role of membrane compartmentalization on the spatial organization of clathrin-coated pits and on the multiplicity of differently structured regulators of actin dynamics at the coated pit.

2.1 Introduction

Membrane dynamics in general and endocytosis in particular reflect and regulate timely alterations to the biophysical and biochemical properties of the plasma membrane. The agility and robustness of the co-regulation of membrane composition and its dynamics result from intertwined feedback loops involving (1) local enzyme-mediated alterations to lipid composition; (2) modifications to the structure and

---

T. Hirschhorn • M. Ehrlich (✉)
Department of Cell Research and Immunology, Tel Aviv University, Tel Aviv, Israel
e-mail: marceloe@post.tau.ac.il

Y. Yarden and G. Tarcic (eds.), *Vesicle Trafficking in Cancer*,
DOI 10.1007/978-1-4614-6528-7_2, © Springer Science+Business Media New York 2013

dynamics of the actin cytoskeleton in the proximity of the membrane; (3) the sensing and manipulation of membrane curvature, tension, and area; and (4) the recruitment and/or removal of regulatory proteins through membrane traffic, lateral diffusion, or cytosol-membrane exchange. These feedback loops are fully integrated into signal transduction cascades and form part of the cellular response to diverse stimuli. Thus, membrane trafficking typically determines the sensitivity of the cell to exogenous stimuli through the determination of the levels and localization of signaling receptors and their second messengers at the plasma membrane [1–3]. Complementarily, the degree of regulation of membrane traffic by signal-mediated cellular programs is exemplified by the sensitivity of clathrin-mediated endocytosis (CME) to perturbations to the expression of numerous kinases [4]. Notably, typical behaviors of tumor cells in general and metastatic cells in particular, such as their adhesiveness, motility, and invasiveness, stem from the derailment of membrane trafficking processes at critical regulatory nodes [5, 6]. In this context, loss of polarity through dedifferentiation and epithelial to mesenchymal transition, and the adhesion of the cell to different surfaces in the course of invasion, may alter the endocytic process and the degree of its dependency on an active actin cytoskeleton (see below).

The overall objective of the present chapter is to describe and discuss mechanisms of involvement of actin in the regulation of endocytosis. Due to the greater degree of mechanistic understanding that has been achieved on CME relative to other endocytic pathways, I center on the roles performed by the actin cytoskeleton and actin dynamics in CME.

Initially, the role of actin in CME was addressed through the observation of the effects of actin-perturbing drugs (such as the G-actin sequester latrunculin, the actin polymerization inhibitor cytochalasin, and the inducer of actin polymerization and stabilization jasplakinolide) on the uptake of ligands and receptors known to be internalized through CME and on the amount and localization of arrested endocytic intermediates [7–13]. These studies identified a variable degree of dependency of CME on the actin cytoskeleton. For example, in polarized cells, CME at the apical membrane depended on actin dynamics while no such dependency was observed at the basolateral membrane. The development of fluorescence-based time-lapse microscopy techniques allowed for experiments addressing the role of actin in "endocytic dynamics." These experiments are now routinely performed at submicron resolution in x-, y-, and z-axis, at ~ microsecond temporal resolution and on large number endocytic events, constituting a critical addition to biochemical and electron microscopy experiments. These technical developments allow for the study of the recruitment kinetics of specific factors to individual coated pits, of the alterations to the dynamic parameters of coated pit assembly upon experimental manipulations, and of the regulation of the spatial distribution of the sites of formation of clathrin-coated structures.

Notable contributions to the understanding of the role of actin dynamics in CME were obtained through the combination of quantitative microscopy techniques with yeast genetics [14–16]. The picture that emerges from multiple studies is of a multistep process, involving ~50 different factors, and three main phases: patch assembly and cargo inclusion, actin-mediated invagination and scission, and vesicle movement along actin cables [14–16]. A modular mode of recruitment and action of regulatory components has also been proposed for the basic endocytic event in mammalian cells. The canonical (also termed terminal) endocytic event is consensually divided into initiation, selection of cargo, invagination and coat assembly, scission, internalization, and uncoating [17–20]. In contrast to the apparent uniformity of the endocytic event in yeast, a considerable degree of heterogeneity in the dynamics, requirement for the actin cytoskeleton, and apparent organization of endocytic structures on the plasma membrane has been revealed in different mammalian model systems. This heterogeneity may stem from inherent variability amongst mammalian cells, where differences in cell type or state lead to alterations to the expression or posttranslational modifications of endocytic regulators. Moreover, mammalian cells also differ in plasma membrane composition and in the relative abundance of membrane subcompartments. Importantly, variability in the endocytic process in mammalian cells may also be reflective of its integration with signal transduction programs, as CME is thought to occur in either constitutive or induced modes. The concept of a constitutive mode of endocytosis has been recently challenged as the internalization of the prototypical constitutive clathrin endocytic cargo, the transferrin receptor, was proposed to depend on the transferrin-induced Src-mediated phosphorylation of dynamin and cortactin, two regulators of the assembly of coated pits and of actin dynamics ([21] and see below). Differences amongst studies may also reflect variations in experimental procedures such as the levels of co- or overexpression of the molecules in study, the identity of the membrane which is observed (dorsal or ventral), and the plating surfaces and conditions of the cells in culture (see below). A recent study employing genome editing to express fluorescent fusion proteins (clathrin light chain and dynamin 2) reported a greater efficiency and uniformity of endocytic events, suggesting that the well-accepted notion of an inherent variability of endocytic events may result from an imbalance in expression levels of different factors [22].

The facultative requirement for actin dynamics in endocytic events in mammalian cells, the inherent parameters of a dynamic cytoskeleton, and the susceptibility of actin dynamics to regulation by different signaling pathways place the actin cytoskeleton as a key candidate mechanism for the integration of modulations to endocytosis in broader cellular programs.

2.2 Spatial Regulation of Endocytosis: Roles of the Actin Cytoskeleton

The actin cytoskeleton may regulate CME through the compartmentalization of the plasma membrane. The notion that the actin cytoskeleton compartmentalizes the plasma membrane is supported by studies on the mode of diffusion of lipids and proteins on the plane of the membrane. These studies revealed a slower diffusion rate of lipids and proteins in biological membranes when compared to artificial membranes or membrane blebs and a hop-like mode of mobility of receptors and lipids in biological membranes. This mode of mobility was characterized by transient confinement into microdomains of differing size (on the nano- and meso-scale, from few to hundreds of nanometers in diameter). Importantly, these movement-confining microdomains were sensitive to manipulations to the structure of the actin cytoskeleton [23–26]. The model resulting from these observations is one of the picket-fence-like organizations of the membrane, in which proteins anchored to the cortical actin cytoskeleton act as pickets in a fence demarking microdomains and retard/confine the movement of lipids and proteins through transient/reversible interactions. The notion of a restriction of the lateral mobility of lipids such as phosphatidylinositol (4,5) biphosphate PI(4,5)P2 by picket-like structures was recently substantiated through biophysical measurements in cells carrying out phagocytosis [27]. The putative functional consequences of such a compartmentalization mechanism were addressed through computerized simulations, where an increase in the efficiency and output levels of chemical reactions at the plasma membrane resulted from the transient focusing of molecules to form clusters through interactions with a contractile cytoskeleton prone to remodeling [28], and the compartmentalization of the membrane was predicted to promote burst-like kinetics of chemical reactions [29]. These phenomena may be of particular importance in overcoming limiting concentration barriers for the formation of multicomponent complexes such as clathrin-coated pits. In a broader context, the importance of cytoskeleton-based nanoclustering in signal transduction is becoming increasingly apparent [30, 31]. This supports the notion of an integration of signaling and trafficking through the similarity of the organizational principles involved in the generation of active regions enriched in the molecular complexes that mediate both of these functions at the plasma membrane.

A nonhomogeneous membrane environment, resulting from cytoskeleton-mediated compartmentalization, is predicted to present areas which differ in their propensity of supporting the formation of clathrin-coated structures or pits. A well-studied example of such compartmentalized membrane regions, specialized in endo- or exocytosis, is observed at the presynaptic nerve terminal. At the active zone, actin and other classical cytoskeletal components, such as tubulin, myosin, and spectrin, form a protein-rich electron-dense matrix that aligns docked vesicles to calcium channels localized at the plasma membrane, ensuring efficient fusion and neurotransmitter release upon calcium entry [32]. Adjacent to the active zone is the periactive zone, a region from which the CME of synaptic vesicle components allows for their efficient recycling. The involvement of actin structures and dynamics has also been shown to couple exocytosis and compensatory endocytosis in additional cellular contexts such as neuroendocrine cells and activated Xenopus eggs [33, 34]; however, a role for CME in these processes may depend on the nature and strength of the inducing stimuli and the kinetics of vesicle retrieval. Interestingly, a spatial coupling between sites of endocytosis and immobile structures, termed eisosomes, was identified in yeast, a cellular context in which clathrin-mediated endocytosis is entirely actin-dependent [35]. However, the functional significance of this membrane compartmentalization mode in yeast to vesicular traffic has been recently questioned [36].

The degree of spatial organization of sites of coated pit formation has also been addressed in fibroblast and epithelial cell lines. Initial studies employing confocal time-lapse microscopy proposed that coated pits occurred at defined sites at the membrane, termed "hot spots," and suggested the tethering of coated pits to a "membrane skeleton" [37]. On similar lines, a co-organization in linear arrays of coated pits containing AP2, the actin cytoskeleton, and nonmuscle myosin was observed in HeLa, NIH-3T3, and CHO cells [38]. This distribution pattern was best observed in the top cortical region of the cells, depended on myosin, and was disrupted by cytochalasin D and by the expression of clathrin hub. Such a pattern of alignment of coated pits to actin was also observed on the bottom surface of live NIH3T3 cells imaged by total internal reflection microscopy (TIRFM) [39]. In addition to the actin cytoskeleton, large and long-lived clathrin structures were also proposed to spatially organize clathrin-coated pit initiation. These clathrin platforms, which may be structurally similar to the clathrin profiles identified by Heuser [40], were proposed to support multiple endocytic events [39, 41, 42], possibly by serving as clathrin reservoirs.

In contrast to the view of fixed sites of coated pit formation, studies employing stably transfected BSC-1 cells, clustering analysis of coated pit formation sites, and computerized simulations identified a predominance of a nucleation-growth mode of coated pit formation, albeit with a deviation from a random distribution of coated pit nucleation sites [18]. These studies proposed a model in which de novo nucleation of coated pits occurred in areas prone to coated pit formation (of ~400 nm in diameter), surrounded by areas in which coated pits did not form. This mode of organization was dependent on cell type and on an assembled actin cytoskeleton [18, 43], which was hypothesized to function as a restriction barrier of factors such as PI(4,5)P2. Depolymerization of the actin cytoskeleton correlated with a reduction in coated pit initiations and with a lack of ability of up-regulation of coated pit nucleations (observed upon the washout phase of the treatment with small primary alcohols) [43]. Similarly, inhibition of actin polymerization with latrunculin inhibited the ability of the β2-adrenergic receptor to induce an enlargement of the coated pit signal following stimulation with isoproterenol in cells expressing GFP-arrestin 3 [44]. Recently, a new pattern of coated pit nucleation sites was proposed [45]. In this model, coated pit nucleations occur either randomly throughout the membrane or within "hot spots" (active sites of 200–270 nm radius that turnover every 160 s), which depend on the presence of the actin cytoskeleton but not on its dynamics. In accord with a barrier-concentration function of the cytoskeleton-delineated clusters, overexpression of type I phosphatidylinositol 4-phosphate 5-kinase-α (PIP5KI-α), exogenous addition of PI(4,5)P2, or the clustering of transferrin receptors diminished the clustering tendencies of nucleation sites while the knockdown of PIP5KI-α or of the μ2 subunit of AP2 increased the clustering of coated pit nucleations [45]. A different type of endocytic hot spots, formed by tubovesicular plaques (2–10 μm in size) which concentrate dynamin 2, AP2, and clathrin, was recently identified in cultured hepatocytes [46]. Surprisingly, these clusters were induced by serum starvation, possibly connecting the distribution pattern of coated pits with the nutritional status of the cell.

In addition to intracellular determinants, the spatial organization of the CME machinery may also be influenced by the cell's surroundings. In this context, clathrin-based structures and CME were proposed to mediate and regulate the interactions of cells with the extracellular matrix. Maupin and Pollard [47] employed electron microscopy and observed structures that resembled a combination of flat and curved clathrin-coated pits and clathrin-coated vesicles which closely apposed the underlying substrate in proximity to focal contacts. Similarly, clathrin sheets are prominent in osteoclast membranes that contact the matrix [48]. A role for clathrin structures and endocytosis in mediating cell matrix interactions is in accord with the requirement for CME in the dismantling of focal adhesions [49], the inhibition of cell spreading that is observed upon dynamin inhibition [50], the role played by the CME adaptors Dab2 and Numb in integrin trafficking and cell motility [51, 52], the enhancement in cell spreading and the enlargement of clathrin assemblies observed upon Dab2 overexpression [53], and the increase in cell motility upon the reduction of clathrin plaques through the expression of mutant clathrin light chain [54]. However, clathrin clusters, possibly related to clathrin sheets or plaques, were also identified in adipocytes, which float due to the buoyancy of the stored lipids [41], and thus may not be strictly dependent on cell-matrix attachment. The mode and degree of influence of cell adhesion on endocytosis may have both methodological and biological implications in different experimental setups and may constitute a possible source of variability in the results obtained in different studies, as these routinely employ model systems based on different cell lines, plating conditions, and microscopy techniques. For example, studies employing TIRFM (e.g., [17, 20, 39, 42, 43, 55–58]) concentrate on the imaging of the membrane that contacts the glass coverslip and are thus prone to a greater degree of influence of cell adhesion to the distribution and dynamics of clathrin assemblies. Moreover, the flat and rigid structure of glass coverslips on which cells are cultured and imaged can influence cell behavior. For example, studies addressing the stress cells exert on matrices report that cells pull harder on stiffer surfaces altering in this manner different aspects of cell morphology and intracellular signaling, which may in turn influence trafficking in general and endocytosis in particular [59]. The contribution of cell adhesion to endocytosis has been addressed through the coating of coverslips with different substances such as bovine serum albumin (inert in terms of cell matrix interactions), polylysine, and fibronectin [60], which lead to differences in the dependence of CME on actin dynamics, possibly reflecting the direct attachment of CME cargo (β1 integrin) to the extracellular matrix and the force required to deform a membrane containing such cargo during clathrin-coated vesicle formation [60]. On micropatterned fibronectin substrates, which induce spatial heterogeneity in the cortical actin architecture within a single cell, the lifetimes of clathrin-coated pits were prolonged within the fibronectin-coated islands [61]. This modulation was proposed to stem from alterations to cortical tension and not from the adhesion per se as it depended on stress-fiber formation and was countered by pharmacological agents that alleviate cortical tension. Interestingly, platting of cells on a nano-environment generated by colloidal lithography suggests that cells sense and react to nanoscale topologies as human fibroblasts accumulate attempts of endocytosis of nano-columns with marked effects on the distribution of clathrin and the organization of the cytoskeleton [62]. A requirement for actin polymerization-mediated force in order to counter membrane tension and allow the membrane curvature necessary for vesicle formation has been proposed to occur in yeast, where there is a need to counter turgor pressure [63], in the apical membranes of polarized cells [64], in cells in which membrane tension has been manipulated through hypo-osmotic medium or physical stretching [64], or when endocytosing cargo of exceptional dimensions [65, 66]. Interestingly, membrane tension has been proposed as a trigger for an exocytotic burst observed in the course of cell spreading and could function as a general mechanism in the establishment of a balance between endo- and exocytosis [67].

The influence of an active cytoskeleton on the motility of clathrin-coated pits was initially addressed by Gaidarov et al., which proposed that coated pits demonstrate a measurable but highly restricted motility (within an area of 0.6–0.8 μm in diameter) [68]. In this study, the sevenfold increase in mobility range upon treatment with latrunculin B led to the conclusion that an actin-based framework restricts this lateral mobility. A possible role for the actin cytoskeleton in the regulation of the lateral mobility of membrane-bound clathrin-coated structures was also addressed in additional cell types [39, 43, 69]. However, a mechanistical understanding of the manner by which a multimeric clathrin structure that binds adaptors, auxiliary proteins, and receptors, and induces the local deformation of the plasma membrane, moves laterally relative to its immediate membrane surroundings is lacking. Indeed, alternative explanations for the observed motility of clathrin structures at the plasma membrane are the coordinated motion of different coated pits through the movement of the membrane itself [43], and the possibility that a portion of the motile clathrin-labeled structures may represent endosomes [69].

2.3 Regulators of the Actin/Endocytosis Interface at the Coated Pit

In this section, I will expand on the structure and mode of action of a subset of the regulators of the actin/endocytosis interface. The intention is not to give a comprehensive list of all such regulators, but rather to exemplify the diversity and complexity of their modes of action.

2.3.1 Phosphatidylinositol (4,5) Biphosphate (PI(4,5)P2)

Regulators of the polymerization, organization, and capping of the actin cytoskeleton and endocytic regulators share an affinity towards PI(4,5)P2 [19, 70], allowing for the coordination of their recruitment, the timely and localized modulation of their enzymatic activity, and the coupling of exocytosis and endocytosis. In accord with the repertoire of endocytic proteins which bind PI(4,5)P2, which includes (but is not limited to) the α-, β2-, and μ2-subunits of the AP2 adaptor complex, epsin, Disabled 2 (Dab2), CALM/AP180, sorting nexin 9 (SNX9), Huntingtin interacting protein 1 and Huntingtin interacting protein 1-related (HIP1/HIP1R), and dynamin [68, 71–78], PI(4,5)P2 has been proposed to regulate all stages of clathrin-mediated endocytosis [79]. This notion has been substantiated by the overexpression or knockdown of PI(4,5)P2 generating or consuming proteins and by procedures that lead to the acute depletion of PI(4,5)P2 from the membrane [45, 80–82]. Interestingly, PI(4,5)P2 also regulates exocytosis [83, 84] and may thus be required for the coupling of exocytosis and endocytosis, at least partially through the regulation of actin dynamics. In this context, by regulating the localization and activity of actin-polymerizing/depolymerizing/modifying proteins, PI(4,5)P2 promotes the formation of actin filament structures beneath the plasma membrane [70]. Typically, binding to PI(4,5)P2 activates inducers of actin filament assembly and inhibits the activity of actin filament disassembly factors [70]. For example, PI(4,5)P2 binds to the N-terminal basic domain of the neural Wiskott–Aldrich syndrome protein (N-WASP) and releases it from its auto-inactivated conformation, allowing it to positively regulate Arp2/3 and promote the assembly of branched actin filaments [70]. Also, profilin, which in the absence of free filament ends acts as an ATP–actin monomer sequestering protein, while promoting actin polymerization in their presence, was proposed to sequester dynamin and inhibit endocytosis in neurons, an inhibitory interaction that was relieved by the binding of profilin to PI(4,5)P2 [85]. PI(4,5)P2 also negatively regulates the activity of the actin depolymerizing factor cofilin, possibly by sequestering it to the membrane [86]. However, in yeast, cofilin was shown to be required for the maintenance of the actin flux and is necessary for the optimal functioning of the endocytic process [87], underscoring the complexity of the regulation of actin dynamics at the coated pit.

2.3.2 BAR-Domain Proteins

BAR (Bin–Amphiphysin–Rvs)-domain (F-BAR and N-BAR) proteins coordinate the induction and/or sensing of alterations to membrane curvature, the recruitment of dynamin, and the onset and maintenance of actin polymerization at the neck of the clathrin-coated pit (recently reviewed in [88, 89]). The coordination of these different functions is made possible by their multidomain structure which integrates membrane-binding modules (such as the BAR domain and amphipathic helices) with protein–protein interaction modules. Examples of these protein–protein interaction modules are Src Homology 3 (SH3) domains, which recruit factors endowed with proline-rich domains (PRDs) such as dynamin and N-WASP; sequences involved in the recruitment of AP2 and clathrin, such as clathrin boxes, DPF/W and FxDxF motifs; and NPF motifs that bind Eps15 homology (EH) domains. The precise function of BAR-domain proteins as well as their degree of specificity or redundancy is an active field of study. In a recent temporal mapping of the kinetics of recruitment of different endocytic factors relative to the scission of clathrin-coated vesicles [17], the recruitment of syndapin2 (endowed with an N-terminal F-BAR domain, NPF motifs, and a C-terminal SH3-domain that mediates its autoinhibition or its activation via conformational changes induced by interactions with PRDs [90]) coincided with the recruitment of N-WASP and preceded the recruitment of N-BAR proteins (endophilin2, BIN1, and amphiphysin1) [17]. This order of recruitment is in line with the differences in curvature sensing/induction by F-BAR and N-BAR domains [91]. The latter, in addition to the curvature-inducing/sensing/stabilizing BAR domain, also contains an amphipathic helix that inserts into the membrane bilayer and generates higher curvature angles.

2.3.3 N-WASP

N-WASP is a ubiquitously expressed member of the mammalian WASP family. This family (comprising WASP, N-WASP, and WAVEs 1–3) is characterized by the VCA region (C-terminal verprolin-homology domain, cofilin-homology domain, and the acidic domain) and by its function in the scaffolding of factors involved in the conversion of signals to localized actin polymerization [92]. Structurally, N-WASP contains binding motifs for regulatory factors such as the WASP-interacting protein (WIP); negatively charged phosphoinositides; small GTPases such as cdc42; SH3-domain containing proteins such as dynamin, cortactin, and N- and F-BAR proteins; actin; and the Arp2/3 complex [92]. N-WASP activation involves the coordinated release from an auto-inhibitory conformation through its binding to phosphoinositides, cdc42, SH3-domain proteins (such as amphiphysin [93], SNX9 [94], syndapins [95], and intersectin [96]), and posttranslational modifications such as phosphorylation by tyrosine kinases of the Src family [92]. The regulatory role for N-WASP in endocytosis is underscored by its timely recruitment to internalizing clathrin-coated pits [17, 57, 97], by the accumulation of the receptors for transferrin [98] or epidermal growth factor (EGF) at the plasma membrane upon N-WASP depletion or deletion [97, 98], and by the interference with transferrin internalization upon the sequestering of N-WASP to the mitochondria or upon the introduction of anti-N-WASP antibodies into cells [95]. Moreover, different N-WASP regulators such as Abi1, WIP, and Nck 1/2 also dynamically localize to coated pits [17, 95, 97, 98].

2.3.4 Dynamin

Dynamin is a multidomain large GTPase (~96 kD) whose function is essential for clathrin-mediated endocytosis [19, 20, 50, 99, 100]. Dynamin comprises an N-terminal catalytic GTPase domain, a middle domain involved in intracellular targeting and oligomeric assembly, a pleckstrin homology (PH) domain that mediates PI(4,5)P2 binding, a GED (GTPase effector domain) that mediates self-assembly and enhances GTPase activity upon assembly, and a C-terminal PRD that binds to multiple SH3-containing protein partners (reviewed in [100, 101]). Dynamin differs from small regulatory GTPases (such as those belonging to the Ras superfamily) as it binds GTP with low affinity and shows a propensity towards self-assembly [100, 101]. The GTPase activity of dynamin is greatly stimulated upon self-assembly into helical arrays on lipid templates [102]. The dynamics of dynamin recruitment to the coated pit reveal a peak of recruitment immediately prior to vesicle release [17, 18, 20, 42, 56, 103] in accord with a connection between its oligomerization, GTPase activity, and biological function in vivo. Further support of this notion comes from the observation, by electron microscopy, of the morphology of the endocytic structures arrested upon the expression of dynamin mutants [99, 104] or upon the acute inhibition of its GTPase activity with dynasore, a specific dynamin GTPase inhibitor [50]. However, based on the recruitment profile of dynamin at early stages of coated pit maturation and on the effects of its acute inhibition or of mutations altering its nucleotide affinity or catalytic activity, additional functions for dynamin in coated pit assembly have been proposed [20, 50, 55, 105]. In a recent analysis of the recruitment profiles of wild-type dynamin, dynamin mutants, actin, and the N-BAR proteins endophilin2 and BIN1, in cells in which actin dynamics were unperturbed or inhibited, a positive feedback loop between the GTPase activity of dynamin and actin dynamics on the recruitment of these different factors and on vesicle scission was proposed [55]. While a broad consensus exists on the centrality of dynamin functions in clathrin-mediated endocytosis, the precise mechanism by which dynamin mediates membrane fission and the involvement of dynamin in earlier steps in coated pit formation remain under debate [20, 50, 100]. Three different (although possibly nonexclusive) modes of action of dynamin were proposed for its execution of membrane fission: (1) conformational changes of oligomerized dynamin, coupled to its GTPase cycle, enabling it to act as a mechanochemical fission apparatus and either constrict, twist, or elongate the lipid neck connecting a deeply invaginated pit to the membrane [100, 102, 106–108]; (2) the recruitment of effectors by GTP-bound dynamin (similarly to small G proteins) [109] (the recruited effectors may then mediate membrane fission through the coupling of the enhancement of membrane curvature and the localized activation of cytoskeletal dynamics); and (3) a model in which oligomerized dynamin, based on its tendencies to bind and sequester PI(4,5)P2, contributes to a lipid phase separation, at the coated pit neck, that drives fission [110].

At clathrin-coated pits and at other intracellular sites of action such as actin comets and tails [111, 112] and podosomes [113], the recruitment and function of dynamin are coupled to actin dynamics. This coupling stems from PRD/SH3 interactions of dynamin with regulators of actin dynamics such as cortactin, profilin, N-WASP, and N-BAR proteins [114] and from a recently identified direct interaction of dynamin and actin [115]. In the context of the latter case, the ability of dynamin to bind and bundle F-actin, the stimulation of the dynamin GTPase activity by short actin filaments, and the ability of oligomerized dynamin to specifically promote the de-capping and elongation of gelsolin-capped actin filaments form a feedback loop and support the regulation of global actin dynamics and cell adhesion by dynamin in podocytes [115]. Interestingly, perturbations to this mode of regulation did not markedly affect dynamin-dependent endocytosis [115].

2.3.5 Cortactin

Cortactin, originally identified as an Src kinase substrate, is a multidomain molecular scaffold involved in the regulation of actin dynamics at various intracellular localizations (including membrane ruffles, podosomes, invadopodia, and clathrin-coated pits). The strategic placement of cortactin, at the crossroads of signal transduction, membrane dynamics, and cytoskeleton remodeling, supports its role as a promoter of tumor pathogenesis. Indeed, cortactin is overexpressed in a number of epithelial carcinomas, including head and neck and breast cancers [116]. Structurally, cortactin contains an N-terminal acidic domain (NTA) that mediates binding and activation of the Arp2/3 complex [117], six complete and one partial tandem repeats segments (cortactin repeats) that bind F-actin, an α-helical domain, a proline-rich region (PRR) containing tyrosine and serine phosphorylation sites involved in its regulation, and an SH3 domain that mediates its interactions with the proline-rich binding sequences of WIP [118], N-WASP [119], dynamin 2 [120], and Hip1R [121]. Cortactin promotes actin assembly through direct binding to Arp2/3 and F-actin, through the activation of N-WASP, and through interactions with WIP [122]. Moreover, binding of cortactin to Arp2/3 branch points stabilizes the branched architecture of the cortical cytoskeleton [123]. A functional role for cortactin in CME is in accord with the inhibition in transferrin uptake observed upon its RNAi-mediated knockdown in HeLa cells and with the decrease in coated vesicle formation upon its depletion in a cell-free system [124]. However, cortactin is only identified in a ~ third of the coated pits at the ventral membrane of cells [56, 121], and in a different study [125] its knockdown did not affect transferrin uptake in HeLa or in NIH3T3 cells. The recruitment of cortactin to clathrin-coated pits presents similar dynamics to the recruitment of actin, Apr2/3 and other actin-binding proteins, and to the late-phase recruitment of dynamin, as it accompanies invagination and peaks just prior to vesicle scission [17, 56]. Dynamin is considered a main interaction partner for cortactin in coated pits. The dynamin–cortactin interaction depends on actin polymerization and is enhanced by the tyrosine phosphorylation of cortactin by Src [124, 126], in accord with the proposed role for Src kinase activity in CME [21]. Moreover, dynamin and cortactin may directly regulate actin dynamics through the bundling/cross-linking of actin filaments [127]. Interestingly, a contrasting function, of inhibition of actin polymerization, was proposed for the cortactin–Hip1R complex [121] (see below).

2.3.6 Hip1R

Hip1R is the mammalian homolog of the yeast Sla2p protein, which performs central functions in endocytosis in yeast. The endocytic function of Hip1R, as a regulator of clathrin-coated pits-actin, is mediated by its structural elements: an amino terminal ENTH domain that binds PI(4,5)P2 [73, 128], the central region containing predicted coil–coils which mediate homo-oligomerization [129] and the interaction with clathrin light chain [128, 130], and the talin-HIP1/R/Sla2p actin-tethering C-terminal homology (THATCH) domain that mediates binding to actin [131]. In accord with a role in mediating clathrin–actin interactions, RNAi-mediated knockdown of Hip1R leads to excessive actin polymerization at coated pits and to a stable association between the endocytic machinery and the actin assembly machinery [132]. Similarly, the knockdown of clathrin light chains causes an overassembly of actin and the accumulation of cargo and Hip1R at the trans-Golgi network [133]. The maximal recruitment of Hip1R to coated pits was shown to coincide with that of clathrin and cortactin [121, 132], and Hip1R was suggested to form part of the dynamin/myosin module [17]. While a role for Hip1R–clathrin light chain interactions in the regulation of the interaction between actin and the coated pit is well accepted, a mechanistical understanding of how this occurs is less clear. The Hip1R–clathrin light chain interaction was proposed to be necessary for the internalization of clathrin-coated plaques (large and long-lived clathrin-based structures which form at the bottom membrane of cultured cells and abruptly internalize in an actin-dependent fashion [54]). Similarly, this interaction was shown to be necessary for the overcoming of membrane tension in the endocytosis of apical clathrin-coated pits, of coated pits of cells in which membrane tension was elevated, but not of coated pits localized to the basolateral membranes of unperturbed polarized cells [64]. A similar requirement was shown for actin dynamics, and in their proposed model the authors draw Hip1R as the connector between the coated pit and the actin filament [64]. However, a Hip1R–cortactin complex was suggested to inhibit actin assembly and block actin filament barbed-end elongation [121]. Moreover, the binding of clathrin light chain to Hip1R induces a compact conformation and attenuates actin binding by their THATCH domains, suggesting that clathrin is a negative regulator of Hip1R–actin interactions [129]. Furthermore, based on studies in yeast, a function of pruning Sla2p-actin attachments in the clathrin lattice was proposed for clathrin light chain [134]. Thus, in accord with the complexity of the molecular environment at the coated pit/actin interface, further studies may be required for a precise understanding of the regulation of actin polymerization by Hip1R and clathrin light chain in mammalian cells.

2.3.7 Myosins

The notion of the involvement of myosins, motor proteins that employ the energy of ATP hydrolysis to move cargo along actin filaments, in the regulation of endocytosis was suggested by the identification of severe endocytic defects and the reduction of the motility of actin patches away from the membrane, in yeast strains deleted for the type I myosins Myo5p and Myo3p [135], and by the defects in fluid-phase uptake in Dictyostellium lacking long-tailed Myo1s (myosins B, C, and D) [136]. Indeed, the analysis of the kinetics of recruitment of actin and Myo5p [14], the localization of Myo5p to both the basement and the tip of invaginated actin patches revealed upon the ultrastructural analysis of the distribution of endocytic regulators [137], and the demonstration that both the actin nucleation promoting and the motor activities of Myo5p were necessary for endocytosis [138]; all favor a central role for type I myosin activity in the endocytosis in lower eukaryotes. Recently, type I myosins have also been proposed to be involved in clathrin-mediated endocytosis in mammalian cells. Thus, the long-tailed myosin 1E was shown to interact with synaptojanin and dynamin [139], the expression of its SH3-domain containing tail inhibited transferrin uptake [139], and a fluorescent myosin 1E-Apple construct presented similar recruitment dynamics to coated pits as dynamin, Hip1R, N-WASP, and myosin VI, peaking just prior to the generation of the coated vesicle [17]. Myosin VI, an unconventional and unique myosin that moves towards the minus end of actin filaments [140], is a candidate mediator of the regulation of clathrin-mediated endocytosis by the actin cytoskeleton in mammalian cells. In accord with a function in endocytosis, myosin VI interacts with PI(4,5)P2 [141], with the endocytic adaptor and coated pit component Dab2 [142], and with the PDZ-domain containing adaptor GIPC, which localizes to uncoated vesicles [143]. This localization pattern, which apparently depends on the alternative splicing of myosin VI, the cell type in which it is expressed, and the levels of expression of its interactors [143], supports the notion of multiple functions for myosin VI in clathrin-mediated endocytosis [144, 145]. Thus, myosin VI has been suggested to mediate the concentration of endocytic cargo to the basis of microvilli, the invagination of coated pits, and the movement of vesicles in regions of the cell characterized by a dense actin cortex [145, 146]. The viability and relatively mild phenotype of the Snell's waltzer mouse, which bears an intragenic deletion in the myosin VI gene that leads to an absence of its expression [147] suggests that either the function of myosin VI is not essential, or that compensatory mechanisms enable viability. Interestingly, a recent analysis of fibroblasts from the Snell's waltzer mouse reported an accumulation of shallow clathrin-coated pits, a decrease in clathrin-coated vesicles and a redirection of the internalization of transferrin receptors to an alternative (caveolar) endocytic pathway [148], suggesting a compensatory mechanism for defects in CME. In addition to its functions in endocytosis, myosin VI has been implicated in diverse cellular functions such as Golgi traffic, cell division and migration [145]. Interestingly, overexpression myosin VI has been associated with prostate cancer [149] and with the dissemination of ovarian cancer cells [150], reinforcing the notion of the importance of trafficking regulators in the signaling output of cells.

2.3.8 Intersectin

Intersectin (ITSN) is a multidomain protein that functions as an adaptor and coordinator of endocytosis and exocytosis (reviewed in [151]). In humans, ITSN is encoded by two genes (*ITSN1* and *ITSN2*) both of which are processed to different alternatively spliced forms that differ in domain composition [151]. In its longest form,

ITSN contains two EH domains which mediate its interactions with NPF-containing endocytic proteins such as epsin and stonin 2; a coiled–coiled domain which interacts with eps15 and the Q-SNARE Snap25; five SH3 domains that interact with the proline-rich regions of multiple proteins such as dynamin, synaptojanin, and N-WASP; DH (disabled homology) and PH domains which function as a guanine-nucleotide exchange factor (GEF) for the actin polymerization regulator cdc42; and a C2 domain that binds Ca^{2+} [151–153]. Based on its repertoire of interactors, ITSN is expected to co-modulate clathrin and actin dynamics. In accord with a role in clathrin endocytosis, a region between the SH3 domains A and B was shown to bind the appendages of the α and β subunits of AP2, an interaction that is important for the synaptic vesicle cycle [154]. ITSN was also shown to be recruited early in the coated pit nucleation cycle by FCHO1/2 and to mediate AP2 recruitment [155]. Alterations to ITSN expression levels by either overexpression or knockdown affect the internalization of cargo of the clathrin-mediated pathway [156, 157]. Moreover, ITSN2 knockdown was shown to affect the maturation kinetics of coated pits [58]. Furthermore, ITSN also regulates exocytosis [151, 153, 158], thus providing a link between the exocytic and endocytic processes. In the tightly organized context of the neuromuscular junction, ITSN was shown to change localization between the synaptic vesicle cluster, a region of exocytic activity, and the periactive endocytic zone after Ca^{2+} influx [159]. ITSN2 was also shown to regulate the internalization of caveolae, possibly through a regulation of the actin cytoskeleton [160]. Recently, connections between ITSN and parameters of tumorigenesis are being revealed as ITSN1 expression was shown to be necessary for neuroblastoma tumorigenesis [74] and low level of ITSN2 expression was associated with poor prognosis of breast cancer patients after chemotherapy [161].

2.3.9 Small G Proteins, Rhos, and Arf6

Small G proteins, such as Arf1 and Sar1, play essential and defined roles in the formation of COPI- and COPII-coated vesicles at intracellular membranes and establish the nucleotide dependency of these vesicular transport events [162]. In these regulatory factors, the rearrangement of the switch I and II regions, following the exchange of GDP for GTP, is accompanied by the displacement of a membrane-interacting amphipathic α-helix, thus effectively coupling activation with membrane translocation, anchorage, and membrane bending [163–165]. Such a defined role for small G proteins is not a characteristic of the process of formation of clathrin-coated vesicles at the plasma membrane. In CME, the recognition and induction of membrane curvature and the coupling of these events to coat assembly, cargo recruitment, membrane deformation, and scission is carried out by multiple factors (other than small G proteins) with partially overlapping functions. Interestingly, the important role recently ascribed to the membrane insertion of the VL1 loop of the PH domain of dynamin is reminiscent of the regulation of membrane curvature by the above-mentioned small G proteins [166] and may connect this function to nucleotide dependency.

A less than clear picture emerges when addressing the action of small G proteins in CME. Initially, activated Rac and Rho were shown to inhibit transferrin uptake [167]. In accord with these results, overexpression of constitutively active Rac induced the membrane translocation of synaptojanin 2, a Rac effector and inositol-polyphosphate 5-phosphatase that uses PI(4,5)P2 as a substrate, resulting in an inhibition of the internalization of the transferrin and EGF receptors [168]. In contrast, Rac1 inhibition with NSC-23766 had a marked inhibitory effect on clathrin dynamics at the apical membrane of polarized MDCK cells, while showing a lesser inhibition of clathrin dynamics at basolateral membranes [64]. Thus, while Rac activity does not seem to be an absolute requirement for the basic endocytic event, a balanced effect of Rac on actin dynamics and on the recruitment of endocytic regulators may be required for optimal endocytosis in different cellular contexts. A similar role was also proposed for Arf6, where inhibition of the Arf-GEF cytohesin with SecinH3 induced similar effects on the clathrin endocytic dynamics of polarized MDCK cells as those described for NSC-23766 [64]. In addition to its putative role as a regulator of the actin cytoskeleton, Arf6, when GTP bound, was shown to interact with AP2 and modulate its recruitment to the membrane [169]. Notably, cdc42 inhibition with Secramine A was devoid of effects on CME at all surfaces of polarized MDCK cells [64]. The minor roles proposed for Rho and Arf GTPases in CME are in sharp contrast to the central roles these enzymes play in clathrin-independent forms of uptake [170], possibly reflecting the facultative role of actin dynamics in most cellular contexts of CME and the greater role the actin cytoskeleton may play in other forms of vesicular transport from the plasma membrane.

2.4 Summarizing Comments

Great progress has been made in the identification of the factors involved in clathrin-mediated endocytic events. The integration of structural information, of the description of the recruitment dynamics of different factors, and of the phenotypes obtained upon genetic manipulations in yeast and mammalian cells supports a modular view of the endocytic event, with clear parallels in different species across evolution. In this context, the roles performed by the actin cytoskeleton, such as the compartmentalization of the plasma membrane, the facultative local application of force, and the initial trafficking of the internalized vesicle, may constitute a basis for the adaptability of the endocytic machinery to diverse biological settings (see scheme, Fig. 2.1). The notion of a flexible endocytic machinery is further supported by the large number of players with partially overlapping functions, which regulate the basic CME internalization event in mammalian cells. The understanding of the individual characteristics and roles of these different factors and the unraveling of the manipulation of their expression and posttranslational modifications by signal transduction pathways is one of the main challenges of the field. Moreover, the dissection of this fine-tuning of the endocytic machinery will require a careful

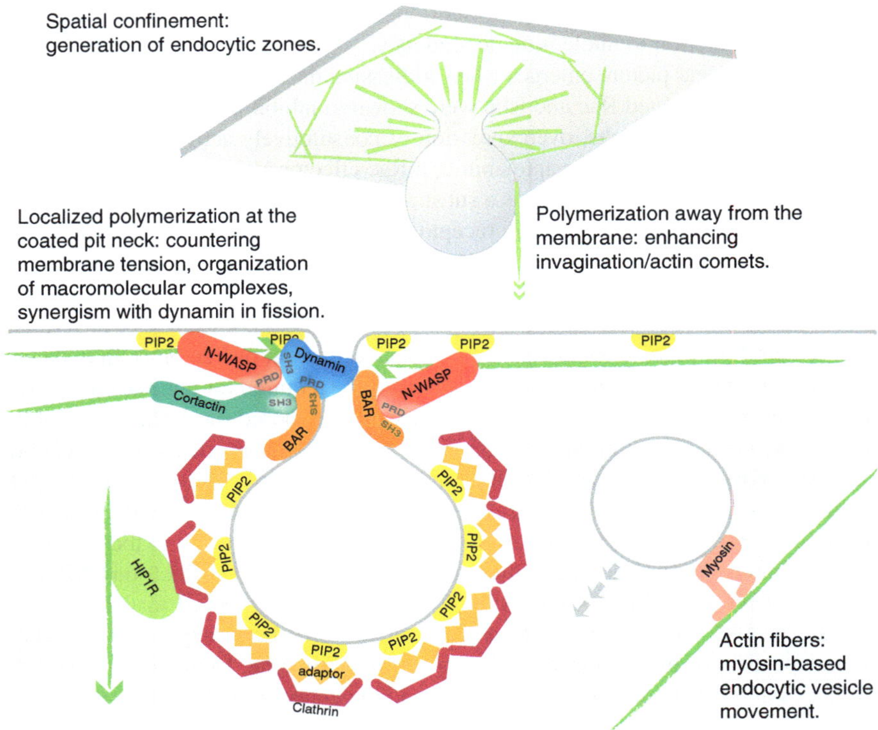

Fig. 2.1 Proposed functions for actin in clathrin-mediated endocytosis. The architecture and dynamic attributes of the actin cytoskeleton form the basis of the different modes of its contribution to clathrin-mediated endocytosis. Thus, the actin cytoskeleton functions as a spatial organizer of the plasma membrane, generating in this manner zones of high propensity for coated pit formations and zones in which coated pit nucleations rarely occur (see Sect. 2.2). Moreover, the timely recruitment of actin polymerization regulators to different localizations of the coated pit via interactions with multidomain regulators of endocytosis (see Sect. 2.3) allows for the contribution of actin dynamics to coated pit invagination and vesicle release. SH3-PRD interactions constitute an important mode of interactions amongst regulators of endocytosis and actin dynamics at the coated pit. Furthermore, the structure of the actin cytoskeleton and the activity of myosins may also contribute to the motility of the internalized vesicle (see Sect. 2.3)

examination of the experimental models to be employed, as the physical and physiological conditions of the cells in study are likely to alter the results which can be obtained in any particular setting. In this context, the membrane trafficking field in general, and the study of the interaction between actin and endocytic dynamics in particular, will certainly benefit from the development of technologies that will enable us to address, with a same degree of precision as the one obtained with the simplified systems of today, systems of a higher complexity, such as cells plated in three dimensional biologically relevant environments and cells challenged with combinations of different stimulatory factors.

References

1. Scita G, Di Fiore PP (2010) The endocytic matrix. Nature 463:464–473
2. Le Roy C, Wrana JL (2005) Clathrin- and non-clathrin-mediated endocytic regulation of cell signalling. Nat Rev Mol Cell Biol 6:112–126
3. Sigismund S, Confalonieri S, Ciliberto A, Polo S, Scita G et al (2012) Endocytosis and signaling: cell logistics shape the eukaryotic cell plan. Physiol Rev 92:273–366
4. Pelkmans L, Fava E, Grabner H, Hannus M, Habermann B et al (2005) Genome-wide analysis of human kinases in clathrin- and caveolae/raft-mediated endocytosis. Nature 436:78–86
5. Mosesson Y, Mills GB, Yarden Y (2008) Derailed endocytosis: an emerging feature of cancer. Nat Rev Cancer 8:835–850
6. Lanzetti L, Di Fiore PP (2008) Endocytosis and cancer: an 'insider' network with dangerous liaisons. Traffic 9:2011–2021
7. Lamaze C, Fujimoto LM, Yin HL, Schmid SL (1997) The actin cytoskeleton is required for receptor-mediated endocytosis in mammalian cells. J Biol Chem 272:20332–20335
8. Gottlieb TA, Ivanov IE, Adesnik M, Sabatini DD (1993) Actin microfilaments play a critical role in endocytosis at the apical but not the basolateral surface of polarized epithelial cells. J Cell Biol 120:695–710
9. Jackman MR, Shurety W, Ellis JA, Luzio JP (1994) Inhibition of apical but not basolateral endocytosis of ricin and folate in Caco-2 cells by cytochalasin D. J Cell Sci 107(pt 9): 2547–2556
10. Altschuler Y, Liu S, Katz L, Tang K, Hardy S et al (1999) ADP-ribosylation factor 6 and endocytosis at the apical surface of Madin-Darby canine kidney cells. J Cell Biol 147:7–12
11. Shurety W, Bright NA, Luzio JP (1996) The effects of cytochalasin D and phorbol myristate acetate on the apical endocytosis of ricin in polarised Caco-2 cells. J Cell Sci 109(pt 12):2927–2935
12. Hyman T, Shmuel M, Altschuler Y (2006) Actin is required for endocytosis at the apical surface of Madin-Darby canine kidney cells where ARF6 and clathrin regulate the actin cytoskeleton. Mol Biol Cell 17:427–437
13. Fujimoto LM, Roth R, Heuser JE, Schmid SL (2000) Actin assembly plays a variable, but not obligatory role in receptor-mediated endocytosis in mammalian cells. Traffic 1:161–171
14. Kaksonen M, Toret CP, Drubin DG (2005) A modular design for the clathrin- and actin-mediated endocytosis machinery. Cell 123:305–320
15. Toret CP, Drubin DG (2006) The budding yeast endocytic pathway. J Cell Sci 119: 4585–4587
16. Robertson AS, Smythe E, Ayscough KR (2009) Functions of actin in endocytosis. Cell Mol Life Sci 66:2049–2065
17. Taylor MJ, Perrais D, Merrifield CJ (2011) A high precision survey of the molecular dynamics of mammalian clathrin-mediated endocytosis. PLoS Biol 9:e1000604
18. Ehrlich M, Boll W, Van Oijen A, Hariharan R, Chandran K et al (2004) Endocytosis by random initiation and stabilization of clathrin-coated pits. Cell 118:591–605
19. McMahon HT, Boucrot E (2011) Molecular mechanism and physiological functions of clathrin-mediated endocytosis. Nat Rev Mol Cell Biol 12:517–533
20. Loerke D, Mettlen M, Yarar D, Jaqaman K, Jaqaman H et al (2009) Cargo and dynamin regulate clathrin-coated pit maturation. PLoS Biol 7:e57
21. Cao H, Chen J, Krueger EW, McNiven MA (2010) SRC-mediated phosphorylation of dynamin and cortactin regulates the "constitutive" endocytosis of transferrin. Mol Cell Biol 30:781–792
22. Doyon JB, Zeitler B, Cheng J, Cheng AT, Cherone JM et al (2011) Rapid and efficient clathrin-mediated endocytosis revealed in genome-edited mammalian cells. Nat Cell Biol 13:331–337
23. Kusumi A, Suzuki KG, Kasai RS, Ritchie K, Fujiwara TK (2011) Hierarchical mesoscale domain organization of the plasma membrane. Trends Biochem Sci 36:604–615

24. Ritchie K, Shan XY, Kondo J, Iwasawa K, Fujiwara T et al (2005) Detection of non-Brownian diffusion in the cell membrane in single molecule tracking. Biophys J 88:2266–2277
25. Fujiwara T, Ritchie K, Murakoshi H, Jacobson K, Kusumi A (2002) Phospholipids undergo hop diffusion in compartmentalized cell membrane. J Cell Biol 157:1071–1081
26. Sako Y, Kusumi A (1995) Barriers for lateral diffusion of transferrin receptor in the plasma membrane as characterized by receptor dragging by laser tweezers: fence versus tether. J Cell Biol 129:1559–1574
27. Golebiewska U, Kay JG, Masters T, Grinstein S, Im W et al (2011) Evidence for a fence that impedes the diffusion of phosphatidylinositol 4,5-bisphosphate out of the forming phagosomes of macrophages. Mol Biol Cell 22:3498–3507
28. Chaudhuri A, Bhattacharya B, Gowrishankar K, Mayor S, Rao M (2011) Spatiotemporal regulation of chemical reactions by active cytoskeletal remodeling. Proc Natl Acad Sci U S A 108:14825–14830
29. Kalay Z, Fujiwara TK, Kusumi A (2012) Confining domains lead to reaction bursts: reaction kinetics in the plasma membrane. PLoS One 7:e32948
30. Plowman SJ, Muncke C, Parton RG, Hancock JF (2005) H-ras, K-ras, and inner plasma membrane raft proteins operate in nanoclusters with differential dependence on the actin cytoskeleton. Proc Natl Acad Sci U S A 102:15500–15505
31. Henis YI, Hancock JF, Prior IA (2009) Ras acylation, compartmentalization and signaling nanoclusters (review). Mol Membr Biol 26:80–92
32. Zhai RG, Bellen HJ (2004) The architecture of the active zone in the presynaptic nerve terminal. Physiology (Bethesda) 19:262–270
33. Yu HY, Bement WM (2007) Control of local actin assembly by membrane fusion-dependent compartment mixing. Nat Cell Biol 9:149–159
34. Villanueva J, Torregrosa-Hetland CJ, Garcia-Martinez V, Del Mar Frances M, Viniegra S et al (2012) The F-actin cortex in chromaffin granule dynamics and fusion: a minireview. J Mol Neurosci 48(2):323–327
35. Walther TC, Brickner JH, Aguilar PS, Bernales S, Pantoja C et al (2006) Eisosomes mark static sites of endocytosis. Nature 439:998–1003
36. Brach T, Specht T, Kaksonen M (2011) Reassessment of the role of plasma membrane domains in the regulation of vesicular traffic in yeast. J Cell Sci 124:328–337
37. Gaidarov I, Santini F, Warren RA, Keen JH (1999) Spatial control of coated-pit dynamics in living cells. Nat Cell Biol 1:1–7
38. Bennett EM, Chen CY, Engqvist-Goldstein AE, Drubin DG, Brodsky FM (2001) Clathrin hub expression dissociates the actin-binding protein Hip1R from coated pits and disrupts their alignment with the actin cytoskeleton. Traffic 2:851–858
39. Yarar D, Waterman-Storer CM, Schmid SL (2005) A dynamic actin cytoskeleton functions at multiple stages of clathrin-mediated endocytosis. Mol Biol Cell 16:964–975
40. Heuser J (1980) Three-dimensional visualization of coated vesicle formation in fibroblasts. J Cell Biol 84:560–583
41. Bellve KD, Leonard D, Standley C, Lifshitz LM, Tuft RA et al (2006) Plasma membrane domains specialized for clathrin-mediated endocytosis in primary cells. J Biol Chem 281:16139–16146
42. Merrifield CJ, Feldman ME, Wan L, Almers W (2002) Imaging actin and dynamin recruitment during invagination of single clathrin-coated pits. Nat Cell Biol 4:691–698
43. Boucrot E, Saffarian S, Massol R, Kirchhausen T, Ehrlich M (2006) Role of lipids and actin in the formation of clathrin-coated pits. Exp Cell Res 312:4036–4048
44. Santini F, Gaidarov I, Keen JH (2002) G protein-coupled receptor/arrestin3 modulation of the endocytic machinery. J Cell Biol 156:665–676
45. Nunez D, Antonescu C, Mettlen M, Liu A, Schmid SL et al (2011) Hotspots organize clathrin-mediated endocytosis by efficient recruitment and retention of nucleating resources. Traffic 12:1868–1878
46. Cao H, Krueger EW, McNiven MA (2011) Hepatocytes internalize trophic receptors at large endocytic "Hot Spots". Hepatology 54:1819–1829

47. Maupin P, Pollard TD (1983) Improved preservation and staining of HeLa cell actin filaments, clathrin-coated membranes, and other cytoplasmic structures by tannic acid-glutaraldehyde-saponin fixation. J Cell Biol 96:51–62
48. Akisaka T, Yoshida H, Suzuki R, Shimizu K, Takama K (2003) Clathrin sheets on the protoplasmic surface of ventral membranes of osteoclasts in culture. J Electron Microsc (Tokyo) 52:535–543
49. Ezratty EJ, Bertaux C, Marcantonio EE, Gundersen GG (2009) Clathrin mediates integrin endocytosis for focal adhesion disassembly in migrating cells. J Cell Biol 187:733–747
50. Macia E, Ehrlich M, Massol R, Boucrot E, Brunner C et al (2006) Dynasore, a cell-permeable inhibitor of dynamin. Dev Cell 10:839–850
51. Teckchandani A, Toida N, Goodchild J, Henderson C, Watts J et al (2009) Quantitative proteomics identifies a Dab2/integrin module regulating cell migration. J Cell Biol 186:99–111
52. Nishimura T, Kaibuchi K (2007) Numb controls integrin endocytosis for directional cell migration with aPKC and PAR-3. Dev Cell 13:15–28
53. Chetrit D, Ziv N, Ehrlich M (2009) Dab2 regulates clathrin assembly and cell spreading. Biochem J 418:701–715
54. Saffarian S, Cocucci E, Kirchhausen T (2009) Distinct dynamics of endocytic clathrin-coated pits and coated plaques. PLoS Biol 7:e1000191
55. Taylor MJ, Lampe M, Merrifield CJ (2012) A feedback loop between dynamin and actin recruitment during clathrin-mediated endocytosis. PLoS Biol 10:e1001302
56. Merrifield CJ, Perrais D, Zenisek D (2005) Coupling between clathrin-coated-pit invagination, cortactin recruitment, and membrane scission observed in live cells. Cell 121:593–606
57. Merrifield CJ, Qualmann B, Kessels MM, Almers W (2004) Neural Wiskott Aldrich Syndrome Protein (N-WASP) and the Arp2/3 complex are recruited to sites of clathrin-mediated endocytosis in cultured fibroblasts. Eur J Cell Biol 83:13–18
58. Mettlen M, Stoeber M, Loerke D, Antonescu CN, Danuser G et al (2009) Endocytic accessory proteins are functionally distinguished by their differential effects on the maturation of clathrin-coated pits. Mol Biol Cell 20:3251–3260
59. Janmey PA, Weitz DA (2004) Dealing with mechanics: mechanisms of force transduction in cells. Trends Biochem Sci 29:364–370
60. Batchelder EM, Yarar D (2010) Differential requirements for clathrin-dependent endocytosis at sites of cell-substrate adhesion. Mol Biol Cell 21:3070–3079
61. Liu AP, Loerke D, Schmid SL, Danuser G (2009) Global and local regulation of clathrin-coated pit dynamics detected on patterned substrates. Biophys J 97:1038–1047
62. Dalby MJ, Berry CC, Riehle MO, Sutherland DS, Agheli H et al (2004) Attempted endocytosis of nano-environment produced by colloidal lithography by human fibroblasts. Exp Cell Res 295:387–394
63. Aghamohammadzadeh S, Ayscough KR (2009) Differential requirements for actin during yeast and mammalian endocytosis. Nat Cell Biol 11:1039–1042
64. Boulant S, Kural C, Zeeh JC, Ubelmann F, Kirchhausen T (2011) Actin dynamics counteract membrane tension during clathrin-mediated endocytosis. Nat Cell Biol 13:1124–1131
65. Cureton DK, Massol RH, Saffarian S, Kirchhausen TL, Whelan SP (2009) Vesicular stomatitis virus enters cells through vesicles incompletely coated with clathrin that depend upon actin for internalization. PLoS Pathog 5:e1000394
66. Cureton DK, Massol RH, Whelan SP, Kirchhausen T (2010) The length of vesicular stomatitis virus particles dictates a need for actin assembly during clathrin-dependent endocytosis. PLoS Pathog 6:e1001127
67. Gauthier NC, Fardin MA, Roca-Cusachs P, Sheetz MP (2011) Temporary increase in plasma membrane tension coordinates the activation of exocytosis and contraction during cell spreading. Proc Natl Acad Sci U S A 108:14467–14472
68. Gaidarov I, Keen JH (1999) Phosphoinositide-AP-2 interactions required for targeting to plasma membrane clathrin-coated pits. J Cell Biol 146:755–764
69. Keyel PA, Watkins SC, Traub LM (2004) Endocytic adaptor molecules reveal an endosomal population of clathrin by total internal reflection fluorescence microscopy. J Biol Chem 279: 13190–13204

70. Saarikangas J, Zhao H, Lappalainen P (2010) Regulation of the actin cytoskeleton-plasma membrane interplay by phosphoinositides. Physiol Rev 90:259–289
71. Jackson LP, Kelly BT, McCoy AJ, Gaffry T, James LC et al (2010) A large-scale conformational change couples membrane recruitment to cargo binding in the AP2 clathrin adaptor complex. Cell 141:1220–1229
72. Yarar D, Waterman-Storer CM, Schmid SL (2007) SNX9 couples actin assembly to phosphoinositide signals and is required for membrane remodeling during endocytosis. Dev Cell 13:43–56
73. Itoh T, Koshiba S, Kigawa T, Kikuchi A, Yokoyama S et al (2001) Role of the ENTH domain in phosphatidylinositol-4,5-bisphosphate binding and endocytosis. Science 291:1047–1051
74. Russo A, O'Bryan JP (2012) Intersectin 1 is required for neuroblastoma tumorigenesis. Oncogene 31(46):4828–4834
75. Vallis Y, Wigge P, Marks B, Evans PR, McMahon HT (1999) Importance of the pleckstrin homology domain of dynamin in clathrin-mediated endocytosis. Curr Biol 9:257–260
76. Yun M, Keshvara L, Park CG, Zhang YM, Dickerson JB et al (2003) Crystal structures of the Dab homology domains of mouse disabled 1 and 2. J Biol Chem 278:36572–36581
77. Ford MG, Pearse BM, Higgins MK, Vallis Y, Owen DJ et al (2001) Simultaneous binding of PtdIns(4,5)P2 and clathrin by AP180 in the nucleation of clathrin lattices on membranes. Science 291:1051–1055
78. Legendre-Guillemin V, Wasiak S, Hussain NK, Angers A, McPherson PS (2004) ENTH/ANTH proteins and clathrin-mediated membrane budding. J Cell Sci 117:9–18
79. Antonescu CN, Aguet F, Danuser G, Schmid SL (2011) Phosphatidylinositol-(4,5)-bisphosphate regulates clathrin-coated pit initiation, stabilization, and size. Mol Biol Cell 22:2588–2600
80. Varnai P, Thyagarajan B, Rohacs T, Balla T (2006) Rapidly inducible changes in phosphatidylinositol 4,5-bisphosphate levels influence multiple regulatory functions of the lipid in intact living cells. J Cell Biol 175:377–382
81. Padron D, Wang YJ, Yamamoto M, Yin H, Roth MG (2003) Phosphatidylinositol phosphate 5-kinase Ibeta recruits AP-2 to the plasma membrane and regulates rates of constitutive endocytosis. J Cell Biol 162:693–701
82. Zoncu R, Perera RM, Sebastian R, Nakatsu F, Chen H et al (2007) Loss of endocytic clathrin-coated pits upon acute depletion of phosphatidylinositol 4,5-bisphosphate. Proc Natl Acad Sci U S A 104:3793–3798
83. Aikawa Y, Martin TF (2003) ARF6 regulates a plasma membrane pool of phosphatidylinositol(4,5)bisphosphate required for regulated exocytosis. J Cell Biol 162:647–659
84. Wen PJ, Osborne SL, Zanin M, Low PC, Wang HT et al (2011) Phosphatidylinositol(4,5) bisphosphate coordinates actin-mediated mobilization and translocation of secretory vesicles to the plasma membrane of chromaffin cells. Nat Commun 2:491
85. Gareus R, Di Nardo A, Rybin V, Witke W (2006) Mouse profilin 2 regulates endocytosis and competes with SH3 ligand binding to dynamin 1. J Biol Chem 281:2803–2811
86. van Rheenen J, Song X, van Roosmalen W, Cammer M, Chen X et al (2007) EGF-induced PIP2 hydrolysis releases and activates cofilin loc.lly in carcinoma cells. J Cell Biol 179:1247–1259
87. Okreglak V, Drubin DG (2007) Cofilin recruitment and function during actin-mediated endocytosis dictated by actin nucleotide state. J Cell Biol 178:1251–1264
88. Qualmann B, Koch D, Kessels MM (2011) Let's go bananas: revisiting the endocytic BAR code. EMBO J 30:3501–3515
89. Rao Y, Haucke V (2011) Membrane shaping by the Bin/amphiphysin/Rvs (BAR) domain protein superfamily. Cell Mol Life Sci 68:3983–3993
90. Kessels MM, Qualmann B (2004) The syndapin protein family: linking membrane trafficking with the cytoskeleton. J Cell Sci 117:3077–3086
91. Henne WM, Kent HM, Ford MG, Hegde BG, Daumke O et al (2007) Structure and analysis of FCHo2 F-BAR domain: a dimerizing and membrane recruitment module that effects membrane curvature. Structure 15:839–852

92. Takenawa T, Suetsugu S (2007) The WASP-WAVE protein network: connecting the membrane to the cytoskeleton. Nat Rev Mol Cell Biol 8:37–48
93. Yamada H, Padilla-Parra S, Park SJ, Itoh T, Chaineau M et al (2009) Dynamic interaction of amphiphysin with N-WASP regulates actin assembly. J Biol Chem 284:34244–34256
94. Shin N, Lee S, Ahn N, Kim SA, Ahn SG et al (2007) Sorting nexin 9 interacts with dynamin 1 and N-WASP and coordinates synaptic vesicle endocytosis. J Biol Chem 282:28939–28950
95. Kessels MM, Qualmann B (2002) Syndapins integrate N-WASP in receptor-mediated endocytosis. EMBO J 21:6083–6094
96. Hussain NK, Jenna S, Glogauer M, Quinn CC, Wasiak S et al (2001) Endocytic protein intersectin-l regulates actin assembly via Cdc42 and N-WASP. Nat Cell Biol 3:927–932
97. Benesch S, Polo S, Lai FP, Anderson KI, Stradal TE et al (2005) N-WASP deficiency impairs EGF internalization and actin assembly at clathrin-coated pits. J Cell Sci 118:3103–3115
98. Innocenti M, Gerboth S, Rottner K, Lai FP, Hertzog M et al (2005) Abi1 regulates the activity of N-WASP and WAVE in distinct actin-based processes. Nat Cell Biol 7:969–976
99. Damke H, Baba T, Warnock DE, Schmid SL (1994) Induction of mutant dynamin specifically blocks endocytic coated vesicle formation. J Cell Biol 127:915–934
100. Ramachandran R (2011) Vesicle scission: dynamin. Semin Cell Dev Biol 22:10–17
101. Praefcke GJ, McMahon HT (2004) The dynamin superfamily: universal membrane tabulation and fission molecules? Nat Rev Mol Cell Biol 5:133–147
102. Stow ell MH, Marks B, Wigge P, McMahon HT (1999) Nucleotide-dependent conformational changes in dynamin: evidence for a mechanochemical molecular spring. Nat Cell Biol 1:27–32
103. Rapp port JZ, Hayman KP, Kendal S, Simon SM (2008) Dynamics of dynamin during clathrin mediated endocytosis in PC12 cells. PLoS One 3:e2416
104. Osaka T, Ikeda K (1983) Possible temperature-dependent blockage of synaptic vesicle recycling induced by a single gene mutation in Drosophila. J Neurobiology 14:207–225
105. Damke H, Bins DD, Ueda H, Schmid SL, Baba T (2001) Dynamin GTPase domain mutants block endocytic vesicle formation at morphologically distinct stages. Mol Biol Cell 12:2578–2589
106. Sweeter SM, Hingham JE (1998) Dynamin undergoes a GTP-dependent conformational change causing gesticulation. Cell 93:1021–1029
107. Roux A, Oyaizu K, Frost A, De Camille P (2006) GTP-dependent twisting of dynamin implicates constriction and tension in membrane fission. Nature 441:528–531
108. Bashkiria PV, Asimov SA, Evasive AI, Schmid SL, Zimmer berg J et al (2008) GTPase cycle of dynamin is coupled to membrane squeeze and release, leading to spontaneous fission. Cell 135:1276–1286
109. Narayanan R, Leonard M, Song BD, Schmid SL, Ram swami M (2005) An internal GAP domain negatively regulates presynaptic dynamin in vivo: a two-step model for dynamin function. J Cell Biol 169:117–126
110. Liu J, Kaksonen M, Drubin DG, Ouster G (2006) Endocytic vesicle scission by lipid phase boundary forces. Proc Natl Acad Sci U S A 103:10277–10282
111. Roth JD, Krueger EW, Cao H, McNiven MA (2002) The large GTPase dynamin regulates actin comet formation and movement in living cells. Proc Natl Acad Sci U S A 99: 167–172
112. Lee E, De Camille P (2002) Dynamin at actin tails. Proc Natl Acad Sci U S A 99:161–166
113. McNiven MA, Baldassarre M, Buccione R (2004) The role of dynamin in the assembly and function of podosomes and invadopodia. Front Biosci 9:1944–1953
114. Schafer DA (2004) Regulating actin dynamics at membranes: a focus on dynamin. Traffic 5:463–469
115. Gu C, Yaddanapudi S, Weins A, Osborn T, Reiser J et al (2010) Direct dynamin-actin interactions regulate the actin cytoskeleton. EMBO J 29:3593–3606
116. Buday L, Downward J (2007) Roles of cortactin in tumor pathogenesis. Biochim Biophys Acta 1775:263–273

117. Higgs HN, Pollard TD (2001) Regulation of actin filament network formation through ARP2/3 complex: activation by a diverse array of proteins. Annu Rev Biochem 70:649–676
118. Kinley AW, Weed SA, Weaver AM, Karginov AV, Bissonette E et al (2003) Cortactin interacts with WIP in regulating Arp2/3 activation and membrane protrusion. Curr Biol 13:384–393
119. Weaver AM, Heuser JE, Karginov AV, Lee WL, Parsons JT et al (2002) Interaction of cortactin and N-WASp with Arp2/3 complex. Curr Biol 12:1270–1278
120. McNiven MA, Kim L, Krueger EW, Roth JD, Cao H et al (2000) Regulated interactions between dynamin and the actin-binding protein cortactin modulate cell shape. J Cell Biol 151:187–198
121. Le Clainche C, Pauly BS, Zhang CX, Engqvist-Goldstein AE, Cunningham K et al (2007) A Hip1R-cortactin complex negatively regulates actin assembly associated with endocytosis. EMBO J 26:1199–1210
122. Ammer AG, Weed SA (2008) Cortactin branches out: roles in regulating protrusive actin dynamics. Cell Motil Cytoskeleton 65:687–707
123. Weaver AM, Karginov AV, Kinley AW, Weed SA, Li Y et al (2001) Cortactin promotes and stabilizes Arp2/3-induced actin filament network formation. Curr Biol 11:370–374
124. Zhu J, Zhou K, Hao JJ, Liu J, Smith N et al (2005) Regulation of cortactin/dynamin interaction by actin polymerization during the fission of clathrin-coated pits. J Cell Sci 118:807–817
125. Barroso C, Rodenbusch SE, Welch MD, Drubin DG (2006) A role for cortactin in Listeria monocytogenes invasion of NIH 3T3 cells, but not in its intracellular motility. Cell Motil Cytoskeleton 63:231–243
126. Zhu J, Yu D, Zeng XC, Zhou K, Zhan X (2007) Receptor-mediated endocytosis involves tyrosine phosphorylation of cortactin. J Biol Chem 282:16086–16094
127. Mooren OL, Kotova TI, Moore AJ, Schafer DA (2009) Dynamin2 GTPase and cortactin remodel actin filaments. J Biol Chem 284:23995–24005
128. Engqvist-Goldstein AE, Warren RA, Kessels MM, Keen JH, Heuser J et al (2001) The actin-binding protein Hip1R associates with clathrin during early stages of endocytosis and promotes clathrin assembly in vitro. J Cell Biol 154:1209–1223
129. Wilbur JD, Chen CY, Manalo V, Hwang PK, Fletterick RJ et al (2008) Actin binding by Hip1 (huntingtin-interacting protein 1) and Hip1R (Hip1-related protein) is regulated by clathrin light chain. J Biol Chem 283:32870–32879
130. Chen CY, Brodsky FM (2005) Huntingtin-interacting protein 1 (Hip1) and Hip1-related protein (Hip1R) bind the conserved sequence of clathrin light chains and thereby influence clathrin assembly in vitro and actin distribution in vivo. J Biol Chem 280:6109–6117
131. Engqvist-Goldstein AE, Kessels MM, Chopra VS, Hayden MR, Drubin DG (1999) An actin-binding protein of the Sla2/Huntingtin interacting protein 1 family is a novel component of clathrin-coated pits and vesicles. J Cell Biol 147:1503–1518
132. Engqvist-Goldstein AE, Zhang CX, Carreno S, Barroso C, Heuser JE et al (2004) RNAi-mediated Hip1R silencing results in stable association between the endocytic machinery and the actin assembly machinery. Mol Biol Cell 15:1666–1679
133. Poupon V, Girard M, Legendre-Guillemin V, Thomas S, Bourbonniere L et al (2008) Clathrin light chains function in mannose phosphate receptor trafficking via regulation of actin assembly. Proc Natl Acad Sci U S A 105:168–173
134. Boettner DR, Friesen H, Andrews B, Lemmon SK (2011) Clathrin light chain directs endocytosis by influencing the binding of the yeast Hip1R homologue, Sla2, to F-actin. Mol Biol Cell 22:3699–3714
135. Galletta BJ, Chuang DY, Cooper JA (2008) Distinct roles for Arp2/3 regulators in actin assembly and endocytosis. PLoS Biol 6:e1
136. Jung G, Wu X, Hammer JA III (1996) Dictyostelium mutants lacking multiple classic myosin I isoforms reveal combinations of shared and distinct functions. J Cell Biol 133:305–323
137. Idrissi FZ, Grotsch H, Fernandez-Golbano IM, Presciatto-Baschong C, Riezman H et al (2008) Distinct acto/myosin-I structures associate with endocytic profiles at the plasma membrane. J Cell Biol 180:1219–1232

138. Sun Y, Martin AC, Drubin DG (2006) Endocytic internalization in budding yeast requires coordinated actin nucleation and myosin motor activity. Dev Cell 11:33–46
139. Krendel M, Osterweil EK, Mooseker MS (2007) Myosin 1E interacts with synaptojanin-1 and dynamin and is involved in endocytosis. FEBS Lett 581:644–650
140. Wells AL, Lin AW, Chen LQ, Safer D, Cain SM et al (1999) Myosin VI is an actin-based motor that moves backwards. Nature 401:505–508
141. Spudich G, Chibalina MV, Au JS, Arden SD, Buss F et al (2007) Myosin VI targeting to clathrin-coated structures and dimerization is mediated by binding to disabled-2 and PtdIns(4,5)P2. Nat Cell Biol 9:176–183
142. Morris SM, Arden SD, Roberts RC, Kendrick-Jones J, Cooper JA et al (2002) Myosin VI binds to and localises with Dab2, potentially linking receptor-mediated endocytosis and the actin cytoskeleton. Traffic 3:331–341
143. Dance AL, Miller M, Seragaki S, Aryal P, White B et al (2004) Regulation of myosin-VI targeting to endocytic compartments. Traffic 5:798–813
144. Aschenbrenner L, Naccache SN, Hasson T (2004) Uncoated endocytic vesicles require the unconventional myosin, Myo6, for rapid transport through actin barriers. Mol Biol Cell 15:2253–2263
145. Buss F, Kendrick-Jones J (2008) How are the cellular functions of myosin VI regulated within the cell? Biochem Biophys Res Commun 369:165–175
146. Hasson T (2003) Myosin VI: two distinct roles in endocytosis. J Cell Sci 116:3453–3461
147. Avraham KB, Hasson T, Steel KP, Kingsley DM, Russell LB et al (1995) The mouse Snell's waltzer deafness gene encodes an unconventional myosin required for structural integrity of inner ear hair cells. Nat Genet 11:369–375
148. Puri C (2009) Loss of myosin VI no insert isoform (NoI) induces a defect in clathrin-mediated endocytosis and leads to caveolar endocytosis of transferrin receptor. J Biol Chem 284:34998–35014
149. Dunn TA, Chen S, Faith DA, Hicks JL, Platz EA et al (2006) A novel role of myosin VI in human prostate cancer. Am J Pathol 169:1843–1854
150. Yoshida H, Cheng W, Hung J, Montell D, Geisbrecht E et al (2004) Lessons from border cell migration in the Drosophila ovary: a role for myosin VI in dissemination of human ovarian cancer. Proc Natl Acad Sci U S A 101:8144–8149
151. Tsyba L, Nikolaienko O, Dergai O, Dergai M, Novokhatska O et al (2011) Intersectin multi-domain adaptor proteins: regulation of functional diversity. Gene 473:67–75
152. Yamabhai M, Hoffman NG, Hardison NL, McPherson PS, Castagnoli L et al (1998) Intersectin, a novel adaptor protein with two Eps15 homology and five Src homology 3 domains. J Biol Chem 273:31401–31407
153. Okamoto M, Schoch S, Sudhof TC (1999) EHSH1/intersectin, a protein that contains EH and SH3 domains and binds to dynamin and SNAP-25. A protein connection between exocytosis and endocytosis? J Biol Chem 274:18446–18454
154. Pechstein A, Bacetic J, Vahedi-Faridi A, Gromova K, Sundborger A et al (2010) Regulation of synaptic vesicle recycling by complex formation between intersectin 1 and the clathrin adaptor complex AP2. Proc Natl Acad Sci U S A 107:4206–4211
155. Henne WM, Boucrot E, Meinecke M, Evergren E, Vallis Y et al (2011) FCHo proteins are nucleators of clathrin-mediated endocytosis. Science 328:1281–1284
156. Sengar AS, Wang W, Bishay J, Cohen S, Egan SE (1999) The EH and SH3 domain Ese proteins regulate endocytosis by linking to dynamin and Eps15. EMBO J 18:1159–1171
157. Martin NP, Mohney RP, Dunn S, Das M, Scappini E et al (2006) Intersectin regulates epidermal growth factor receptor endocytosis, ubiquitylation, and signaling. Mol Pharmacol 70:1643–1653
158. Malacombe M, Ceridono M, Calco V, Chasserot-Golaz S, McPherson PS et al (2006) Intersectin-1L nucleotide exchange factor regulates secretory granule exocytosis by activating Cdc42. EMBO J 25:3494–3503
159. Evergren E, Gad H, Walther K, Sundborger A, Tomilin N et al (2007) Intersectin is a negative regulator of dynamin recruitment to the synaptic endocytic zone in the central synapse. J Neurosci 27:379–390

160. Klein IK, Predescu DN, Sharma T, Knezevic I, Malik AB et al (2009) Intersectin-2L regulates caveola endocytosis secondary to Cdc42-mediated actin polymerization. J Biol Chem 284:25953–25961
161. Specht K, Harbeck N, Smida J, Annecke K, Reich U et al (2009) Expression profiling identifies genes that predict recurrence of breast cancer after adjuvant CMF-based chemotherapy. Breast Cancer Res Treat 118:45–56
162. Kirchhausen T (2000) Three ways to make a vesicle. Nat Rev Mol Cell Biol 1:187–198
163. Antonny B (2006) Membrane deformation by protein coats. Curr Opin Cell Biol 18: 386–394
164. Beck R, Sun Z, Adolf F, Rutz C, Bassler J et al (2008) Membrane curvature induced by Arf1-GTP is essential for vesicle formation. Proc Natl Acad Sci U S A 105:11731–11736
165. Lee MC, Orci L, Hamamoto S, Futai E, Ravazzola M et al (2005) Sar1p N-terminal helix initiates membrane curvature and completes the fission of a COPII vesicle. Cell 122: 605–617
166. Ramachandran R, Pucadyil TJ, Liu YW, Acharya S, Leonard M et al (2009) Membrane insertion of the pleckstrin homology domain variable loop 1 is critical for dynamin-catalyzed vesicle scission. Mol Biol Cell 20:4630–4639
167. Lamaze C, Chuang TH, Terlecky LJ, Bokoch GM, Schmid SL (1996) Regulation of receptor-mediated endocytosis by Rho and Rac. Nature 382:177–179
168. Malecz N, McCabe PC, Spaargaren C, Qiu R, Chuang Y et al (2000) Synaptojanin 2, a novel Rac1 effector that regulates clathrin-mediated endocytosis. Curr Biol 10:1383–1386
169. Paleotti O, Macia E, Luton F, Klein S, Partisani M et al (2005) The small G-protein Arf6GTP recruits the AP-2 adaptor complex to membranes. J Biol Chem 280:21661–21666
170. Donaldson JG, Porat-Shliom N, Cohen LA (2009) Clathrin-independent endocytosis: a unique platform for cell signaling and PM remodeling. Cell Signal 21:1–6

Chapter 3
Oncogenic Signaling from the Plasma Membrane

Eli Zamir, Nachiket Vartak, and Philippe I.H. Bastiaens

Abstract Signaling reactions on membranes play an important role in extracellular information processing by cells. The amount of signaling proteins on the plasma membrane is dynamically maintained by cells to tightly control the qualitative response properties of the signaling system. When oncogenic mutations occur in signaling proteins that are associated with the plasma membrane, the ensemble behavior of signaling molecules can change to a completely different response regime that changes the phenotype of the cell. In order to illuminate the relevance of this spatial dimension in signaling systems, we will first describe how the concentration of signaling proteins determines the qualitative response properties of simple reaction cycles in homogenous protein solutions. From there, we discuss how this concentration parameter is determined by the spatial distribution of proteins in cells and expand this to explain how the translocation of signaling proteins to membrane surfaces elicits a signaling response by changing their local concentration. Within this framework we then describe how an oncogene product's interaction with its wild-type variant can lead to qualitatively different signaling behaviors that depend on their local concentration at membranes as maintained by spatially organizing reactions. We then argue that spatially organizing reaction systems provide an interesting target for cancer therapy.

E. Zamir • N. Vartak
Department of Systemic Cell Biology, Max Planck Institute of Molecular Physiology, Otto-Hahn-Strasse 11, 44227 Dortmund, Germany

P.I.H. Bastiaens (✉)
Department of Systemic Cell Biology, Max Planck Institute of Molecular Physiology, Otto-Hahn-Strasse 11, 44227 Dortmund, Germany

Technische Universität Dortmund, Otto-Hahn-Strasse 6, 44227 Dortmund, Germany
e-mail: philippe.bastiaens@mpi-dortmund.mpg.de

Y. Yarden and G. Tarcic (eds.), *Vesicle Trafficking in Cancer*,
DOI 10.1007/978-1-4614-6528-7_3, © Springer Science+Business Media New York 2013

3.1 Time and Space in Information Processing by Protein Networks

The second law of thermodynamics implies that isolated systems tend to reach a thermodynamic equilibrium in which all energy is thermal and no energy is left available for work (i.e., maximal entropy)—a state that does not allow life. Living systems must maintain themselves far away from thermodynamic equilibrium to generate asymmetries in their organization by using free energy from the environment. At the cellular level, these processes lead to dynamically maintained steady states in the localization and conformational state of signaling proteins. Such steady states reflect a net balance between rates of opposing biochemical reactions such as phosphorylation as catalyzed by kinases and dephosphorylation as catalyzed by phosphatases. Binding of a growth factor to a cognate receptor will induce a change in the relative rates of these reactions (e.g., by upregulating the activity of a kinase) and thereby alter the steady-state concentration of a reaction state of signaling molecules in response to signals. Such a dynamically maintained network state that corresponds to a specific cell phenotype has been termed "cytoplasmic state" [1].

Cellular phenotypes emerge as a system property of an interconnected ensemble of proteins by changing the dynamically maintained network state. In order to grasp the logic of such a dynamic information processing system, it is helpful to describe it in terms of causality between the active states of their components. Thus, a protein (e.g., an active kinase) that enhances the rate of conversion of another protein to its active state (e.g., phosphorylated state) has a positive causal effect on it. Importantly, causal connections between proteins make their states depend on each other and therefore constrain the possible response properties of the system [2].

We will elaborate on a few simple hypothetical biochemical reaction schemes to clarify these concepts discussed above. Consider a reaction cycle in which a kinase phosphorylates a substrate that is then subsequently dephosphorylated by a phosphatase (Fig. 3.1a). Both enzymes operate in the saturatable Michaelis–Menten kinetic regime. The steady-state concentration of the phosphorylated substrate in this ATP-consuming reaction cycle exhibits a sigmoidal response curve to the concentration of the input signal. This provides a reversible switch-like dose–response to kinase activity. It is noteworthy that the output of such systems can be reversibly turned on and off as the input signal crosses the switching threshold. Many cellular functions, such as cell-fate decisions and memory, require irreversible switches. Achieving this property requires the addition of a fundamental causal topology motif into this system—a positive feedback loop (e.g., Fig. 3.1b). Such a system can get locked in an "on" state once the input signal crosses a threshold [3]. In contrast to responses reaching a fixed-point steady state, many important cellular functions require a stable, but dynamic oscillatory response to signal. Negative feedback loops can generate damped oscillations and autoregulated homeostasis, while their combination with positive feedback loops can give rise to stable oscillations [3].

While the causal topology of a protein network confines the possible types of responses to a signal, the actual response that takes place depends on the kinetic

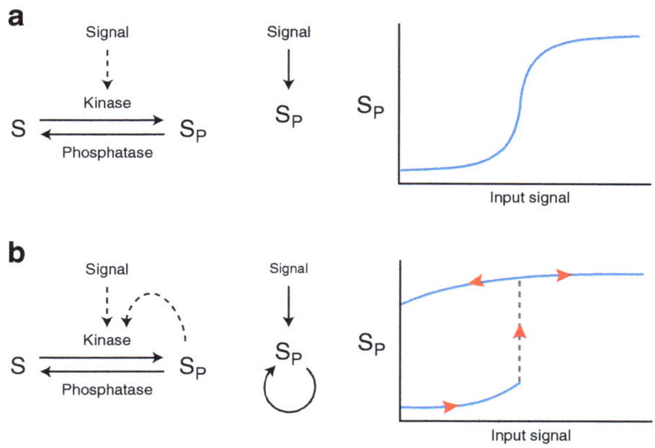

Fig. 3.1 The causal topology of biochemical networks shapes their response properties. (**a**) A biochemical reaction scheme of enzymatic phosphorylation and dephosphorylation of a substrate (S and Sp, respectively) in which the kinase gets further activated by a signal while the phosphatase has a constant constitutive activity level (*left*). If we regard the phosphorylated form of the substrate, Sp, as the active form (hence the "output"), there is a positive causal effect of the signal on the output (*middle*). Given enzymes with Michaelis–Menten kinetics, such a system exhibits a sigmoidal dose–response behavior (*right*). (**b**) The same biochemical system as above except that the phosphorylated form of the substrate (Sp) activates the kinase (*left*). This generates a positive feedback of Sp on itself (*middle*). Due to the positive feedback, the system can become bistable, exhibiting hysteresis to the point of irreversibility and thereby memory. Once the signal passes a certain threshold level the output switches to a high (on) level. This steady state is maintained even if the input signal is no longer present (*right*)

parameters and concentrations of the reactants. Here, it is important to realize that a change in the total concentration of any of the reactants in the system can completely alter the response profile of the system. Let's consider a reaction scheme of a receptor tyrosine kinase (RTK) and its inactivating phosphatase (PTP) that now contains two feedback loops (Fig. 3.2a, b): (1) The RTK is autocatalytic (phosphorylating itself) and (2) the PTP that inhibits the phosphorylation of the kinase is inhibited by the kinase activity. This type of network architecture has been postulated to explain the response properties of epidermal growth factor receptor [4–6]. An important system parameter is the ratio between the expression levels of the kinase and the phosphatase, *R* [4]. By plotting the level of the phosphorylated substrate against *R*, one obtains a bifurcation diagram (Fig. 3.2c), indicating that the system can exhibit three possible behaviors (Fig. 3.2d): (1) At high *R* values (PTP dominates) the system has a single steady state and responds in an analog fashion to input signals that increase the kinase activity (Fig. 3.2d, right). (2) At intermediate *R* values (balance of activities) the system has three steady states of which one is instable. This is the bistable (all-or-none) regime in which a stimulus that raises the kinase activity above a threshold will trigger the full activation of the kinase by its autocatalytic loop and thereby lead to the phosphorylation of most of the kinase (Fig. 3.2d, middle). This state has interesting emergent properties that allow it to

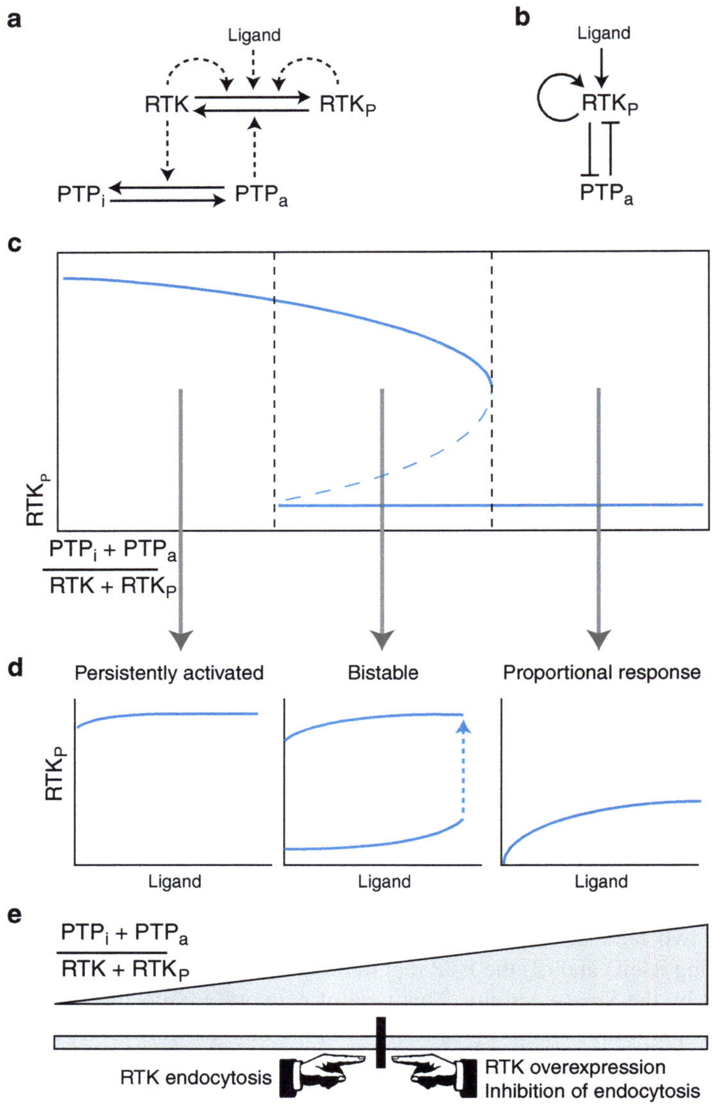

Fig. 3.2 The relative concentration of proteins can determine the response properties of the network. (**a**) A hypothetical biochemical network containing RTKs and PTPs. *Horizontal arrowheads* indicate conversions while *vertical arrowheads* indicate influences on the conversion kinetics. (**b**) The same system as in (**a**) with *arrows* indicating causal relations. (**c**) A bifurcation diagram showing the steady state of the phosphorylated RTK output (RTKp) as a function of the ratio of total PTP level (active+inactive PTP, PTPa and PTPi, respectively) over total RTK level (RTK+RTKp), *R*. (**d**) Dose–response curves showing the steady-state output level (RTKp) as a function of its ligand concentration. The three curves represent different ranges of the bifurcation parameter *R*. (**e**) Normal and pathological processes tune the concentration ratio parameter *R* and accordingly affect the qualitative response properties of the system

propagate signals through space much faster than, for example, by simple diffusion of the kinase. (3) At low R values (RTK dominates) the system has a single (permanently active) steady state that does not respond to input signals (Fig. 3.2d, left).

The example discussed above demonstrates that one can switch the qualitative response properties of a kinase/phosphatase system by changing the ratio of protein concentrations (Fig. 3.2e). Of relevance to our discussion on spatial regulation is that this ratio can be locally regulated in the cell, especially on membranes. For example, the level of RTKs at the plasma membrane is determined by the net balance between the rates of delivery of newly synthesized receptors from the ER to the cell surface, the internalization of the receptors from the surface by endocytosis, and their recycling back to it. In the canonical example of EGFR [7], the internalization rate constant of the receptor in the absence of ligand is slower (~0.02–0.05 min) than its recycling rate (>0.2 min). Therefore, under basal conditions a substantial amount of EGFR is at the plasma membrane at steady-state levels that correlates with its overall expression level. Binding of EGF to EGFR activates the receptor but in parallel also enhances its internalization rate by an order of magnitude (~0.2–0.4 min). While endocytosed inactive EGFRs get recycled rapidly to the plasma membrane, endocytosed active EGFRs are directed to the lysosome to be degraded. Overall, this mechanism thus changes locally the ratio between overall RTK versus PTP activity levels, keeping it low enough to maintain it in a regime that exhibits responsiveness to ligand. On the other hand, genetic aberrations that cause overexpression of EGFR or prevent its downregulation by internalization and degradation bring the system into a permanently active regime on the bifurcation diagram that is unresponsive to ligand and thereby leads to oncogenicity [8–10].

3.2 Dimensionality Reduction and Binding Affinities at the Plasma Membrane

Many signaling paths involve the translocation of cytosolic proteins to the plasma membrane where they interact with other proteins. In some cases these interactions are with proteins that have to be at the plasma membrane for their function, like transmembrane receptors. But in many other cases the plasma membrane hosts interactions between cytosolic proteins that are transiently located there. A fundamental implication of this phenomenon is the reduction of the space dimensionality at which the reaction occurs from the 3D cytosol to the 2D membrane. The effect of space dimensionality reduction on the kinetics and thermodynamics of biochemical reactions is profound and had been explored in several studies. One of these [11] described the binding rate of membrane-associated ligands to the extracellular portion of their receptors. In this model, reduction of dimensionality is generated by nonspecific adsorption of the ligand to the cell surface, followed by 2D surface diffusion to the receptor. As a mirrored analogy, we will adopt here the same formalism to discuss the effect of translocation of cytosolic proteins to the inner aspect of the plasma membrane.

The kinetics and steady states of a reversible protein-binding reaction is determined by the balance between its forward (association) and backward (dissociation) rates. The association rate of proteins is the product of the rate of their random collisions and the probability that a collision will lead to binding. While it is hard to dissect these parameters experimentally, the law of mass action factorizes the association rate to the concentrations of the unbound proteins and a remaining factor, k_{on}:

$$\mathrm{d}C_{AB} / \mathrm{d}t = \underbrace{\overbrace{k_{on} C_A C_B}^{\text{association}} - \overbrace{k_{off} C_{AB}}^{\text{dissociation}}}_{\text{mass action}} \propto \underbrace{(D_A + D_B)\chi_{AB}}_{\propto k_{on}} C_A C_B - k_{off} C_{AB}$$

The concentration of a protein (C) determines the probability to find one of its particles at a certain time and place. Accordingly, the multiplication of the concentrations of two proteins determines the probability that their particles coincide in space. For a given concentration of two proteins, the probability that their particles will be in a colliding distance depends also on their space occupancy (surface area for 3D and circumference length for 2D) as determined by their molecular radius (R_a). Importantly, the rate of collisions depends not only on the concentration of the two proteins but also on how fast they change their positions and thereby integrate random coincidences. Therefore k_{on} has to encapsulate the diffusion speed of the proteins (D). In addition, k_{on} also accounts for the probability (χ) that a collision will lead to binding. Hence, the association rate (i.e., forward rate, F) of a binding reaction in 2D and 3D is proportional to the product of C, D, and χ (to simplify the formalism we consider a binding reaction between two copies of the same protein) [11]:

$$F_2 \approx \chi_2 R_a D_2 C_2 / \sigma_2 \quad F_3 \approx \chi_3 R_a^2 D_3 C_3 / \sigma_3$$

where σ is the free-run distance of the proteins between their collisions with solvent molecules and approximated to be comparable in the membrane and the cytosol ($\sigma_2 \approx \sigma_3$)[11].

To understand what makes association rates at the plasma membrane and in the cytosol different, we can factorize their ratio (F_2/F_3) to ratios between concentrations, diffusion speeds, space occupancy, and probability of a collision to be successful:

$$\frac{F_2}{F_3} \approx \left(\frac{\chi_2}{\chi_3}\right) \cdot \left(\frac{D_2}{D_3}\right) \cdot \left(\frac{C_2}{C_3}\right) \cdot \left(\frac{R_a}{R_a^2}\right) \cdot \left(\frac{\sigma_3}{\sigma_2}\right)$$

1. Changes in concentrations (C_2/C_3) and space occupancies (R_a/R_a^2) (Fig. 3.3 right)—the concentration of a protein with a given copy number, ρ, located inside a sphere with a radius r or on its surface will be $C_3 = \rho/(4\pi r^3/3)$ and $C_2 = \rho/(4\pi r^2)$, respectively. Therefore, assuming spherical cell geometry, recruitment of all protein molecules from the cytosol to the plasma membrane will increase their effective concentrations by a factor of $C_2/C_3 = r/3$. Since this ratio has a dimension of distance, it is useful to combine it with the ratio of the space occupancies (R_a/R_a^2), which has a dimension of distance. The product of these two components equals $r/3R_a$, and is dimensionless.

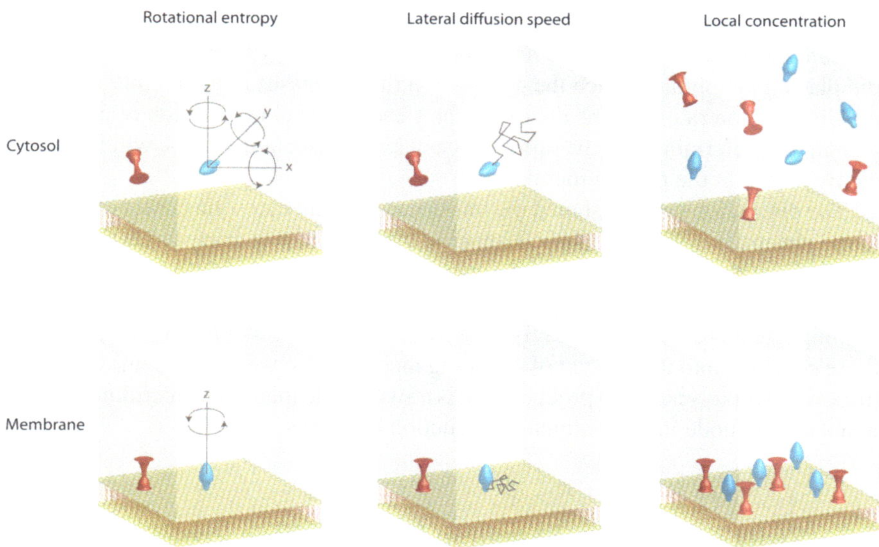

Fig. 3.3 Decomposing the influences of dimensionality reduction on protein-binding rates. At the plasma membrane the rotational entropy of proteins is reduced due to partial confinement of two rotational angles (*left*). This reduction in the degrees of freedom is likely to increase the probability that the proteins that collide will be in an orientation that enables binding. The higher viscosity at the plasma membrane reduces the frequency of collisions by slowing down diffusion (*middle*). On the other hand, the local concentration of proteins gets dramatically higher when they are recruited from the cytosol to the plasma membrane, which increases the frequency of collisions (*right*)

2. Change in diffusion speed (D_2/D_3)—the translocation of proteins to the plasma membrane changes their diffusion speed, roughly reducing it by ten- to hundred-fold due to the higher viscosity of the membrane in comparison to the cytosol (Fig. 3.3 middle).
3. Change in the probability for a productive collision (χ_2/χ_3)—the probability that two proteins will encounter each other in orientation that enables binding depends on their rotational degrees of freedom (i.e., rotational entropy) [12]. In 3D all three Euler rotational angles are unrestricted, while in ideal 2D situation only one rotational angle is unrestricted. If two proteins evolved to bind each other while being anchored to the plasma membrane, their anchoring should allow them to orient properly for interaction between their binding sites. Therefore, any confinement that membrane anchoring of proteins imposes on their rotational diffusion will increase the probability that their collision will lead to binding (Fig. 3.3, left).

Overall, both the gain in concentration and reduction of rotational entropy increase the binding rate between proteins, while the decrease in the diffusion speed counters this effect. Quantitatively, since $r \gg R_a$, it appears that for a wide range of parameters the gain in concentration obtained by the dimensionality reduction is by far the most dominant factor. Therefore, the translocation of proteins to the plasma membrane will lead to a dramatic increase in their association rates. In contrast, as a first-order reaction the dissociation of a protein complex is independent on its concentration and diffusion and hence not expected to vary between the cytosol and the plasma membrane. Since the association rate increases at the membrane, while the dissociation rate remains the same, the steady-state concentration of proteins in a complex, determined by the ratio of association/dissociation rates, will be significantly higher at the plasma membrane.

The discussion so far assumed the translocation of all cytosolic copies of a protein to the plasma membrane. Live cell imaging indicates that in many cases the translocation of fluorescently tagged proteins to the membrane is indeed dramatic. Yet, even if just a small fraction (α) of a protein translocates to the plasma membrane, its binding rate there will significantly increase, as long as $\alpha \gg (R_a/r)$. In order to put these concepts into the context of signaling mechanisms, we describe one of the first discovered proto-oncogene products that is a switchable membrane recruitment factor and a central node in early signal transduction networks.

3.3 Switching on Ras Signaling by Recruiting Proteins to the Plasma Membrane

The Ras small GTPases transmit growth factor signals from the plasma membrane to which they are anchored by posttranslational lipid modifications at their C-terminal hypervariable region (HVR). Several Ras molecules are known that can be distinguished by the posttranslational modification at the HVR that determines their distribution on membranes. Ras is activated by the exchange of bound GDP for GTP mediated by the interaction with guanine nucleotide exchange factors (GEFs) that expose the guanine nucleotide-binding pocket for effective exchange with the surplus of cytoplasmic GTP. Ras proteins are themselves inefficient GTPases requiring interaction with catalytic residues provided by GTPase activating proteins (GAPs) for hydrolyzing GTP to GDP to switch back off. In the active GTP state the conformation of the two so-called switch regions enables the interaction with downstream effectors that harbor Ras-binding (RB) or Ras association (RA) domains. For example, in the canonical Raf–Mek–Erk MAPK cascade, activated Ras-GTP can bind and recruit the cytoplasmic effector Raf to the plasma membrane. This Ras–Raf interaction alone is however not sufficient to fully activate the Raf protein. It is the Ras-mediated enrichment of Raf on the plasma membrane that enables it to be efficiently phosphorylated on several residues by a plethora of plasma membrane-associated kinases (such as Src, protein kinase C (PKC), protein kinase A, and Akt) to become fully active [13] (Fig. 3.4). Ras is in essence a switchable recruitment factor that brings its effector proteins to the plasma membrane and thereby elicits a dimensionality reduction in the space where they can diffuse. This translocation increases their effective concentration by at least an order of magnitude and thereby enhances their reactivity towards each other as discussed above. The amount of Ras that is maintained at the plasma membrane by spatially organizing reaction cycles determines the concentration of translocated effectors that react at the plasma mebrane

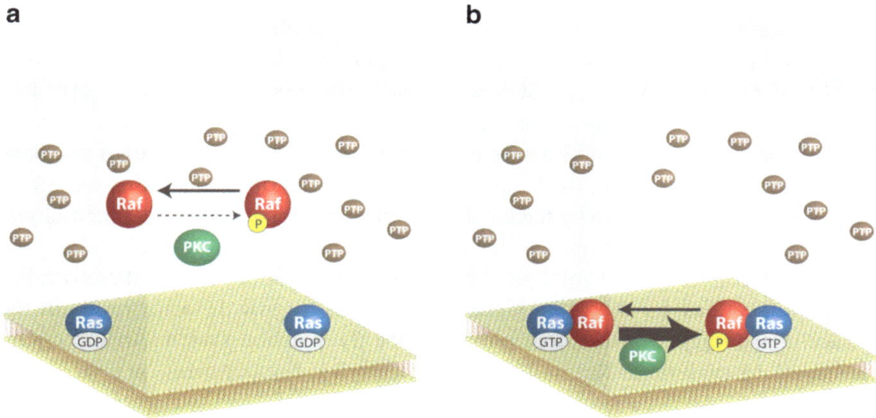

Fig. 3.4 Activation of Raf through dimensionality reduction. (**a**) In the absence of input signal Raf is in the cytosol. In this compartment Raf is maintained inactivated due to its low concentration and that of its activating kinase PKC. On the other hand, the low Km of PKC's phosphatases enables an efficient inactivation. (**b**) Upon growth factor signal, both Raf and PKC, but not the phosphatases, get recruited to the plasma membrane. This shifts the phosphorylation–dephosphorylation balance to promote net phosphorylation and thereby leads to Raf activation

and therefore has a major effect on Ras signaling output. In the cytoplasm, Raf concentrations are in the sub-micromolar range, which is far below the Michaelis–Menten constant (K_m) for an activating kinase such as PKC (Fig. 3.4). Furthermore, phosphatase activity that operates at low K_m will maintain a dephosphorylated state of Raf in this cytoplasmic space [14]. The cytoplasm can thus be perceived as a signaling inactivating space or milieu, in contrast to the signaling activating space of the plasma membrane. The recruitment of effectors to the same cellular compartment where the activating kinases are located not only increases their reactivity towards the activating kinases but also integrates the output of various signaling pathways to make the response more robust to spurious activation events in any of the individual signaling components. For example, if PKC is the activating kinase of Raf, propagation of growth factor signals would require the simultaneous activation of the RasGEF "Son of *sevenless*" (Sos)—to activate Ras and translocate Raf—and phospholipase Cγ, to generate the signaling lipid diacylglycerol to translocate PKC to the plasma membrane [15].

When the effector itself is a regulator of Ras activity, the efficiency of plasma membrane recruitment of such an effector by Ras will also determine the Ras activity response profile. Sos has been extensively studied in that context [16, 17]. Sos is normally recruited to the plasma membrane via its interaction with the adapter protein Grb2, containing SH2 domains that bind phosphotyrosines on activated growth factor receptors. This recruitment of Sos to the plasma membrane is sufficient to activate Ras, representing a straightforward mechanism for downstream Ras signaling to continue. However, Sos also possesses an allosteric binding site to which activated Ras-GTP can bind and enhance its GEF activity. Active Ras-GTP can also directly recruit Sos to the plasma membrane through this allosteric interaction, thereby increasing its concentration on the plasma membrane. Although the accessibility of the allosteric site exhibits another level of regulation via the interaction of the histone and PH domains on SOS with lipids on the plasma membrane [18, 19], it clearly constitutes a node for positive feedback regulation of Ras activity. As explained above, such a system can generate an all-or-none (bistable) response dependent on the strength of the feedback connection [3, 20]. Because the strength of the feedback on Sos is given by the local Ras concentration on the plasma membrane, the reaction systems that maintain Ras on the plasma membrane will also affect the response of Ras to growth factors. The spatial organization of Ras therefore has major implications for the cell's signaling response to growth factor levels in the extracellular milieu. We therefore now describe the reaction systems that maintain the spatial distribution of Ras molecules on membrane compartments in cells.

3.4 The Spatial Organization of Ras GTPases

From where Ras transmits signals in the cell depends on a variety of posttranslational modifications (PTMs) at the C-terminal HVR of Ras. Irreversible prenylation via a thioether bond at the C-terminal cysteine of Ras proteins, which include either addition of a 15-C chain (farnesylation) or a 20-C chain (geranylgeranylation), is an important prerequisite for these proteins to form weak associations with any membrane in the cell [21]. Some Ras family proteins—typically but not exclusively H/NRas—undergo reversible S-palmitoylation via a thioester bond on cysteines at the C-terminus. A subset of the 25 members of DHHC-cysteine-rich domain proteins that are encoded in the human genome catalyzes this palmitoyltransferase (PAT) reaction with broad substrate specificity [22, 23]. Palmitoylation further increases the hydrophobicity of these farnesylated Ras molecules and hence stabilizes the interaction with any membrane in the cell. The specific plasma membrane distribution of palmitoylated Ras proteins is maintained by an acylation cycle that works in concert with the secretory functionality of the Golgi [24, 25]. De novo synthesized, farnesylated, but still un-palmitoylated Ras proteins undergo rapid intermembrane diffusion leading to an unspecific partitioning over the extensive and densely packed (endo-)membrane systems of an eukaryotic cell. Upon chance encounter with the Golgi, the Golgi-localized PAT activity ensures efficient palmitoylation of Ras trapping it at the Golgi [26]. From there palmitoylated Ras is transferred to the plasma membrane by the vesicular transport in the secretory pathway [25, 27]. The highly concentrated palmitoylated Ras on the Golgi and the plasma membrane will tend to slowly redistribute to all membranes in the cell by both passive diffusion and active vesicular transport. This entropy-driven randomization of palmitoylated Ras to all membranes is countered by ubiquitously distributed acyl protein thioesterase (APT) activity [28] that depalmitoylates Ras in order to increase its diffusion speed and thereby enhance the kinetics of re-trapping farnesylated Ras at the Golgi by re-palmitoylation to start a new cycle (Fig. 3.5a).

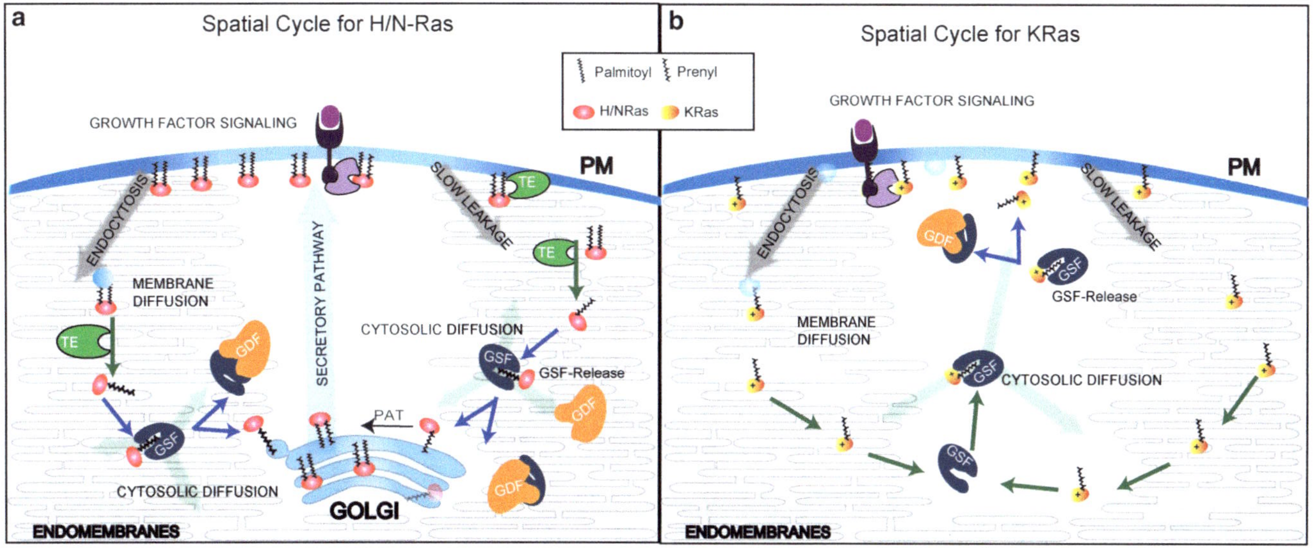

Fig. 3.5 (**a**) the spatial cycle for H/NRas—solely farnesylated HRas tends to accumulate on endomembranes at equilibrium owing to its weak membrane affinity and the large surface area of the endomembrane compartment. The acylation cycle is a reaction–diffusion cycle wherein farnesylated HRas is palmitoylated on the Golgi apparatus, substantially increasing its hydrophobicity. The resultant increased affinity for membranes allows it to be transferred to the plasma membrane via the secretory pathway. Entropic and membrane mixing processes transfer PM-localized Ras to endomembranes. Cell-wide depalmitoylation by thioesterases converts HRas to the weak membrane affinity state and allows interaction with GSFs that solubilize HRas. GDFs release depalmitoylated HRas from GSFs on the Golgi where it can be re-palmitoylated, completing a single turn of the acylation cycle. (**b**) the spatial cycle for KRas—farnesylated KRas tends to dwell on the endomembrane compartment due to its large surface area. GSFs solubilize KRas, allowing it to diffuse faster in the cytoplasm to gain access to the plasma membrane. GDFs release KRas from GSFs at the PM where electrostatic interactions between the polybasic stretch and negatively charged phospholipids stabilize KRas on the PM. Entropic and membrane fission processes transfer KRas to endomembranes where it may once again interact with GSFs to explore the cell volume by the increased diffusion speed for access to the PM—thus completing a turn of the KRas spatial cycle

A central factor in the spatial organization of palmitoylated Ras isoforms is the chaperone that aids in enhancing the diffusion speed of depalmitoylated Ras in the cytoplasm. The 17 kDa delta subunit PDEδ of phosphodiesterase 6 (PDE6) was identified as this factor that specifically binds the farnesyl moiety and thereby solubilizes Ras that is depalmitoylated by thioesterase activity [29]. We have termed the functionality of PDEδa "GDI-like solubilizing factor (GSF)" because it bears structural similarity with guanine nucleotide dissociation inhibitors (GDIs) but does not exhibit preferential binding to the GDP state of G-proteins, as is the case for Rho-GDIs [30]. Knockdown of PDEδ by RNA interference leads to randomized distributions of not only palmitoylated H/NRas but also polycationic KRas to all membranes in the cell, which emphasizes its prominent role as a general solubilizer of farnesylated Ras proteins [29]. The proto-oncogene product KRas4B (hence referred to as KRas) is a prominent member of the class of Ras proteins that bear a polybasic stretch instead of palmitoylatable cysteines in the HVR. This positive charge stabilizes the KRas interaction specifically with the dense negatively charged inner leaflet of the plasma membrane [31, 32]. Consistent with previous findings that KRas reaches the plasma membrane via a non-vesicular, diffusional mechanism [33], newly synthesized, solely farnesylated endomembrane-bound KRas undergoes an intermembrane transfer to be enriched at the plasma membrane by the favorable electrostatic interaction. Because the rate of intermembrane transfer of prenylated Ras is a slow process [34], its interaction with PDEδ facilitates its diffusional exploration of the cytoplasm and thereby increases the kinetics of trapping at the negatively charged plasma membrane (Fig. 3.5b). Fission and fusion processes that maintain the size and shape of the plasma membrane continuously perturb the resulting KRas distribution. Energy-driven processes such as various forms of endocytosis create a flux of KRas on vesicles towards endomembranes where its interaction is weakened by lack of the electrostatic component. Here, KRas is limited to 2D diffusion and low-frequency detachment from membranes where the probability of encountering another endomembrane is far greater than encountering the plasma membrane. This presents a substantial kinetic barrier to plasma membrane (re-) binding leading to a persistent dispersive distribution of KRas on endomembranes. The solution to this kinetic barrier occurs in the form of the GSF, PDEδ, that binds the lipophilic prenyl group of Ras, thus shielding its hydrophobicity from the aqueous cytosol. The soluble GSF-KRas complex increases the diffusional mobility of KRas by an order of magnitude resulting in higher frequency of plasma membrane encounter, where it can rebind to reinstate equilibrium (Fig. 3.5b).

Despite that KRas leaves the plasma membrane at a much lower rate as compared to endomembranes, this difference in off-rate is offset by the excess of binding sites on the large surface of the endomembrane system that is at least an order of magnitude greater than that of the plasma membrane [35]. The GSF-mediated rapidly equilibrated KRas distribution would thus still be biased towards the surface of endomembranes. To obtain the experimentally observed five to tenfold KRas enrichment at the plasma membrane [32], an additional energy input is necessary that drives the release of PDEδ-bound KRas specifically at the plasma membrane. This may well be the function of the GDI displacement factor (GDF), Arl2/3 [36]. The allosteric release of KRas from PDEδ by Arl2/3-GTP at the plasma membrane would generate an out-of-equilibrium population of KRas specifically at the plasma membrane as driven by the energy of GTP hydrolysis on Arl2/3. So far, Arl3 has been identified on the Golgi in interphase cells as well as to a much lesser extent on the plasma membrane [37]. This might open up the interesting possibility that KRas is also transported via vesicular transport to reach the plasma membrane. However, further experiments need to be done to map the activity profile of Arl2/3 in the cell since it is ARL2/3-GTP that interacts with PDEδ and thereby releases Ras at the right membrane.

3.5 Targeting the Systems that Maintain Ras Organization

Oncogenic, gain-of-function mutations in the components of signaling networks such as in the RTK–Ras–ERK cascade have an effect that the cell becomes uncoupled from the composition of the extracellular milieu that normally guides its social behavior within the tissue. Oncogene addiction [38, 39] has been described in these terms where the cancer cell becomes dependent on strong intrinsic survival signals that emanate from the signal network in which the oncogene product resides to counter the intrinsic apoptotic signals that occur due to the chromosome aberrations and lack of extrinsic survival signals from the extracellular milieu. Ras proteins occupy a central node in early signal transduction from which both proliferative/survival Raf–Mek–Erk signals and survival/anabolism signals from the PI3K-Akt axis emanate [40]. This central position in the transmission of proliferative and survival signals from extracellular growth factors is reflected in the 30 % overall frequency of Ras mutations in cancer. Strikingly, up to 90 % of pancreatic cancers have mutations in KRas [41], and a high incidence of HRas mutations occurs in other malignancies [42] and in Costello syndrome patients [43]. Oncogenic mutations in position G12 (G12V or G12D) or Q61 (Q61K) render Ras in a permanently active, GTP-bound state by abolishing its intrinsic GTP-hydrolyzing activity as accelerated by the catalytic residues of interacting GAPs. Developing inhibitors for constitutively active signaling on the level of Ras has proven extremely challenging, as is evidenced by the complete lack of potent direct Ras inhibitors [44].

As discussed above, the signaling activity of Ras is tightly coupled to its association with specific membranes in the cell. It is therefore to be expected that the signaling activity of the Ras oncogene product is also affected by the oncogene product's spatial distribution in the cell which makes the molecular systems that maintain Ras localization lucrative targets for the modulation of oncogenic Ras signaling. Here, it is important to consider that the signaling output of wild-type Ras is determined by the regulatory GEF and GAP activities as well as its interaction with downstream effectors, whereas the signaling activity of permanently GTP-bound oncogenic RasG12V is a result of its interaction with downstream effectors only. The first level at which the distribution of oncogenic Ras in the cell affects its signaling output is therefore the interaction with its effectors at the plasma membrane such as for Raf and the catalytic P110 subunit of PI3K [40]. Active Ras-GTP recruits these factors to the plasma membrane, but it is the other protein activities that are localized to the plasma membrane that fully activate these kinases. Any interference with the systems that maintain oncogenic Ras at the plasma membrane, or the post-translational modifications that let it get there in the first place should therefore reduce its oncogenic signaling activity by uncoupling recruitment of signaling factors by Ras from plasma membrane-localized growth factor-induced activating activities. Farnesyl transferase inhibitors are based on this principle but suffer on the one hand from their obvious pleiotropic effects on cell signaling by completely inhibiting signaling from farnesylated molecules [9, 45], and on the other hand from the compensatory mechanism of geranylgeranylation that reinstates Ras plasma membrane distributions that take over in case of farnesyl transferase inhibition [46].

Before we move on to discuss approaches to affect the spatial distribution of Ras proteins, we need to consider another layer of Ras signal output regulation that is dependent on its localization at the plasma membrane. For this, it is important to realize that oncogenic mutations occur mostly on a single allele. This implies that an oncogene product such as RasG12V can coexist with its wild-type variant in the cell, if it is not lost due to genetic instability. If the activity of the oncogene product and wild-type GTPase would be completely uncoupled, the oncogene product would cause an offset in the activity state of downstream signaling molecules such as Erk. Given that some of the gene expression machinery behaves in accordance with Weber's law [47], not responding to absolute levels of signaling activity but to fold changes in activity relative to the background, as for example shown for MAPK and Wnt signaling [48], the wild-type Ras population might still provide the switchable activity that transmits changes in the composition of the extracellular milieu to the gene expression machinery and thereby control the phenotype of the cell. However, the guanine nucleotide binding state of oncogenic and wild-type Ras are coupled via positive feedbacks as described above for the RasGEF, Sos [49]. The activation of Sos, and thus Ras, via this feedback mechanism is dependent on the localization of active Ras at the plasma membrane. Oncogenic RasG12V-GTP can thus activate the exchange activity of Sos at the plasma membrane and thereby switch the wild-type Ras population to the active GTP-bound state. This activation is dependent on the dose of oncogenic RasG12V that resides at the plasma membrane because the interactions between Sos and Ras are driven by their high local concentrations there. By lowering the amount of Ras at the plasma membrane below a threshold for effective feedback coupling to Sos, the remaining wild-type Ras population could reacquire its GEF-regulated switching ability that is dependent on growth factor receptor activation by extracellular ligands.

The objective would therefore be to develop pharmacological agents that lower the dose of oncogenic Ras at the plasma membrane to down-modulate oncogenic signaling from Ras by uncoupling it from its effectors. Ubiquitous depalmitoylation plays a crucial role in the maintenance of palmitoylated Ras localization and the inhibition of thioesterase activity is not lethal to cells [28]. Based on these premises, the thioesterase inhibitor Palmostatin B was recently developed and was shown to inhibit the Ras-depalmitoylating enzyme APT. Most strikingly, treatment of cells with Palmostatin B leads to a random distribution of fully palmitoylated Ras over all membrane systems in the cell. The net result of thioesterase inhibition is the reduction of Ras at the plasma membrane and thereby the envisioned effect of the reduction of its interaction with effectors. On the phenotypic level, thioesterase inhibition by Palmostatin B caused a partial reversion of a tumor-like phenotype to a normal phenotype in H-RasG12V transformed MDCK-f3 cells. Thioesterase inhibition is still in its infancy as an approach to affect the phenotype of cancer cells and much research needs to be done to prove its efficacy in tumor models. The fact is that it will only have a chance of working in cancer cells that bear oncogenic palmitoylated Ras.

To also affect the spatial distribution of polybasic stretch containing oncogenic KRas, one needs to interfere with the spatially organizing systems of Ras proteins at a more fundamental level. The universal functionality that maintains the spatial organization of farnesylated Ras proteins by enhancing their diffusion in the cytoplasm is the GSF PDEδ. The importance of PDEδ in maintaining the spatial organization of Ras proteins is apparent from the loss of the plasma membrane partitioning of both palmitoylated H/NRas as well as polybasic stretch containing KRas in MDCK cells in which PDEδ had been knocked down by RNA interference [29]. The loss of the spatial organization of Ras in these cells was paralleled by a reduction in EGF-induced Erk activity, and ectopic expression of PDEδ showed a several fold increase in EGF-induced Erk activity. Reinstating PDEδ by ectopic expression in human hepatocarcinoma cells that do not express PDEδ also reinstates the plasma membrane enrichment of Ras as well as EGF-induced Erk signaling. These experiments clearly demonstrate that effective coupling of Ras to its effectors indeed occur at the plasma membrane. Evidence that PDEδ also affects oncogenic Ras signaling by maintaining its spatial distribution came from PDEδ knockdown experiments in both HRasG12V-transformed fibroblasts and KRasG12D knock-in models for pancreatic adenocarcinoma cells [29, 50]. PDEδ downregulation resulted in a randomized distribution of endogenous oncogenic Ras and strongly reduced cell proliferation/survival as observed by clonogenic assays. The reduction in proliferation was paralleled by a reduced pErk phosphorylation within 72 h of PDEδ knockdown. Erk activity down-modulation was lost after longer times of PDEδ knockdown, possibly reflecting a selection process in the clonogenic assay that overcomes the lack of PDEδ-mediated Ras signaling from the plasma membrane by switching to a Ras-independent signaling pathway. Despite the general problem that genetically instable cancer cells can become resistant to therapy by selection processes under chemotherapeutic pressure, the approach of inhibiting the PDEδ-Ras interaction to affect the spatial organization of Ras has another boon to it that makes its pursuit worthwhile. Similar to thioesterase inhibition, the inhibition of PDEδ-Ras interaction will not completely remove Ras from the plasma membrane, and the residual wild-type Ras at the plasma membrane in "normal" cells could still respond to growth factors and thereby maintain their viability. This is reflected by the fairly normal development of PDEδ knockout mice that have 20–30 % less body weight, apart from the retinal degeneration that occurs because of the role of PDEδ in transporting GRK1 and the catalytic subunit of PDE6 to the outer segment of photoreceptors [51]. PDEδ is a target with two sites for which small molecules could be developed. The farnesyl-binding pocket offers a site for competitive inhibition of Ras binding, and the allosteric Arl2/3 site offers the possibility of locking PDEδ in either the closed or open conformation that will interfere with the delivery of Ras at the right membrane trap.

Tuning signaling network states by the pharmacological manipulation of the spatial organization of oncogenes as described here might well be a more general approach to reinstate some social behavior in cancer cells with a minimal impact on the functioning of normal tissue.

References

1. Niethammer P, Kronja I, Kandels-Lewis S, Rybina S, Bastiaens P, Karsenti E (2007) Discrete states of a protein interaction network govern interphase and mitotic microtubule dynamics. PLoS Biol 5(2):e29
2. Zamir E, Bastiaens PI (2008) Reverse engineering intracellular biochemical networks. Nat Chem Biol 4(11):643–647
3. Tyson JJ, Chen KC, Novak B (2003) Sniffers, buzzers, toggles and blinkers: dynamics of regulatory and signaling pathways in the cell. Curr Opin Cell Biol 15(2):221–231
4. Reynolds AR, Tischer C, Verveer PJ, Rocks O, Bastiaens PI (2003) EGFR activation coupled to inhibition of tyrosine phosphatases causes lateral signal propagation. Nat Cell Biol 5(5):447–453
5. Tischer C, Bastiaens PI (2003) Lateral phosphorylation propagation: an aspect of feedback signalling? Nat Rev Mol Cell Biol 4(12):971–974
6. Verveer PJ, Wouters FS, Reynolds AR, Bastiaens PI (2000) Quantitative imaging of lateral ErbB1 receptor signal propagation in the plasma membrane. Science 290(5496):1567–1570
7. Sorkin A, Goh LK (2008) Endocytosis and intracellular trafficking of ErbBs. Exp Cell Res 314(17):3093–3106
8. Mosesson Y, Mills GB, Yarden Y (2008) Derailed endocytosis: an emerging feature of cancer. Nat Rev Cancer 8(11):835–850
9. Levkowitz G, Waterman H, Zamir E, Kam Z, Oved S, Langdon WY et al (1998) c-Cbl/Sli-1 regulates endocytic sorting and ubiquitination of the epidermal growth factor receptor. Genes Dev 12(23):3663–3674
10. Bache KG, Slagsvold T, Stenmark H (2004) Defective downregulation of receptor tyrosine kinases in cancer. EMBO J 23(14):2707–2712
11. Axelrod D, Wang MD (1994) Reduction-of-dimensionality kinetics at reaction-limited cell surface receptors. Biophys J 66(3 Pt 1):588–600
12. Wu Y, Vendome J, Shapiro L, Ben-Shaul A, Honig B (2011) Transforming binding affinities from three dimensions to two with application to cadherin clustering. Nature 475(7357): 510–513
13. Chong H, Vikis HG, Guan K-L (2003) Mechanisms of regulating the Raf kinase family. Cell Signal 15(5):463–469
14. Kolch W (2000) Meaningful relationships: the regulation of the Ras/Raf/MEK/ERK pathway by protein interactions. Biochem J 351(Pt 2):289–305
15. Slater SJ (2002) Regulation of PKCalpha activity by C1-C2 domain interactions. J Biol Chem 277:15277–15285
16. Freedman TS, Sondermann H, Friedland GD, Kortemme T, Bar-Sagi D, Marqusee S et al (2006) A Ras-induced conformational switch in the Ras activator son of sevenless. Proc Natl Acad Sci U S A 103(45):16692–16697
17. Gureasko J, Galush WJ, Boykevisch S, Sondermann H, Bar-Sagi D, Groves JT et al (2008) Membrane-dependent signal integration by the Ras activator son of sevenless. Nat Struct Mol Biol 15(5):452–461

18. Gureasko J, Kuchment O, Makino DL, Sondermann H, Bar-Sagi D, Kuriyan J (2010) Role of the histone domain in the autoinhibition and activation of the Ras activator Son of Sevenless. Proc Natl Acad Sci U S A 107(8):3430–3435
19. Zhao C, Du G, Skowronek K, Frohman MA, Bar-Sagi D (2007) Phospholipase D2-generated phosphatidic acid couples EGFR stimulation to Ras activation by Sos. Nat Cell Biol 9(6): 706–712
20. Sabouri-Ghomi M, Ciliberto A, Kar S, Novak B, Tyson JJ (2008) Antagonism and bistability in protein interaction networks. J Theor Biol 250(1):209–218
21. Hougland JL, Fierke CA (2009) Getting a handle on protein prenylation. Nat Chem Biol 5(4):197–198
22. Raymond FL, Tarpey PS, Edkins S, Tofts C, O'Meara S, Teague J et al (2007) Mutations in ZDHHC9, which encodes a palmitoyltransferase of NRAS and HRAS, cause X-linked mental retardation associated with a Marfanoid habitus. Am J Hum Genet 80(5):982–987
23. Hou H, John Peter AT, Meiringer C, Subramanian K, Ungermann C (2009) Analysis of DHHC acyltransferases implies overlapping substrate specificity and a two-step reaction mechanism. Traffic 10(8):1061–1073
24. Rocks O, Peyker A, Kahms M, Verveer PJ, Koerner C, Lumbierres M et al (2005) An acylation cycle regulates localization and activity of palmitoylated Ras isoforms. Science 307(5716): 1746–1752
25. Rocks O, Gerauer M, Vartak N, Koch S, Huang ZP, Pechlivanis M et al (2010) The palmitoylation machinery is a spatially organizing system for peripheral membrane proteins. Cell 141(3):458–471
26. Dudler T, Gelb MH (1996) Palmitoylation of Ha-Ras facilitates membrane binding, activation of downstream effectors, and meiotic maturation in Xenopus oocytes. J Biol Chem 271: 11541–11547
27. Misaki R (2010) Palmitoylated Ras proteins traffic through recycling endosomes to the plasma membrane during exocytosis. J Cell Biol 191:23–29
28. Dekker FJ, Rocks O, Vartak N, Menninger S, Hedberg C, Balamurugan R et al (2010) Small-molecule inhibition of APT1 affects Ras localization and signaling. Nat Chem Biol 6(6): 449–456
29. Chandra A, Grecco HE, Pisupati V, Perera D, Cassidy L, Skoulidis F et al (2012) The GDI-like solubilizing factor PDEdelta sustains the spatial organization and signalling of Ras family proteins. Nat Cell Biol 14(2):148–158
30. Nancy V (2002) The delta subunit of retinal rod cGMP phosphodiesterase regulates the membrane association of Ras and Rap GTPases. J Biol Chem 277(17):15076–15084
31. Leventis R, Silvius JR (1998) Lipid-binding characteristics of the polybasic carboxy-terminal sequence of K-ras4B. Biochemistry 37:7640–7648
32. Gomez GA, Daniotti JL (2007) Electrical properties of plasma membrane modulate subcellular distribution of KRas. FEBS J 274:2210–2228
33. Apolloni A (2000) H-ras but not K-ras traffics to the plasma membrane through the exocytic pathway. Mol Cell Biol 20:2487
34. Silvius JR (2006) K-ras4B and prenylated proteins lacking "second signals" associate dynamically with cellular membranes. Mol Biol Cell 17:192–202
35. Shibata Y, Shemesh T, Prinz WA, Palazzo AF, Kozlov MM, Rapoport TA (2010) Mechanisms determining the morphology of the peripheral ER. Cell 143(5):774–788
36. Ismail SA, Chen YX, Rusinova A, Chandra A, Bierbaum M, Gremer L et al (2011) Arl2-GTP and Arl3-GTP regulate a GDI-like transport system for farnesylated cargo. Nat Chem Biol 7(12):942–949
37. Zhou C, Cunningham L, Marcus AI, Li Y, Kahn RA (2006) Arl2 and Arl3 regulate different microtubule-dependent processes. Mol Biol Cell 17(5):2476–2487
38. Sharma SV, Settleman J (2007) Oncogene addiction: setting the stage for molecularly targeted cancer therapy. Genes Dev 21(24):3214–3231
39. Hanahan D, Weinberg RA (2000) The hallmarks of cancer. Cell 100(1):57–70
40. Aksamitiene E (2012) Cross-talk between mitogenic Ras/MAPK and survival PI3K/Akt pathways: a fine balance. Biochem Soc Trans 40:139–146

41. Deramaudt T, Rustgi AK (2005) Mutant KRAS in the initiation of pancreatic cancer. Biochim Biophys Acta 1756:97–101
42. Luwor RB, Lu Y, Li X, Liang K, Fan Z (2011) Constitutively active Harvey Ras confers resistance to epidermal growth factor receptor-targeted therapy with cetuximab and gefitinib. Cancer Lett 306(1):85–91
43. Aoki Y (2005) Germline mutations in HRAS proto-oncogene cause Costello syndrome. Nat Genet 37:1038–1040
44. Verdine GL, Walensky LD (2007) The challenge of drugging undruggable targets in cancer: lessons learned from targeting BCL-2 family members. Clin Cancer Res 13(24):7264–7270
45. Brunner TB, Hahn SM, Gupta AK, Muschel RJ, McKenna WG, Bernhard EJ (2003) Farnesyltransferase inhibitors. Cancer Res 63(18):5656–5668
46. Downward J (2003) Targeting RAS, signalling pathways in cancer therapy. Nat Rev Cancer 3(1):11–22
47. Ferrell JE Jr (2009) Signaling motifs and Weber's law. Mol Cell 36(5):724–727
48. Goentoro L, Kirschner MW (2009) Evidence that fold-change, and not absolute level, of beta-catenin dictates Wnt signaling. Mol Cell 36(5):872–884
49. Groves JT, Kuriyan J (2010) Molecular mechanisms in signal transduction at the membrane. Nat Struct Mol Biol 17:659–665
50. Guerra C, Collado M, Navas C, Schuhmacher AJ, Hernandez-Porras I, Canamero M et al (2011) Pancreatitis-induced inflammation contributes to pancreatic cancer by inhibiting oncogene-induced senescence. Cancer Cell 19(6):728–739
51. Zhang H, Li S, Doan T, Rieke F, Detwiler PB, Frederick JM et al (2007) Deletion of PrBP/delta impedes transport of GRK1 and PDE6 catalytic subunits to photoreceptor outer segments. Proc Natl Acad Sci U S A 104(21):8857–8862

Chapter 4
Endocytosis in the Spatial Control of Polarised Cell Functions

Giorgio Scita, Andrea Disanza, Emanuela Frittoli, Andrea Palamidessi, and Sara Sigismund

Abstract Endocytosis, originally thought of as a device to transport nutrients and membrane-associated molecules across the plasma membrane through vesicles, is emerging as a connectivity infrastructure (which we have called "the endocytic matrix") of different cellular networks necessary for the execution of various cellular programmes. A primary role of the endocytic matrix is the delivery of space- and time-resolved signals to the cell, and it is thus essential for the execution of polarised functions. Here, by discussing paradigmatic cases, we intend to outline emerging concepts of how the endocytic wiring system functions as a highly interconnected intracellular highway that mobilises membrane and signalling molecules, ensuring polarised compartmentalisation of signals. We will specifically focus on two exemplar cases: the impact of the endocytic matrix on cell migration and on asymmetric cell division. In each of these cases, endocytosis and recycling have been shown to ensure the asymmetric distribution of biological molecules, which, in turn, is crucial for proper polarised cellular functions.

The authors Andrea Disanza, Emanuela Frittoli, Andrea Palamidessi have contributed equally. Part of the text and Figure 2 have been reprinted with modifications from *Physiol Rev January 2012 vol. 92 no. 1 273-366*, by Sara Sigismund, Stefano Confalonieri, Andrea Ciliberto, Simona Polo, Giorgio Scita and Pier Paolo Di Fiore, entitled "Endocytosis and Signaling: Cell Logistic Shape The Eukaryotic Cell Plan", Copyright (2012), with permission from Rightslink, American Physiological Society.

G. Scita (✉)
IFOM Foundation, Institute FIRC of Molecular Oncology,
Via Adamello 16, 20139 Milan, Italy

Dipartimento di Medicina, Chirurgia ed Odontoiatria, Università degli
Studi di Milano, Milan, Italy
e-mail: giorgio.scita@ifom.eu

A. Disanza • E. Frittoli • A. Palamidessi • S. Sigismund
IFOM Foundation, Institute FIRC of Molecular Oncology, Via Adamello 16, 20139 Milan, Italy

Y. Yarden and G. Tarcic (eds.), *Vesicle Trafficking in Cancer*,
DOI 10.1007/978-1-4614-6528-7_4, © Springer Science+Business Media New York 2013

4.1 Introduction

Cells of unicellular and multicellular organisms must sense and decode spatial information. This is achieved by adapting cytoskeletal and membrane components and signalling machineries so as to acquire and maintain an asymmetric architectural organisation and a polarised distribution of cellular structures and signalling molecules whose output, thus, becomes spatially restricted.

One powerful tool to confer spatial dimensions to signalling is through endocytosis and recycling of membrane-bound receptors and effector molecules. Endocytosis, originally viewed as a device to transport nutrients across the plasma membrane, appears today as a pervasive programme that permeates every aspect of cell physiology, with a primary role in the delivery of space- and time-resolved signals to the cell in an interpretable format [1, 2]. Endocytic internalisation of plasma membrane-localised receptor, for example, has emerged as built-in mechanisms required for chemotactic gradient sensing and directed cell migration. In addition, internalisation of membrane and membrane-associated proteins is frequently accompanied by recycling of these factors back to the plasma membrane. These cycles of endo-/exocytosis (or EEC), on the one hand, function to replenish ligand-free receptor for the next round of signalling and transport. On the other hand, EEC can also serve either as a mean to redirect and confine signalling molecules to specialised and distinct areas of the plasma membrane or as a positive feedback mechanism capable of maintaining the polarisation state of critical signalling molecules. Endocytic-dependent spatial restriction of signalling is, thus, considered as a critical device for the execution of polarised cellular functions, including directed cell migration and cell invasion of metastatic cells, epithelial cell polarisation during tissue morphogenesis, and asymmetric cell division (ACD) [2]. Here, we will discuss paradigmatic examples that support this notion with the explicit intent to conceptualise, rather than being comprehensive, how spatially directed trafficking of membrane and membrane-associated signalling molecules critically contributes to the initiation and maintenance of polarised cellular functions. Due to space limitation, we will primarily focus on the impact of trafficking networks, which we have recently termed "endocytic matrix" (to indicate a complex network of intersections between input and output functioning as a decoder of cell signalling) on directed cell migration and cell fate determination. However, a variety of additional polarity functions, including epithelial morphogenesis and neuronal cell organisation, have also been shown to rely on endocytic-dependent spatial restriction of cell signalling.

4.2 Endocytosis Spatially Restricts Signalling and Processes During Directed Cell Migration

One process that depends on the intertwined connection between signalling and localisation is cell motility, where the precise perception of extracellular cues in a three-dimensional setting is remarkably complex, particularly when cells move towards chemoattractants in a polarised fashion. Under these conditions, cells must reorient directionally by polarising key plasma membrane proteins according to the direction of travel. Additionally, coordination between membrane traffic, cell substrate adhesion, and actin remodelling is required to generate propulsive forces responsible for the protrusive activity at the leading edge of motile cells. Not surprisingly, multiple mechanisms have been revealed in recent years through which trafficking of membrane and membrane-associated motogenic transducers directly impinges on cell migration.

4.2.1 Endocytosis Extinguishes the Signal at the Proper Time and Location

The endocytic process is commonly regarded as a critical means to attenuate extracellular ligand-induced responses. This is of particular relevance for those soluble signals that promote directed cell migration. This is intuitively achieved by direct clearance of motogenic receptors, which function as the first-line sensors of the cellular microenvironment from the plasma membrane limiting the intensities of the ensuing signalling events. In addition to this, internalised activated receptors are often destined to trafficking routes that lead to their degradation, thus hampering the ability of cells to respond to the continuous presence of extracellular cues by directly reducing the number of responsive receptors. A number of plasma membrane receptors, for example, receptor tyrosine kinases and G protein-coupled receptor (GPCR) function as motogenic sensors, which especially during developmental morphogenesis respond to gradients of chemotactic factors that guide the migration of cells to their final destination. These cells, in addition to migrate directionally, must also be capable to arrest at their target sites, where the concentration of the chemotactic factors is the highest [3] (Fig. 4.1a). Thus, a ligand-dependent internalisation/sorting mechanism that drives a motogenic receptor towards a degradative pathway may be critical to switch off the migratory signal and the ensuing actin polymerisation/depolymerisation cycles when appropriate. A scenario of this kind has been, for instance, demonstrated to operate during the migration of primordial germ cells (PGCs) towards the gonads in *zebrafish* development [3]. PGCs express the chemokine receptor CXCR4b and directionally migrate towards sites in the embryo at which the ligand SDF-1 is expressed [4]. Binding of SDF-1 elicits also internalisation of the receptor promoting its spatial redistribution and restricting its signalling, two factors that are required for proper chemotaxis. Indeed an internalisation defective receptor led to aberrantly elevated signals and increased time spent "running", preventing cells to reach their final target while promoting ectopic cell migration [5] (Fig. 4.1a). Recently, an additional non-cell-autonomous endocytic mechanism has been also shown to contribute to the directed migration of PGC [6]. The somatic cells surrounding the germ cells were found to express another SDF-1 receptor, CXCR7. Binding of the "somatic" receptor to SDF-1 was shown not to enhance cell signalling, but rather to clear the ligand from the extracellular environment ensuring

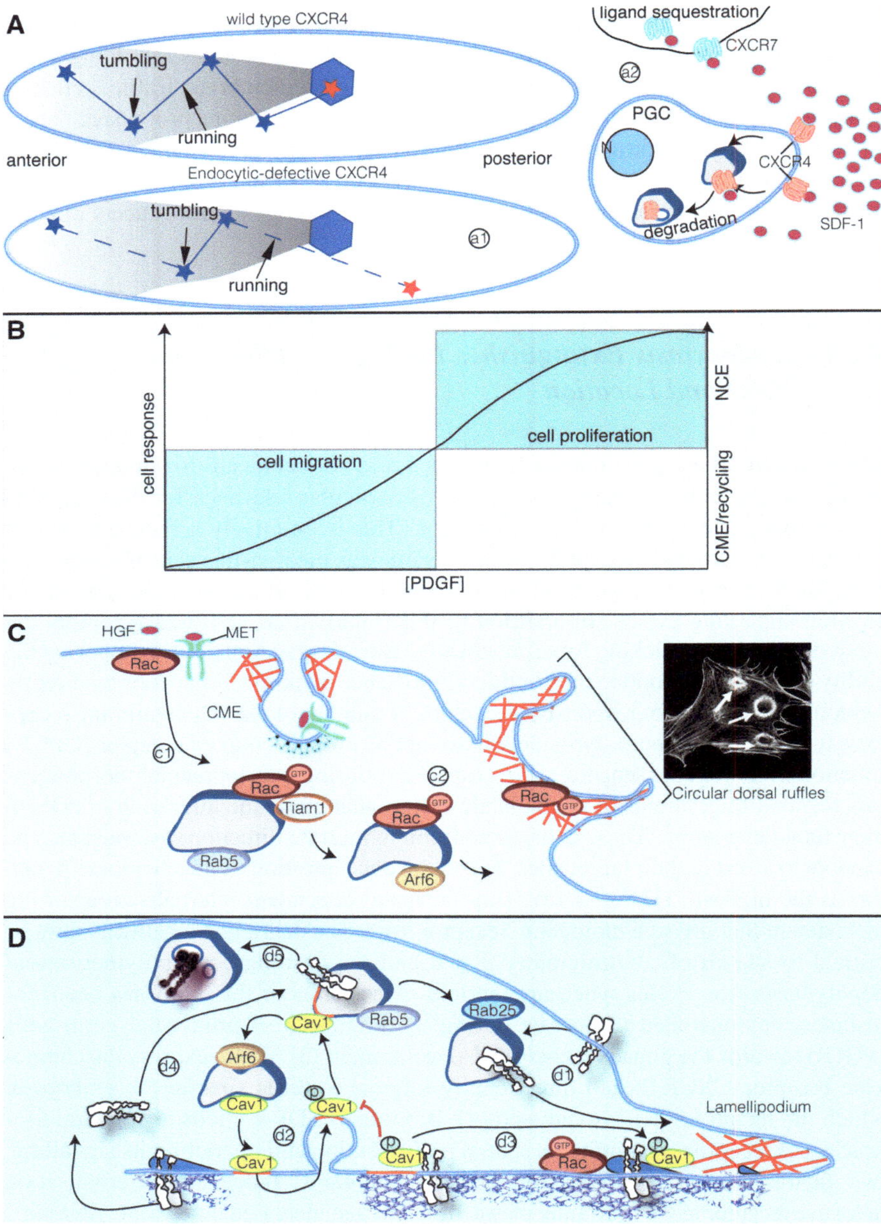

Fig. 4.1 Endocytosis and directed cell migration. (**A**) Endocytic control of chemotaxis. In *zebrafish*, PGCs, the progenitors of the gametes, migrate from the (anterior) position where they are specified towards the (posterior) region where the gonad develops in response to soluble chemokine, such as SDF-1 [3]. Gradients of SDF-1 are sensed by the chemokine receptor CXCR4. Activation of CXCR4 leads to a cyclic migratory behaviour composed of "running" phase in the direction of the SDF-1 gradient, followed by pausing, during which cells loose polarity and tumble, before resuming their correct migration towards the gonads. In endocytic defective CXCR4 mutants, the time spent

endomembrane organelles, which thus become bona fide signalling platforms influencing not only the time and amplitude of the resulting signal but also its specificity [11]. Consistent with this view, more and more signal transduction pathways are reported to require an active endocytic machinery or strikingly to originate from various types of endosomes. A variation and an extension of this latter concept has further emerged with the recognition that plasma membrane receptors are internalised through different pathways, e.g. clathrin-mediated endocytosis (CME) or non-clathrin endocytosis (NCE), which have been shown to control directly the biological outcome of their signalling [12, 13]. During the endocytic transport, molecules undergo a discrete set of route-dependent posttranslational modifications, such as phosphorylation/dephosphorylation and ubiquitination, that directly influence the composition of the signalling cascade that is being activated [11]. Thus, the biological output might be controlled not only through compartmentalisation of signalling into endosomal platforms but also by the routes through which molecules reach the different compartments.

Fig. 4.1 (continued) running and tumbling is altered and PGC fails to arrest at gonads (**a1**). The SDF-1 gradient is also maintained by a decoy receptor CXCR7 present on somatic cells that surround germ cells (**a2**). (**B**) Endocytic routes regulate the signalling response to soluble factors. RTK can either be internalised through a CME or NCE [12]. The different internalisation routes elicit different signalling responses. For example, fibroblasts stimulated with low doses of PDGF responds by acquiring migratory properties in a CME-dependent manner. Mitogenesis and cell proliferation are, instead, preferentially induced by higher concentration of PDGF, a response that requires a functional raft/caveolin, NCE pathway [14]. (**C, D**) Spatial restriction of RAC and integrin signal by endo-/exocytic trafficking. Coordination between membrane traffic, cell substrate adhesion and actin remodelling is required for the formation of polarised cell protrusions, such as lamellipodia and circular dorsal ruffles (CDR) (*inset* shows a fibroblast moving in a mesenchymal mode, stained for filamentous actin to highlight the presence of PDGF-induced polarised, apical CDR). Two endocytic/signalling networks implicated in polarised migration are shown. In the first (**C**), in response to stimulation of RTKs, such as HGF stimulation of the MET receptor, CME and RAB5 activation promote the internalisation of RAC and its GEF, TIAM1, into early endosomes (**c1**). Activated GTP-bound RAC is subsequently recycled through the ARF6 endosomal pathway (**c2**) to PM regions, where actin polymerisation supports the formation of CDR. In the second network (**D**), the trafficking of integrins, through CME and raft-dependent NCE, enables sustained and polarised signalling as well as precise coordination of integrin activity with the changing dynamics of focal adhesions. Integrins, such as $\alpha5\beta1$, are continuously internalised (**d1**) and recycled to the PM, through RAB25 endosomes that are compartmentalised at the leading edge of cells for lamellipodial extension. Coordination of integrin adhesion and lipid raft endocytosis and recycling is also crucial to integrate RAC and integrin activation (**d2**). Lipid rafts are endocytosed through caveolin-1 (CAV1) containing caveolae (**d3**). Lipid rafts are also binding sites for RAC (**d3**). Integrin signalling blocks lipid raft internalisation by promoting CAV1 phosphorylation and its retention in focal adhesions at the PM (**d3**). Thus, when integrins are engaged by the ECM, RAC binding sites at the PM become available. Cell detachment abrogates integrin activation and extinguishes RAC signalling at the PM, by enabling the relocalisation and subsequent caveolae-mediated internalisation of CAV1 and lipid rafts. Recycling of CAV1, as well as of RAC and integrins may be coordinated by ARF6. Mature, integrin-containing focal adhesions at the rear of the cells need to be disassembled to enable effective cell locomotion. This process involves dynamin and CME (**d4**). Finally, internalised integrin, bound to ECM ligands such as FN, may be specifically directed to lysosomes for degradation, through a mechanism involving integrin ubiquitination and recognition by the ESCRT machinery (**d5**)

In keeping with this latter notion, it has been recently shown that the stimulation of one of the most potent fibroblastic motogenic factors, platelet-derived growth factor (PDGF) and its cognate receptors, can elicit either proliferation or cell migration depending on the ligand concentrations and the routes taken by the activated receptor. Cells stimulated with low doses of PDGF responded by rearranging their cytoskeleton and acquiring migratory properties in a CME-dependent manner. Mitogenesis and cell proliferation were, instead, preferentially induced by higher concentrations of PDGF, a response that required a functional raft/caveolin pathway [14] (Fig. 4.1b). In this case, the precise mechanisms and molecular targets that control actin-based machineries responsible for directional motility are yet to be fully defined. It is, however, tempting to speculate that a low dose of PDGF stimulation, in analogy to EGF [13], may preferentially promote recycling of PDGFR or of its key signal transducers back to the plasma membrane, triggering positive feedback loops that ensure localised and sustained signalling to actin-based dynamics (Fig. 4.1b). Within this context, the members of the RhoGTPases, and in particular RAC proteins, which are known to control multiple critical events (e.g. de novo actin nucleation through the nucleation promoting factors (NPF), WAVEs, Capping proteins, and F-actin depolymerisation, through ADF/cofilin [15]) concomitantly required for actin dynamics, may represent ideal cargos that hitchhike a trafficking ride back to the plasma membrane resulting in spatial restriction and polarisation of their signalling output.

Endocytic-dependent signal dysregulation has also been shown to account for the acquisition of oncogenic and prometastatic properties of protumorigenic mutants of the c-MET [16]. Certain tumour-associated c-MET activating mutations exhibit increased endocytosis/recycling activity and decreased levels of degradation, leading to accumulation on endosomes, hyperactivation of the GTPase RAC1 and increased levels of cell migration and invasion. Blockade of endocytosis hampered mutants anchorage-independent growth, in vivo tumorigenesis and metastasis while maintaining their activation [16]. Thus, prolonged endosomal signalling or increased recycling of oncogenic c-MET receptor contributes to aberrant invasive migration leading to augmented tumour dissemination.

4.2.3 EEC and Membrane Flow

A default mechanism linked to EEC that potentially has direct consequences on polarity phenotypes is the generation of membrane flow. By analogy with actin tread-milling, the flow of internalised and recycled membranes was proposed more than a decade ago either to generate forces for the extension of migratory protrusions [17] or to promote the rearward movement of molecules bound to the surface of these protrusions during cell motility. Results consistent with membrane flow have been obtained in various cell types [17], although, for other motile cell types, a number of experiments failed to detect any significant rearward membrane flow [18]. Thus, membrane flow may not be a universal property of moving cells, though it may be important for some of them. This notwithstanding, the requirement for a continuous flow of membranes propelled by endocytic molecules is essential for the highly dynamic changes of cell shape that occur during directional, chemotactic migration of the amoeba *Dictyostelium discoideum*, a professional mover [19, 20]. Clathrin-null *Dictyostelium* mutants, in addition to displaying dysfunctional cytokinesis, are characterised by increased roundness, defective polarity, reduced cell velocity and inefficient chemotaxis [19]. This was originally proposed to be due to an impaired ability to extend polarised cell protrusions at the front of the cell. More recent evidence suggests, instead, that an intact clathrin-dependent EEC of membrane is necessary for a moving cell to adjust its cell surface area to match changes in cell shape. Lack of this adaptation system thus severely impairs cell locomotion [20].

4.2.4 EEC Spatially Restricts Signals for Directional Migration

One additional way to achieve signal polarisation and directional motility is through localised redistribution, via EEC, of signalling molecules in response to extracellular cues. The first genetic evidence in support of this concept was produced in *Drosophila melanogaster*. Disruption of typical endocytic regulators, such as the E3 ubiquitin ligase CBL, the RAB5 activator SPRINT (the homologue of mammalian RIN1) or the recycling regulators RAB11 [21], resulted in aberrant cell migration in response to stimulation [22], by affecting the EEC of the motogenic RTKs of the EGFR and PVR (PDGF/VEGF receptor) families. Thus, endocytic pathways, particularly those impinging on RAB5, are required to ensure the spatial resolution of chemotactic signalling emanating from different RTKs, in order to regulate actin-based, polarised protrusive activity and motility.

There is evidence that a similar circuitry also operates in mammalian cells to modulate polarised cellular function. Endocytic trafficking of RAC, and its recycling to the PM, is required for the transduction and spatial resolution of information emanating from motogenic stimuli [23]. As occurs in *Drosophila*, an endocytic RAB5-based circuitry is pivotal. By activating endocytosis, RAB5 causes internalisation of RAC, its activation in recycling endosomes and its subsequent delivery through ARF6-dependent routes to specific regions of the PM. Once redelivered to the membrane, polarised RAC-dependent functions take place, leading to the formation of migratory protrusions that promote a mesenchymal mode of cell motility [23] (Fig. 4.1c). Notably, ARF6-dependent recycling controls also the spatial and polarised distribution of CDC42 in migratory cells [24].

The importance of endosomal recycling routes for directional migration is highlighted by various studies in different mammalian cells. There is evidence, for example, that inhibition of the slow recycling pathway by expression of dominant negative RAB11 or truncated myosin Vb or RAB11-FIP, an effector of RAB11, impairs cell migration [25] and chemotaxis of basophilic leukaemia cells [26]. These latter results have been recently confirmed in epithelial PtK1 cells, where, however, interference with the RAB11 recycling pathway increased random motility and impaired directional and persistent migration, possibly as a consequence of the delocalised formation of protrusive lamellipodia [27]. Thus, polarised endosomal recycling is not required for cell locomotion per se, but rather, it appears to be critical for the maintenance of the polarity of cell migration, which when disrupted leads to disorganised motility. The relevance of RAB11 recycling for polarised cell migration has recently been extended also in border cells in *Drosophila*. In this system, a trafficking loop between the plasma membrane and the RAB11-recycling endosome was demonstrated to be essential for collective cell migration in vivo. This loop regulates the spatial restriction of active RTKs at the leading edge of the cell cluster, transforming the extracellular gradient into a robust intracellular polarity [21].

A similar endo-/exocytic cycle appears to control the cellular trafficking of integrins. These major cell surface adhesion receptors play a critical role in cell migration. Several different mechanisms control their activity, including expression and subunit heterodimerisation patterns, clustering and lateral diffusion in the plane of the PM and interaction with the actin cytoskeleton and the inside of cells (reviewed in [28]). In addition to this, many integrins are continually internalised from the PM into endosomal compartments and are subsequently recycled, prompting the proposal that spatially polarised EEC of these adhesion receptors is essential to control various aspects of cell locomotion (reviewed in [29]). Consistent with this view, for instance, the blockade of integrin $\alpha5\beta1$ recycling by functional interference with the integrin-associated RAB25, a member of the RAB11 family of proteins that control endosomal recycling, impaired the formation of "pseudopodal protrusions" (mesenchymal motility) and directional motility during 3D cell migration [30] (Fig. 4.1d). Recently, it has been shown that in addition to recycling from late RAB25 endosome, $\alpha5\beta1$ can also be targeted to lysosomes and multivesicular bodies, which concentrate towards the rear of a migratory cells. Importantly, active $\alpha5\beta1$ integrin, instead of being degraded within these organelles as most cargo would be, is retrogradely transported and recycled to the plasma membrane at the back of invading cells via a pathway requiring the Chloride Intracellular Channel Protein 3 (CLIC3). Integrin retrograde recycling is thought to maintain active Src signalling and contribute to cell invasion and tumour dissemination [31].

Mechanistically, one important question that these findings raise is how signalling molecules are recycled to specific regions of the PM (as opposed to the bulk PM) to execute spatially restricted signalling. In the case of RAC and integrins, one possible answer came from recent studies connecting localised RAC activation with integrin-mediated adhesion and lipid raft internalisation (Fig. 4.1c). These studies suggested that RAC positioning at and trafficking from and to specific locations of the PM may be regulated through raft-dependent endocytosis. This process is needed, in turn, to specify the localisation of RAC activity for the execution of relevant biological processes. Thus, upon activation of integrins, sites of high RAC affinity become available on the PM preventing RAC internalisation, which only occurs following cell detachment in a dynamin and caveolin-1-dependent manner [32]. Indeed, caveolin-1-deficient cells show increased RAC activation, which however is not spatially confined, leading to loss of directional migration [33]. The RAC/integrin EEC and targeting circuitry appears to require the coordinated action of two different routes of endocytosis (clathrin-dependent [23] and raft/caveolar-mediated [34]). Within this context, ARF6-dependent recycling appears to be the critical factor controlling not only the redelivery of RAC [23] and integrins [32, 34] but also of lipid rafts, back to the PM, ultimately coordinating RAC signalling and directional migration with adhesion-dependent cell growth [35] (Fig. 4.1d). Alternatively, other phospholipids, such as diacylglycerol kinase-α (DGK-α)-generated phosphatidic acid (PA), may serve as local, pseudopodia tip-restricted anchor points for either RCP-integrin complex [36] or RhoGDI, a key regulator of RAC proteins [37].

One additional attractive hypothesis to account for how various endocytic routes may promote directional migration in a coordinated fashion is based on observations that caveolar-based endocytosis frequently occurs only at the trailing edges of migrating cells [38], while CME, coupled to fast recycling, is restricted to the advancing leading edges [39]. These findings suggest that polarised locomotion may be facilitated by a front-rear distribution of diverse endocytic routes. This notion has recently been extended to include clathrin- and raft-independent endocytic routes as well as macropinocytosis. The clathrin-independent carrier (CLIC) internalisation pathway has been shown to be responsible for the vast majority of bulk endocytosis in lamellipodia and to be required for directional cell migration by promoting rapid non-clathrin-mediated EEC of focal adhesion cargoes [40]. Conversely, macropinocytosis induced by PDGF was shown to promote the rapid redistribution of both β1 and β3 integrins to circular dorsal ruffles, their subsequent internalisation through macropinosomes and redelivery to nascent focal adhesions at the leading edge of migratory fibroblasts [41], ultimately promoting cell locomotion.

In the case of integrin trafficking, questions that still remain to be addressed are the following: (1) whether integrins that undergo EEC are the active ones bound to their extracellular matrix (ECM) ligand and (2) whether their activation status affects their endocytic routes and intracellular fate. One recent report shed lights on these issues, further providing evidence in support of the notion that proper targeting of activated integrins to lysosomal degradation is required for cell motility [42] (Fig. 4.1d). A sizable fraction of internalised Fibronectin (FN)-bound α5β1 integrin dimers are specifically directed to lysosomes for degradation through a mechanism involving integrin ubiquitination and recognition by the ESCRT machinery. Cells expressing an α5β1 integrin mutant, which could no longer be ubiquitinated, were severely impaired in cell migration, suggesting that FN-integrin complex turnover is essential for locomotion. Since FN degradation is also required for cell migration [43], it is possible that the FN-integrin complex must be degraded, instead of being continuously recycled, to avoid the formation of dysfunctional adhesion sites that would result in increased adhesion and build-up of ECM, both of which would hinder cell migration. Alternatively, degradation, as opposed to or in equilibrium with recycling, may be required for the proper attenuation of integrin signalling in order to have an impact on migration.

An additional trafficking mechanism that impinges on the ability of cells to sense and respond to shallow differences in the concentrations of soluble gradient during development has been recently unveiled. Morphogen gradients orchestrate developmental processes, such as differentiation and cell migration. A case in point is represented by FGF8 gradients during zebrafish gastrulation. FGF8 is expressed and secreted by cells positioned at the embryonic margin during gastrulation. FGF8 then diffuses away from this source, forming a concentration gradient across the neighbouring tissue to induce graded expression of target genes that promote differentiation. FGF8 binds to the receptor tyrosine kinase FGFR1 (fibroblast growth factor receptor 1) on the surface of target cells, and this triggers endocytosis of the ligand-receptor complex, which then subsequently progresses through early endosomes towards MVBs and lysosomes for degradation. The conjugation of ubiquitin to FGFR1 cytoplasmic domains, which is key to this progression, is catalysed by the E3 ubiquitin ligase CBL. To explore the impact of trafficking on FGF8 sensing, Nowak et al. [44] implanted a source of exogenous FGF8 (FGF8-loaded beads) into embryos and monitored the amount of endosomal FGF8 in cells at various distances from the beads. The amount of endosomal FGF8 was high near the bead and decreased with increasing distance from the source. However, after interference with CBL function, the differential between the amounts of endosomal FGF8 in cells at various distances from the bead was markedly reduced. This indicated that the rate at which internalised receptors are transported to lysosomes dictates the relationship between the morphogen concentration and the signalling response of the cell. Normal rates of endosome to lysosome transport yield a close relationship between FGF8 dose and the cellular response (e.g. high FGF8 concentrations give a high response and low FGF8 concentrations yield a low response), whereas inhibition of lysosomal targeting flattens this relationship such that the extent of FGFR1 signalling differs less over a broader range of distances from the source. Thus, the time spent by motogenic receptors en route to lysosomal degradation is critical for the accurate and precise sensing of morphogen gradients.

4.2.5 Endocytosis Acts Locally to Regulate Focal Adhesion Turnover

A key aspect of directional migration of adherent cells is the establishment of transient attachments to the ECM through integrin clusters that form plaques known as focal adhesions. Focal adhesions establish a connection between the ECM and the actin cytoskeleton and serve as points of traction for the cell. The contraction of focal adhesion-associated actin stress fibres is thought to propel the cell body forward. During migration, there is a constant turnover of focal adhesions that form at the leading edge, often as focal complexes that mature into focal adhesions as tension builds up, and that are then disassembled, allowing for tail retraction, and integrin detachment from the ECM [45]. While the mechanisms of adhesion assembly have been largely defined, focal adhesion disassembly still remains unclear. Given the importance of integrins in adhesion and the role of integrin trafficking in migration, a prevailing idea is that the formation and disassembly of focal adhesions during cell migration are coupled to the endo-/exocytic cycles of integrins [29]. In keeping with this notion, focal adhesion disassembly was shown to be dependent on the activity of dynamin, which can form a complex with the kinase FAK and the adaptor GRB2, and is essential for microtubule-dependent focal adhesion disassembly [46, 47]. Additionally, clathrin and various clathrin accessory proteins can accumulate at focal adhesion sites where, following targeting by microtubules, they promote the localised internalisation of integrin and focal adhesion disassembly [48, 49] (Fig. 4.1d). Thus, while integrin EEC may globally serve as a device to maintain a spatially confined front-to-back gradient of adhesion receptors, focal adhesion-restricted CME may terminate mechanosignalling, suggesting that membrane trafficking is a versatile system for the temporal and spatial control of motogenic inputs.

4.2.6 Crosstalk Between PM Receptors Within the Endocytic Network

An additional emerging level through which PM motogenic receptors, including mechanosensors, such as integrins, and canonical signal transducers, such as RTKs, influence cell migration is by exerting a reciprocal control over their trafficking routes.

It is well established that RTK and integrin signalling are inextricably linked in such a way that full activation of various RTK pathways can be achieved only if cell adhesion is engaged, while inside-out integrin activation is frequently promoted by growth factors in a variety of cellular processes ranging from cell spreading, epithelial cell morphogenesis and cell migration (reviewed in [50]).

One mechanism to initiate inside-out signalling is through the mobilisation of the endosomal pool of integrin heterodimers for rapid redelivery to the PM [29]. PDGF selectively promotes the recycling of integrin $\alpha v\beta 3$, but not of integrin $\alpha 5\beta 1$, through RAB4-dependent endosomal routes, enhancing cell adhesion and spreading [51]. Endocytic and signalling pathways are deeply integrated as indicated by the observation that integrin $\alpha v\beta 3$ primarily activates RAC, which is essential for the formation of lamellipodia and focal complexes, and which drives directional cell migration, while integrin $\alpha 5\beta 1$ controls RHOA-dependent stress fibre formation and cell contraction [52]. The selective activation of integrin heterodimers, therefore, influences the balance of their signalling to RHO-GTPases, ultimately controlling the mode of cell motility. Fibroblast growth factor receptor 1 (FGFR1) and cell-cell adhesion molecules display a similar mode of interaction. For example, Neural Cell Adhesion Molecule (NCAM) associates with FGFR1 [53]. FGF induces endocytosis and degradation of FGFR1, while NCAM instead promotes stabilisation of the receptor, which is recycled to the cell surface in a RAB11- and SRC-dependent manner, resulting in sustained signalling. This promotes NCAM-induced cell migration, and presumably also accounts for the NCAM pro-invasive role during tumour progression.

There is accumulating molecular evidence of interactions between different integrin heterodimers along the endocytic routes and of integrin regulation of RTK trafficking (reviewed in [29]). A specific case in point is provided by studies linking integrins αvβ3 and α5β1 with EGFR trafficking networks. Upon αvβ3 ECM ligand engagement, the heterodimer is actively routed to RAB11-recycling compartments through its interaction with RAB-coupling protein, RCP [54]. Disruption of αvβ3 function causes RCP to dissociate from αvβ3 integrin and to bind α5β1 instead. This mechanism results in efficient rerouting of integrin α5β1 back to the PM, effectively accomplishing an integrin heterodimer switch that enhances RHOA signalling with concomitant enhanced turnover of lamellipodial extension and increased random cell migration. Importantly, disruption of αvβ3 by cyclic peptidomimetic drugs not only drives the recruitment of RCP to the cytoplasmic tail of β1 integrin but also enables RCP to associate with EGFR [54]. α5β1 integrin and EGFR thus coordinately recycle to the PM, with a striking effect on EGFR and integrin signalling that enhances the invasiveness of various epithelial tumour cells [54]. The "endocytic interaction" of integrins with RTKs is not limited to this specific example. In endothelial cells, pharmacological inactivation of αvβ3 results in enhanced VEGFR2 recycling through a RAB4-dependent pathway, diverting this receptor away from degradation while boosting VEGFR2 cell surface levels, endothelial sprouting and tubulation, ultimately resulting in neoangiogenesis [55]. This is the opposite effect to that originally predicted for αvβ3 interfering drugs, such as Cilengitide, highlighting the importance of understanding endocytic networks in regulating complex pathophysiological processes.

4.3 Endocytosis and the Determination of Cell Fate

While the majority of cell divisions generate two identical daughter cells, in a number of cases the two progenies assume different fates; in these instances one of the two daughter cells might retain the mother cell fate, or both daughters can assume fates that are different from the mother cell and from each other. These events are crucial in development and in the maintenance of stem cell (SC) compartments in adult life. Furthermore their subversion is thought to play a central role in cancer (reviewed in [56]).

The phenomenon defined as ACD sits at the heart of the process of cell fate determination. ACD can be influenced by intrinsic and extrinsic mechanisms. In the former case, the unequal partitioning of molecular machinery at mitosis gives raise to daughter cells that are intrinsically different. In the latter, the influence of external stimuli, for instance, an SC niche, imparts different cues to the two progenies, helping shaping their fates. Endocytosis plays a paramount role in both intrinsic and extrinsic mechanisms, and it is one of the major programmes (arguably, the major) through which ACD, and ensuing cell specification, is achieved.

4.3.1 Learning From Genetics: The SOP System

The most advanced mechanistic knowledge of how endocytosis impacts on ACD derives from studies in the fruit fly and in particular on the bristle sensory organ. This organ is formed by four cells that originate from a mother cell (the sensory organ precursor—SOP—cell) through a pattern of ACDs. The SOP cell (Fig. 4.2) divides asymmetrically along the anteroposterior axis to originate an anterior pIIb cell and a posterior pIIa cell. The pIIb and pIIa cells give rise, through further ACDs, to the final four cells of the organ, the sheath cell and the neuron (from the pIIb), and the socket and shaft (hair) cells (from pIIa) (for details, see [56]). Although the SOP cell does not constitute a bona fide SC compartment, being devoid of self-renewal ability, the study of the divisions it undergoes to form the pIIa and pIIb cells has enormously advanced our knowledge of the intrinsic mechanism of ACD.

It has been known for many years that a protein called NUMB partitions asymmetrically at one of the poles of the dividing SOP, thereby being inherited almost exclusively by the pIIb cell and imparting cell specification (for extensive reviews of the mechanism of the asymmetric partitioning of NUMB and of several other cell fate determinants identified in Drosophila, see [56]). Genetically, NUMB counteracts the action of the signalling receptor NOTCH [57, 58]. It has also been known for several years that, in *Drosophila* neurogenesis, the signalling function of NOTCH requires dynamin, and therefore presumably endocytosis [59]. These two observations, however, were not rationalised together until the discovery that NUMB was an endocytic protein [60]. It was then discovered that NUMB binds to the major endocytic adaptor AP-2 and determines its asymmetrical segregation in the pIIb cell (Fig. 4.2). A flurry of papers subsequently defined how a series of differential endocytic/recycling events, taking place in the pIIb and pIIa cell, create sufficient asymmetry in the repertoire of signalling molecules, at the PM and in intracellular signalling compartment, to allow directional signalling from the pIIb cell (which behaves as a signal-sending cell) to the pIIa cell (the signalling-receiving cell). At least four endocytic-based circuitries concur to create this asymmetry (Fig. 4.2):

(a) NUMB (and/or AP-2)-dependent endocytosis in the pIIb cell of NOTCH or of SANPODO, a positive regulator of NOTCH signalling during ACDs in *Drosophila* [61]. It is of note that recent evidence in mammalian cells argues for a role of NUMB as an inhibitor of NOTCH recycling to the PM [62], rather than as a positive modulator of internalisation, suggesting that the function of NUMB in the pIIb cell might be that of preventing NOTCH recycling to the PM, favouring its commitment to degradation. Whatever the case, the presence of NUMB determines a functional NOTCH-null (or attenuated) situation in the pIIb cell (Fig. 4.2).
(b) DELTA activation by neuralised-dependent endocytosis in the pIIb cell. The UB-ligase neuralised is also asymmetrically segregated during SOP cell mitosis, being preferentially partitioned into the pIIb cell, where it ubiquitinates the

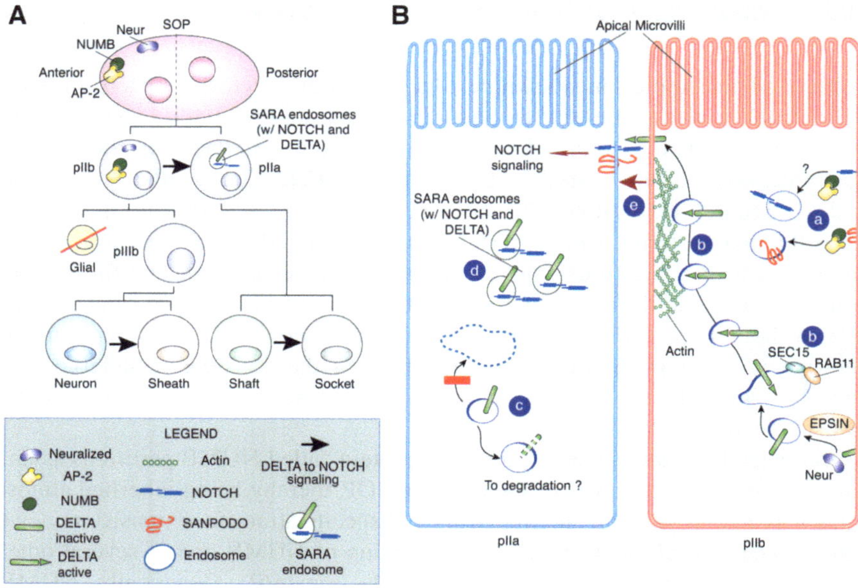

Fig. 4.2 Endocytosis in asymmetric cell division (ACD). (**A**) ACDs in the SOP lineage of *Drosophila*. The SOP lineage is shown. All divisions are asymmetric with directional (DELTA to NOTCH) signalling between daughter cells (*red arrows*) [56]. The first division (from SOP to pIIa and pIIb) is shown in detail. The plane of division with orientations and asymmetrically partitioned molecules (NUMB, AP-2 and SARA endosomes) are indicated [73]. (**B**) Endocytosis generates asymmetry in pIIa and pIIb cells. (**a**) NOTCH is non-functional in pIIb cells, because it is internalised and degraded or because SANPODO is internalised [56]. While the internalisation of SANPODO is established, it is not clear whether NOTCH is preferentially internalised/degraded in the pIIb cell. Evidence from mammals [62] indicates that in the pIIb cell, the function of NUMB may be to prevent NOTCH recycling to the PM, favouring its degradation. (**b**) DELTA-related events in pIIb. The E3 ligase neuralised is asymmetrically partitioned in pIIb, allowing endocytosis of DELTA [63, 64]. DELTA is trafficked by epsin to a RAB11/SEC15 endosome [66] that is directed, for cargo release, along actin networks to a microvillar-dense region of the apical membrane of the pIIb [69]. (**c**) DELTA-related events in pIIa. DELTA is also internalised in the pIIa cell through a neuralised and Ub-independent mechanism. In this cell the recycling to the PM is blocked and DELTA is destined to degradation, because the RAB11-positive endosomal compartment cannot form, possibly because a critical RAB11 partner (Nuclear fallout/Arfophilin 1) is inactivated [66]. DELTA might also be internalised before mitosis of the SOP cell; in pIIb it could be recycled to the PM, whereas in pIIa it might be destined to a degradative pathway. (**d**) Asymmetric distribution of SARA endosomes. In the SOP cell, both NOTCH and DELTA are trafficked to SARA endosomes before cell division [72]. These endosomes are then transported asymmetrically to the nascent pIIa cell [72]. The described events are not necessarily "all or none" situations. They might occur in both cells, with a cell-specific bias in favour of one of them that is further amplified through reinforcement/extinction events that lead from a quasi-symmetric situation to the final DELTA/NOTCH asymmetry needed for directional signalling (**e**)

NOTCH ligand DELTA, thereby promoting its internalisation [63, 64]. In this case, internalisation functions as an "activation" strategy, as it is known that the internalisation of DELTA and its recycling to the PM (through as yet unknown molecular mechanisms) are necessary for the ability of DELTA to engage NOTCH on neighbouring cells [65]. Neuralised might actually promote endocytosis of DELTA through a particular pIIb-specific pathway. Indeed, DELTA is trafficked differently in pIIa and pIIb cells. In the latter, it is routed through RAB11-positive recycling endosomes and probably recycled to the PM. In pIIa cells, conversely, RAB11 endosomes do not form, and DELTA cannot be recycled and is presumably destined to degradation [66]. This mechanism would ensure that the expression of DELTA (and possibly of "activated" DELTA) at the PM is skewed towards the pIIb (signal-sending) cell. These results reinforce the notion of recycling as an important aspect of the mechanisms of cell fate specification, as also supported by the involvement of SEC15, a component of the exocyst (a complex involved in tethering and spatially targeting exocytic and recycling vesicles to the PM), in the ACD of SOP cells [67] (Fig. 4.2).

(c) Recycling- and actin-dependent topological segregation of signalling molecules. In *NUMB* or *a-ADAPTIN* SOP mutants, SANPODO is enriched at the pIIa-pIIb cell interface [68]. In addition, it has been recently reported that the apical surface of pIIa and pIIb cells display actin-rich microvillar structures, to which DELTA is preferentially recy led. The formation of these structures depends on the presence of the actin nucleator complex ARP2/3, and the presence of ARP3 is required in the signal-sending pIIb cell for fate specification [69]. In a system in which both ligands and receptors are membrane-tethered, the PM region within the area of cell-to-cell contact is clearly the most relevant for directional signalling. This suggests that the overall PM level of effector molecules (SANPODO, NOTCH or DELTA) might not matter as much as their levels within defined signalling domains, which in the case of pIIa and pIIb might be represented by microvillar structures that would greatly amplify the surface area available for cell-cell contacts (Fig. 4.2).

(d) Asymmetric partitioning of smad anchor for receptor activation (SARA) endosomes. The asymmetric partitioning of entire endocytic compartments also plays an important role in cell fate specification. Unequal distribution of endosomes between daughter cells is observed frequently during ACD, for instance, at the first cleavage of the *Caenorhabditis elegans* embryo [70], or during ACD of mammalian hematopoietic SCs [71]. In SOP cells, both NOTCH and DELTA are trafficked to SARA endosomes before ACD [72] (Fig. 4.2). These endosomes are then directionally transported to the nascent pIIa cell [72]. This is functionally important, since mistargeting of SARA endosomes to the pIIb cell causes ectopic activation of NOTCH in that cell [72]. These findings define a mechanism, operating in the pIIa cell that acts synergistically with the other described mechanisms towards the generation of asymmetry.

4.3.2 Endocytosis and Stem Cells

ACD is crucial in the maintenance of adult SC compartments, in which SCs divide asymmetrically to give rise to a daughter cell that retains the mother fate (i.e. becomes an SC and withdraws into quiescence) and to a daughter, the progenitor, that undergoes multiple rounds of cell division to generate a vast progeny that eventually differentiates. Most of what we know about the role of endocytosis and SC compartments revolves, not surprisingly, around the role of NUMB as an intrinsic determinant of ACD in SCs. Such a role has been demonstrated, both in lower organisms and in mammals, in neuroblasts, in muscle satellite cells, in hematopoietic SCs and in mammary SCs [73]. At the mechanistic level, one issue deserves additional comments, since it has been shown that in neuroblasts, NUMB might couple with different signalling pathways, in a context-dependent manner. One such pathway involves ACBD3, a NUMB-interacting Golgi protein, which undergoes changes in its subcellular distribution during the cell cycle [74]. When ACBD3 is redistributed in the cytosol after Golgi fragmentation at mitosis, it acts synergistically with NUMB in specifying an SC fate, whereas, when it is associated with the Golgi during interphase, it can promote neuronal differentiation in post-mitotic neurons [74].

4.4 Summary

Our view of the functional implications of endocytosis and recycling has significantly changed in the course of the past 15 years. A variety of experimental evidence points to the notion that EEC networks are intimately intertwined with signalling pathways and represents communications and supply routes ("the endocytic matrix" [2]) of the cell. Indeed, a more precise definition of endocytosis is that of a vast programme, deeply ingrained in the cellular master plan and inextricably intertwined with signalling, which constitutes the major communications infrastructure of the cell. As such, it governs almost all aspects of the relationships of the cell with the extracellular environment and of intracellular communication. Its evolution constitutes, arguably, the major driving force in the evolution of prokaryotic to eukaryotic organisms.

Within this framework it is not surprising that the endocytic matrix is pivotal in processes that require the interpretation and transduction of signals in a spatially defined and temporally controlled manner, ultimately promoting the acquisition of polarised cellular functions and phenotypes. Cell migration represents the perfect case in point as endo-/exocytic trafficking pervasively regulates virtually all aspects of cell motility. This control extends from cell autonomous interpretation of soluble cues and physical constraints imposed by the ECM via the dynamically controlled positioning of motogenic receptors and mechanosensory device, to the precise distribution and spatially restricted activation of signalling adaptors. It is the generation of a dynamic and flexible cell endocytic circuitry that plastically adapts cellular cytoskeletal forces, intracellular hydrostatic pressure and plasma membrane tensions with the ever-changing conditions of the extracellular environment, enabling the choice of the optimal mode of cell locomotion or impacting on cell fate decision. Our understanding, however, of the molecular and regulatory circuitry controlling the trafficking of cargos and membrane through the intracellular road network is still in its infancy. As the molecular players come into focus, it will be possible to build integrated cellular maps of signalling cascades and trafficking routes. This will be critical to unravel the mechanisms controlling physiological and pathological processes that must respond to spatial cues, such as tissue morphogenesis or cell migration, or subvert spatial confinements, such as tumour invasion.

Acknowledgments We apologise to all those colleagues whose primary references or important discoveries could not be properly acknowledged for lack of space. The authors of this review are supported by grants from AIRC (Associazione Italiana Ricerca sul Cancro. N. 10168 and N. 8678) to GS and AD, European Research Council (ERC-Adv-2010-268836) and PRIN2009 (progetti di ricerca di interesse nazionale-prot. 2009X23L78), and Cariplo Fondation (Projc. 2010-0737) and International Association of Cancer Research (AICR) to GS.

References

1. Sigismund S, Confalonieri S, Ciliberto A, Polo S, Scita G et al (2012) Endocytosis and signaling: cell logistics shape the eukaryotic cell plan. Physiol Rev 92:273–366
2. Scita G, Di Fiore PP (2010) The endocytic matrix. Nature 463:464–473
3. Eggert US, Kiger AA, Richter C, Perlman ZE, Perrimon N et al (2004) Parallel chemical genetic and genome-wide RNAi screens identify cytokinesis inhibitors and targets. PLoS Biol 2:e379
4. Raz E (2004) Guidance of primordial germ cell migration. Curr Opin Cell Biol 16:169–173
5. Minina S, Reichman-Fried M, Raz E (2007) Control of receptor internalization, signaling level, and precise arrival at the target in guided cell migration. Curr Biol 17:1164–1172
6. Boldajipour B, Raz E (2007) What is left behind—quality control in germ cell migration. Sci STKE 2007:pe16
7. Pollard TD (2003) The cytoskeleton, cellular motility and the reductionist agenda. Nature 422:741–745
8. Charras GT, Yarrow JC, Horton MA, Mahadevan L, Mitchison TJ (2005) Non-equilibration of hydrostatic pressure in blebbing cells. Nature 435:365–369
9. Matsumura F (2005) Regulation of myosin II during cytokinesis in higher eukaryotes. Trends Cell Biol 15:371–377
10. Blaser H, Reichman-Fried M, Castanon I, Dumstrei K, Marlow FL et al (2006) Migration of zebrafish primordial germ cells: a role for myosin contraction and cytoplasmic flow. Dev Cell 11:613–627
11. Sorkin A, von Zastrow M (2009) Endocytosis and signaling: intertwining molecular networks. Nat Rev Mol Cell Biol 10:609–622
12. Le Roy C, Wrana JL (2005) Signaling and endocytosis: a team effort for cell migration. Dev Cell 9:167–168
13. Sigismund S, Argenzio E, Tosoni D, Cavallaro E, Polo S et al (2008) Clathrin-mediated internalization is essential for sustained EGFR signaling but dispensable for degradation. Dev Cell 15:209–219

14. De Donatis A, Comito G, Buricchi F, Vinci MC, Parenti A et al (2008) Proliferation versus migration in platelet-derived growth factor signaling: the key role of endocytosis. J Biol Chem 283:19948–19956
15. Stradal TE, Scita G (2006) Protein complexes regulating Arp2/3-mediated actin assembly. Curr Opin Cell Biol 18:4–10
16. Joffre C, Barrow R, Menard L, Calleja V, Hart IR et al (2011) A direct role for Met endocytosis in tumorigenesis. Nat Cell Biol 13:827–837
17. Bretscher MS (1996) Moving membrane up to the front of migrating cells. Cell 85:465–467
18. Lee J, Gustafsson M, Magnusson KE, Jacobson K (1990) The direction of membrane lipid flow in loc.moting polymorphonuclear leukocytes. Science 247:1229–1233
19. Wessels D, Reynolds J, Johnson O, Voss E, Burns R et al (2000) Clathrin plays a novel role in the regulation of cell polarity, pseudopod formation, uropod stability and motility in Dictyostelium. J Cell Sci 113(pt 1):21–36
20. Traynor D, Kay RR (2007) Possible roles of the endocytic cycle in cell motility. J Cell Sci 120:2318–2327
21. Assaker G, Ramel D, Wculek SK, Gonzalez-Gaitan M, Emery G (2010) Spatial restriction of receptor tyrosine kinase activity through a polarized endocytic cycle controls border cell migration. Proc Natl Acad Sci U S A 107:22558–22563
22. Jekely G, Sung HH, Luque CM, Rorth P (2005) Regulators of endocytosis maintain loc.lized receptor tyrosine kinase signaling in guided migration. Dev Cell 9:197–207
23. Palamidessi A, Frittoli E, Garre M, Faretta M, Mione M et al (2008) Endocytic trafficking of Rac is required for its activation and for the spatial restriction of signaling in cell migration. Cell 134:135–147
24. Osmani N, Peglion F, Chavrier P, Etienne-Manneville S (2010) Cdc42 localization and cell polarity depend on membrane traffic. J Cell Biol 191:1261–1269
25. Mammoto A, Ohtsuka T, Hotta I, Sasaki T, Takai Y (1999) Rab11BP/Rabphilin-11, a downstream target of rab11 small G protein implicated in vesicle recycling. J Biol Chem 274:25517–25524
26. Fan GH, Lapierre LA, Goldenring JR, Sai J, Richmond A (2004) Rab11-family interacting protein 2 and myosin Vb are required for CXCR2 recycling and receptor-mediated chemotaxis. Mol Biol Cell 15:2456–2469
27. Prigozhina NL, Waterman-Storer CM (2006) Decreased polarity and increased random motility in PtK1 epithelial cells correlate with inhibition of endosomal recycling. J Cell Sci 119:3571–3582
28. Calderwood DA (2004) Integrin activation. J Cell Sci 117:657–666
29. Margadant C, Monsuur HN, Norman JC, Sonnenberg A (2011) Mechanisms of integrin activation and trafficking. Curr Opin Cell Biol 23:607–614
30. Caswell PT, Spence HJ, Parsons M, White DP, Clark K et al (2007) Rab25 associates with alpha5beta1 integrin to promote invasive migration in 3D microenvironments. Dev Cell 13:496–510
31. Dozynkiewicz MA, Jamieson NB, Macpherson I, Grindlay J, van den Berghe PV et al (2012) Rab25 and CLIC3 collaborate to promote integrin recycling from late endosomes/lysosomes and drive cancer progression. Dev Cell 22:131–145
32. del Pozo MA, Balasubramanian N, Alderson NB, Kiosses WB, Grande-Garcia A et al (2005) Phospho-caveolin-1 mediates integrin-regulated membrane domain internalization. Nat Cell Biol 7:901–908
33. Grande-Garcia A, Echarri A, de Rooij J, Alderson NB, Waterman-Storer CM et al (2007) Caveolin-1 regulates cell polarization and directional migration through Src kinase and Rho GTPases. J Cell Biol 177:683–694
34. del Pozo MA, Alderson NB, Kiosses WB, Chiang HH, Anderson RG et al (2004) Integrins regulate Rac targeting by internalization of membrane domains. Science 303:839–842
35. Balasubramanian N, Scott DW, Castle JD, Casanova JE, Schwartz MA (2007) Arf6 and microtubules in adhesion-dependent trafficking of lipid rafts. Nat Cell Biol 9:1381–1391

36. Rainero E, Caswell PT, Muller PA, Grindlay J, McCaffrey MW et al (2012) Diacylglycerol kinase alpha controls RCP-dependent integrin trafficking to promote invasive migration. J Cell Biol 196:277–295
37. Chianale F, Rainero E, Cianflone C, Bettio V, Pighini A et al (2010) Diacylglycerol kinase alpha mediates HGF-induced Rac activation and membrane ruffling by regulating atypical PKC and RhoGDI. Proc Natl Acad Sci U S A 107:4182–4187
38. Parat MO, Anand-Apte B, Fox PL (2003) Differential caveolin-1 polarization in endothelial cells during migration in two and three dimensions. Mol Biol Cell 14:3156–3168
39. Fletcher SJ, Rappoport JZ (2010) Moving forward: polarised trafficking in cell migration. Trends Cell Biol 20:71–78
40. Howes MT, Kirkham M, Riches J, Cortese K, Walser PJ et al (2010) Clathrin-independent carriers form a high capacity endocytic sorting system at the leading edge of migrating cells. J Cell Biol 190:675–691
41. Gu Z, Noss EH, Hsu VW, Brenner MB (2011) Integrins traffic rapidly via circular dorsal ruffles and macropinocytosis during stimulated cell migration. J Cell Biol 193:61–70
42. Lobert VH, Brech A, Pedersen NM, Wesche J, Oppelt A et al (2010) Ubiquitination of alpha 5 beta 1 integrin controls fibroblast migration through lysosomal degradation of fibronectin-integrin complexes. Dev Cell 19:148–159
43. Hocking DC, Chang CH (2003) Fibronectin matrix polymerization regulates small airway epithelial cell migration. Am J Physiol Lung Cell Mol Physiol 285:L169–L179
44. Nowak M, Machate A, Yu SR, Gupta M, Brand M (2011) Interpretation of the FGF8 morphogen gradient is regulated by endocytic trafficking. Nat Cell Biol 13:153–158
45. Ridley AJ (2011) Life at the leading edge. Cell 145:1012–1022
46. Ezratty EJ, Partridge MA, Gundersen GG (2005) Microtubule-induced focal adhesion disassembly is mediated by dynamin and focal adhesion kinase. Nat Cell Biol 7:581–590
47. Wang Y, Cao H, Chen J, McNiven MA (2011) A direct interaction between the large GTPase Dynamin2 and FAK regulate focal adhesion dynamics in response to active Src. Mol Biol Cell 22(9):1529–1538
48. Ezratty EJ, Bertaux C, Marcantonio EE, Gundersen GG (2009) Clathrin mediates integrin endocytosis for focal adhesion disassembly in migrating cells. J Cell Biol 187:733–747
49. Chao WT, Ashcroft F, Daquinag AC, Vadakkan T, Wei Z et al (2010) Type I PIP kinase beta regulates focal adhesion disassembly by promoting beta1 integrin endocytosis. Mol Cell Biol 30(18):4463–4479
50. Luo BH, Carman CV, Springer TA (2007) Structural basis of integrin regulation and signaling. Annu Rev Immunol 25:619–647
51. Roberts M, Barry S, Woods A, van der Sluijs P, Norman J (2001) PDGF-regulated rab4-dependent recycling of alphavbeta3 integrin from early endosomes is necessary for cell adhesion and spreading. Curr Biol 11:1392–1402
52. Danen EH, van Rheenen J, Franken W, Huveneers S, Sonneveld P et al (2005) Integrins control motile strategy through a Rho-cofilin pathway. J Cell Biol 169:515–526
53. Francavilla C, Cattaneo P, Berezin V, Bock E, Ami D et al (2009) The binding of NCAM to FGFR1 induces a specific cellular response mediated by receptor trafficking. J Cell Biol 187:1101–1116
54. White DP, Caswell PT, Norman JC (2007) alpha v beta3 and alpha5beta1 integrin recycling pathways dictate downstream Rho kinase signaling to regulate persistent cell migration. J Cell Biol 177:515–525
55. Reynolds AR, Hart IR, Watson AR, Welti JC, Silva RG et al (2009) Stimulation of tumor growth and angiogenesis by low concentrations of RGD-mimetic integrin inhibitors. Nat Med 15:392–400
56. Neumuller RA, Knoblich JA (2009) Dividing cellular asymmetry: asymmetric cell division and its implications for stem cells and cancer. Genes Dev 23:2675–2699
57. Guo M, Jan LY, Jan YN (1996) Control of daughter cell fates during asymmetric division: interaction of Numb and Notch. Neuron 17:27–41

58. Spana EP, Kopczynski C, Goodman CS, Doe CQ (1995) Asymmetric localization of numb autonomously determines sibling neuron identity in the Drosophila CNS. Development 121:3489–3494
59. Seugnet L, Simpson P, Haenlin M (1997) Requirement for dynamin during Notch signaling in Drosophila neurogenesis. Dev Biol 192:585–598
60. Santolini E, Puri C, Salcini AE, Gagliani MC, Pelicci PG et al (2000) Numb is an endocytic protein. J Cell Biol 151:1345–1352
61. Berdnik D, Torok T, Gonzalez-Gaitan M, Knoblich JA (2002) The endocytic protein alpha-Adaptin is required for numb-mediated asymmetric cell division in Drosophila. Dev Cell 3:221–231
62. McGill MA, Dho SE, Weinmaster G, McGlade CJ (2009) Numb regulates post-endocytic trafficking and degradation of notch1. J Biol Chem 284(39):26427–26438
63. Lai EC, Deblandre GA, Kintner C, Rubin GM (2001) Drosophila neuralized is a ubiquitin ligase that promotes the internalization and degradation of delta. Dev Cell 1:783–794
64. Le Borgne R, Schweisguth F (2003) Unequal segregation of neuralized biases Notch activation during asymmetric cell division. Dev Cell 5:139–148
65. Parks AL, Klueg KM, Stout JR, Muskavitch MA (2000) Ligand endocytosis drives receptor dissociation and activation in the Notch pathway. Development 127:1373–1385
66. Emery G, Hutterer A, Berdnik D, Mayer B, Wirtz-Peitz F et al (2005) Asymmetric Rab 11 endosomes regulate delta recycling and specify cell fate in the Drosophila nervous system. Cell 122:763–773
67. Jafar-Nejad H, Andrews HK, Acar M, Bayat V, Wirtz-Peitz F et al (2005) Sec15, a component of the exocyst, promotes notch signaling during the asymmetric division of Drosophila sensory organ precursors. Dev Cell 9:351–363
68. Tong X, Zitserman D, Serebriiskii I, Andrake M, Dunbrack R et al (2010) Numb independently antagonizes Sanpodo membrane targeting and Notch signaling in Drosophila sensory organ precursor cells. Mol Biol Cell 21:802–810
69. Rajan A, Tien AC, Haueter CM, Schulze KL, Bellen HJ (2009) The Arp2/3 complex and WASp are required for apical trafficking of Delta into microvilli during cell fate specification of sensory organ precursors. Nat Cell Biol 11:815–824
70. Andrews R, Ahringer J (2007) Asymmetry of early endosome distribution in C. elegans embryos. PLoS One 2:e493
71. Beckmann J, Scheitza S, Wernet P, Fischer JC, Giebel B (2007) Asymmetric cell division within the human hematopoietic stem and progenitor cell compartment: identification of asymmetrically segregating proteins. Blood 109:5494–5501
72. Coumailleau F, Furthauer M, Knoblich JA, Gonzalez-Gaitan M (2009) Directional Delta and Notch trafficking in Sara endosomes during asymmetric cell division. Nature 458:1051–1055
73. Pece S, Confalonieri S, Romano PR, Di Fiore PP (2011) NUMB-ing down cancer by more than just a NOTCH. Biochim Biophys Acta 1815:26–43
74. Zhou Y, Atkins JB, Rompani SB, Bancescu DL, Petersen PH et al (2007) The mammalian Golgi regulates numb signaling in asymmetric cell division by releasing ACBD3 during mitosis. Cell 129:163–178

Chapter 5
Aberrant Vesicular Trafficking Contributes to Altered Polarity and Metabolism in Cancer

Shreya Mitra and Gordon B. Mills

Abstract Epithelial cancers demonstrate loss of cell polarity, hyperproliferation, and altered cellular metabolism due to acquisition of a suite of genomic aberrations as well as a consequence of metabolic challenges in the tumor microenvironment. Whether these neoplastic properties represent a coordinate process leading to tumor initiation and progression is still poorly understood. In this review, we posit that abnormal vesicular trafficking targets cellular metabolism not just by altering trafficking of receptor tyrosine kinases and nutrient transporters but also by disrupting tight junctions and cell polarity. Apical–basal polarity is required for the formation of normal cellular structures that maintain cellular junctions as well as to regulate asymmetric division of stem cells; disruption of these processes contributes to tumor initiation and progression. Indeed, derailed endocytosis and subsequent aberrations in targeting of vesicles and their cargoes to the correct intracellular compartments is an emerging hallmark of cancer. This chapter will review existing literature to highlight the vicious nexus between trafficking, polarity, and metabolism in order to identify potential "Achilles heels" that can be exploited therapeutically.

5.1 Introduction

Polarized holographic structures are a recurring theme in living organisms [1]. Indeed loss of polarity and breakdown of normal cell–cell interactions in polarized epithelium are required to bypass cellular competition, a process that limits the consequence of cellular damage and oncogenic stress [2]. These checkpoints must be overcome for a cell to adopt a number of malignant behaviors including proliferation, motility, and invasion [3–9].

S. Mitra (✉) • G.B. Mills
The Department of Systems Biology, University of Texas MD Anderson Cancer Center, 7435 Fannin Street, Unit 950, Houston, TX 77054, USA
e-mail: smitra@mdanderson.org

Y. Yarden and G. Tarcic (eds.), *Vesicle Trafficking in Cancer*,
DOI 10.1007/978-1-4614-6528-7_5, © Springer Science+Business Media New York 2013

The stability of the "polarized state" is susceptible to effects of mutations in multiple oncogenes including Ras, phosphatidylinositide 3-kinase (PI3K), Notch [10–13], as well as tumor suppressor genes such as LKB1, p53, and Scribble [14–20], making aberrant polarity both an emerging hallmark and a precondition for tumor initiation and progression. Not surprisingly, during tumor progression, epithelial polarity pathways are hijacked leading to collapse of structured epithelium, breakdown of cell–cell junctions, and cell–extracellular matrix (ECM) interactions, allowing the metastatic cascade to progress [21–24]. However, the nature of the initial trigger leading to loss of epithelial polarity has remained largely unknown.

In the epithelium, a uniform polarized architecture leads to efficient cell–cell and cell–matrix communication, which is critical for tissue homeostasis [6, 7, 25]. This homeostasis is sensitive to energetic stress resulting from microenvironmental fluctuations in nutrient and oxygen levels [26], indicating that microenvironmental cues leading to metabolic changes in the tumor could influence cell polarity and subsequently altered polarity would also impact cellular metabolic state. In normal tissues, the complexity of the endosomal recycling network allows efficient compensation for resource scarcity by recalibrating uptake and usage of cellular resources during energy and nutrient stress [27, 28]. Thereby, the trafficking machinery is equipped to maintain polarized states by spatiotemporal regulation of import, sorting, trafficking, and export of growth factors, nutrients, integrins, polarity, and junction proteins [29–34]. Polarity and vesicular trafficking networks are thus intricately intertwined, with defects in one process having marked consequence on the other (Fig. 5.1).

Moreover, spatially localized intra-tumoral mutational heterogeneity as well as adaptive responses of cancer cells to mutational stress and changes in the microenvironment further contribute to the challenge of integrating the complex genomic aberrations in the tumor with metabolic diversity, growth profiles, and microenvironmental stresses [35]. While the actors in the recycling vesicle pathways linking polarity and metabolism are beginning to be identified and functionally characterized, there is as yet insufficient information to provide an integrated systems biology approach to accurately model the network of trafficking systems and how they impinge on accompanying changes in the cellular metabolic profile and polarity.

In this chapter, we will present evidence for potential mechanisms by which aberrant vesicular trafficking could lead to a loss of cell polarity with consequent metabolic changes allowing survival and progression of cancers. Given the complex interactions within these homeostatic networks, determining the cause and effect between altered trafficking, loss of polarity, and changes in metabolism has been and remains challenging.

5.2 Interrogating the Interrelationship Between Vesicular Trafficking and Cell Polarity

Multiple polyvalent interactions, involving signaling pathways, cytoskeletal systems, ECM, and most importantly the trafficking machinery, coordinately establish and regulate cell polarity [27, 36–38]. Polarity is cell type specific and dynamic [7, 9, 39]. It is evident that loss of polarity precedes epithelial–mesenchymal transition

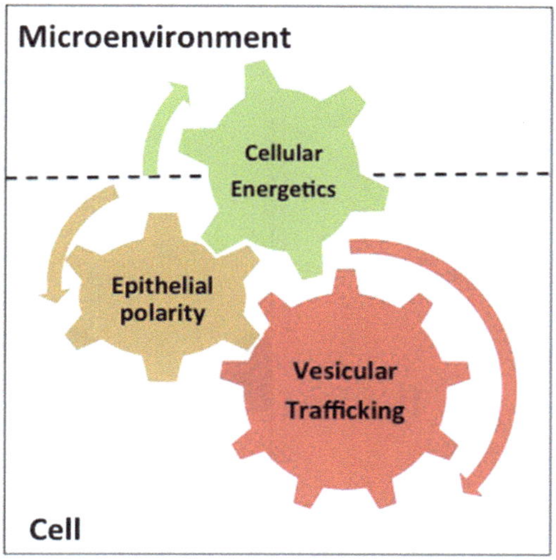

Fig. 5.1 Vesicular trafficking integrates epithelial polarity proteins and cellular energetic sensors to maintain tissue homeostasis. The highly complex vesicular trafficking network in eukaryotes is critical for transportation of polarity proteins as well as junctional proteins, integrins, growth factors and nutrient transporters. Polarity tool kit itself cross talks with metabolic pathways to communicate cellular energy status across the epithelium. The metabolic and bioenergetics pathways maintain equilibrium across the tissue based on its interactions with junctional proteins and polarity checkpoints. Thus the three pathways are intimately linked to each other. Deregulation of any of these three wheels would alter tissue homeostasis and lead disease including malignancy

(EMT) [12, 40] and several key polarity proteins (e.g., LKB1, Scribble) are potent tumor suppressors [15, 19, 21, 41–43].

5.2.1 Endosomal Trafficking Machinery Is the Architect of Dynamic Cell Polarity

Vesicular trafficking plays a pivotal role in establishment of the two most prevalent forms of polarity observed in normal and malignant cells (Fig. 5.2). In the following section, we will highlight how vesicular trafficking systems comprising of small GTPases of Rab and Rho superfamily are key integrators of the information processes linking vesicle dynamics, polarity, and metabolism in multicellular organisms [37, 44, 45].

5.2.1.1 Establishing "Front–Back" Polarity to Favor Motility in Response to Environmental Cues

Front–rear asymmetry facilitates rapid response to environmental cues and represents the earliest polarity state that evolved in unicellular organisms. By forming a

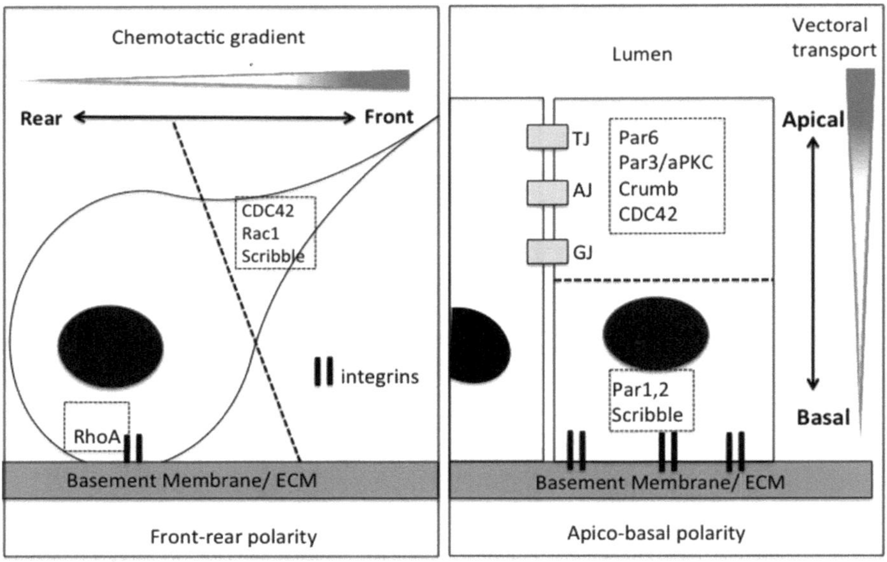

Fig. 5.2 Front-rear polarity and apico-basal polarity. The *left side box* is a schematic representation of a mesenchymal cell with typical front-back polarity. The axis of polarity is dictated by the chemotactic gradient, with the leading edge, place for high actin turn over, geared to move towards nutrient rich microenvironment. The front part of the cell is marked by polarity proteins such as Scribble and GTPases namely CDC42 and Rac1. The GTPases communicate with the actin cytoskeleton and facilitate the formation of lamellipodia and filopodias. Integrins are rapidly endocytosed from the rear and recycled to the front by Rab25 and Rab11 to enhance migration. The rear end (marked by *dotted line*) of the cell remains attached to the basement membrane and enriched in Rho GTPases. The *right box* represents 2 adjacent polarized epithelial cells. The axis of polarity is vertical, apico-basal type where the apical end faces the lumen and the basal end adheres to the basement membrane. Cellular junctions demarcate the apical part of the cell along with polarity proteins, Par6, Par3/aPKC and Crumb. CDC42 is the main GTPase found in the apical region. Par1, 2 and Scribble are present in the typo basal part (marked by dotted line). Growth factors, integrins, nutrients are available to the cell from the basement membrane and the vasculature feeding the ECM. The apicobasal polarity facilitates vectoral transports and cellular wastes are released into the lumen. The cellular junctions maintain limited paracellular permeability and help to keep receptors and their mitogenic ligands separated to achieve regulated cell proliferation

leading edge and a lagging end, an internal directional compass sensitizes the cytoskeleton towards an external chemotactic gradient. Following installation of polarity proteins, microtubule tracts are laid out to carry bulk cargo towards the leading edge allowing directed migration [46].

Polarized trafficking of integrins (cell–matrix-focal adhesion interactions) and RTKs (chemotactic response) are integral to the creation of the front–rear polarity axis [47–50]. In a migrating cell, these receptors are internalized and trafficked via Rab GTPase decorated vesicles to the front of the cell to fuel actin polymerization [51–55] generating differential cytoskeletal activity, especially actin turnover and remodeling at cellular poles. The internalization, sorting, and fate of vesicle cargos are determined by the dynamic exchange of specific Rab family members with vesicles. Furthermore, the polarized activity of actin is mediated by localized activation of specific Rho family small GTPases [56–58]. For example, in migrating cells, Cdc42 and Rac1 populate the expanding front membrane where they increase actin assembly by promoting Arp2/3-mediated actin nucleation [59]. As the expanding plasma membrane forms an anterior lamellipodia, integrins from the rear of the cell are endocytosed in Rab coated vesicles, dissolving focal adhesions [60, 61]. Coordinate, RhoA-mediated assembly and activation of actin–myosin contractile networks, detaches the rear pole of the cell from the ECM, allowing forward motion [62].

Epithelial cancer cells co-opt processes that normally promote front–rear polarity to facilitate cell migration during development and response to injury as well as to mediate the early steps of invasion and metastasis. Consistent with a major role in tumorigenesis, components of the front–back polarity regulatory process, such as small GTPases of the Rab and Rho family and integrins, undergo marked copy number and expression changes in addition to trafficking alterations during oncogenic progression [34, 53, 56, 63–69].

5.2.1.2 Establishing "Apicobasal" Polarity to Enable Vectorial Transport

During phylogeny, epithelium-specific polarized, compact organization of cells arose from "coalesced" aggregation and subsequent epithelial differentiation of migratory mesenchymal cells [46]. The modification of the existing "front–back" polarization mechanics of the mesenchymal cells to a columnar (vertical) apicobasal axis was developmentally advantageous since it synchronized and unified responses to signaling inputs and outputs of a larger mass of cells within the tissue. Indeed, evolution of this social network proved critical for survival of multicellular organisms. A sophisticated and complex polarity system evolved as sheets of cells formed an interface between the organism and its environment or folded into ducts and villi along a luminal axis [23]. Importantly, the adhesive side remained attached to a basement membrane (BM) and eventually acquired perfusion with blood vessels and a coordinated interaction with a mitogen enriched stroma, while the other end faced the lumen and was in direct contact with the microenvironment of the organism [22]. Subsequently, junctions developed in these epithelial cells to ensure adhesion. Cellular junctions selectively regulates permeability of the epithelial sheets to specific ions and small molecules [22]. However, there are several variations to this theme that allow normal cells to initiate cytokinesis as well as cell migration during tissue repair. Many of these processes are maintained in tumor cells, but rather than being used for the social benefit of the whole organism, are exploited for the growth and metastasis of the tumor [21, 22, 70, 71].

The basal membrane allows cell-BM connections via integrins [22]. Strikingly the loss of attachment of normal epithelial cells to the BM leads to programmed cell death through anoikis, again providing a form of tissue surveillance of cellular state that must be overcome during epithelial tumorigenesis. The "normal" mutual exclusion of proteins to the apical or baso-lateral domains is frequently lost early in oncogenic transformation [22].

Functionally, maintenance of the apico-basal axis is a key component of cellular differentiation as well as the surveillance of cellular state and removal of damaged or mutated cells [2]. Studies on cooperative tumorigenesis (e.g., Ras and Notch) and cellular competition suggest that endocytic machinery and apico-basal polarity toolkit collaborate to protect against neoplastic development [13]. In drosophila, aberrant Scribble increases endocytosis of JNK activators, leading to JNK-driven apoptosis. Tumorigenesis requires bypass of JNK activation as a consequence of endocytosis. Similarly, in polarized mammalian cell complexes, cellular competition can prevent the outgrowth of cells with oncogenic mutations providing an additional hurdle to tumor initiation and progression [2]. Thus deregulation of the "polarity checkpoint" is required to bypass apoptosis and cell elimination [72].

Rab and Rho GTPases not only transport and stabilize polarity complexes but also remove aberrant apical proteins from the baso-lateral membrane [73]. Notably, growth factors receptors such as Her2 and EGFR normally localize strictly to the basolateral membrane in polarized cells [74]. In contrast, their ligands are restricted to apical sides. A disruption of apico-basal polarity either due to injury or a malignant transformation allows ready access of growth factors to their cognate receptors. Oligomerized Caveolin-1 [75] and EPS8 (Epidermal Growth Factor Receptor Substrate 8) [76], a scaffold interacting with Rab5 GAP [77] and connecting Ras to Rac [78, 79], are key molecular guards that restrict EGFR and other receptors to the basolateral membrane [78, 80] in normal cells. A loss of Caveolin-1 or changes in EPS8-binding interactions would disrupt membrane segregation and cause excessive receptor activation and increased motility and proliferation.

The segregation of receptor and ligand ensure that apico-basal polarized cells remain quiescent but provide the opportunity for a rapid and efficient response to injury. In tumor cells, loss of apico-basal polarity allows access of growth factor to their cognate receptors acting to propagate aspects of the malignant phenotype.

5.2.2 Polarity Checkpoint Proteins and Their Spatiotemporal Regulation by Endosomal Trafficking Machinery

Initially identified via fly screens but highly conserved throughout metazoans, the polarity toolkit includes the Scribble complex (Scribble, DLG, and LGL), the PARtition (PAR) complex (Par3/Par6/atypical protein kinase C [aPKC] and Par4/LKB1), and the transmembrane Crumbs complex (Crumbs, PALS1, and PATJ) [81, 82]. These complexes are common mediators engendering both apico-basal polarity and front–tail polarity. Indeed, the use of the same complexes in both apico-basal and front–tail polarities are energy conserving and also ensures coordinate transition from one to the other state. The PAR-aPKC system converts initial cues to establish complementary membrane domains along the polarity axis [83]. Par6 facilitates functional junctions leading to an initial segregation of the Par3/aPKC complex at the apical end of the cell with Par1 or 2 and the Scribble complex at the baso-lateral (BL) end of the cell [84] (Fig. 5.2). Subsequent mutual exclusion occurs with Scribble in the BL surface suppressing apical membrane identity by inhibiting the function of the Par3 complex. In the apical surface, recruitment of Crumb to Par3 locally activates the polarity checkpoint [85].

Intriguingly, the polarity complexes appear to act both upstream and downstream of small GTPases. For example, Scribble complexes with a guanine exchange factor (GEF) for Cdc42/Rac1, namely, βPIX (PAK-interacting exchange factor), and with GIT1 (G-protein-coupled receptor interacting protein1) [86], during Cdc42 activation and localization [87], which facilitates its relocalization to the leading edge. LGL, also a part of this complex, then enables vesicle fusion with the plasma membrane via SNARE proteins [88]. In epithelial cells, interaction of GTP bound CDC42 with Par6 recruits CDC42 to the apical cortex [89, 90] and increases the activity of both aPKC [89, 91] and Crumbs [92]. CDC42, by virtue of its dynamic interactions with Par6, regulates polarity in migrating cells [56, 93, 94], and Par6 itself mediates alignment of centrosome and nucleus with respect to the direction of migration in astrocytes and fibroblasts [95, 96]. Again, the dynamic roles of CDC2, aPKC, and Crumbs coordinate the transitions between apico-basal and front–tail polarity.

TGFβ induces EMT in breast cancer cells when TGFβ receptors, present at tight junctions of breast cancer cells, bind to and phosphorylate Par6 (on Ser 345), recruiting Smurf1 to the complex [97]. Presence of Smurf destroys RhoA at tight junctions and reverses junction assembly leading to loss of apico-basal polarity with consequent hyperproliferation and EMT [98]. In another instance, aPKC proteins, specifically PKCι, is deregulated as part of the 3q26.2 amplicon that is present in multiple tumor lineages [99]. Overexpression of PKCι as a consequence of the 3q26.2 amplicon alters polarity leading to increased expression of cyclin E and subsequent cellular proliferation. Indeed, coordinate elevation of cyclin E and PKCι is associated with worsened outcomes in human cancers. aPKC also cross talks with the GTPase system via T-cell lymphoma invasion and metastasis1 (TIAM1) [100], a GEF for Rac1 [101]. The apical recycling endosome protein Rab11a associates with PDK1, that phosphorylates the T-loop of aPKC, to maintain steady state levels of active aPKC [102]. This coordinates the effects of cell signaling (PDK1), vesicle recycling (Rab11a), and cell polarity (aPKC) and potentially cell proliferation (cyclin E). It is important to note that aPKC also interacts with metabolic proteins as discussed later suggesting that it may represent a key node integrating and coordination of transmission of information between polarity and metabolism.

5.3 Interrogating the Interrelationship Between Polarity Proteins and Bioenergetic Stress Sensors

Polarity proteins are highly sensitive to oncogenic insults as well as microenvironmental stress, including bioenergetic stress. For example, aPKC, which as described above is a key regulator of polarity, appears to represent a nodal point linking polarity and metabolism. Phosphorylation of glycogen synthase kinase-3β (GSK3β), by aPKC when sequestered in vesicles inactivates GSK3β, potentially altering cellular metabolism and energy storage as GSK3β determines the rate of synthesis of glycogen. Phosphorylation of GSK3β also alters APC (Adenomatous Polyposis Coli) phosphorylation, resulting in destabilization of microtubules at the plasma membrane [103]. If aPKC and GSK3β are not sequestered into complexes, their functionality becomes limited. Thus aPKC functions as a key coordinator of information determining polarity, vesicle recycling, and metabolic fates of cells.

Strikingly, the demonstration that the Par4/LKB1 polarity protein regulates cellular metabolism through the adenosine monophosphate (AMP)-activated protein kinase (AMPK) [104], a key sensor of cellular energy status [105] under stress conditions, uncovered an intriguing but powerful link between loss of polarized architecture and altered metabolism [106, 107]. Importantly, germ line inactivation of LKB1 in the Peutz Jaegers syndrome [108–111] and somatic inactivation of LKB1 in many tumor lineages underlines a potential critical link between polarity, metabolism and tumor progression. LKB1 regulates metabolic control of cell growth, mitosis, cell polarization, and asymmetric cell division [111] while coordinated effects of LKB1 and AMPK maintain epithelial cell polarity under energetic stress. LKB1 potently inhibits motility and invasion pathways and downregulates the expression of several mesenchymal marker proteins such as ZEB1, a transcriptional repressor for E-cadherin and an EMT inducer in lung cancer [112]. Additionally, kinase activity of LKB1 affects cellular metabolites such as lactate by inducing apical trafficking of lactate transporters such as Sln (a homolog of the MCT lactate transporter) in polarized cells in drosophila. LKB1-mediated localization of the lactate transporter inhibits apoptosis and expands the repertoire of usable fuels for survival [113, 114]. Thus, in recent years, multiple reports showcasing a major role of LKB1 in regulating metabolic balance through both AMPK and non-AMPK mechanisms show great promise in unraveling the link between altered polarity and cancer cell metabolism.

5.4 Vesicular Trafficking Ensures Epithelial Homeostasis via Maintenance of Cell–Cell Interactions

Sequential establishment of the different types of cellular junctions, namely, tight junctions (TJ), adherent junctions (AJ), and gap junctions (GJ), is necessary for the establishment of overall polarization. Oncogenic insults, such as Ras and TGFβ, by derailing the trafficking machinery alters the molecular composition of junctions, leading to their collapse [22]. Key junction proteins, namely, cadherins and claudins, are replaced by integrins (especially β1 integrins) [115] allowing a switch from an apical–basal polarity to a front–back polarity, facilitating EMT and cell motility.

5.4.1 *Formation of Tight Junctions and Their Role in Oncogenesis*

TJ, the trademark feature of apico-basal polarity, fences off the two protein pools [116–118] from apical and basal membranes. Furthermore TJs form a selective paracellular barrier to ion and macromolecule transit. TJ comprises cytoplasmic zona occludens (ZO), membrane spanning claudins, and occludins macromolecular modules [85, 119] as well as cytoskeletal proteins and signaling molecules [118]. Extracellular homotypic interactions with adjacent cells stabilize TJ while the cytoplasmic ends of the transmembrane proteins interact with growth regulatory pathways [120–122]. There is emerging evidence that the formation and dissolution of TJ as well as the intercellular cross talk that they mediate, both regulate and are regulated by the metabolic status of the cell as well as by nutrient availability in the microenvironment.

Expression analysis of multiple epithelial cancer lineages shows deregulation of TJ proteins associated with EMT, invasion, and metastasis [123, 124]. Junction proteins such as claudins are gaining relevance as novel clinical targets [125] and also as cancer biomarkers [126]. In fact, frequent loss of claudins 3, 4, and 7 in breast carcinomas identifies a new intrinsic subtype of breast cancer called the "claudin-low" tumors [127, 128], featuring low levels of E-cadherin, along with a unique mutational landscape consistent with being an independent lineage from basal-like tumors [128]. The claudin-low tumors are poorly differentiated, enriched in stem cell markers, with high resistance to chemotherapy and with poor prognosis [127].

A number of Rab family members, in particular Rab13 and to a lesser extent Rab3B, Rab8, and Rab14, are implicated in the formation and dissolution of TJ in polarized normal epithelial cells [129]. Rab13 regulates protein kinase A (PKA) during TJ assembly [130] and also recycles occludin and claudin 1 by forming a complex with its TJ-specific effector, junctional Rab13-binding protein, and MICAL-L2 [131]. Rab13 and Rab8 bind to the same complex albeit at spatially distinct locations compatible with a collaborative role in TJ and AJ formation [131]. Rab5a, an early endosomal protein, is also indirectly involved with polarity and TJ. Rab5a drives EGFR trafficking and sorting with a consequent regulation of the EGFR scaffold protein, Gab1 [132]. Gab1, in turn, contributes to cell polarity by acting as a PAR protein scaffold [133], linking Rab5a to polarity regulation. In endothelial cells, Rab5a also regulates localization of claudin-1 in a proteasome-dependent manner [134]. This suggests that Rab5a-coated endosomes serve as a hub where EGFR signaling modules, polarity regulators, and TJ proteins intersect. Furthermore, in migrating cells, Rab14 and its effector form an endocytic-recycling pathway that traffic ADAM protease and regulates cell–cell junctions [123].

The Rab11 family, comprising Rab 25 (aka Rab11c), and its related family member, Rab11a, are both implicated in formation of TJ in normal and cancer models. In normal MDCK cells, phosphorylation of Rab11Fip2, a target of Rab11a, at Ser227 is essential for the formation of TJ and AJ [135]. This phosphorylation allows retention of occludin and specific claudins at TJs in a Rab11a-independent manner [135].

Interestingly, we found that claudin-low tumors have greatly reduced levels of Rab25 (unpublished) suggesting a potential link between Rab25 and expression of junction proteins. A role for recycling endosomes in regulation of polarity is further supported by the observation that Rab25 is necessary for claudin 4 expression and localization at TJ [32, 136]. As Rab25 is a key regulator of cellular metabolism [28, 137], this supports a coordinate regulation of vesicle recycling, cellular polarity, and cellular metabolism. Based on our unpublished data, we speculate that Rab25 may be the key Rab11 family member that interacts with the effector Rab11Fip2 to regulate the composition of TJ.

The experimental evidence discussed here supports a role for abnormal vesicular trafficking in loss of TJ as an early event during metastatic progression. Would detection of aberrant trafficking provide an earlier window of intervention or serve as a predictive marker for tumor aggressiveness? Towards that end, our laboratory and others are working to unravel the mechanisms of TJ regulation by endocytic trafficking proteins.

5.4.2 Formation of Adherent Junctions: The Critical Nexus Between E-cadherin and the Recycling Endosomes

AJ designate the contact points between adjacent epithelial cells serving as sensors for contact inhibition thereby limiting the growth potential of a cellular aggregate. Cell–cell adhesion is mediated by homophilic interactions (*trans* and *cis*) of cadherins, which are connected to the cytoskeleton via catenins (p120). E-cadherin is a calcium-dependent adhesion receptor present at the AJ of epithelial cells [138]. The dynamic cellular distribution of E-cadherin facilitates response to biochemical and mechanical stimuli during remodeling and stabilization of AJ. In mature junctions, membrane E-cadherin renewal is regulated by small GTPase-mediated endocytosis and trafficking [139]. Upon cellular contact, E-cadherin is trafficked via a Rab11a module consisting of a ternary complex of Rab11a, myosin Vb, and Rab11/FIP2 [140] from the trans-Golgi network (TGN) to the lateral membrane where it senses the mechanical forces generated by contact with the adjacent cell. E-cadherin transmits a "contact inhibition" message to F-actin, which in turn stabilizes a nonmotile cytoskeletal configuration and suppresses cell proliferation pathways [140]. In the absence of stable cell–cell contacts, a limited pool of surface E-cadherin is maintained via constant endocytosis and recycling [141].

E-cadherin is a classical epithelial cell maker that is lost during EMT, typically postulated due to transcriptional deregulation. However, new evidence suggests that sorting endosomes may be the dominant determinant of the fate of cellular E-cadherin [142]. The transmembrane domains of signaling proteins within sorting compartments possess targeting information for active sorting to the apical surface. Sorting endosomes with multiple tubular extensions form sub-compartments that provide a platform for distinct combinations of transport carriers and their specific cargo (e.g., MET vs.

transferrin receptor) interactions. In an oncogenic setting, prolonged exposure to growth factors, in particular EGF or TGFβ, leads to internalization of cell surface E-cadherin [143] which is then trafficked for degradation via Rab7/Rab9 or recycled via Rab11a vesicles. EGF stimulation of ER-positive breast cancer cells induces Rac1-modulated macropinocytosis of the E-cadherin–catenin complex into early endosomal complexes (EEA1), decorated with the sorting nexin, SNX1 [144]. SNX1 extracts E-cadherin out of EEA1 and channels it to Rab11a coated vesicles, thus bypassing lysosomal degradation and returning E-cadherin to the membrane. The absence of nexin thus leads to cytoplasmic accumulation and degradation of E-cadherin and loss of the adhesion junction [144]. In the absence of functional adhesion junctions, the polarity axis is weakened facilitating EMT. Since Rab11a vesicles carry E cadherin to a basolateral membrane address, Rab11a mutants also perturb E-cadherin [145] function. Errors in the early endosomal sorting machinery could misregulate transport of AJ to appropriate Rab11a-coated recycling vesicles and relegate E-cadherin cargo for degradation via Rab7 vesicles. In fact there is strong evidence that E-cadherin accumulates in Rab7-coated vesicles during Src-mediated EMT [146].

5.4.3 Cell–ECM Interactions Are Constructed by Polarized Vesicular Trafficking and Are Deregulated in Cancers

Cell–ECM interaction occurs primarily through the integrins. Integrin trafficking is a key consequence of polarization and also necessary for maintenance of cell polarity. These transmembrane proteins function as mechano-transducers between actin and the BM and engage with the BM, forming focal adhesion at the basolateral end of a polarized cell. Endocytic trafficking by Rab family members, including Rab4A, Rab5, Rab8A, Rab11, Rab21, Rab25, and Rab11FIP1, critically regulate the integrated vesicle cascade leading to integrin loc.lization [32, 147–151]. Essentially, endocytosis redistributes integrins from the rear to the front of the cell or restricts them to the baso-lateral domain, in the case of apico-basally polarized epithelial cells.

In migrating cells, integrins are relocated to the leading edge via endocytic trafficking from the rear of the cell. Multiple Rab family members and effectors are involved in the coordinate relocalization of integrin complexes. Rab5 and Rab21 regulate endocytosis and recycling of β integrins, while Rab25 facilitates transport of α5β1 integrin to actin hotspots [51, 55]. Further Rab25 facilitates the formation of functional EGFR-α5β1 integrin complexes at the leading edge of the cell (Fig. 5.3). Specifically, short loop Rab4-decorated recycling vesicles carry integrin α5β3 and transferrin receptor while long loop perinuclear recycling endosomes traffic components required for actin polymerization, via Rab11, and its two effectors Fip2 and Myosin Vb. Once at the leading edge, other Rab11 effectors, namely, Fip3 and Fip4, bind to Rab11 and Arf and promote nucleation required during actin polymerization [152].

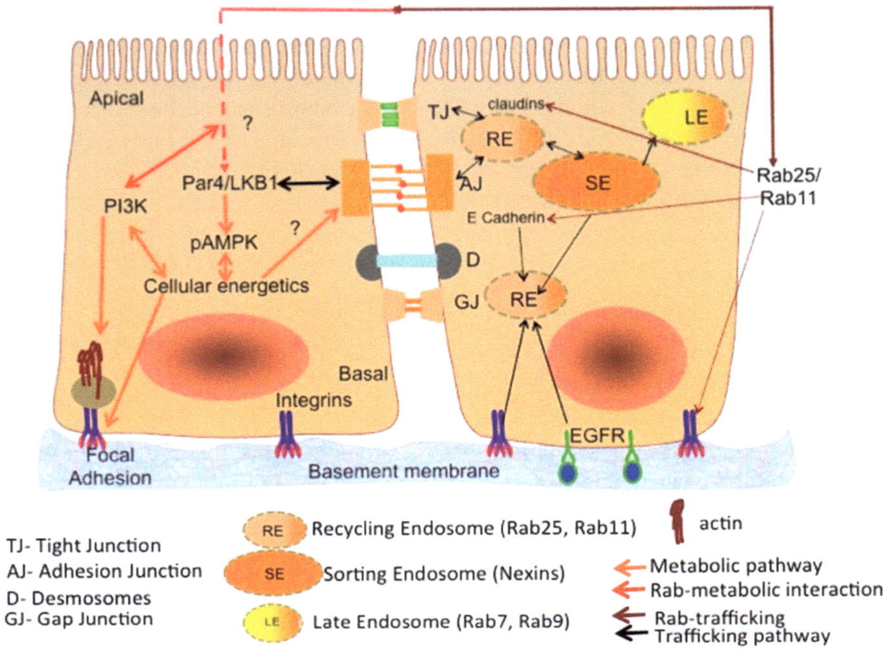

Fig. 5.3 Endocytic machinery as a link between polarity and cellular energetics. A schematic representation of known and suggested interactions between vesicular trafficking by Rab GTPases, key junctional proteins, and energy sensors in (apico-basal) polarized epithelial cells. Both TJ and AJ are trafficked via sorting endosome and recycling endosomes (Rab11a, Rab25 and Rab4). E-cadherin is trafficked to basolateral region via recycling endosome. During oncogenic progression, aberrant trafficking could misplace TJ and AJ proteins into late endosome/lysosome vesicles where the junctions are degraded leading to EMT. Alternatively, loss of Rab11a, or Rab25 could prevent recycling of critical junctional proteins favoring metastasis. In contrast, integrins and growth factors, could be increasingly sorted into recycling compartment rather than into late endosomal compartment, enhancing mitogenic signaling. This would break the epithelial homeostasis that prevents uncontrolled growth and metastasis. In parallel, cross talk between polarity protein LKB1 with intact cellular junction is necessary for activation of energy sensor AMPK. The Rab11a/Rab25 decorated recycling vesicle could be a potential platform where this critical interaction takes place. PI3K pathway, which dictates mitogenic signaling as well as bioenegetic pathways, is another node where trafficking proteins such as Rab25 can cross talk with metabolic pathways

Integrin expression and function are typically deregulated in cancers. Rab25 and its effector Rab11Fip1, which promotes tumor invasion and migration via association with $\alpha5\beta1$ in 3D culture models and tumors, underscore the contribution of vesicular trafficking in regulating integrin function [55, 149]. Rab25-decorated vesicles in collaboration with its effector, Rab-coupling protein (RCP), physically promote interactions between beta-l integrin and EGFR within the cytosolic compartment, promoting cellular motility [51, 55]. That both Rab25 and Rab11Fip1 are aberrant in a large number of cancers and correlate with poor patient outcomes is consistent with being drivers of tumor behavior, at least in part, through altering integrin function.

The BM provides a directional context for polarized integrin function. Therefore basal restriction of BM is essential for epithelial polarization. Rho kinase (ROCK), an effector of RhoA, by restricting the microtubule stabilizer, Par1B (MARK2) to the outer basal periphery, correctly positions the BM. Interestingly, in non-epithelial fibroblasts, the rigidity of the membrane contributes significantly to its polarization. Tensile strength of matrix and its contributions to tumorigenesis is gaining prominence as a likely regulator of function of the HIPPO tumor suppressor-signaling cascade [8, 153, 154]. Further, the ability of transmembrane molecules in the apico-basal complex to sense tension and act as mechano-transducers contributes to the regulation of HIPPO function. Whether internalization and fate of integrin trafficking is altered based on adhesion to matrices of different tensile strength remains to be tested.

5.5 Polarity and Metabolism Pathways Intersect at Endosomal Trafficking Platforms at Cellular Junctions and During Cytokinesis

Perturbations resulting from internal cellular energetics and external microenvironmental cues alter the GTPase trafficking networks to generate operational "systems." By corollary, multiple systems are in constant competition to develop and enforce stable steady states. Outcomes of this competition could be decisive for maintenance or loss of polarized organization and promote hyperproliferation (aberrant cell division) and EMT (loss of cell junctions). Indeed, in various models, the Rab11a-driven recycling system is a major driver and determinant of epithelial morphogenesis [142, 145]. However, if instead of a Rab that is sensitive to the effects of GAPs and GEFs such as Rab11a, there is an overexpression of a constitutively active Rab such as Rab25, or if GAP vs. GEF ratios were altered, a different endpoint would result. Indeed Rab25, which does not appear to be regulated by GAPs and GEFs and not Rab11a that is regulated by GAPs and GEFs, is frequently aberrant in cancer accompanied by an upregulation of EMT markers [155].

5.5.1 Cell–Cell Junctions: Recycling Endosome as a Hub for LKB1, E-cadherin Cross Talk

As presented previously, vesicular trafficking, especially the recycling endosome, is critical for delivery of junction proteins like E-cadherin to the appropriate membrane compartment. Interestingly, in polarized epithelial cells, E-cadherin-mediated localization of LKB1/STRAD complex at AJ is required for AMPK activation at threonine 172 [156]. Conversely, LKB1 is also essential for maturation of AJ [106, 107, 156, 157].

Although it is known that LKB1 has to complex with the pseudokinase STRAD and scaffolding protein MO25 to achieve its functional role at the junction, further studies are required to uncover the upstream factors that mediate LKB1/STRAD interaction with E-cadherin and how they are transported to AJ. Since recycling endosome Rab11a recycles E-cadherin to plasma membrane AJ, it is possible that LKB1 is also transported in the same vesicular compartment. Overall, the involvement of AMPK at cellular junctions underscores a tight link between cellular energy status and TJ function [104, 158].

In general, AMPK activation exerts cytostatic effects by inhibiting biosynthetic pathways thus decreasing cellular ATP consumption. Whether the activation of AMPK and alterations in biosynthetic pathways as a result of metabolic stress drives contact inhibition via AJ needs to be tested. Indeed, aberrant E-cadherin trafficking could block contact inhibition-dependent activation of AMPK and thus lead to increased mTOR activity, protein synthesis, and increased proliferation. Alternatively, as found in ovarian cancers, PI3K/mTOR signaling could downregulate E-cadherin and increase proliferation by increasing transcriptional represses of E-cadherin, namely, Snail and Slug [159].

5.5.2 Cytokinesis and Regulation Phosphoinositide Trafficking by GTPases

Arguably the biggest collaboration between energetics, trafficking machinery, and the polarity toolbox occurs during cytokinesis [160]. In both front–rear and apico-basal polarized cells, the recycling endosome directs the formation of furrows in the mid-body of the cell in a Rab11a and Crumbs-dependent manner [161]. This step is especially critical during asymmetric division necessary to sustain a subpopulation of stem cells. In several studies, Rab11 effectors, Fip3 and Fip4, along with Arf6 and the exocyst are implicated in this critical abscission step of cytokinesis [161]. Whether aberrations in this process contribute to tumor initiation or progression remains to be fully elucidated.

Furthermore, during cytokinesis, differential distribution of endogenous phosphoinositides, second messengers for various signaling modules, defines surface microdomains. Phosphatidylinositide-3 phosphates (PIP3) are generated by the activity of phosphatidylinositol 3-kinase (PI3K) and are trafficked to specific membranes locations via binding to vesicular trafficking proteins containing FYVE domains (PIKFYVE). Dephosphorylation of PIP3 by PTEN is a prime tumor-suppressive mechanism. Indeed, demonstrating the importance of this dynamic process, the PI3K–PTEN pathway is more commonly mutationally activated than any other pathway in tumors. PIP2, generated by dephosphorylation of PIP3 by PTEN, is enriched at the apical surfaces in a Par3-dependent process, while PIP3 species dominate basolateral surfaces [74] and AJ [90, 162, 163]. In fact, PTEN actively localizes to the apical surface to degrade PIP3 locally, resulting in the accumulation of PIP2 [164]. Enrichment of PIP2 at the apical membrane is responsible for the recruitment of a suite of proteins that bind PIP2 such as the CDC42–annexin 2 complex [164], triggering downstream signaling networks. Since Par6 and aPKC are recruited subsequent to CDC42, localization of PIP2 provides an critical early step for organization of key polarity proteins [74]. Thus PTEN loss/inactivation, one of the most frequent aberrations in cancer, contributes directly to polarity defects [118, 165–169] resulting in abnormal cell division.

Rab GTPases coordinate with the PI3K pathway to achieve precise trafficking of PI3K activation [170] involving not just the plasma membrane but also major intracellular compartments such as early and late endosomes, Golgi, autophagosomes, TGN, and endoplasmic reticulum, where phosphoinositides and specific Rabs co-localize [170]. In the context of cytokinesis, Rab11 and Rab21 are critical, with Rab11 being recruited to mid zone to facilitate furrowing and asymmetric division.

The p85α, regulatory subunit of PI3K, contains a Rho GAP domain proposed to interact with Rab4 and Rab5 trafficking pathways [171–173]. Although this interaction of p85α with Rab GTPases is linked to regulation of RTK signaling [172], it could potentially effect PIP2 localization by sequestering CDC42 [174] as well as regulate other functions of p85α. As p85 is the key adapter and regulator of PI3K, its localization determines the site of production of PIP3 and subsequent conversion to PIP2. This coordinates localization of the components regulating cell signaling and polarity. Furthermore since p85α homodimers bind and regulate PTEN stability [16], the recruitment of p85α to particular subcellular compartments could further regulate cellular polarity and vesicle recycling.

Strikingly, during front–rear polarity, activation of PI3K at the leading edge of cells combined with exclusion of PTEN from this domain with continued PTEN activity throughout the remainder of the cell, concentrates PIP3 at the leading edge [81, 90, 164, 166]. The localization of PI3K activity could be CDC42-dependent through recruitment of p85α and the associated p110 PI3K catalytic subunit [174]. Indeed, since p85α binding to p110 and PTEN are mutually exclusive, this could further contribute to increased PIP3 levels at the leading edge of cells. Thus, oncogenic mutations in PI3K that enhance its binding to CDC42 could contribute to early changes in apico-basal polarity that could contribute to tumorigenesis as well as to EMT. Indeed, while the PI3K pathway is implicated in cellular motility, invasion and metastasis, a coordinated mathematical model based on systems approaches able to account for the multiple functions of the PI3K pathway including potent direct roles in regulation of polarity remains beyond the availability of high-quality integrative data.

Cellular energetics dictates cell growth and proliferation. The presence of active AMPK (Thr172), which is a key sensor of metabolic states and in particular ATP/AMP ratios at mitotic structures reflects a need for an in situ energy sensor that communicates the bioenergetic state of the cell to the chromosomal and cytoskeletal dimensions of the mitotic system [175–177]. How AMPK is trafficked to the mitotic apparatus remains unknown. Nevertheless, the presence of LKB1 and MO25 at mitotic spindles in *C. elegans* and Drosophila models supports the need to extensively explore the interactions between trafficking and metabolic pathways in human cancers [178].

5.6 Derailed Endocytosis as a Cause and Consequence of Energetic Stress

Solid tumors, especially at the center, are frequently challenged by low nutrients, growth factors, and oxygen levels and high lactate levels, resulting in a hypoxic and acidic microenvironment, which alters trafficking networks, TJ, and subsequently disrupt polarity [179].

Protein trafficking by Rab and Arf family small GTPases in response to amino acid availability regulates mTOR activation. A number of Rab GTPases including Rab5, Rab7, Rab11, and Rab31, when bound to GTP, selectively blocks mTORC1 activation in response to amino acids [180]. This again emphasizes the link between energy balance and vesicle trafficking. Constitutive activation of Rab5 and Arf1 strongly inhibits mTORC1 activity in mammalian systems post amino acid stimulation via Rag GTPases. Endocytic trafficking and in particular the integrity of the late endosome is essential for the ability of nutrients to regulate mTORC1. Indeed, blocking early (Rab5) to late endosomal (Rab7) conversion stalls insulin- and amino acid-stimulated mTORC1 activation resulting in restriction of mTORC1 to hybrid early/late endosomes [181].

The loss of cell polarity and alterations in protein localization due to aberrant trafficking are two key non-transcriptional systemic and rapid responses to nutrient depletion, especially low glucose levels [182–184]. Activation of AMPK by low-energy balance results in phosphorylation AS160, a Rab GAP, and triggers its association with 14-3-3 proteins and its consequent dissociation from glucose transporter, Glut4, carrying vesicles [185]. This facilitates transfer of Glut4 from intracellular storage vesicles to the plasma membrane and increases glucose uptake, alleviating the metabolic stress [185–187]. Again, this suggests that a tight coupling between polarity, trafficking, and metabolism is integral to normal cellular function and is co-opted by tumor cells to allow responses to the microenvironment.

Recently, our laboratory identified Rab25, which is frequently genomically amplified in multiple tumor lineages, as a key regulator of cellular bioenergetics and autophagy [28, 137] in epithelial cancers. The Rab25 transcriptome is highly enriched in metabolic genes suggesting a role in cellular metabolism. Indeed Rab25 enhances survival during nutrient stress by preventing apoptosis and autophagy in a PI3K-dependent manner. In addition to activating AKT, Rab25 associates with glucose transporters and translocates Glut1 to the membrane, increasing uptake of glucose leading to accumulation of glycogen in epithelial cancer cells. The action of Rab25 in elevating basal ATP levels and increasing glycogen stores allows cancer cells to transiently accommodate to energy stress [28] and favors cell survival in the toxic tumor microenvironment and during anoikis stress.

An unexpected molecular link between nutrient levels and polarity was reported in yeast, where glucose starvation transiently inhibited translation initiation, cell polarity, and trafficking networks. The alterations in trafficking were PKA-dependent and temporarily delocalized clathrin adaptors from the PM. It appears that activity of PKA prior to glucose depletion dictates adaptive responses of endosomal and TGN trafficking [188].

In ischemia models (tumor hypoxia), depletion of cellular ATP results in disruption of the epithelial TJ barrier, followed by loss of membrane polarity. At the molecular level, low ATP favors a stable complex of TIAM1 and Par3, thus sequestering the activating GEF away from Rac and reducing aPKC-Par3 complex formation. With subsequent reduction in Rac and aPKC activity, the TJ disassembles [189], destroying cellular polarity. Now growth factors can access to their cognate receptors increasing glucose uptake as well as inducing expression of cell survival mediators.

As described above, nutrient stress disrupts apico-basal polarity. However, rather than immediately reinstituting polarity, once nutrients become available, cancer cells with compromised apico-basal polarity, benefit from transitioning to a front–rear state, favoring their migration machinery. High glucose uptake that occurs following nutrient stress increases Rac1 activity, priming cells for migration.

Interestingly MARK2/Par1, a microtubule stabilizing protein, which is phosphorylated and activated by Par4/LKB1, may play an important role during an apico-basal to front–rear transition. Such transition could be prompted by accumulating microenvironmental stresses such as hypoxia and/or altered energetics, potentially collapsing the polarity axis. Moreover, changes in mechanical forces at work between adjacent tumor cells and between tumors cells and the tissue matrix can contribute to cytoskeletal alterations towards a more migratory phenotype. MARK2/Par1 dictates the alignment of microtubules either in columnar or horizontal organization [190]. In a simplified scenario, a change in the axis of alignment of microtubules would also change the direction of travel of Rab GTPase-coated vesicles. Once again, Rab11a emerges as a major player since its effector, Rab11FIP2, is a substrate of MARK2, and contributes to the establishment of cell–cell junctions [140]. Other Rabs, such as Rab 20, which regulates mitochondrial homeostasis under hypoxic conditions, could play a key role in response to cellular energy changes [191], are deregulated in various inherited and acquired disorders, notably in pancreatic carcinomas.

Thus microenvironmental stress, leading to energy crisis, promotes the evolution of a heterogeneous population of genetically altered cells [5, 6, 192–196]. Systems involved in maintaining uniform epithelial tissue organization are outcompeted by systems promoting alternative metabolic pathways with subsequent changes in polarity and increased cellular motility. As suggested above, the increased cellular motility may be a phylogenetic memory of the response of unicellular organisms to stress that is co-opted for the function of mammalian cells or alternatively a residual function that is not beneficial to the mammalian cells. The physical and cytoskeletal changes accompanying loss of apico-basal polarity and function of cellular junctions subsequently stimulate transcriptional programs that translate to a more glycolytic phenotype, i.e., the Warburg effect [197, 198].

Overall, we have snapshots of how polarity proteins are transported by Rho and Rab GTPases but further studies are needed to determine (1) whether vesicular trafficking is a fundamental non-transcriptional regulator of polarity proteins that effects tumor progression, (2) how Rab GTPases communicate junctional status to polarity checkpoints and energy sensors, and vice versa, (3) whether altered trafficking precede changes in polarity and metabolism, (4) at what stage of tumorigenesis do aberrations in the Rab GTPases and transport of polarity proteins occur, and (5) how polarized trafficking alter anabolic and catabolic processes in the cell.

A systems biology approach, incorporating a detailed analysis of DNA copy number, methylation and mutation, mRNA and miRNA levels, protein levels as well as dynamic changes in protein complexes and localization as a consequence of perturbations in trafficking and polarity pathways, will likely be necessary to integrate the functional effects of the complex cross talk between members of the trafficking and polarity programs.

5.7 Conclusions

Cell "polarization" is essential for creating the gradients within the cell that drive biochemical reactions, cellular energetics, motility, and cell–cell interactions. It appropriates allocation of cellular components during cytokinesis and distributes plasma membrane with specific external interacting ligands. The vesicular trafficking system, as the master manager and distributor of cellular resources including polarity checkpoint proteins, defines various functional hotspots in the cell. Therefore, oncogenes target polarity proteins to disrupt glandular organization during tumor initiation and progression.

Oncogenic transformation engages the "polarity and trafficking systems" that normally function during wound healing to bring about EMT. Tumor acidosis resulting from the release of lactate as a component of the Warburg glycolytic metabolic phenotype found in the majority of cancer cells can compromise BM and cell–BM adhesion. Through a series of unknown events, the orientation of the cytoskeletal axis changes, as the adhesion junctions are degraded, slowly depolarizing the cell. The cell then gains a new front–rear polarity, which allows growth factor ligands to activate their cognate receptors contributing to cell survival as well as motility. This can lead to directional motility towards the vasculature due to a gradient of lactate and pH, resulting in cancer cells coming into contact with the vasculature, a critical step in the metastatic cascade. Considering that cytoskeletal motors and various ion pumps consume a large fraction of cellular energy, it is worth evaluating how this energy consumption pattern is altered in highly glycolytic cancers, where cellular energy usage is focused on driving biosynthetic processes.

Aberrations in multiple members of the polarity complex are found in human tumors and animal models of cancer. For example, PKCι, which is critical for apico-basal polarity, is genomically amplified with increased protein expression in serous ovarian cancers. Importantly, PKCι is also mislocalized in these tumors. The aberrant expression and mislocalization of this polarity protein elevates cyclin E expression and increases proliferation [99].

In normal systems Rab11a is arguably the single most important Rab GTPase in terms of polarized transport [49]. However, it is important to note that Rab25 (aka Rab11c) a close homolog of Rab11a has overlapping as well as competing functions and is implicated in the pathophysiology of an increasing array of cancers. Stress inputs like hypoxia that leads to the expression of mesenchymal gene signatures favoring invasion and migration also favor "resource conservation" routes of vesicle trafficking. Indeed, during hypoxia, Rab11a reinstates α6β4 integrin expression by recycling these proteins to the cell membrane instead of to the degradation pathway. Elevated levels of recycled α6β4 results in a microtubule configuration that facilitates migration and cytokinesis. The diversity of Rab11 family effectors implies that their spatial distribution and their dynamic association with vesicles as they transit the cell are important matrices for their function [199].

While the PI3K signaling pathway represents a critical intersection point between the polarity and metabolic pathways, the geographical coordinate of this interaction is possibly the recycling endosome. Although oncogenic activation of the recycling endosomal Rab11a is relatively rare in tumors [155], its homolog, Rab25, is frequently amplified and is regarded as a driver of tumorigenicity in a variety of cancers. Our laboratory has also shown that oncogenic activities of Rab25 are PI3K-dependent and include altered metabolism and increased metastasis [28, 167]. However while overexpression of Rab25 promotes metastasis in various cancer models, it also appears to function as a tumor suppressor in a context-dependent manner in the claudin-low subset of breast cancers [200], reportedly with a relatively increased glycolytic index [201]. The observed increase in mesenchymal gene signatures and loss of TJ [127, 128] in claudin-low cell lines suggests a complex role for Rab25 in "epithelial polarity programs" in cancer cells. With recent findings of our group that Rab25 regulates cancer cell metabolism [28], we are using the multiple functions of Rab25 to probe the integration of polarity and metabolism in epithelial cancers.

Cancer is a product of evolution and it systemically hijacks latent developmental programs to its advantage. Loss of polarity and altered metabolism, which are intimately related to vesicle trafficking, are established hallmarks of cancer [202, 203]. The evidence presented herein calls for the inclusion of "derailed endocytosis" as a hallmark of cancer. Indeed the vesicular trafficking system forms a regulatory link between cellular energetics and polarity states, which presents a novel option for clinical intervention.

Funding Funding sources include Cancer Center Support Grant P20CA16672, PO1CA099031 to GBM and Susan G Komen Postdoctoral Fellowship to SM.

References

1. Bryant DM, Mostov KE (2008) From cells to organs: building polarized tissue. Nat Rev Mol Cell Biol 9:887–901. doi:10.1038/nrm2523
2. Leung CT, Brugge JS (2012) Outgrowth of single oncogene-expressing cells from suppressive epithelial environments. Nature 482:410–413. doi:10.1038/nature10826
3. Wodarz A, Nathke I (2007) Cell polarity in development and cancer. Nat Cell Biol 9: 1016–1024. doi:10.1038/ncb433

4. Wang F et al (2002) Phenotypic reversion or death of cancer cells by altering signaling pathways in three-dimensional contexts. J Natl Cancer Inst 94:1494–1503
5. Muranen T et al (2012) Inhibition of PI3K/mTOR leads to adaptive resistance in matrix-attached cancer cells. Cancer Cell 21:227–239. doi:10.1016/j.ccr.2011.12.024
6. Bissell MJ, Radisky DC, Rizki A, Weaver VM, Petersen OW (2002) The organizing principle: microenvironmental influences in the normal and malignant breast. Differentiation 70:537–546. doi:10.1046/j.1432-0436.2002.700907.x
7. Bissell MJ, Bilder D (2003) Polarity determination in breast tissue: desmosomal adhesion, myoepithelial cells, and laminin 1. Breast Cancer Res 5:117–119
8. Xu R, Boudreau A, Bissell MJ (2009) Tissue architecture and function: dynamic reciprocity via extra- and intra-cellular matrices. Cancer Metastasis Rev 28:167–176. doi:10.1007/s10555-008-9178-z
9. Godde NJ, Galea RC, Elsum IA, Humbert PO (2010) Cell polarity in motion: redefining mammary tissue organization through EMT and cell polarity transitions. J Mammary Gland Biol Neoplasia 15:149–168. doi:10.1007/s10911-010-9180-2
10. Yan Z, Chen M, Perucho M, Friedman E (1997) Oncogenic Ki-ras but not oncogenic Ha-ras blocks integrin beta1-chain maturation in colon epithelial cells. J Biol Chem 272:30928–30936
11. Rossi F, Gonzalez C (2012) Synergism between altered cortical polarity and the PI3K/TOR pathway in the suppression of tumour growth. EMBO Rep 13:157–162. doi:10.1038/embor.2011.230
12. Larue L, Bellacosa A (2005) Epithelial-mesenchymal transition in development and cancer: role of phosphatidylinositol 3′ kinase/AKT pathways. Oncogene 24:7443–7454. doi:1209091[pii]10.1038/sj.onc.1209091
13. Brumby AM, Richardson HE (2003) Scribble mutants cooperate with oncogenic Ras or Notch to cause neoplastic overgrowth in Drosophila. EMBO J 22:5769–5779. doi:10.1093/emboj/cdg548
14. Muller PA et al (2009) Mutant p53 drives invasion by promoting integrin recycling. Cell 139:1327–1341. doi:10.1016/j.cell.2009.11.026
15. Partanen JI et al (2012) Tumor suppressor function of liver kinase B1 (Lkb1) is linked to regulation of epithelial integrity. Proc Natl Acad Sci USA 109:E388–E397. doi:10.1073/pnas.1120421109
16. Cheung LW et al (2011) High frequency of PIK3R1 and PIK3R2 mutations in endometrial cancer elucidates a novel mechanism for regulation of PTEN protein stability. Cancer Discov 1:170–185. doi:10.1158/2159-8290.CD-11-0039
17. Rainero E et al (2012) Diacylglycerol kinase alpha controls RCP-dependent integrin trafficking to promote invasive migration. J Cell Biol 196:277–295. doi:10.1083/jcb.201109112
18. Pearson HB et al (2011) SCRIB expression is deregulated in human prostate cancer, and its deficiency in mice promotes prostate neoplasia. J Clin Invest 121:4257–4267. doi:10.1172/JCI58509
19. Doggett K, Grusche FA, Richardson HE, Brumby AM (2011) Loss of the Drosophila cell polarity regulator Scribbled promotes epithelial tissue overgrowth and cooperation with oncogenic Ras-Raf through impaired Hippo pathway signaling. BMC Dev Biol 11:57. doi:10.1186/1471-213X-11-57
20. Dow LE et al (2008) Loss of human Scribble cooperates with H-Ras to promote cell invasion through deregulation of MAPK signalling. Oncogene 27:5988–6001. doi:10.1038/onc.2008.219
21. Ellenbroek SI, Iden S, Collard JG (2012) Cell polarity proteins and cancer. Semin Cancer Biol 22(3):208–215. doi:S1044-579X(12)00048-X [pii]10.1016/j.semcancer.2012.02.012
22. Martin-Belmonte F, Perez-Moreno M (2012) Epithelial cell polarity, stem cells and cancer. Nat Rev Cancer 12:23–38. doi:10.1038/nrc3169nrc3169 [pii]
23. Marongiu F, Doratiotto S, Sini M, Serra MP, Laconi E (2012) Cancer as a disease of tissue pattern formation. Prog Histochem Cytochem 47(3):175–207. doi:10.1016/j.proghi.2012.08.001

24. Doratiotto S, Marongiu F, Faedda S, Pani P, Laconi E (2009) Altered growth pattern, not altered growth per se, is the hallmark of early lesions preceding cancer development. Histol Histopathol 24:101–106
25. Bissell MJ, Aggeler J (1987) Dynamic reciprocity: how do extracellular matrix and hormones direct gene expression? Prog Clin Biol Res 249:251–262
26. Bartsch H, Nair J (2004) Oxidative stress and lipid peroxidation-derived DNA-lesions in inflammation driven carcinogenesis. Cancer Detect Prev 28:385–391. doi:10.1016/j.cdp.2004.07.004
27. Cao X, Surma MA, Simons K (2012) Polarized sorting and trafficking in epithelial cells. Cell Res 22:793–805. doi:10.1038/cr.2012.64 cr201264 [pii]
28. Cheng KW et al (2012) Rab25 increases cellular ATP and glycogen stores protecting cancer cells from bioenergetic stress. EMBO Mol Med 4:125–141. doi:10.1002/emmm.201100193
29. Iden S, Collard JG (2008) Crosstalk between small GTPases and polarity proteins in cell polarization. Nat Rev Mol Cell Biol 9:846–859. doi:10.1038/nrm2521
30. Cao X, Surma MA, Simons K (2012) Polarized sorting and trafficking in epithelial cells. Cell Res 22:793–805. doi:10.1038/cr.2012.64
31. Duncan MC, Peifer M (2008) Regulating polarity by directing traffic: Cdc42 prevents adherens junctions from crumblin' aPart. J Cell Biol 183:971–974. doi:10.1083/jcb.200811057
32. Stenmark H (2009) Rab GTPases as coordinators of vesicle traffic. Nat Rev Mol Cell Biol 10:513–525. doi:10.1038/nrm2728
33. Kang RS, Folsch H (2009) An old dog learns new tricks: novel functions of the exocyst complex in polarized epithelia in animals. F1000 Biol Rep 1:83. doi:10.3410/B1-83
34. Mosesson Y, Mills GB, Yarden Y (2008) Derailed endocytosis: an emerging feature of cancer. Nat Rev Cancer 8:835–850. doi:10.1038/nrc2521
35. Eilon T, Barash I (2008) Different gene-expression profiles for the poorly differentiated carcinoma and the highly differentiated papillary adenocarcinoma in mammary glands support distinct metabolic pathways. BMC Cancer 8:270. doi:10.1186/1471-2407-8-270
36. Duncan MC, Peifer M (2008) Regulating polarity by directing traffic: Cdc42 prevents adherens junctions from crumblin' aPart. J Cell Biol 183:971–974. doi:jcb.200811057 [pii]10.1083/jcb.200811057
37. Shivas JM, Morrison HA, Bilder D, Skop AR (2010) Polarity and endocytosis: reciprocal regulation. Trends Cell Biol 20:445–452. doi:10.1016/j.tcb.2010.04.003
38. Mellman I, Nelson WJ (2008) Coordinated protein sorting, targeting and distribution in polarized cells. Nat Rev Mol Cell Biol 9:833–845. doi:10.1038/nrm2525
39. Apodaca G (2010) Opening ahead: early steps in lumen formation revealed. Nat Cell Biol 12:1026–1028. doi:10.1038/ncb1110-1026
40. Moreno-Bueno G, Portillo F, Cano A (2008) Transcriptional regulation of cell polarity in EMT and cancer. Oncogene 27:6958–6969. doi:onc2008346 [pii]10.1038/onc.2008.346
41. Hezel AF, Bardeesy N (2008) LKB1; linking cell structure and tumor suppression. Oncogene 27:6908–6919. doi:10.1038/onc.2008.342
42. Shorning BY, Griffiths D, Clarke AR (2011) Lkb1 and Pten synergise to suppress mTOR-mediated tumorigenesis and epithelial-mesenchymal transition in the mouse bladder. PLoS One 6:e16209. doi:10.1371/journal.pone.0016209
43. Elsum I, Yates L, Humbert PO, Richardson HE (2012) The Scribble-Dlg-Lgl polarity module in development and cancer: from flies to man. Essays Biochem 53:141–168. doi:10.1042/bse0530141
44. Gao N, Kaestner KH (2010) Cdx2 regulates endo-lysosomal function and epithelial cell polarity. Genes Dev 24:1295–1305. doi:10.1101/gad.1921510
45. Golachowska MR, Hoekstra D, van IJzendoorn SC (2010) Recycling endosomes in apical plasma membrane domain formation and epithelial cell polarity. Trends Cell Biol 20:618–626. doi:10.1016/j.tcb.2010.08.004
46. Nelson WJ (2009) Remodeling epithelial cell organization: transitions between front-rear and apical-basal polarity. Cold Spring Harb Perspect Biol 1:a000513. doi:10.1101/cshperspect.a000513

47. Weaver VM et al (2002) Beta4 integrin-dependent formation of polarized three-dimensional architecture confers resistance to apoptosis in normal and malignant mammary epithelium. Cancer Cell 2:205–216
48. Weaver VM et al (1997) Reversion of the malignant phenotype of human breast cells in three-dimensional culture and in vivo by integrin blocking antibodies. J Cell Biol 137:231–245
49. Jing J, Prekeris R (2009) Polarized endocytic transport: the roles of Rab11 and Rab11-FIPs in regulating cell polarity. Histol Histopathol 24:1171–1180
50. Parachoniak CA, Park M (2012) Dynamics of receptor trafficking in tumorigenicity. Trends Cell Biol 22:231–240. doi:10.1016/j.tcb.2012.02.002
51. Caswell P, Norman J (2008) Endocytic transport of integrins during cell migration and invasion. Trends Cell Biol 18:257–263. doi:S0962-8924(08)00107-4 [pii]10.1016/j.tcb.2008.03.004
52. Caswell PT, Vadrevu S, Norman JC (2009) Integrins: masters and slaves of endocytic transport. Nat Rev Mol Cell Biol 10:843–853. doi:nrm2799 [pii]10.1038/nrm2799
53. Clark EA, King WG, Brugge JS, Symons M, Hynes RO (1998) Integrin-mediated signals regulated by members of the rho family of GTPases. J Cell Biol 142:573–586
54. Veevers-Lowe J, Ball SG, Shuttleworth A, Kielty CM (2011) Mesenchymal stem cell migration is regulated by fibronectin through alpha5beta1-integrin-mediated activation of PDGFR-beta and potentiation of growth factor signals. J Cell Sci 124:1288–1300. doi:10.1242/jcs.076935
55. Caswell PT et al (2007) Rab25 associates with alpha5beta1 integrin to promote invasive migration in 3D microenvironments. Dev Cell 13:496–510. doi:S1534-5807(07)00331-0 [pii]10.1016/j.devcel.2007.08.012
56. Hall A, Nobes CD (2000) Rho GTPases: molecular switches that control the organization and dynamics of the actin cytoskeleton. Philos Trans R Soc Lond B Biol Sci 355:965–970. doi:10.1098/rstb.2000.0632
57. Amano M, Nakayama M, Kaibuchi K (2010) Rho-kinase/ROCK: a key regulator of the cytoskeleton and cell polarity. Cytoskeleton 67:545–554. doi:10.1002/cm.20472
58. Spiering D, Hodgson L (2011) Dynamics of the Rho-family small GTPases in actin regulation and motility. Cell Adh Migr 5:170–180
59. Higgs HN, Pollard TD (2000) Activation by Cdc42 and PIP(2) of Wiskott-Aldrich syndrome protein (WASp) stimulates actin nucleation by Arp2/3 complex. J Cell Biol 150:1311–1320
60. Kawauchi T (2012) Cell adhesion and its endocytic regulation in cell migration during neural development and cancer metastasis. Int J Mol Sci 13:4564–4590. doi:10.3390/ijms13044564
61. Shin S, Wolgamott L, Yoon SO (2012) Integrin trafficking and tumor progression. Int J Cell Biol 2012:516789. doi:10.1155/2012/516789
62. Ridley AJ et al (2003) Cell migration: integrating signals from front to back. Science 302:1704–1709. doi:10.1126/science.1092053
63. Mills GB, Jurisica I, Yarden Y, Norman JC (2009) Genomic amplicons target vesicle recycling in breast cancer. J Clin Invest 119:2123–2127. doi:10.1172/JCI4025640256 [pii]
64. Sen A et al (2012) Paxillin mediates extranuclear and intranuclear signaling in prostate cancer proliferation. J Clin Invest 122:2469–2481. doi:10.1172/JCI62044
65. Sattler M, Pisick E, Morrison PT, Salgia R (2000) Role of the cytoskeletal protein paxillin in oncogenesis. Crit Rev Oncog 11:63–76
66. Sachdev S, Bu Y, Gelman IH (2009) Paxillin-Y118 phosphorylation contributes to the control of Src-induced anchorage-independent growth by FAK and adhesion. BMC Cancer 9:12. doi:1471-2407-9-12 [pii]10.1186/1471-2407-9-12
67. Chia WJ, Tang BL (2009) Emerging roles for Rab family GTPases in human cancer. Biochim Biophys Acta 1795:110–116
68. Deakin NO, Turner CE (2011) Distinct roles for paxillin and Hic-5 in regulating breast cancer cell morphology, invasion, and metastasis. Mol Biol Cell 22:327–341. doi:mbc.E10-09-0790 [pii]10.1091/mbc.E10-09-0790
69. Heasman SJ, Ridley AJ (2008) Mammalian Rho GTPases: new insights into their functions from in vivo studies. Nat Rev Mol Cell Biol 9:690–701. doi:nrm2476 [pii] 10.1038/nrm2476

70. Asnacios A, Hamant O (2012) The mechanics behind cell polarity. Trends Cell Biol 22(11): 584–591. doi:10.1016/j.tcb.2012.08.005
71. Tervonen TA et al (2011) Faulty epithelial polarity genes and cancer. Adv Cancer Res 111:97–161. doi:B978-0-12-385524-4.00003-9 [pii]10.1016/B978-0-12-385524-4.00003-9
72. Igaki T, Pagliarini RA, Xu T (2006) Loss of cell polarity drives tumor growth and invasion through JNK activation in Drosophila. Curr Biol 16:1139–1146. doi:10.1016/j.cub.2006.04.042
73. Mellman I, Warren G (2000) The road taken: past and future foundations of membrane traffic. Cell 100:99–112
74. Tanos B, Rodriguez-Boulan E (2008) The epithelial polarity program: machineries involved and their hijacking by cancer. Oncogene 27:6939–6957. doi:10.1038/onc.2008.345
75. Lajoie P et al (2007) Plasma membrane domain organization regulates EGFR signaling in tumor cells. J Cell Biol 179:341–356. doi:10.1083/jcb.200611106
76. Castagnino P et al (1995) Direct binding of eps8 to the juxtamembrane domain of EGFR is phosphotyrosine- and SH2-independent. Oncogene 10:723–729
77. Lanzetti L et al (2000) The Eps8 protein coordinates EGF receptor signalling through Rac and trafficking through Rab5. Nature 408:374–377. doi:10.1038/35042605
78. Offenhauser N et al (2004) The eps8 family of proteins links growth factor stimulation to actin reorganization generating functional redundancy in the Ras/Rac pathway. Mol Biol Cell 15:91–98. doi:10.1091/mbc.E03-06-0427
79. Scita G et al (1999) EPS8 and E3B1 transduce signals from Ras to Rac. Nature 401:290–293. doi:10.1038/45822
80. Stetak A et al (2006) Cell fate-specific regulation of EGF receptor trafficking during Caenorhabditis elegans vulval development. EMBO J 25:2347–2357. doi:10.1038/sj.emboj.7601137
81. McCaffrey LM, Macara IG (2009) Widely conserved signaling pathways in the establishment of cell polarity. Cold Spring Harb Perspect Biol 1:a001370. doi:10.1101/cshperspect.a001370
82. McCaffrey LM, Macara IG (2009) The Par3/aPKC interaction is essential for end bud remodeling and progenitor differentiation during mammary gland morphogenesis. Genes Dev 23:1450–1460. doi:10.1101/gad.1795909
83. Suzuki A, Ohno S (2006) The PAR-aPKC system: lessons in polarity. J Cell Sci 119: 979–987. doi:10.1242/jcs.02898
84. Totong R, Achilleos A, Nance J (2007) PAR-6 is required for junction formation but not apicobasal polarization in C. elegans embryonic epithelial cells. Development 134: 1259–1268. doi:10.1242/dev.02833
85. Goldstein B, Macara IG (2007) The PAR proteins: fundamental players in animal cell polarization. Dev Cell 13:609–622. doi:10.1016/j.devcel.2007.10.007
86. Audebert S et al (2004) Mammalian Scribble forms a tight complex with the betaPIX exchange factor. Curr Biol 14:987–995. doi:10.1016/j.cub.2004.05.051
87. Osmani N, Vitale N, Borg JP, Etienne-Manneville S (2006) Scrib controls Cdc42 localization and activity to promote cell polarization during astrocyte migration. Curr Biol 16:2395–2405. doi:10.1016/j.cub.2006.10.026
88. Klezovitch O, Fernandez TE, Tapscott SJ, Vasioukhin V (2004) Loss of cell polarity causes severe brain dysplasia in Lgl1 knockout mice. Genes Dev 18:559–571. doi:10.1101/gad.1178004
89. Atwood SX, Chabu C, Penkert RR, Doe CQ, Prehoda KE (2007) Cdc42 acts downstream of Bazooka to regulate neuroblast polarity through Par-6 aPKC. J Cell Sci 120:3200–3206. doi:10.1242/jcs.014902
90. Martin-Belmonte F et al (2007) PTEN-mediated apical segregation of phosphoinositides controls epithelial morphogenesis through Cdc42. Cell 128:383–397. doi:10.1016/j.cell.2006.11.051
91. Yamanaka T et al (2001) PAR-6 regulates aPKC activity in a novel way and mediates cell-cell contact-induced formation of the epithelial junctional complex. Genes Cells 6:721–731
92. Margolis B, Borg JP (2005) Apicobasal polarity complexes. J Cell Sci 118:5157–5159. doi:10.1242/jcs.02597

93. Joberty G, Petersen C, Gao L, Macara IG (2000) The cell-polarity protein Par6 links Par3 and atypical protein kinase C to Cdc42. Nat Cell Biol 2:531–539. doi:10.1038/35019573
94. Lin D et al (2000) A mammalian PAR-3-PAR-6 complex implicated in Cdc42/Rac1 and aPKC signalling and cell polarity. Nat Cell Biol 2:540–547. doi:10.1038/35019582
95. Etienne-Manneville S, Hall A (2001) Integrin-mediated activation of Cdc42 controls cell polarity in migrating astrocytes through PKCzeta. Cell 106:489–498
96. Schlessinger K, McManus EJ, Hall A (2007) Cdc42 and noncanonical Wnt signal transduction pathways cooperate to promote cell polarity. J Cell Biol 178:355–361. doi:10.1083/jcb.200701083
97. Ozdamar B et al (2005) Regulation of the polarity protein Par6 by TGFbeta receptors controls epithelial cell plasticity. Science 307:1603–1609. doi:10.1126/science.1105718
98. Viloria-Petit AM et al (2009) A role for the TGFbeta-Par6 polarity pathway in breast cancer progression. Proc Natl Acad Sci USA 106:14028–14033. doi:10.1073/pnas.0906796106
99. Eder AM et al (2005) Atypical PKCiota contributes to poor prognosis through loss of apical-basal polarity and cyclin E overexpression in ovarian cancer. Proc Natl Acad Sci USA 102:12519–12524. doi:10.1073/pnas.0505641102
100. Nishimura T et al (2005) PAR-6-PAR-3 mediates Cdc42-induced Rac activation through the Rac GEFs STEF/Tiam1. Nat Cell Biol 7:270–277. doi:10.1038/ncb1227
101. Pegtel DM et al (2007) The Par-Tiam1 complex controls persistent migration by stabilizing microtubule-dependent front-rear polarity. Curr Biol 17:1623–1634. doi:10.1016/j.cub.2007.08.035
102. Mashukova A, Forteza R, Wald FA, Salas PJ (2012) PDK1 in apical signaling endosomes participates in the rescue of the polarity complex atypical PKC by intermediate filaments in intestinal epithelia. Mol Biol Cell 23:1664–1674. doi:10.1091/mbc.E11-12-0988
103. Zumbrunn J, Kinoshita K, Hyman AA, Nathke IS (2001) Binding of the adenomatous polyposis coli protein to microtubules increases microtubule stability and is regulated by GSK3 beta phosphorylation. Curr Biol 11:44–49
104. Zhang L, Li J, Young LH, Caplan MJ (2006) AMP-activated protein kinase regulates the assembly of epithelial tight junctions. Proc Natl Acad Sci USA 103:17272–17277. doi:10.1073/pnas.0608531103
105. Hardie DG, Ross FA, Hawley SA (2012) AMPK: a nutrient and energy sensor that maintains energy homeostasis. Nat Rev Mol Cell Biol 13:251–262. doi:10.1038/nrm3311
106. Mirouse V, Billaud M (2011) The LKB1/AMPK polarity pathway. FEBS Lett 585:981–985. doi:10.1016/j.febslet.2010.12.025
107. Mirouse V, Swick LL, Kazgan N, St Johnston D, Brenman JE (2007) LKB1 and AMPK maintain epithelial cell polarity under energetic stress. J Cell Biol 177:387–392. doi:10.1083/jcb.200702053
108. Forcet C et al (2005) Functional analysis of Peutz-Jeghers mutations reveals that the LKB1 C-terminal region exerts a crucial role in regulating both the AMPK pathway and the cell polarity. Hum Mol Genet 14:1283–1292. doi:10.1093/hmg/ddi139
109. Baas AF, Smit L, Clevers H (2004) LKB1 tumor suppressor protein: PARtaker in cell polarity. Trends Cell Biol 14:312–319. doi:10.1016/j.tcb.2004.04.001
110. Baas AF et al (2003) Activation of the tumour suppressor kinase LKB1 by the STE20-like pseudokinase STRAD. EMBO J 22:3062–3072. doi:10.1093/emboj/cdg292
111. Alessi DR, Sakamoto K, Bayascas JR (2006) LKB1-dependent signaling pathways. Annu Rev Biochem 75:137–163. doi:10.1146/annurev.biochem.75.103004.142702
112. Roy BC et al (2010) Involvement of LKB1 in epithelial-mesenchymal transition (EMT) of human lung cancer cells. Lung Cancer 70:136–145. doi:10.1016/j.lungcan.2010.02.004
113. Jang C, Lee G, Chung J (2008) LKB1 induces apical trafficking of Silnoon, a monocarboxylate transporter, in Drosophila melanogaster. J Cell Biol 183:11–17. doi:10.1083/jcb.200807052
114. Martinez-Outschoorn UE et al (2011) Cancer cells metabolically "fertilize" the tumor microenvironment with hydrogen peroxide, driving the Warburg effect: implications for PET imaging of human tumors. Cell Cycle 10:2504–2520

115. Chen YS et al (2011) Proteomics profiling of Madin-Darby canine kidney plasma membranes reveals Wnt-5a involvement during oncogenic H-Ras/TGF-beta-mediated epithelial-mesenchymal transition. Mol Cell Proteomics 10:M110.001131. doi:10.1074/mcp.M110.001131
116. Zahraoui A (2004) [Tight junctions, a platform regulating cell proliferation and polarity]. Med Sci (Paris) 20:580–585. doi:10.1051/medsci/2004205580
117. Cereijido M, Valdes J, Shoshani L, Contreras RG (1998) Role of tight junctions in establishing and maintaining cell polarity. Annu Rev Physiol 60:161–177. doi:10.1146/annurev.physiol.60.1.161
118. Itoh M, Bissell MJ (2003) The organization of tight junctions in epithelia: implications for mammary gland biology and breast tumorigenesis. J Mammary Gland Biol Neoplasia 8: 449–462. doi:10.1023/B:JOMG.0000017431.45314.07
119. Giepmans BN, van Ijzendoorn SC (2009) Epithelial cell-cell junctions and plasma membrane domains. Biochim Biophys Acta 1788:820–831. doi:10.1016/j.bbamem.2008.07.015
120. Guillemot L, Paschoud S, Pulimeno P, Foglia A, Citi S (2008) The cytoplasmic plaque of tight junctions: a scaffolding and signalling center. Biochim Biophys Acta 1778:601–613. doi:10.1016/j.bbamem.2007.09.032
121. Matter K, Balda MS (2003) Signalling to and from tight junctions. Nat Rev Mol Cell Biol 4:225–236. doi:10.1038/nrm1055
122. Tsukita S, Furuse M, Itoh M (1999) Structural and signalling molecules come together at tight junctions. Curr Opin Cell Biol 11:628–633
123. Linford A et al (2012) Rab14 and its exchange factor FAM116 link endocytic recycling and adherens junction stability in migrating cells. Dev Cell 22:952–966. doi:10.1016/j.devcel.2012.04.010
124. Soini Y (2012) Tight junctions in lung cancer and lung metastasis: a review. Int J Clin Exp Pathol 5:126–136
125. Szabo I, Kiss A, Schaff Z, Sobel G (2009) Claudins as diagnostic and prognostic markers in gynecological cancer. Histol Histopathol 24:1607–1615
126. Sobel G et al (2006) Claudin 1 differentiates endometrioid and serous papillary endometrial adenocarcinoma. Gynecol Oncol 103:591–598. doi:10.1016/j.ygyno.2006.04.005
127. Kamel D, Brady B, Tabchy A, Mills GB, Hennessy B (2012) Proteomic classification of breast cancer. Curr Drug Targets 13(12):1495–1509
128. Prat A, Perou CM (2011) Deconstructing the molecular portraits of breast cancer. Mol Oncol 5:5–23. doi:10.1016/j.molonc.2010.11.003
129. Yamamoto Y et al (2003) Distinct roles of Rab3B and Rab13 in the polarized transport of apical, basolateral, and tight junctional membrane proteins to the plasma membrane. Biochem Biophys Res Commun 308:270–275
130. Kohler K, Louvard D, Zahraoui A (2004) Rab13 regulates PKA signaling during tight junction assembly. J Cell Biol 165:175–180. doi:10.1083/jcb.200312118
131. Yamamura R, Nishimura N, Nakatsuji H, Arase S, Sasaki T (2008) The interaction of JRAB/MICAL-L2 with Rab8 and Rab13 coordinates the assembly of tight junctions and adherens junctions. Mol Biol Cell 19:971–983. doi:10.1091/mbc.E07-06-0551
132. Kostenko O, Tsacoumangos A, Crooks D, Kil SJ, Carlin C (2006) Gab1 signaling is regulated by EGF receptor sorting in early endosomes. Oncogene 25:6604–6617. doi:10.1038/sj.onc.1209675
133. Yang Z et al (2012) The signaling adaptor gab1 regulates cell polarity by acting as a par protein scaffold. Mol Cell 47:469–483. doi:10.1016/j.molcel.2012.06.037
134. Asaka M, Hirase T, Hashimoto-Komatsu A, Node K (2011) Rab5a-mediated localization of claudin-1 is regulated by proteasomes in endothelial cells. Am J Physiol Cell Physiol 300:C87–C96. doi:10.1152/ajpcell.00565.2010
135. Lapierre LA et al (2012) Phosphorylation of Rab11-FIP2 regulates polarity in MDCK cells. Mol Biol Cell 23:2302–2318. doi:10.1091/mbc.E11-08-0681
136. Senga K, Mostov KE, Mitaka T, Miyajima A, Tanimizu N (2012) Grainyhead-like 2 regulates epithelial morphogenesis by establishing functional tight junctions through the organization

of a molecular network among claudin3, claudin4, and Rab25. Mol Biol Cell 23:2845–2855. doi:10.1091/mbc.E12-02-0097
137. Liu Y et al (2012) Knockdown of RAB25 promotes autophagy and inhibits cell growth in ovarian cancer cells. Mol Med Report 6:1006–1012. doi:10.3892/mmr.2012.1052
138. Boggetti B, Niessen CM (2012) Adherens junctions in mammalian development, homeostasis and disease: lessons from mice. Subcell Biochem 60:321–355. doi:10.1007/978-94-007-4186-7_14
139. de Beco S, Amblard F, Coscoy S (2012) New insights into the regulation of E-cadherin distribution by endocytosis. Int Rev Cell Mol Biol 295:63–108. doi:10.1016/B978-0-12-394306-4.00008-3
140. Ducharme NA et al (2006) MARK2/EMK1/Par-1Balpha phosphorylation of Rab11-family interacting protein 2 is necessary for the timely establishment of polarity in Madin-Darby canine kidney cells. Mol Biol Cell 17:3625–3637. doi:10.1091/mbc.E05-08-0736
141. Le TL, Yap AS, Stow JL (1999) Recycling of E-cadherin: a potential mechanism for regulating cadherin dynamics. J Cell Biol 146:219–232
142. Desclozeaux M et al (2008) Active Rab11 and functional recycling endosome are required for E-cadherin trafficking and lumen formation during epithelial morphogenesis. Am J Physiol Cell Physiol 295:C545–C556. doi:00097.2008 [pii] 10.1152/ajpcell.00097.2008
143. Janda E et al (2006) Raf plus TGFbeta-dependent EMT is initiated by endocytosis and lysosomal degradation of E-cadherin. Oncogene 25:7117–7130. doi:10.1038/sj.onc.1209701
144. Bryant DM et al (2007) EGF induces macropinocytosis and SNX1-modulated recycling of E-cadherin. J Cell Sci 120:1818–1828. doi:10.1242/jcs.000653
145. Lock JG, Stow JL (2005) Rab11 in recycling endosomes regulates the sorting and basolateral transport of E-cadherin. Mol Biol Cell 16:1744–1755. doi:10.1091/mbc.E04-10-0867
146. Palacios F, Tushir JS, Fujita Y, D'Souza-Schorey C (2005) Lysosomal targeting of E-cadherin: a unique mechanism for the down-regulation of cell-cell adhesion during epithelial to mesenchymal transitions. Mol Cell Biol 25:389–402. doi:10.1128/MCB.25.1.389-402.2005
147. Eva R et al (2010) Rab11 and its effector Rab coupling protein contribute to the trafficking of beta 1 integrins during axon growth in adult dorsal root ganglion neurons and PC12 cells. J Neurosci 30:11654–11669. doi:10.1523/JNEUROSCI.2425-10.2010
148. Caswell PT et al (2007) Rab25 associates with alpha5beta1 integrin to promote invasive migration in 3D microenvironments. Dev Cell 13:496–510. doi:10.1016/j.devcel.2007.08.012
149. Dozynkiewicz MA et al (2012) Rab25 and CLIC3 collaborate to promote integrin recycling from late endosomes/lysosomes and drive cancer progression. Dev Cell 22:131–145. doi:10.1016/j.devcel.2011.11.008
150. Mai A et al (2011) Competitive binding of Rab21 and p120RasGAP to integrins regulates receptor traffic and migration. J Cell Biol 194:291–306. doi:10.1083/jcb.201012126
151. Roberts M, Barry S, Woods A, van der Sluijs P, Norman J (2001) PDGF-regulated rab4-dependent recycling of alphavbeta3 integrin from early endosomes is necessary for cell adhesion and spreading. Curr Biol 11:1392–1402
152. Anitei M, Hoflack B (2012) Bridging membrane and cytoskeleton dynamics in the secretory and endocytic pathways. Nat Cell Biol 14:11–19. doi:10.1038/ncb2409
153. Pan D (2010) The hippo signaling pathway in development and cancer. Dev Cell 19:491–505. doi:10.1016/j.devcel.2010.09.011
154. Wojciak-Stothard B, Ridley AJ (2003) Shear stress-induced endothelial cell polarization is mediated by Rho and Rac but not Cdc42 or PI 3-kinases. J Cell Biol 161:429–439. doi:10.1083/jcb.200210135jcb.200210135 [pii]
155. Mills GB, Jurisica I, Yarden Y, Norman JC (2009) Genomic amplicons target vesicle recycling in breast cancer. J Clin Invest 119:2123–2127. doi:10.1172/JCI40256
156. Sebbagh M, Santoni MJ, Hall B, Borg JP, Schwartz MA (2009) Regulation of LKB1/STRAD localization and function by E-cadherin. Curr Biol 19:37–42. doi:10.1016/j.cub.2008.11.033
157. Lee JH et al (2007) Energy-dependent regulation of cell structure by AMP-activated protein kinase. Nature 447:1017–1020. doi:10.1038/nature05828
158. Zheng B, Cantley LC (2007) Regulation of epithelial tight junction assembly and disassembly by AMP-activated protein kinase. Proc Natl Acad Sci USA 104:819–822. doi:10.1073/pnas.0610157104

159. Lau MT, Leung PC (2012) The PI3K/Akt/mTOR signaling pathway mediates insulin-like growth factor 1-induced E-cadherin down-regulation and cell proliferation in ovarian cancer cells. Cancer Lett 326(2):191–198. doi:10.1016/j.canlet.2012.08.016
160. Hehnly H, Doxsey S (2012) Polarity sets the stage for cytokinesis. Mol Biol Cell 23:7–11. doi:10.1091/mbc.E11-06-0512
161. Fielding AB et al (2005) Rab11-FIP3 and FIP4 interact with Arf6 and the exocyst to control membrane traffic in cytokinesis. EMBO J 24:3389–3399. doi:10.1038/sj.emboj.7600803
162. Rivard N (2009) Phosphatidylinositol 3-kinase: a key regulator in adherens junction formation and function. Front Biosci 14:510–522. doi:3259 [pii]
163. Thapa N et al (2012) Phosphoinositide signaling regulates the exocyst complex and polarized integrin trafficking in directionally migrating cells. Dev Cell 22:116–130. doi:10.1016/j.devcel.2011.10.030
164. Martin-Belmonte F, Mostov K (2007) Phosphoinositides control epithelial development. Cell Cycle 6:1957–1961
165. Langlois MJ et al (2010) The PTEN phosphatase controls intestinal epithelial cell polarity and barrier function: role in colorectal cancer progression. PLoS One 5:e15742. doi:10.1371/journal.pone.0015742
166. Comer FI, Parent CA (2007) Phosphoinositides specify polarity during epithelial organ development. Cell 128:239–240. doi:10.1016/j.cell.2007.01.010
167. Cheng KW et al (2004) The RAB25 small GTPase determines aggressiveness of ovarian and breast cancers. Nat Med 10:1251–1256. doi:10.1038/nm1125
168. Liu H, Radisky DC, Wang F, Bissell MJ (2004) Polarity and proliferation are controlled by distinct signaling pathways downstream of PI3-kinase in breast epithelial tumor cells. J Cell Biol 164:603–612. doi:10.1083/jcb.200306090
169. Reddy P et al (2005) Formation of E-cadherin-mediated cell-cell adhesion activates AKT and mitogen activated protein kinase via phosphatidylinositol 3 kinase and ligand-independent activation of epidermal growth factor receptor in ovarian cancer cells. Mol Endocrinol 19:2564–2578. doi:me.2004-0342 [pii]10.1210/me.2004-0342
170. Jean S, Kiger AA (2012) Coordination between RAB GTPase and phosphoinositide regulation and functions. Nat Rev Mol Cell Biol 13:463–470. doi:10.1038/nrm3379
171. Chamberlain MD, Berry TR, Pastor MC, Anderson DH (2004) The p85alpha subunit of phosphatidylinositol 3′-kinase binds to and stimulates the GTPase activity of Rab proteins. J Biol Chem 279:48607–48614. doi:10.1074/jbc.M409769200
172. Chamberlain MD et al (2010) Deregulation of Rab5 and Rab4 proteins in p85R274A-expressing cells alters PDGFR trafficking. Cell Signal 22:1562–1575. doi:10.1016/j.cellsig.2010.05.025
173. Mellor P, Furber LA, Nyarko JN, Anderson DH (2012) Multiple roles for the p85alpha isoform in the regulation and function of PI3K signalling and receptor trafficking. Biochem J 441:23–37. doi:10.1042/BJ20111164
174. Zheng Y, Bagrodia S, Cerione RA (1994) Activation of phosphoinositide 3-kinase activity by Cdc42Hs binding to p85. J Biol Chem 269:18727–18730
175. Hardie DG (2004) The AMP-activated protein kinase pathway—new players upstream and downstream. J Cell Sci 117:5479–5487. doi:10.1242/jcs.01540
176. Menendez JA, Vazquez-Martin A (2012) AMPK: a bona fide resident of the mitotic spindle midzone. Cell Cycle 11:841–842. doi:10.4161/cc.11.5.19461
177. Vazquez-Martin A, Oliveras-Ferraros C, Menendez JA (2009) The active form of the metabolic sensor: AMP-activated protein kinase (AMPK) directly binds the mitotic apparatus and travels from centrosomes to the spindle midzone during mitosis and cytokinesis. Cell Cycle 8:2385–2398
178. Wei C, Bhattaram VK, Igwe JC, Fleming E, Tirnauer JS (2012) The LKB1 tumor suppressor controls spindle orientation and localization of activated AMPK in mitotic epithelial cells. PLoS One 7:e41118. doi:10.1371/journal.pone.0041118
179. Balkovetz DF, Chumley P, Amlal H (2009) Downregulation of claudin-2 expression in renal epithelial cells by metabolic acidosis. Am J Physiol Renal Physiol 297:F604–F611. doi:10.1152/ajprenal.00043.2009

180. Li L et al (2010) Regulation of mTORC1 by the Rab and Arf GTPases. J Biol Chem 285:19705–19709. doi:10.1074/jbc.C110.102483
181. Flinn RJ, Yan Y, Goswami S, Parker PJ, Backer JM (2010) The late endosome is essential for mTORC1 signaling. Mol Biol Cell 21:833–841. doi:10.1091/mbc.E09-09-0756
182. Zaman S, Lippman SI, Schneper L, Slonim N, Broach JR (2009) Glucose regulates transcription in yeast through a network of signaling pathways. Mol Syst Biol 5:245. doi:10.1038/msb.2009.2
183. De Wever V, Reiter W, Ballarini A, Ammerer G, Brocard C (2005) A dual role for PP1 in shaping the Msn2-dependent transcriptional response to glucose starvation. EMBO J 24: 4115–4123. doi:10.1038/sj.emboj.7600871
184. Faulhammer F et al (2007) Growth control of Golgi phosphoinositides by reciprocal localization of sac1 lipid phosphatase and pik1 4-kinase. Traffic 8:1554–1567. doi:10.1111/j.1600-0854.2007.00632.x
185. Ramm G, Larance M, Guilhaus M, James DE (2006) A role for 14-3-3 in insulin-stimulated GLUT4 translocation through its interaction with the RabGAP AS160. J Biol Chem 281:29174–29180. doi:10.1074/jbc.M603274200
186. Stockli J et al (2008) Regulation of glucose transporter 4 translocation by the Rab guanosine triphosphatase-activating protein AS160/TBC1D4: role of phosphorylation and membrane association. Mol Endocrinol 22:2703–2715. doi:10.1210/me.2008-0111
187. Martin S et al (2000) Biogenesis of insulin-responsive GLUT4 vesicles is independent of brefeldin A-sensitive trafficking. Traffic 1:652–660
188. Aoh QL, Graves LM, Duncan MC (2011) Glucose regulates clathrin adaptors at the trans-Golgi network and endosomes. Mol Biol Cell 22:3671–3683. doi:10.1091/mbc.E11-04-0309
189. Gopalakrishnan S, Hallett MA, Atkinson SJ, Marrs JA (2007) aPKC-PAR complex dysfunction and tight junction disassembly in renal epithelial cells during ATP depletion. Am J Physiol Cell Physiol 292:C1094–C1102. doi:10.1152/ajpcell.00099.2006
190. Cohen D, Brennwald PJ, Rodriguez-Boulan E, Musch A (2004) Mammalian PAR-1 determines epithelial lumen polarity by organizing the microtubule cytoskeleton. J Cell Biol 164:717–727. doi:10.1083/jcb.200308104
191. Hackenbeck T et al (2011) The GTPase RAB20 is a HIF target with mitochondrial localization mediating apoptosis in hypoxia. Biochim Biophys Acta 1813:1–13. doi:10.1016/j.bbamcr.2010.10.019
192. Petersen OW, Ronnov-Jessen L, Howlett AR, Bissell MJ (1992) Interaction with basement membrane serves to rapidly distinguish growth and differentiation pattern of normal and malignant human breast epithelial cells. Proc Natl Acad Sci USA 89:9064–9068
193. Salem AF et al (2012) Two-compartment tumor metabolism: autophagy in the tumor microenvironment and oxidative mitochondrial metabolism (OXPHOS) in cancer cells. Cell Cycle 11:2545–2556. doi:10.4161/cc.20920
194. Sotgia F et al (2012) Caveolin-1 and cancer metabolism in the tumor microenvironment: markers, models, and mechanisms. Annu Rev Pathol 7:423–467. doi:10.1146/annurev-pathol-011811-120856
195. Martinez-Outschoorn UE et al (2011) Stromal-epithelial metabolic coupling in cancer: integrating autophagy and metabolism in the tumor microenvironment. Int J Biochem Cell Biol 43:1045–1051. doi:10.1016/j.biocel.2011.01.023
196. Lisanti MP et al (2010) Understanding the "lethal" drivers of tumor-stroma co-evolution: emerging role(s) for hypoxia, oxidative stress and autophagy/mitophagy in the tumor microenvironment. Cancer Biol Ther 10:537–542. doi:10.4161/cbt.10.6.13370
197. Carito V et al (2012) Metabolic remodeling of the tumor microenvironment: migration stimulating factor (MSF) reprograms myofibroblasts toward lactate production, fueling anabolic tumor growth. Cell Cycle 11:3403–3414. doi:10.4161/cc.21701
198. Guido C et al (2012) Metabolic reprogramming of cancer-associated fibroblasts by TGF-beta drives tumor growth: connecting TGF-beta signaling with "Warburg-like" cancer metabolism and L-lactate production. Cell Cycle 11:3019–3035. doi:10.4161/cc.21384

199. Horgan CP, McCaffrey MW (2009) The dynamic Rab11-FIPs. Biochem Soc Trans 37: 1032–1036. doi:10.1042/BST0371032
200. Cheng JM, Ding M, Aribi A, Shah P, Rao K (2006) Loss of RAB25 expression in breast cancer. Int J Cancer 118:2957–2964. doi:10.1002/ijc.21739
201. Tong SY et al (2012) Correlation between FDG uptake by PET/CT and the expressions of glucose transporter type 1 and hexokinase II in cervical cancer. Int J Gynecol Cancer 22: 654–658. doi:10.1097/IGC.0b013e31824864e6
202. Hanahan D, Weinberg RA (2000) The hallmarks of cancer. Cell 100:57–70
203. Hanahan D, Weinberg RA (2011) Hallmarks of cancer: the next generation. Cell 144: 646–674. doi:10.1016/j.cell.2011.02.013

Chapter 6
Endocytosis and the Regulation of Cell Signaling, Cell Adhesion, and Epithelial to Mesenchymal Transition in Cancer

Crislyn D'Souza-Schorey and Guangpu Li

Abstract Endosomes play key roles in the control and execution of such diverse and spatially restricted processes as cell signaling, epithelial to mesenchymal transitions as well as cell adhesion and migration. The endocytosis of growth factor receptors and adhesion molecules, such as cadherins and integrins, is coming into focus as a major mechanism in the regulation of cellular processes that govern cell growth, differentiation, survival, and motility. Subversion of these pathways accompanies disease progression, especially cancer. We suggest that endosomes are multifunctional dynamic platforms on which unique sets of molecular components are assembled and sorted to adapt to different environmental and cellular cues. A better understanding of how endosomes can function as conduits for the acquisition of oncogenic phenotypes will lead to more specific therapeutic approaches to combat cancer progression.

6.1 Introduction

Although endocytosis has long been regarded as a conduit for the internalization and degradation of nutrients and cell surface receptors, it is now accepted that the endosomal membrane system plays a vital role in the control and execution of spatially restricted functions, such as cell adhesion and motility. Cell adhesion is

C. D'Souza-Schorey (✉)
Department of Biological Sciences, University of Notre Dame,
Notre Dame, IN 46556-0369, USA
e-mail: cdsouzas@nd.edu

G. Li
Department of Biochemistry and Molecular Biology, University of Oklahoma
Health Sciences Center, Oklahoma City, OK 73104, USA
e-mail: guangpu-li@ouhsc.edu

Y. Yarden and G. Tarcic (eds.), *Vesicle Trafficking in Cancer*,
DOI 10.1007/978-1-4614-6528-7_6,

essential for the maintenance of multicellularity in living organisms. Intercellular adhesion and cell-to-extracellular matrix (ECM) adhesion are a result of the assembly and functional integrity of "adhesion complexes" at sites of cell–cell or cell–ECM contacts, respectively [34]. These complexes consist of transmembrane adhesion molecules coupled to intracellular scaffold or signaling proteins and the cytoskeleton. Cadherin family cell adhesion molecules and their associated scaffold proteins, the catenins, are major components of cell–cell adhesive contacts, namely, the adherens junctions and desmosomes [34]. The major transmembrane protein at cell–ECM adhesive contacts (i.e., focal adhesions and hemidesmosomes) is the heterodimeric integrin receptors [34]. These cell–cell and cell–ECM adhesion complexes are linked to and stabilized by the actin or intermediate filaments but undergo significant remodeling during the acquisition of migratory and invasive phenotypes in tumor cells [93, 94]. In the initial stages of tumor progression, tumor cells sometimes undergo epithelial–mesenchymal transition, which requires the disruption of cell–cell adhesions [5]. The trafficking of cadherins along the endocytic pathway is now accepted as an important mechanism involved in the remodeling of cell–cell adhesions [39]. Some cancer cells maintain or reestablish cell–cell adhesions during metastasis and their collective migration requires that both cell–cell and cell–ECM adhesions stay intact [25]. Mesenchymal tumor cells will migrate and invade into the basement membrane and surrounding stromal tissues [54]. To facilitate migration and invasion, tumor cells continuously form new cell–ECM adhesions at the leading edge, whereas focal adhesions at the trailing edge are disrupted by endocytosis.

Besides serving as carriers and sorting stations for the trafficking of adhesion molecules, emerging evidence suggests that endosomal compartments are also essential sites of signal transduction [53, 77, 84]. Activated receptors can function in endosomes, and certain signaling components are localized, even exclusively, to endosomes. Signals transmitted from endosomes are robust and typically long-lived, different from those that arise from the plasma membrane. Endosomal signaling is widespread across species and regulates essential processes including growth and differentiation in addition to cell adhesion and motility. Subversion of the mechanisms involved is predicted to play an important role in several human diseases and, most especially, cancer. Pharmacological agents that target receptor signaling at the plasma membrane have proved to be effective therapeutics for some cancers [53]. Thus, selective disruption of receptor signaling in endosomes, which can be accomplished by targeting endosomal-specific signaling pathways that are altered in cancers, could also provide novel therapies for tumor progression.

In this review chapter, we discuss current evidence coupling endocytosis and the regulation of signaling pathways in cells and how altered regulation of these pathways can lead to acquisition of oncogenic phenotypes. We also describe how endocytosis and recycling of adhesion molecules, such as cadherins and integrins, is coming into focus as a major mechanism in the regulation of adhesive and migratory properties of cells.

6.2 Endocytosis Regulates Signal Transduction

Endocytosis plays an important role in the duration and extent of signal transduction via controlling the number and accessibility of cell surface receptors [52, 82, 84]. Signal transduction starts at the cell surface when ligands such as growth factors and hormones bind to their cognate receptors, which recruit and activate signaling molecules such as adaptors and enzymes (e.g., kinases) that further activate downstream effectors to amplify the signal transduction processes ultimately leading to regulation of gene expression, cell proliferation, or cell differentiation. Malfunction of the signal transduction processes is a major cause of cancer.

Receptor tyrosine kinases (RTKs) constitute a large family of receptors whose signal transduction processes promote cell proliferation or differentiation. A well-characterized example is the epidermal growth factor receptor (EGFR). Ligand (EGF) binding causes conformational changes, dimerization, and tyrosine phosphorylation of EGFR, which activates phosphoinositide 3-kinase (PI3-K), phospholipase Cγ (PLCγ), and the mitogen-activated protein (MAP) kinase pathways at the cell surface. The ligand also induces endocytosis of EGFR into endosomes and eventually intraluminal vesicles (ILVs) and lysosomes for degradation, which reduces the number of active EGFR molecules and attenuates the signaling in the so-called "downregulation" process. This process is delayed under hypoxia conditions characteristic of cancer cells where hypoxia-inducible factor HIF1α blocks Rabaptin-5 expression, leading to reduced Rab5-mediated endosome fusion and endocytosis [91]. As a result, EGFR signaling is prolonged with increased cell proliferation.

EGFR in endosomes, before delivery to ILVs and lysosomes, may remain engaged to the ligand, and its cytoplasmic domain may continue signaling. Indeed, the first evidence of endosomes serving as a platform for EGFR signaling inside the cell came from a study that showed reduced phosphorylation/activation of MAP kinases in the cells where endocytosis was blocked by a dynamin mutant [90]. This concept was later generalized as the signaling endosome hypothesis [32]. Signaling endosomes not only continue the signaling processes initiated at the cell surface but also gain access to new signaling molecules and start new signaling processes. For example, EGFR can activate the small GTPase Rab5 via RIN1 on endosomes [85]. Rab5-GTP and certain receptors can directly recruit APPL1 to endosomes, which activates Akt and regulates its substrate specificity for GSK-3β, a process critical for cell survival and embryonic development in zebrafish and possibly other vertebrates [76]. On the other hand, an intriguing recent observation suggests that accumulation of activated EGFR on endosomes via blocking ILV formation triggers apoptosis [72], although the signaling pathway is yet to be established. It is clear that the endosomes provide a membrane environment distinct from that of the plasma membrane and their mobility inside cells allows access to additional signaling molecules leading to new functional consequences.

Endosomal transport is critical for retrograde signaling by nerve growth factor (NGF) and its receptor (TrkA) in neuronal survival, migration, axon growth, and target cell innervation [3, 10]. In this case, target-derived NGF binding at the tip of the axon induces endocytosis of the NGF–TrkA complex into the signaling endosomes that start signaling locally inside the axon and undergo retrograde transport along microtubules towards the cell body/soma. The signaling endosomes carry signaling molecules such as the activated TrkA, ERKs, and Akt [28, 32] as well as specific transcription factors (e.g., CREB) translated in the axon [11] to activate specific nuclear gene expression in the soma, including a positive feedback loop of increased expression of TrkA [48]. The functional consequences are the aforementioned neurotrophic effects.

Endosomal membrane is enriched with PI3P, which recruits PI3P-binding signaling molecules such as the FYVE domain-containing proteins important for TGFβ receptor- and G protein-coupled receptor (GPCR)-mediated signal transduction processes. In the case of TGFβ signaling, the TGFβ receptor is endocytosed into endosomes where it interacts with SMAD anchor for receptor activation (SARA), a FYVE domain-containing adaptor protein that recruits SMAD2 to the endosomes for phosphorylation by the TGFβ receptor [88]. The phosphorylated SMAD2 then forms a complex with SMAD4, followed by translocation to the nucleus for activation of gene expression. In the case of GPCR signaling, activated G protein on endosomes can activate the PI3K Vps34 to produce PI3P that in turn recruits FYVE domain-containing proteins to promote MAPK and Cdc42 signaling pathways [84].

Endocytosis also plays an important role in Notch-mediated signal transduction and neuronal cell proliferation [22, 39]. Both Notch and its ligands (the Delta/Serrate/Lag2 (Dsl) domain-containing proteins) are transmembrane proteins on apposing cells. Ligand binding leads to two consecutive proteolytic cleavages (S1 and S2) in the ectodomain of Notch. The C-terminal fragment of Notch is then endocytosed, followed by another cleavage (S3) in the transmembrane domain by γ secretase to release the cytoplasmic intracellular domain that translocates to the nucleus and activates expression of target genes. Interestingly, the ligand itself, e.g., Delta, requires endocytosis and recycling to concentrate at specific regions on the plasma membrane for efficient binding and activation of Notch on the signal-receiving cells [22, 39].

An important mechanism that controls the endocytosis and/or endosomal sorting of the signaling receptors involves ubiquitination at their cytoplasmic domains. While polyubiquitination with the Lys-48 linkage marks the substrate for degradation by the proteosome, polyubiquitination with the Lys-63 linkage and monoubiquitination may regulate other functions of the substrate including endocytosis and endosomal sorting [84]. Ubiquitination is necessary for the endocytosis of Ste2, a GPCR, in yeast [29], but it is not essential for other GPCRs and RTKs in animal cells [20, 79]. In the latter case, ubiquitination may still increase the interaction of these signaling receptors with components of clathrin-coated pits to facilitate their endocytosis [2, 36]. Importantly, ubiquitination is critical for subsequent sorting of signaling receptors into ILVs/MVBs [20, 79], via interaction with the ESCRT complexes, to terminate the signal [35, 81, 82]. Consistently, ubiquitination-deficient EGFR shows increased signaling activity. A well-documented E3 ubiquitin ligase, Cbl, is responsible for the ubiquitination of various RTKs. Mutations that abrogate Cbl ubiquitin ligase activity are known to cause cancer such as myeloid leukemia and lung cancer [74, 86], suggesting that ubiquitination-mediated RTK sequestration in ILVs/MVBs and signal termination are critical for normal signal transduction processes and cell growth. In the case of Notch, non-ligand-bound Notch is ubiquitinated by the E3 ligase ITCH, endocytosed and delivered to lysosomes for degradation [8].

In addition to signaling receptors, ubiquitination of downstream signaling molecules on the plasma membrane, such as the Ras GTPases (H- and N-Ras), leads to endocytosis and signal downregulation in both mammalian cells and *Drosophila* where the system controls organ development [96, 97]. The E3 ubiquitin ligase responsible for Ras ubiquitination is Rabex-5 [96, 97], which was originally identified as a GEF for activation of the endosome-associated Rab5 and endosome fusion [31]. Rabex-5 targets early endosomes and plasma membrane by binding to Rab22 [99, 100] or by forming a complex with Rabaptin-5 that in turn binds to Rab5 [31]. The K-Ras isoform, on the other hand, is not ubiquitinated, but a recent study shows that a fraction of K-Ras is endocytosed via the clathrin-dependent pathway and follows the conventional endocytic pathway to early endosomes, late endosomes/MVBs, and lysosomes [46]. Interestingly, K-Ras is able to recruit Raf1 to elicit signal transduction on late endosomes [46], which normally degrade endocytosed cargoes and reduce signaling.

Endocytosed receptors are not always destined to degradation in late endosomes and lysosomes; instead they can be recycled to the plasma membrane for reutilization and additional rounds of signaling, depending on the types of ligands and endocytic portals. For example, TGFα also interacts with EGFR but its affinity is lower than EGF and readily dissociates from the receptor in early endosomes [19, 24]. As a result, the receptor is sorted back to the plasma membrane, due to insufficient ubiquitination [45], which likely contributes to the high potency of TGFα in promoting tumor cell growth [56]. In addition, receptors can be endocytosed via clathrin-dependent or clathrin-independent pathways or both [17, 81]. In the case of EGFR, low concentrations of EGF induce clathrin-dependent endocytosis while high concentrations of EGF promote clathrin-independent endocytosis [81]. The former pathway largely recycles EGFR back to the plasma membrane for continual signaling and the latter pathway promotes EGFR traffic to late endosomes and lysosomes for degradation and signal attenuation [81]. In addition to the RTKs, some GPCRs such as the β2 adrenergic receptor (β2AR) require endocytosis and recycling to be resensitized for sustained signaling. In this case, β2AR can be inactivated by phosphorylation at the plasma membrane and endocytosis allows β2AR to gain access to the endosome-associated phosphatase 2A for dephosphorylation and resensitization [66]. The active β2AR is then recycled back to the plasma membrane, via a Rab4-dependent fast recycling pathway, for ligand binding and new rounds of signaling [21, 59, 78].

6.3 EMT in Cancer

Most solid tumors are epithelial in origin. A loss of epithelial cell markers and concomitant acquisition of mesenchymal cell markers have been observed in some epithelial tumors, such as non-small cell lung cancer (NSCLC) and pancreatic, colorectal, and hepatocellular cancers particularly at the invasive front [37, 87]. This profound phenotypic conversion, referred to as epithelial to mesenchymal transition (EMT), is orchestrated by integrated networks of signal transduction pathways that direct marked alterations in cell adhesion and motility. Although EMT is best known for its role in embryonic development, in the adult, several oncogenic pathways (growth factors, Src, Ras, Wnt/beta-catenin, and Notch) may induce EMT [87].

Although there are accumulating data to suggest a critical role for EMT in cancer progression, the demonstration of this process in human cancer has been controversial [67]. Moreover, the acquisition of mesenchymal phenotypes is not a prerequisite for cell migration/invasion and in many cases appears to be characteristic of only a few cells at the invasive fronts in a tumor mass. The evidence to date suggests that the induction of EMT depends on the tumor type and its genetic alterations as well as on its interaction with the extracellular matrix [67]. Over the last decade, however, it's been shown that some cancer cells reactivate EMT in an effort to escape their normal boundaries [67]. The loss of epithelial cell markers (e.g., E-cadherin) is associated with disease progression and metastatic potential of a tumor. There is accruing evidence that cancer cells can dedifferentiate through activation of specific biological pathways associated with EMT, thereby gaining the ability to migrate and invade. Hence, what has been observed experimentally regarding EMT and normal embryonic development is also thought to apply in the progression of solid tumors—a cellular reprogramming process whereby epithelial tumor cells lose cell polarity and cell junction proteins and at the same time gain signal transduction activities associated with cell invasion and survival in an anchorage-independent environment. Mesenchymal-like tumor cells gain migratory capacity at the expense of proliferative potential. Cellular changes resulting in EMT in cancer are thought to play a major role in disease progression and have been associated with poor prognosis in patients [37].

6.4 Endocytic Trafficking of Cadherins in EMT

A critical molecular event underpinning the dissolution of cell–cell contacts during EMT is the loss of the cell–cell adhesion molecule, E-cadherin, a key component of the adherens junctions [87]. EMT and metastatic progression are most often associated with a reversible downregulation of E-cadherin (encoded by *CDH1*) involving either hypermethylation of the *CDH1* promoter or repression by EMT-inducing transcription factors [5, 87]. In particular, EMT is accompanied by the activation of two related zinc finger-containing transcription factors, Snail and Slug. The basic helix–loop–helix proteins, Twist 1 and Twist 2, and the ZEB family proteins have also been shown to induce EMT via transcription regulation. Notably, however, in addition to transcriptional downregulation, posttranscriptional regulation of adhesive structures can also markedly influence the progression of EMT [12]. The endocytosis and lysosomal degradation of E-cadherin as described below is one such cellular mechanism that can have a profound impact on the initial stages of EMT.

The cytoplasmic domain of E-cadherin contains a dileucine motif, which is a binding site for clathrin adaptor complexes, and mutations in the motif inhibit E-cadherin endocytosis [50]. The E-cadherin dileucine motif also binds to p120ctn, which in polarized poorly motile epithelia masks the dileucine motif to prevent the endocytosis of E-cadherin [51]. Cellular depletion of p120ctn results in the internalization of cadherins and loss of cell–cell contacts [9, 14, 95]. The binding of an adaptor molecule, Numb, to p120ctn negates the p120ctn-mediated suppression of E-cadherin endocytosis since Numb recruits the AP-2 clathrin adaptor complex promoting endocytosis [75]. Numb can also interact with the NVYYY motif on E-cadherin, which in turn can hinder the p120ctn–E-cadherin interaction [92].

Growth factors such as HGF, as well as oncogenic v-Src, can also induce epithelial–mesenchymal transition (EMT) in part by promoting the endocytosis of cadherin molecules [12, 94]. Src-mediated phosphorylation of the NVYYY motif on E-cadherin induces the dissociation of p120ctn and recruits a c-Cbl-related E3 ubiquitin ligase, Hakai [26]. Hakai induces the ubiquitination of E-cadherin and subsequently its endocytosis and lysosomal degradation, which require the activation of the Rab small GTP-binding proteins, Rab5 and Rab7 [63]. HGF treatment or v-Src expression activates ARF6, an ARF family small GTP-binding protein, which enhances E-cadherin endocytosis [60]. In Madin–Darby Canine Kidney (MDCK) cells, ARF6-GTP recruits the nucleoside diphosphate kinase, Nm23-H1, initiating a downregulation of Rac1 activity and promoting the clathrin-dependent endocytosis of E-cadherin to the early endosome, both of which facilitate the disassembly of the adherens junctions [62]. Furthermore, upon Src-induced loss of cell–cell contacts, internalized E-cadherin is targeted to the lysosome for degradation in an ARF6-dependent manner so that it cannot be recycled to the plasma membrane, thus ensuring that cell–cell contacts will not be reformed. The TBC/RabGAP Armus is thought to play a role in this process, bridging signaling between ARF6, Rac1, and Rab7 [23]. Armus was shown to bind Rac1 and locally facilitate lysosome biogenesis and the degradation of E-cadherin. Inhibition of ARF6 activity, by expression of a dominantly interfering mutant, ARF6(T27N), enhances the epithelial phenotype by preventing the internalization of E-cadherin into endosomal compartments and thereby blocking hepatocyte growth factor (HGF) and Src-induced cell scattering [60, 61]. In addition, the downregulation of ARF6 activity, established by a positive feedback loop between EphA2 and E-cadherin, has been shown to enhance E-cadherin-based adhesion and the maturation of apical–basal polarity in MDCK cells [49]. The formation of stable adherens junctions during epithelial cell polarization is dependent on the spatially regulated activation of ARF6, which is modulated by the formation of a complex between FRMD4A/GRSP-1/PAR3 and the ARF6-GEF cytohesin-1 at cell junctions [33]. HGF also induces Ras-mediated activation of RIN2, an activator for Rab5, which also facilitates the endocytosis of E-cadherin [38].

The aforementioned effects of ARF6 on the internalization of E-cadherin and other cell surface receptors have also been shown to impact epithelial glandular organization in 3D cell cultures [89]. In this regard, sustained ARF6 activation in basement membrane cultures of epithelial cysts, a structural unit of epithelial glandular organs, leads to the internalization of E-cadherin as well as growth factor receptors by ARF6-regulated pathways. Sustained signaling from endosomes (described further below), in turn, leads to the formation of aberrant glandular morphologies, reminiscent of tumorigenic phenotypes seen in vivo [16].

In addition to clathrin-mediated endocytosis, caveolin-mediated endocytosis is also involved in E-cadherin internalization in some cell types [1, 47]. EGF treatment of MCF-7 cells, on the other hand, promotes E-cadherin into endosomes via clathrin-independent pathways [64]. While multiple internalization routes have been implicated in E-cadherin endocytosis, collectively these findings show that the cycling of E-cadherin along the endosomal pathway can markedly impinge on the dynamics of the adherens junctions of epithelial tissues and loss of the epithelial phenotype.

6.5 Endocytic Trafficking of Integrins and Acquisition of Motile Phenotypes

Migrating and invading cells display an increased internalization and recycling of integrins from the retracting edge of the cell to the leading edge and at sites of cell invasion. β1 integrin receptors have been localized to clathrin-coated pits [15] where the endocytic adaptors Numb and Dab2 bind to integrin receptors to facilitate internalization [57]. HS1-associated protein X-1 (HAX-1) has been shown to bind to the cytoplasmic tail of β6 integrin to facilitate the clathrin-mediated internalization of αvβ6 integrin receptors, which in turn enhances migration and invasion of oral sqaumous carcinoma cell lines [69]. While the above is one example, integrin trafficking has been associated with migratory potential of several tumor cells lines [54, 70]. Integrin endocytosis is also thought to facilitate the uptake of ECM proteins such as fibronectin and vitronectin and transport of these molecules to the lysosomes [43, 80].

In actively migrating cells, ARF6 and Rab family GTPases have been linked to the trafficking of integrin receptors. In this regard, β1 integrin has been shown to localize to ARF6-regulated recycling endosomes [6, 68]. Expression of dominant-negative ARF6 inhibits the cell's ability to recycle this endosomal compartment to the plasma membrane leading to a decrease in receptors at the migrating edge [6, 68]. In addition, the recycling of β1 integrin is regulated by an ARF6-GAP, ACAP1; the inhibition of ACAP1 or its phosphorylation by Akt inhibits integrin recycling and cell migration [41]. The ARF6 GEF, GEP100, which is upregulated in breast cancers, has also been implicated in the trafficking of β1 integrin receptors [18, 73]. The migratory ability of cells is also modulated by engagement with the extracellular matrix and requires ARF6-mediated activation of Rac1 for the formation of protrusive structures such as lamellipodia at the leading edge [13]. The formation of an α4 integrin–paxillin–Arf-GAP complex at the trailing edge of the cell assists in directional migration by inhibiting ARF6 activity, thereby blocking adhesion-dependent Rac1 activation and the extension of lamellipodia [58]. Slit2-Robo signaling can block the ARF6 and Rac1 activation induced by integrin engagement, important in pathologies of endothelial cell protrusive activity [44].

A subset of Rab GTPases on the plasma membrane, early and recycling endosomes control the integrin trafficking, and altered expression levels of these Rabs are often associated with various types of cancer [40]. Rab5 and Rab21 are associated with the plasma membrane and early endosomes and regulate the internalization of β1 integrins via direct interaction with their α subunits [65]. Reduction of Rab5 or Rab21 in carcinoma-associated fibroblasts can decrease α5 integrin at the plasma membrane and remodeling of cell–extracellular matrix interaction, which is required for the invasion of squamous cell carcinoma [30]. Indeed, Rab5 is highly expressed in lung adenocarcinoma and hepatocellular carcinoma [27, 42]. The Rab11 subfamily members (Rab11 and Rab25) are localized in a perinuclear recycling compartment that controls a slow recycling pathway important for trafficking of integrins back to the plasma membrane at the leading edge. Rab11 appears to collaborate with ARF6 to control the exit of integrins from the perinuclear recycling compartment [68]. Importantly, Rab11 controls α6β4 integrin recycling involved in hypoxia-induced breast cancer cell invasion [98]. Rab25 is related to Rab11 and directly binds to the β1 subunit of α5β1 integrin to facilitate its recycling to the leading edge of the plasma membrane, a process implicated in cancer cell invasion and metastasis [7]. Overexpression of Rab25 is well documented in many types of cancers [40] including ovarian cancer, breast cancer, testicular tumor, Wilms tumor, and bladder and hepatocellular carcinomas. Interestingly, Rab25 expression is decreased in colorectal adenocarcinomas, which correlates with poor prognosis of colon cancer patients [55]. In this regard, mouse models of colon cancer show that loss of Rab25 can increase colonic tumor formation [55]. It is not yet clear why different types of cancers require overexpression and loss of Rab25, respectively, to promote cell invasion and aggressiveness. In contrast to Rab11/Rab25-mediated slow recycling, Rab4 is localized in a fast recycling compartment and facilitates the fast recycling of certain integrins (e.g., αvβ3) to the cell surface in response to the platelet-derived growth factor (PDGF) [71].

6.6 Concluding Remarks

Endocytosis has emerged as an important regulatory mechanism in signal transduction/cell proliferation, cell polarity/EMT, and cell adhesion/invasion, in addition to its conventional role in the uptake and digestion of nutrients. The regulatory function of endocytosis in these processes involves internalization and sorting of signaling receptors, cell junction molecules (e.g., cadherins), and integrins to degradative or recycling pathway. Aberrant endocytic machinery and unbalanced endocytosis can contribute to uncontrolled cell proliferation, EMT, and aggressive cell invasion, which are hallmarks of cancer cells. Indeed, altered expression and/or mutations in endocytic genes are frequently found in tumors [82]. One well-documented example is Cbl, an E3 ubiquitin ligase involved in ubiquitination and endocytosis of signaling receptors. Inactivating mutations in Cbl are associated with myeloid malignancies [74]. In addition, mutations in the cytoplasmic domain of signaling receptors may abrogate their endocytosis and reduce degradation, contributing to tumorigenesis. For example, ErbB-2, an endocytosis-defective variant of EGFR, shows strong transforming effect and is highly expressed in breast cancer [4, 83]. A better understanding of the endocytic machinery that regulates the degradation and recycling of signaling receptors, cadherins, and integrins should help develop novel and specific therapeutics against cancer.

Acknowledgements The authors' research programs are supported in part by NIH grants, CA115316 to CDS and GM074692 to GL.

References

1. Akhtar N, Hotchin NA (2001) RAC1 regulates adherens junctions through endocytosis of E-cadherin. Mol Biol Cell 12:847–862
2. Arevalo JC, Waite J, Rajagopal R, Beyna M, Chen ZY, Lee FS, Chao MV (2006) Cell survival through Trk neurotrophin receptors is differentially regulated by ubiquitination. Neuron 50:549–559
3. Ascano M, Bodmer D, Kuruvilla R (2012) Endocytic trafficking of neurotrophins in neural development. Trends Cell Biol 22:266–273
4. Baulida J, Kraus MH, Alimandi M, Di Fiore PP, Carpenter G (1996) All ErbB receptors other than the epidermal growth factor receptor are endocytosis impaired. J Biol Chem 271:5251–5257
5. Berx G, van Roy F (2009) Involvement of members of the cadherin superfamily in cancer. Cold Spring Harb Perspect Biol 1:a003129
6. Brown FD, Rozelle AL, Yin HL, Balla T, Donaldson JG (2001) Phosphatidylinositol 4,5-bisphosphate and Arf6-regulated membrane traffic. J Cell Biol 154:1007–1017
7. Caswell PT, Spence HJ, Parsons M, White DP, Clark K, Cheng KW, Mills GB, Humphries MJ, Messent AJ, Anderson KI, McCaffrey MW, Ozanne BW, Norman JC (2007) Rab25 associates with alpha5beta1 integrin to promote invasive migration in 3D microenvironments. Dev Cell 13:496–510
8. Chastagner P, Israel A, Brou C (2008) AIP4/Itch regulates Notch receptor degradation in the absence of ligand. PLoS One 3:e2735
9. Chen X, Kojima S, Borisy GG, Green KJ (2003) p120 catenin associates with kinesin and facilitates the transport of cadherin-catenin complexes to intercellular junctions. J Cell Biol 163:547–557
10. Cosker KE, Courchesne SL, Segal RA (2008) Action in the axon: generation and transport of signaling endosomes. Curr Opin Neurobiol 18:270–275
11. Cox LJ, Hengst U, Gurskaya NG, Lukyanov KA, Jaffrey SR (2008) Intra-axonal translation and retrograde trafficking of CREB promotes neuronal survival. Nat Cell Biol 10:149–159
12. D'Souza-Schorey C (2005) Disassembling adherens junctions: breaking up is hard to do. Trends Cell Biol 15:19–26
13. D'Souza-Schorey C, Chavrier P (2006) ARF proteins: roles in membrane traffic and beyond. Nat Rev Mol Cell Biol 7:347–358
14. Davis MA, Ireton RC, Reynolds AB (2003) A core function for p120-catenin in cadherin turnover. J Cell Biol 163:525–534
15. De Deyne PG, O'Neill A, Resneck WG, Dmytrenko GM, Pumplin DW, Bloch RJ (1998) The vitronectin receptor associates with clathrin-coated membrane domains via the cytoplasmic domain of its beta5 subunit. J Cell Sci 111(Pt 18):2729–2740
16. Debnath J, Brugge JS (2005) Modelling glandular epithelial cancers in three-dimensional cultures. Nat Rev Cancer 5:675–688

17. Di Guglielmo GM, Le Roy C, Goodfellow AF, Wrana JL (2003) Distinct endocytic pathways regulate TGF-beta receptor signalling and turnover. Nat Cell Biol 5:410–421
18. Dunphy JL, Moravec R, Ly K, Lasell TK, Melancon P, Casanova JE (2006) The Arf6 GEF GEP100/BRAG2 regulates cell adhesion by controlling endocytosis of beta1 integrins. Curr Biol 16:315–320
19. Ebner R, Derynck R (1991) Epidermal growth factor and transforming growth factor-alpha: differential intracellular routing and processing of ligand-receptor complexes. Cell Regul 2:599–612
20. Eden ER, Huang F, Sorkin A, Futter CE (2011) The role of EGF receptor ubiquitination in regulating its intracellular traffic. Traffic 13:329–337
21. Filipeanu CM, Zhou F, Lam ML, Kerut KE, Claycomb WC, Wu G (2006) Enhancement of the recycling and activation of beta-adrenergic receptor by Rab4 GTPase in cardiac myocytes. J Biol Chem 281:11097–11103
22. Fortini ME, Bilder D (2009) Endocytic regulation of Notch signaling. Curr Opin Genet Dev 19:323–328
23. Frasa MA, Maximiano FC, Smolarczyk K, Francis RE, Betson ME, Lozano E, Goldenring J, Seabra MC, Rak A, Ahmadian MR, Braga VM (2010) Armus is a Rac1 effector that inactivates Rab7 and regulates E-cadherin degradation. Curr Biol 20:198–208
24. French AR, Sudlow GP, Wiley HS, Lauffenburger DA (1994) Postendocytic trafficking of epidermal growth factor-receptor complexes is mediated through saturable and specific endosomal interactions. J Biol Chem 269:15749–15755
25. Friedl P, Hegerfeldt Y, Tusch M (2004) Collective cell migration in morphogenesis and cancer. Int J Dev Biol 48:441–449
26. Fujita Y, Krause G, Scheffner M, Zechner D, Leddy HE, Behrens J, Sommer T, Birchmeier W (2002) Hakai, a c-Cbl-like protein, ubiquitinates and induces endocytosis of the E-cadherin complex. Nat Cell Biol 4:222–231
27. Fukui K, Tamura S, Wada A, Kamada Y, Igura T, Kiso S, Hayashi N (2007) Expression of Rab5a in hepatocellular carcinoma: possible involvement in epidermal growth factor signaling. Hepatol Res 37:957–965
28. Grimes ML, Zhou J, Beattie EC, Yuen EC, Hall DE, Valletta JS, Topp KS, LaVail JH, Bunnett NW, Mobley WC (1996) Endocytosis of activated TrkA: evidence that nerve growth factor induces formation of signaling endosomes. J Neurosci 16:7950–7964
29. Hicke L, Riezman H (1996) Ubiquitination of a yeast plasma membrane receptor signals its ligand-stimulated endocytosis. Cell 84:277–287
30. Hooper S, Gaggioli C, Sahai E (2010) A chemical biology screen reveals a role for Rab21-mediated control of actomyosin contractility in fibroblast-driven cancer invasion. Br J Cancer 102:392–402
31. Horiuchi H, Lippe R, McBride HM, Rubino M, Woodman P, Stenmark H, Rybin V, Wilm M, Ashman K, Mann M, Zerial M (1997) A novel Rab5 GDP/GTP exchange factor complexed to Rabaptin-5 links nucleotide exchange to effector recruitment and function. Cell 90:1149–1159
32. Howe CL, Mobley WC (2004) Signaling endosome hypothesis: A cellular mechanism for long distance communication. J Neurobiol 58:207–216
33. Ikenouchi J, Umeda M (2010) FRMD4A regulates epithelial polarity by connecting Arf6 activation with the PAR complex. Proc Natl Acad Sci U S A 107:748–753
34. Juliano RL (2002) Signal transduction by cell adhesion receptors and the cytoskeleton: functions of integrins, cadherins, selectins, and immunoglobulin-superfamily members. Annu Rev Pharmacol Toxicol 42:283–323
35. Katzmann DJ, Odorizzi G, Emr SD (2002) Receptor downregulation and multivesicular-body sorting. Nat Rev Mol Cell Biol 3:893–905
36. Kazazic M, Bertelsen V, Pedersen KW, Vuong TT, Grandal MV, Rodland MS, Traub LM, Stang E, Madshus IH (2009) Epsin 1 is involved in recruitment of ubiquitinated EGF receptors into clathrin-coated pits. Traffic 10:235–245
37. Keshamouni VG, Schiemann WP (2009) Epithelial-mesenchymal transition in tumor metastasis: a method to the madness. Future Oncol 5:1109–1111

38. Kimura T, Sakisaka T, Baba T, Yamada T, Takai Y (2006) Involvement of the Ras-activated Rab5 guanine nucleotide exchange factor RIN2-Rab5 pathway in the hepatocyte growth factor-induced endocytosis of E-cadherin. J Biol Chem 281:10598–10609
39. Kopan R, Ilagan MX (2009) The canonical Notch signaling pathway: unfolding the activation mechanism. Cell 137:216–233
40. Li G (2011) Rab GTPases, membrane trafficking and diseases. Curr Drug Targets 12:1188–1193
41. Li J, Ballif BA, Powelka AM, Dai J, Gygi SP, Hsu VW (2005) Phosphorylation of ACAP1 by Akt regulates the stimulation-dependent recycling of integrin beta1 to control cell migration. Dev Cell 9:663–673
42. Li Y, Meng X, Feng H, Zhang G, Liu C, Li P (1999) Over-expression of the RAB5 gene in human lung adenocarcinoma cells with high metastatic potential. Chin Med Sci J 14:96–101
43. Lobert VH, Brech A, Pedersen NM, Wesche J, Oppelt A, Malerod L, Stenmark H (2010) Ubiquitination of alpha 5 beta 1 integrin controls fibroblast migration through lysosomal degradation of fibronectin-integrin complexes. Dev Cell 19:148–159
44. London NR, Li DY (2011) Robo4-dependent Slit signaling stabilizes the vasculature during pathologic angiogenesis and cytokine storm. Curr Opin Hematol 18:186–190
45. Longva KE, Blystad FD, Stang E, Larsen AM, Johannessen LE, Madshus IH (2002) Ubiquitination and proteasomal activity is required for transport of the EGF receptor to inner membranes of multivesicular bodies. J Cell Biol 156:843–854
46. Lu A, Tebar F, Alvarez-Moya B, Lopez-Alcala C, Calvo M, Enrich C, Agell N, Nakamura T, Matsuda M, Bachs O (2009) A clathrin-dependent pathway leads to KRas signaling on late endosomes en route to lysosomes. J Cell Biol 184:863–879
47. Lu Z, Ghosh S, Wang Z, Hunter T (2003) Downregulation of caveolin-1 function by EGF leads to the loss of E-cadherin, increased transcriptional activity of beta-catenin, and enhanced tumor cell invasion. Cancer Cell 4:499–515
48. Miller FD, Mathew TC, Toma JG (1991) Regulation of nerve growth factor receptor gene expression by nerve growth factor in the developing peripheral nervous system. J Cell Biol 112:303–312
49. Miura K, Nam JM, Kojima C, Mochizuki N, Sabe H (2009) EphA2 engages Git1 to suppress Arf6 activity modulating epithelial cell-cell contacts. Mol Biol Cell 20:1949–1959
50. Miyashita Y, Ozawa M (2007) A dileucine motif in its cytoplasmic domain directs beta-catenin-uncoupled E-cadherin to the lysosome. J Cell Sci 120:4395–4406
51. Miyashita Y, Ozawa M (2007) Increased internalization of p120-uncoupled E-cadherin and a requirement for a dileucine motif in the cytoplasmic domain for endocytosis of the protein. J Biol Chem 282:11540–11548
52. Mosesson Y, Mills GB, Yarden Y (2008) Derailed endocytosis: an emerging feature of cancer. Nat Rev Cancer 8:835–850
53. Murphy JE, Padilla BE, Hasdemir B, Cottrell GS, Bunnett NW (2009) Endosomes: a legitimate platform for the signaling train. Proc Natl Acad Sci U S A 106:17615–17622
54. Nagano M, Hoshino D, Koshikawa N, Akizawa T, Seiki M (2012) Turnover of focal adhesions and cancer cell migration. Int J Cell Biol 2012:310616
55. Nam KT, Lee HJ, Smith JJ, Lapierre LA, Kamath VP, Chen X, Aronow BJ, Yeatman TJ, Bhartur SG, Calhoun BC, Condie B, Manley NR, Beauchamp RD, Coffey RJ, Goldenring JR (2010) Loss of Rab25 promotes the development of intestinal neoplasia in mice and is associated with human colorectal adenocarcinomas. J Clin Invest 120:840–849
56. Nicholson RI, Gee JM, Harper ME (2001) EGFR and cancer prognosis. Eur J Cancer 37 (Suppl 4):S9–S15
57. Nishimura T, Kaibuchi K (2007) Numb controls integrin endocytosis for directional cell migration with aPKC and PAR-3. Dev Cell 13:15–28
58. Nishiya N, Kiosses WB, Han J, Ginsberg MH (2005) An alpha4 integrin-paxillin-Arf-GAP complex restricts Rac activation to the leading edge of migrating cells. Nat Cell Biol 7:343–352

59. Odley A, Hahn HS, Lynch RA, Marreez Y, Osinska H, Robbins J, Dorn GW 2nd (2004) Regulation of cardiac contractility by Rab4-modulated beta2-adrenergic receptor recycling. Proc Natl Acad Sci U S A 101:7082–7087
60. Palacios F, D'Souza-Schorey C (2003) Modulation of Rac1 and ARF6 activation during epithelial cell scattering. J Biol Chem 278:17395–17400
61. Palacios F, Price L, Schweitzer J, Collard JG, D'Souza-Schorey C (2001) An essential role for ARF6-regulated membrane traffic in adherens junction turnover and epithelial cell migration. EMBO J 20:4973–4986
62. Palacios F, Schweitzer JK, Boshans RL, D'Souza-Schorey C (2002) ARF6-GTP recruits Nm23-H1 to facilitate dynamin-mediated endocytosis during adherens junctions disassembly. Nat Cell Biol 4:929–936
63. Palacios F, Tushir JS, Fujita Y, D'Souza-Schorey C (2005) Lysosomal targeting of E-cadherin: a unique mechanism for the down-regulation of cell-cell adhesion during epithelial to mesenchymal transitions. Mol Cell Biol 25:389–402
64. Paterson AD, Parton RG, Ferguson C, Stow JL, Yap AS (2003) Characterization of E-cadherin endocytosis in isolated MCF-7 and Chinese hamster ovary cells: the initial fate of unbound E-cadherin. J Biol Chem 278:21050–21057
65. Pellinen T, Arjonen A, Vuoriluoto K, Kallio K, Fransen JA, Ivaska J (2006) Small GTPase Rab21 regulates cell adhesion and controls endosomal traffic of beta1-integrins. J Cell Biol 173:767–780
66. Pitcher JA, Payne ES, Csortos C, DePaoli-Roach AA, Lefkowitz RJ (1995) The G-protein-coupled receptor phosphatase: a protein phosphatase type 2A with a distinct subcellular distribution and substrate specificity. Proc Natl Acad Sci U S A 92:8343–8347
67. Polyak K, Weinberg RA (2009) Transitions between epithelial and mesenchymal states: acquisition of malignant and stem cell traits. Nat Rev Cancer 9:265–273
68. Powelka AM, Sun J, Li J, Gao M, Shaw LM, Sonnenberg A, Hsu VW (2004) Stimulation-dependent recycling of integrin beta1 regulated by ARF6 and Rab11. Traffic 5:20–36
69. Ramsay AG, Keppler MD, Jazayeri M, Thomas GJ, Parsons M, Violette S, Weinreb P, Hart IR, Marshall JF (2007) HS1-associated protein X-1 regulates carcinoma cell migration and invasion via clathrin-mediated endocytosis of integrin alphavbeta6. Cancer Res 67:5275–5284
70. Ramsay AG, Marshall JF, Hart IR (2007) Integrin trafficking and its role in cancer metastasis. Cancer Metastasis Rev 26:567–578
71. Roberts M, Barry S, Woods A, van der Sluijs P, Norman J (2001) PDGF-regulated rab4-dependent recycling of alphavbeta3 integrin from early endosomes is necessary for cell adhesion and spreading. Curr Biol 11:1392–1402
72. Rush JS, Quinalty LM, Engelman L, Sherry DM, Ceresa BP (2012) Endosomal accumulation of the activated epidermal growth factor receptor (EGFR) induces apoptosis. J Biol Chem 287:712–722
73. Sabe H, Hashimoto S, Morishige M, Hashimoto A, Ogawa E (2008) The EGFR-GEP100-Arf6 pathway in breast cancer: Full invasiveness is not from the inside. Cell Adh Migr 2:71–73
74. Sanada M, Suzuki T, Shih LY, Otsu M, Kato M, Yamazaki S, Tamura A, Honda H, Sakata-Yanagimoto M, Kumano K, Oda H, Yamagata T, Takita J, Gotoh N, Nakazaki K, Kawamata N, Onodera M, Nobuyoshi M, Hayashi Y, Harada H, Kurokawa M, Chiba S, Mori H, Ozawa K, Omine M, Hirai H, Nakauchi H, Koeffler HP, Ogawa S (2009) Gain-of-function of mutated C-CBL tumour suppressor in myeloid neoplasms. Nature 460:904–908
75. Sato K, Watanabe T, Wang S, Kakeno M, Matsuzawa K, Matsui T, Yokoi K, Murase K, Sugiyama I, Ozawa M, Kaibuchi K (2011) Numb controls E-cadherin endocytosis through p120 catenin with aPKC. Mol Biol Cell 22:3103–3119
76. Schenck A, Goto-Silva L, Collinet C, Rhinn M, Giner A, Habermann B, Brand M, Zerial M (2008) The endosomal protein App l1 mediates Akt substrate specificity and cell survival in vertebrate development. Cell 133:486–497
77. Scita G, Di Fiore PP (2010) The endocytic matrix. Nature 463:464–473
78. Seachrist JL, Anborgh PH, Ferguson SS (2000) beta 2-adrenergic receptor internalization, endosomal sorting, and plasma membrane recycling are regulated by rab GTPases. J Biol Chem 275:27221–27228

79. Shenoy SK, McDonald PH, Kohout TA, Lefkowitz RJ (2001) Regulation of receptor fate by ubiquitination of activated beta 2-adrenergic receptor and beta-arrestin. Science 294:1307–1313
80. Shi F, Sottile J (2008) Caveolin-1-dependent beta1 integrin endocytosis is a critical regulator of fibronectin turnover. J Cell Sci 121:2360–2371
81. Sigismund S, Argenzio E, Tosoni D, Cavallaro E, Polo S, Di Fiore PP (2008) Clathrin-mediated internalization is essential for sustained EGFR signaling but dispensable for degradation. Dev Cell 15:209–219
82. Sigismund S, Confalonieri S, Ciliberto A, Polo S, Scita G, Di Fiore PP (2012) Endocytosis and signaling: cell logistics shape the eukaryotic cell plan. Physiol Rev 92:273–366
83. Sorkin A, Di Fiore PP, Carpenter G (1993) The carboxyl terminus of epidermal growth factor receptor/erbB-2 chimerae is internalization impaired. Oncogene 8:3021–3028
84. Sorkin A, von Zastrow M (2009) Endocytosis and signalling: intertwining molecular networks. Nat Rev Mol Cell Biol 10:609–622
85. Tall GG, Barbieri MA, Stahl PD, Horazdovsky BF (2001) Ras-activated endocytosis is mediated by the Rab5 guanine nucleotide exchange activity of RIN1. Dev Cell 1:73–82
86. Tan YH, Krishnaswamy S, Nandi S, Kanteti R, Vora S, Onel K, Hasina R, Lo FY, El-Hashani E, Cervantes G, Robinson M, Hsu HS, Kales SC, Lipkowitz S, Karrison T, Sattler M, Vokes EE, Wang YC, Salgia R (2010) CBL is frequently altered in lung cancers: its relationship to mutations in MET and EGFR tyrosine kinases. PLoS One 5:e8972
87. Thiery JP, Sleeman JP (2006) Complex networks orchestrate epithelial-mesenchymal transitions. Nat Rev Mol Cell Biol 7:131–142
88. Tsukazaki T, Chiang TA, Davison AF, Attisano L, Wrana JL (1998) SARA, a FYVE domain protein that recruits Smad2 to the TGFbeta receptor. Cell 95:779–791
89. Tushir JS, Clancy J, Warren A, Wrobel C, Brugge JS, D'Souza-Schorey C (2010) Unregulated ARF6 activation in epithelial cysts generates hyperactive signaling endosomes and disrupts morphogenesis. Mol Biol Cell 21:2355–2366
90. Vieira AV, Lamaze C, Schmid SL (1996) Control of EGF receptor signaling by clathrin-mediated endocytosis. Science 274:2086–2089
91. Wang Y, Roche O, Yan MS, Finak G, Evans AJ, Metcalf JL, Hast BE, Hanna SC, Wondergem B, Furge KA, Irwin MS, Kim WY, Teh BT, Grinstein S, Park M, Marsden PA, Ohh M (2009) Regulation of endocytosis via the oxygen-sensing pathway. Nat Med 15:319–324
92. Wang Z, Sandiford S, Wu C, Li SS (2009) Numb regulates cell-cell adhesion and polarity in response to tyrosine kinase signalling. EMBO J 28:2360–2373
93. Weber GF, Bjerke MA, DeSimone DW (2011) Integrins and cadherins join forces to form adhesive networks. J Cell Sci 124:1183–1193
94. Wu WJ, Hirsch DS (2009) Mechanism of E-cadherin lysosomal degradation. Nat Rev Cancer 9:143; author reply 143
95. Xiao K, Allison DF, Buckley KM, Kottke MD, Vincent PA, Faundez V, Kowalczyk AP (2003) Cellular levels of p120 catenin function as a set point for cadherin expression levels in microvascular endothelial cells. J Cell Biol 163:535–545
96. Xu L, Lubkov V, Taylor LJ, Bar-Sagi D (2010) Feedback regulation of Ras signaling by Rabex-5-mediated ubiquitination. Curr Biol 20:1372–1377
97. Yan H, Jahanshahi M, Horvath EA, Liu HY, Pfleger CM (2010) Rabex-5 ubiquitin ligase activity restricts Ras signaling to establish pathway homeostasis in Drosophila. Curr Biol 20:1378–1382
98. Yoon SO, Shin S, Mercurio AM (2005) Hypoxia stimulates carcinoma invasion by stabilizing microtubules and promoting the Rab11 trafficking of the alpha6beta4 integrin. Cancer Res 65:2761–2769
99. Zhu H, Liang Z, Li G (2009) Rabex-5 is a Rab22 effector and mediates a Rab22-Rab5 signaling cascade in endocytosis. Mol Biol Cell 20:4720–4729
100. Zhu H, Qian H, Li G (2010) Delayed onset of positive feedback activation of rab5 by rabex-5 and rabaptin-5 in endocytosis. PLoS One 5:e9226

Chapter 7
Efficient Enhancement of Signaling Capacity: Signaling Endosomes

Iwona Pilecka and Marta Miaczynska

Abstract The concept of signaling endosomes originated from the observations that receptors activated at the plasma membrane can continue signaling after their endocytic internalization into endosomal compartments. In this chapter we discuss how the unique features of endosomes, such as their biochemical properties, regulated cytoskeleton-mediated transport or heterogeneity, can be exploited to modulate signal transduction. While endosomes may regulate the magnitude, kinetics, and specificity of the signals, they can also control the intracellular localization of signaling molecules and spatial signal propagation, thus contributing to cell polarization. In particular, we describe known mechanisms by which early or late endosomes act as platforms for signal propagation, diversification, or sequestration. Finally, we review some examples of how signaling endosomes contribute to the development or physiology of multicellular organisms and how aberrations in these processes may lead to pathologies.

7.1 The Signaling–Endocytosis Nexus

The mutually dependent relationship between signal transduction and endocytosis is an important feature that determines functioning of many types of transmembrane signaling receptors. Generally speaking, the process of endocytosis controls signal transduction by integrating different signaling cascades on the surface of plasma membrane and of endocytic vesicles which affects the duration, intensity, specificity, and distribution of signaling events. Frequently endocytosis is stimulated by the activation of receptors, resulting in ligand-dependent internalization of receptors and associated molecules. Obviously, endocytosis controls the receptor availability at the

I. Pilecka (✉) • M. Miaczynska
International Institute of Molecular and Cell Biology, Laboratory of Cell Biology,
4 Ks. Trojdena Street, 02-109 Warsaw, Poland
e-mail: ipilecka@iimcb.gov.pl; miaczynska@iimcb.gov.pl

Y. Yarden and G. Tarcic (eds.), *Vesicle Trafficking in Cancer*,
DOI 10.1007/978-1-4614-6528-7_7,

cell surface, thus providing a regulatory loop that prevents excessive ligand-induced activation. Eventually the internalized receptors get delivered to lysosomes for degradation (leading to the irreversible signal termination) or become recycled back to the plasma membrane (e.g., for the next round of activation). According to this classical schema that was discovered in the late 1970s and prevailed for over two decades, endocytosis plays a mostly negative but also regulatory role in signal transduction. However, since the 1990s, there has been accumulating evidence for the presence of activated receptors and associated signaling complexes on endosomal membranes in the cytoplasm. Thanks to the development of advanced resources and technologies, nowadays we gained certain insights into a complex and largely positive role of endocytosis in signaling. First, receptors trafficked through the endosomal compartments may retain or even gain signaling abilities, resulting in signal prolongation or diversification. Second, signaling molecules residing in/on endosomes are efficiently delivered to distant areas of the cell thanks to actin- and microtubule-based vesicle transport. Third, the orchestra of molecules that coordinate endocytic transport participates directly in a variety of signaling complexes, regulating such diverse processes as gene transcription, cellular division, and differentiation. Thus, endosomes can be viewed as platforms for sustained and specific signaling that add temporal and spatial dimensions to signaling downstream the plasma membrane receptors. In this chapter we will discuss in more detail various aspects of endosome-specific signaling events (recently reviewed also in [58, 65, 71, 80]).

7.2 Distinct Internalization Routes

Efficient signal transduction requires precise regulation at multiple levels. At the plasma membrane, receptor signaling is controlled by the availability of ligands and receptors as well as by distribution of signaling components. Upon ligand binding, signaling receptors residing at the plasma membrane become internalized via clathrin-mediated endocytosis or via several types of clathrin-independent endocytosis and eventually arrive in early endosomes. These represent a highly dynamic compartment from where receptors can be recycled back to the plasma membrane or sorted to late endosomes/multivesicular bodies (MVBs) and eventually degraded in lysosomes (Fig. 7.1). The critical sorting steps that coordinate the trafficking via a fast and slow recycling route or towards late endosomes are dependent on specific members of the Rab GTPase family [82]. In addition, sorting towards degradation is coordinated by the ubiquitylation and endosomal sorting complexes required for transport (ESCRT-0, ESCRT-I, ESCRT-II, and ESCRT-III) [34, 67]. The numerous examples of how the route of internalization affects the biological outcome of receptor activation have been recently reviewed [22, 65, 71]. Each internalization route promotes formation of particular signaling complexes and favors distinct signaling events. For example, it was proposed that clathrin-mediated endocytosis preferentially directs the epidermal growth factor receptor (EGFR) towards recycling and sustained signaling, while clathrin-independent pathways of endocytosis destine it for degradation [78].

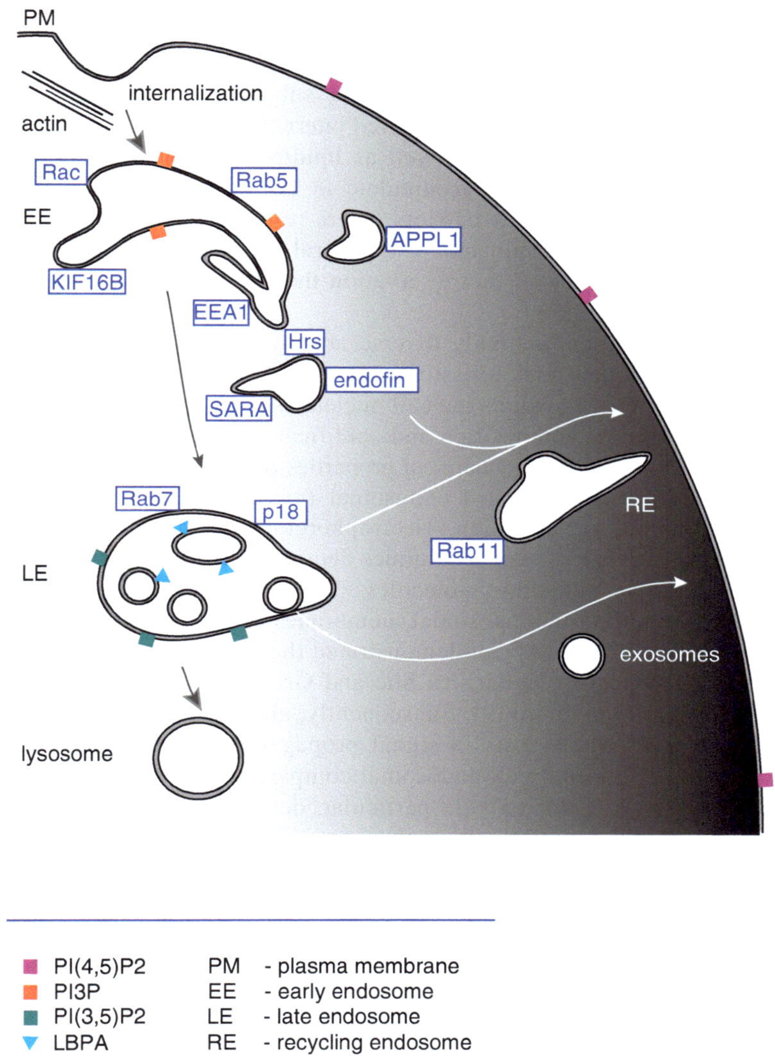

Fig. 7.1 Overview of compartments in the endocytic network with their lipid and protein markers. Selected signaling components described in the text are also indicated

Similarly, clathrin-mediated endocytosis promotes signaling of transforming growth factor-β (TGF-β), while the lipid raft–caveolar internalization pathway preferentially induces the receptor turnover [20]. The endocytosis route may determine signal specificity, as exemplified by Wnt signaling, where caveolin-dependent and clathrin-dependent internalization activate or inhibit the canonical β-catenin pathway, respectively [91]. In addition, clathrin-mediated endocytosis contributes to the noncanonical Wnt signaling in the planar cell polarization (PCP) pathway [29, 60].

7.3 Unique Features of Endosomes

Upon internalization, receptor signaling is subjected to additional control mechanisms which depend on the preferred internalization route and require the specific biochemical properties as well as lipid/protein composition of endosomes. As activated receptors accumulate in endosomes, these intracellular compartments serve as motile carriers where receptor signaling may be sustained, reactivated, and terminated. The general definition of "signaling endosomes" is a consequence of the observation that receptors continue to signal after internalization.

Endosomes represent a highly diverse and dynamic endomembrane compartment, with frequent fusion/fission events, where a general membrane flow from the cell periphery towards the cell nucleus is accompanied by progressive changes in membrane composition and acidification of endosomal lumen [21, 38]. Thanks to the distinct biochemical properties of endosomal subpopulations; it is feasible to obtain enriched endosomal fractions using density gradient ultracentrifugation, gel filtration, electrophoretic, and immunoaffinity techniques [28, 57]. Usage of such techniques demonstrated the existence of endosome-bound active signaling molecules. The early evidence for signaling complexes assembling on endosomal membranes came from subcellular fractionation studies of rat liver that demonstrated the presence of activated EGFR and associated signaling molecules Shc and Grb2 on endosomal membranes following EGF stimulation [19]. Subsequently, abundant evidence established a positive role of endocytosis in signal propagation and diversification and demonstrated that virtually all endosomal compartments are signaling competent (reviewed in [22, 54, 65]). In particular, development of high-resolution microscopy provided multiple examples of signaling functions of endosomes (for review, see [52]). On the other hand, signaling of activated receptors from endosomes may not always be an absolute requirement for their biological function, as recent reports show that the transcriptional response as well as MAPK and AKT activation are primarily mediated by activated EGFR at the cell surface [5, 81].

The important features of endosomes that specifically affect signaling are (a) unique lipid and protein composition in comparison to the plasma membrane; (b) progressive acidification of endosomal lumen occurring during endosomal maturation; (c) small volume and limited surface area that favor assembly of multiprotein complexes containing active receptors, adaptors and signaling molecules, as well as support multiple weak intermolecular interactions; (d) intracellular actin- and microtubule-mediated transport that enables rapid and controlled transmission of endosome-derived signals; (e) existence of heterogeneous endosomal populations in terms of morphology, localization, composition, and function. Below we will present selected examples showing the importance of two of these unique features.

7.3.1 Specific Lipid/Protein Composition of Endosomes

Endosomal membranes are differentially enriched in particular phosphoinositides that contribute to compartmental identity and anchor a number of phosphoinositide-binding proteins [31, 41, 51] (Fig. 7.1). Thanks to its association with several phosphoinositide kinases and phosphatases, Rab5 GTPase regulates both generation and turnover of these phospholipids and thus coordinates sorting and signaling events [77]. Dynamic changes in lipid composition of endosomal membranes occurs either via fusion with preexisting compartments or through vesicle maturation. Typically, early endosomes are enriched in phosphatidylinositol 3-phosphate (PI3P), which in turn becomes converted to phosphatidylinositol 3,5-bisphosphate (PI(3,5)P_2), prevalent in late endosomes. In early endosomes, the Rab5 effector phosphatidylinositol 3-kinase (PI3K) VPS34 is responsible for PI3P generation. The presence of PI3P is required for recruitment of EEA1, Rabenosyn-5, sorting nexins (SNXs), and the plus-end-directed kinesin-3 motor KIF16B to early endosomes [41]. The PI3P level is controlled through the lipid kinase PIKFyve and the lipid phosphatase MTM1 [41]. Vesicles maturing to late endosomes contain PI(3,5)P_2, which is necessary for retrograde membrane trafficking to the trans-Golgi network, but not required for degradative sorting of receptors [70]. Additionally, the internal membranes of late endosomes contain lysobisphosphatidic acid (LBPA), which controls membrane invagination and endosome biogenesis [50].

7.3.2 Endosomal pH and Ionic Content

Endosomal transport depends on the gradual acidification of endosomal compartments, ranging from 6.8–6.1 in early endosomes to 6.0–4.8 in late endosomes and 4.5 in lysosomes [38]. The acidic pH of late endosomes affects the fate of EGFR in a ligand-specific manner [69]. Ligands such as EGF, heparin-binding EGF-like growth factor (HB-EGF), and betacellulin (BTC) remain bound to the receptor, which targets EGFR for lysosomal degradation. In contrast, acidic pH causes dissociation of TGF-α from EGFR already in early endosomes, which enables recycling of unoccupied receptors to the surface and prolonged receptor phosphorylation and signaling. Importantly, the ability of certain ligands to induce receptor recycling rather than degradation affects downstream signaling and can partially determine oncogenic potential of the various EGFR ligands [69].

Simultaneously but independently of acidification, endosomal changes in the luminal concentrations of chloride, sodium, or potassium ions may directly affect the regulation and function of endocytic processes [75]. The ionic environment can regulate the membrane curvature and fusion events and eventually affect the fate of the transported cargo.

7.4 Multifaceted Signaling from Endosomes

Over the past years, substantial experimental evidence proved the involvement of endosomes in the regulation of signal magnitude, kinetics and specificity, as well as in the control of spatial signal propagation and cell polarity. Below we will summarize the current knowledge on endosomal contribution to signal transduction on the molecular level.

7.4.1 Signal Propagation

The collective action of the essential signaling molecules, which are shared between the plasma membrane- and endosome-originating complexes, ensures the proper duration and intensity of the signal. There are multiple examples of receptors from the GPCR (G protein-coupled receptor) and RTK (receptor tyrosine kinase) families that can continue signaling from endosomal compartments. Certain GPCRs remain associated with β-arrestins at the plasma membrane and endosomes, with the contribution of endosomal signaling proportional to the residence time in endosomes [7, 94]. Similarly, the ligand-bound and phosphorylated EGFR has been detected on early and late endosomes, together with all the components of the ERK/MAPK signaling cascades [61]. Of note, different cellular responses may be triggered by apparently the same signaling events originating with different kinetics from the cell surface and from intracellular sites. Growth factors induce acute MAPK activation from the plasma membrane and more sustained from endosomes. In PC12 cells, stimulation with nerve growth factor (NGF) leads to different outcomes depending on the location of Trk receptors: activated receptors at the cell surface promote prolonged AKT activation and proliferation, while catalytically active receptors on endosomes initiate neuronal differentiation [93]. Also, thyroid-stimulating hormone (TSH) receptor and parathyroid hormone (PTH) receptor continue to signal to $G\alpha_s$ and adenylyl cyclases after internalization, leading to persistent cAMP production [8, 26] (Fig. 7.2a). Signaling from internalized TSH receptors, likely located in the perinuclear endosomal recycling compartment, is required for VASP phosphorylation and actin rearrangement [8].

7.4.2 Signal Diversification

Specific signaling pathways originating from various endosomal populations were studied for several receptor systems, including RTKs, GPCRs, tumor necrosis factor receptor (TNFR1), TGF-β receptor family, Toll-like receptors, and Notch [58]. Below we will describe some available examples of unique signals originating from early and late endosomes. Importantly, these endosomal compartments are largely heterogeneous, often with particular proteins acting as markers of separate subpopulations of endosomes, trafficking intermediates, or membrane subdomains.

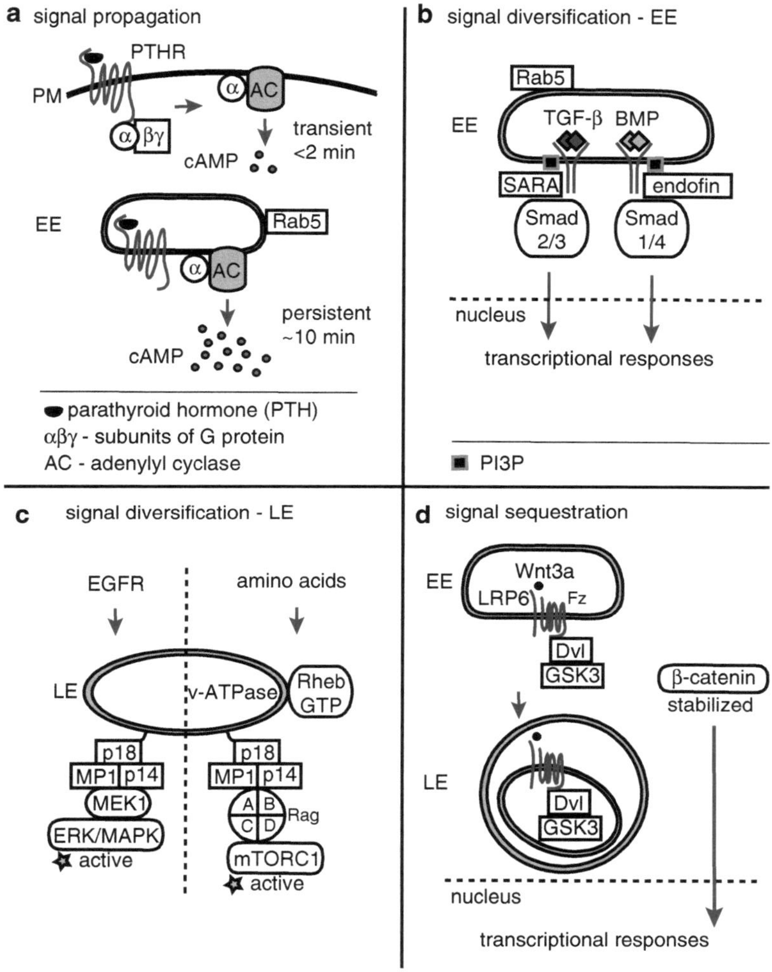

Fig. 7.2 Selected examples of endosomal contribution to signal propagation (**a**), signal diversification in early (**b**) or late endosomes (**c**), and signal sequestration (**d**). For details, see the text

7.4.2.1 Early Endosomes

Early endosomes serve as the first sorting compartment for receptors internalized via clathrin-dependent and clathrin-independent pathways. Characteristically, the membrane of early endosomes is enriched in Rab5 and PI3P, the latter recruiting proteins that contain FYVE and PX domains [46].

One example of the signaling role of FYVE domain-containing proteins is a family of endosome-associated ubiquitin ligases CARPs. They have been implicated in diverse signaling events, such as regulation of endosomal recycling [16], degradation of p53 [92], and negative regulation of TNF-induced NF-κB activation [47].

Another example is represented by three FYVE domain-containing proteins, SARA, endofin, and Hrs, that are positive regulators of signaling stimulated by TGF-β and related growth factor family, including activin and bone morphogenetic proteins (BMPs). Endosomal SARA, endofin, and Hrs have been shown to facilitate formation of complexes containing type I and II receptors and Smads and to promote signaling after stimulation with TGF-β [12, 55, 87] (Fig. 7.2b). As a consequence, clathrin-mediated endocytosis is required for the phosphorylation and thus activation of Smad2/3, nuclear accumulation of Smads, and efficient Smad-dependent transcriptional responses [11]. Also, affecting Rab5 function by overexpression of Rab5-S34N, a GDP-bound dominant-negative mutant, increases the expression of a Smad-dependent promoter in a ligand-independent manner, while the GTP-bound Rab5-Q79L mutant attenuates the transcriptional activity induced by activin [63]. On the other hand, blocking clathrin-mediated endocytosis, mislocalization of SARA, or disturbing the binding between SARA and Smad2 do not inhibit Smad2 activation [48].

A subpopulation of Rab5-positive early endocytic vesicles enriched in adaptor protein APPL1, termed APPL endosomes, is endowed with signaling properties [18, 53] (Fig. 7.1). These endosomes are involved in signaling from a variety of receptors, including RTKs, and serve as a scaffold for AKT2 and PI3K subunits [18]. APPL1 is required for proper activation of the anti-apoptotic AKT–GSK3β signaling axis [74], for GLUT4 translocation and glucose uptake after stimulation by insulin [72], and for the cross talk between adiponectin and insulin signaling pathways [49]. Eventual enrichment in PI3P causes dissociation of APPL1, recruitment of EEA1 protein, and further maturation of such vesicles into late endosomes [68, 96].

Cumulatively, available data position early endosomes among the multiple complementary stations that contribute to the maximal activation of vital signaling systems.

7.4.2.2 Late Endosomes

Late endosomes or MVBs have characteristic morphology and are marked by the presence of Rab7, LAMP-1, and LBPA [32].

MAPK signaling is sustained in late endosomes, thanks to the endosome-specific lipid raft adaptor protein p18 that anchors the p14–MP1–MEK1 complex [59, 90] (Fig. 7.2c). EGF-dependent activation of MEK was significantly suppressed in $p18^{-/-}$ cells, since approximately half of the total MEK–ERK activity in cells engages the endosomal p18–p14–MP1 scaffold. Knockdown of p14 or MP1 in cultured cells decreased the basal activity of MEK1/2 and ERK1/2 [86]. The endosomal MAPK signaling may have regulatory function in late endosome-specific membrane dynamics [85]. Interestingly, the endosomal p18–p14–MP1 complex has been also implicated in amino acid-specific activation of the mTORC1 pathway [73] (Fig. 7.2c). Termed Ragulator, this trimeric protein complex recruits the Rag GTPases and mTORC1 to the lysosomal surface, which promotes cell growth in response to amino acids [73]. Interestingly, amino acids must accumulate in the lysosomal lumen to initiate signaling, and the vacuolar H^+-adenosine triphosphatase ATPase (v-ATPase) is required for Rag and mTORC1 activation [95].

Also, the sustained activation of Rap1 small G protein on late endosomes prolongs the activation time of the MAPK cascade and results in the upregulation of gene expression. Stimulation of TrkA neurotrophin receptor with NGF results in initial activation of Rap1 on early endosomes via a specific GDP/GTP exchange factor called C3G and subsequent transport of GTP-Rap1 in complex with TrkA receptor to late endosomes, where it recruits PDZ-GEF1 [35]. Late endosomes provide platform for the formation of a tetrameric complex containing TrkA receptor, PDZ-GEF1, synaptic scaffolding molecule (S-SCAM), and ankyrin repeat-rich membrane spanning protein (ARMS). Such complex induces sustained activation of Rap1 and ERK that leads to neurite outgrowth [35].

7.4.3 Signal Termination

Endocytosis can attenuate signaling at various levels. First, internalization may separate the receptors from signaling mediators localized at the plasma membrane and absent in endosomes (e.g., phosphatidylinositol 4,5-bisphosphate or phospholipase Cγ). In the case of receptor trafficking to late endosomes, prior to their lysosomal degradation, their signaling capabilities are terminated after they are sorted into the lumen of the MVBs by the coordinated action of the ESCRT complexes and the ubiquitin system [67].

7.4.4 Sequestration of Signaling Components

In contrast to the above-mentioned sequestration of activated receptors within the MVB lumen leading to signal termination, analogous sequestration of an inhibitor would lead to enhanced signaling. This has been elegantly demonstrated for the canonical Wnt pathway, where upon stimulation by Wnt, its receptors Frizzled and LRP6 become internalized and associate with the GSK3 kinase. In the absence of Wnt ligands, the cytosolic GSK3 phosphorylates β-catenin and promotes its degradation, thus preventing signaling. Upon stimulation, sequestration of GSK3 together with Wnt-bound receptor complexes into internal vesicles of late endosomes in an ESCRT-dependent manner results in the accumulation of β-catenin, thus activating signaling [83] (Fig. 7.2d).

7.4.5 Intracellular Transport

Retrograde transport of signaling endosomes enables effective transmission of information from its origin at the plasma membrane to its effectors in the cell nucleus. Specific features of endosomes rely not only on a unique protein and lipid composition and intraluminal pH but also on their localization within the cell. Altered residence of signaling molecules within distinct endocytic compartments could underlie pathologies involving aberrant activation of a variety of signaling pathways.

According to the large-scale simulations of diffusing STAT3 molecules coupled with probabilistic modeling of dephosphorylation kinetics, signal propagation over the distances greater than 200 nm from the plasma membrane preferentially utilizes vesicular transport rather than spreading of signaling molecules by free diffusion [37]. Experimental work on localization of STAT3 and tyrosine phosphorylated STAT3 supports this model, pointing to early endosomes serving as a carrier delivering STAT3 from the plasma membrane to the cell nucleus [76]. In another study, transcriptional activity of STAT3 downstream of hepatocyte growth factor (HGF) requires endosomal trafficking of active Met receptors to the perinuclear region [43]. Overall, endosomes provide a regulatory role enhancing relatively weak signals initiated by some stimuli (i.e., interleukin-6 or HGF), while strong activation of STAT3 via the cytokine oncostatin M occurs independently of endosomal transport [43].

By modulating the level of the PX domain-containing kinesin KIF16B it was possible to relocate early endosomes to the cell periphery, which slowed down degradation of EGFR, or to cluster them in the perinuclear region, which accelerated EGFR degradation [36]. Interestingly, forced localization of late endosomes to the cell periphery or the perinuclear region also affects the kinetics of EGF receptor degradation and signaling [84]. In particular, the peripheral translocation of late endosomes resulted in prolonged EGFR activation, sustained ERK and p38 signaling, and hyperactivation of the nuclear target Elk-1. Conversely, induced clustering of late endosomes in the perinuclear region delayed EGFR degradation and sustained MAPK signaling but reduced the activation of nuclear targets [84]. In conclusion, the spatial organization of endosomes affects EGFR trafficking and degradation kinetics and thus fine-tunes MAPK signaling.

7.4.6 Exosomes and Intercellular Communication

Exosomes are extracellular vesicles (40–100 nm diameter) that origin from intraluminal vesicles of MVBs and can be secreted upon fusion of MVBs with the plasma membrane (reviewed in [27, 79]). Exosomal release may be an alternative to lysosomal processing but may also provide means of cell–cell communication. Exosomes can directly fuse with the plasma membrane or can be internalized by the recipient cell. They are able to propagate genetic material [88] as well as oncogenes and their associated transforming phenotypes [2]. Furthermore, exosomes participate in modulation of the immune response, dissemination of viral particles and prions, and pathogenesis of neurodegenerative diseases [79].

7.4.7 Endocytic Proteins in Nuclear Signaling

Membrane sorting and cargo transport within the endocytic system is governed by a multitude of adaptor and accessory proteins. Intriguingly, some of these proteins have been observed in the cell nucleus, where they may regulate the activity of various nuclear factors. Endosomal proteins can act as transcriptional coactivators and corepressors through at least two mechanisms (reviewed in [64, 66]): (1) acting at the regulatory step of transcription initiation by affecting the activity (e.g., TSG101), stability (e.g., HIP1), or localization (e.g., β-arrestins) of transcription factors and (2) functioning in chromatin modification/remodeling (i.e., APPL1/2 and ESCRT-III proteins). Such functions of endocytic adaptors may imply that they can act as scaffold platforms not only in endocytic compartments but also in the nucleus, recruiting transcription factors, and cofactors. Still, the functional interdependence between the nuclear and endocytic roles of the same protein is in most cases not clear.

7.5 Physiological Relevance of Signaling Endosomes

Mutual dependence between endocytosis and signaling has functional consequences at the level of individual cells, affecting their differentiation, proliferation, survival, motility, and metabolism. Proper coordination of these processes warrants correct development and functioning of tissues and whole organisms, while any aberrations may lead to pathologies. Among the best described examples of physiological processes affected by endosomal signaling are embryogenesis, cell migration, neurotrophin action within the nervous system, and metabolic regulation (Fig. 7.3) which we will review below. Dysfunction of each of these processes can result in diseases. Aberrations in embryonic development or cell migration may lead to tumorigenesis and tumor invasion, abnormal neuronal signaling—to neurodegeneration, while improper metabolic regulation may cause metabolic diseases, including diabetes (Fig. 7.3). Importantly, some of these pathologies can be at least in part attributed to the disruption of signaling cascades which at the cellular level take place on endosomes. In the following paragraphs we will present selected paradigms illustrating the role of signaling endosomes at the level of a cell and an organism, with relevance to pathogenesis.

7.5.1 Embryonic Development and Cell Fate Determination

A striking example of how signaling endosomes are exploited to determine cell fate in embryonic development is provided by an asymmetric distribution of SARA endosomes observed during division of sensory organ precursors (SOP) cells in Drosophila [17]. SARA endosomes, named after their marker protein, can carry various signaling molecules such as TGF-β and its receptors (termed Dpp and Tkv

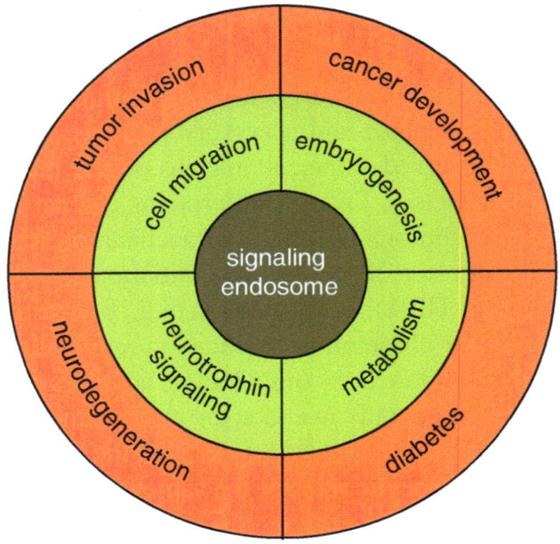

Fig. 7.3 Schematic representation of the physiologically relevant functions of signaling endosomes (in *green*) and example pathologies associated with their dysfunction (in *red*). More details in the text

in flies, respectively) [4] or Delta–Notch complexes [17]. SARA endosomes can associate with the mitotic spindle during mitosis which regulates their inheritance by the daughter cells. While in the developing wing epithelial cells, SARA endosomes carrying Dpp-Tkv are partitioned symmetrically to provide equal amounts of signaling molecules to the daughters [4]; during the sensory organ biogenesis, SARA endosomes ferrying Delta–Notch complexes are distributed asymmetrically [17]. Specifically, during the division of the SOP pI cell, SARA endosomes are inherited only by one of the daughter cells, termed pIIa. This mechanism leads to the activation of Notch signaling in the pIIa and its suppression in the pIIb daughter cell, thus guiding their further differentiation. Experimental mistargeting of SARA endosomes causes abnormalities in the development of sensory organ, arguing that signaling endosomes contribute to cell fate determination. Importantly, asymmetric cell division is crucial for maintenance of stem cells, including cancer stem cells, and therefore shifting the balance between the asymmetric and symmetric cell divisions was shown to contribute to cancer development in humans [13].

7.5.2 Cell Migration and Invasion in Tumorigenesis

Endocytosis and endocytic compartments may contribute to various stages of tumorigenesis in a number of ways which are comprehensively covered in some excellent recent reviews [1, 45, 56]. In brief, endocytic processes regulate establishment and maintenance of cell polarity, cell–cell contacts and adhesion, cell migration, and proliferation and cell fate determination, all of which are frequently aberrant in tumorigenesis. In particular, the significance of signaling endosomes has been well documented in the regulation of cell migration and invasion. Signaling endosomes carrying active Rac and their polarized delivery to the plasma membrane have been shown to direct localized formation of actin-rich protrusions and drive migration, also in cancer cells [62]. Similarly, polarized trafficking of late endosomes carrying active integrins to the rear of migrating cells is crucial for invasion of cancer cells [23].

7.5.3 Neurotrophin Signaling and Neurodegeneration

Some of the earliest examples of signaling endosomes refer to the compartments transporting neurotrophins, such as NGF, in neurons [30]. In these cells, NGF binds its receptor TrkA in axon terminals and undergoes internalization into endosomes. Subsequently, endosomes ferrying active NGF–TrkA complexes are transported retrogradely along the microtubules to the cell body, to elicit pro-survival signaling [15, 37]. Intriguingly, the ability to generate signaling endosomes depends on the neurotrophin ligand rather than the TrkA receptor itself. Neurotrophin NT3 also binds TrkA but it does not induce retrograde endosomal transport and pro-survival signaling. As explained recently, NT3–TrkA complexes are dissociated in the acidic pH of endosomes, while NGF–TrkA complexes remain stable and induce Rac1–cofilin signaling, leading to actin depolymerization around the endosomes and their retrograde transport [33].

Intriguingly, aberrations in the endosomal system are frequently observed in neurodegenerative syndromes, such as Alzheimer's disease, Huntington's disease, or Charcot–Marie–Tooth neuropathies (reviewed in [6]). Among others, endosomes contribute to processing of amyloid precursor protein (APP), as one of the secretases responsible for APP cleavage (β-site APP-cleaving enzyme or BACE) is localized to endosomes [25, 89]. Indeed, a number of reports document abnormally enlarged early endosomes in patients with Alzheimer's disease or Down syndrome [9, 10, 39, 40]. While it is difficult to separate the roles of endosomes in trafficking and in signaling, it remains possible that aberrant endosomal signaling contributes to some neuropathologies.

7.5.4 Metabolic Regulation

The endosomal system is obviously linked to various aspects of metabolic regulation. As already mentioned under Sect. 7.4.2.2, late endosomes and lysosomes serve as amino acid-sensing platforms for mTORC1 signaling in response to the nutrient status of the cell [73, 95]. Moreover, mTOR is a key cellular regulator of autophagy and lysosomes participate in the degradation of autophagic substrates [42]. Intriguingly, nutrient- and pH-dependent intracellular positioning of lysosomes was recently shown to coordinate mTOR signaling and autophagy [44]. Under nutrient rich conditions, lysosomes are localized peripherally, concomitant with mTORC1 activation and suppression of autophagy. Under starvation, increase in the intracellular pH induces clustering of lysosomes in the perinuclear region, promoting generation of autophagosomes and their fusion with lysosomes.

Another aspect of metabolic regulation involving endosomal organelles is insulin signaling. Insulin stimulates translocation of GLUT4 glucose transporters from endosomal storage vesicles to the cell surface to promote glucose uptake, followed afterwards by GLUT4 internalization [3]. Cycling of GLUT4-containing vesicles requires coordination between exocytosis, endocytosis, and recycling which is regulated by PI3K, AKT, and atypical protein kinase C (aPKC) signaling [24]. Ongoing efforts to decipher the exact targets of insulin signaling on GLUT4 endosomal vesicles will contribute to our understanding of diabetes and related pathologies.

Finally, many more regulatory links between metabolic signaling and endocytosis are bound to be discovered. In a recent genome-wide siRNA screen, the loss of function of many metabolic enzymes exhibited endocytic phenotypes, arguing that they impact, directly or indirectly, endocytic processes [14]. It remains an interesting possibility that some of the metabolic processes are compartmentalized on endosomes, in a way similar to mTORC1 signaling from lysosomes.

7.6 Conclusions and Perspectives

In the "signaling endosome" hypothesis, endocytic compartments provide intracellular platforms that generate quantitative and qualitative differences in signals. While it is likely that not under all circumstances the signaling from endosomes is crucial, it provides additional means to ensure the specificity of signaling. This is particularly important considering that several pathways, although evoking different responses, employ many common signaling effectors. If compartmentalized on different populations of endosomes, such effectors face different molecular environments and certain interactions with other membrane-localized molecules (both proteins and lipids) may be specifically favored. Moreover, the types of endosomes and their composition vary in different cell types or in the course of development and tissue differentiation. Such functional diversity of endosomal compartments can be therefore exploited to modulate tissue-specific or developmental signaling. We anticipate that future development of supra-resolution imaging techniques and specific endosome-localized probes will deliver further examples for the contribution of the endosomal network to the regulation of intracellular signaling. Considering a very broad involvement of endocytosis in multiple aspects of physiology and pathology, it is conceivable that endosomal signaling may be exploited as a future target for selective therapeutic approaches.

References

1. Abella JV, Park M (2009) Breakdown of endocytosis in the oncogenic activation of receptor tyrosine kinases. Am J Physiol Endocrinol Metab 296(5):E973–E984. doi:10.1152/ajpendo.90857.2008
2. Al-Nedawi K, Meehan B, Micallef J, Lhotak V, May L, Guha A, Rak J (2008) Intercellular transfer of the oncogenic receptor EGFRvIII by microvesicles derived from tumour cells. Nat Cell Biol 10(5):619–624. doi:10.1038/ncb1725
3. Bogan JS (2012) Regulation of glucose transporter translocation in health and diabetes. Annu Rev Biochem 81:507–532. doi:10.1146/annurev-biochem-060109-094246
4. Bokel C, Schwabedissen A, Entchev E, Renaud O, Gonzalez-Gaitan M (2006) Sara endosomes and the maintenance of Dpp signaling levels across mitosis. Science 314(5802):1135–1139. doi:10.1126/science.1132524
5. Brankatschk B, Wichert SP, Johnson SD, Schaad O, Rossner MJ, Gruenberg J (2012) Regulation of the EGF transcriptional response by endocytic sorting. Sci Signal 5(215):ra21. doi:10.1126/scisignal.2002351
6. Bronfman FC, Escudero CA, Weis J, Kruttgen A (2007) Endosomal transport of neurotrophins: roles in signaling and neurodegenerative diseases. Dev Neurobiol 67(9):1183–1203. doi:10.1002/dneu.20513
7. Bunnett NW, Cottrell GS (2010) Trafficking and signaling of G protein-coupled receptors in the nervous system: implications for disease and therapy. CNS Neurol Disord Drug Targets 9(5):539–556
8. Calebiro D, Nikolaev VO, Gagliani MC, de Filippis T, Dees C, Tacchetti C, Persani L, Lohse MJ (2009) Persistent cAMP-signals triggered by internalized G-protein-coupled receptors. PLoS Biol 7(8):e1000172. doi:10.1371/journal.pbio.1000172
9. Cataldo A, Rebeck GW, Ghetri B, Hulette C, Lippa C, Van Broeckhoven C, van Duijn C, Cras P, Bogdanovic N, Bird T, Peterhoff C, Nixon R (2001) Endocytic disturbances distinguish among subtypes of Alzheimer's disease and related disorders. Ann Neurol 50(5):661–665
10. Cataldo AM, Petanceska S, Terio NB, Peterhoff CM, Durham R, Mercken M, Mehta PD, Buxbaum J, Haroutunian V, Nixon RA (2004) Abeta localization in abnormal endosomes: association with earliest Abeta elevations in AD and Down syndrome. Neurobiol Aging 25(10):1263–1272. doi:10.1016/j.neurobiolaging.2004.02.027
11. Chen YG (2009) Endocytic regulation of TGF-beta signaling. Cell Res 19(1):58–70. doi:10.1038/cr.2008.315
12. Chen YG, Wang Z, Ma J, Zhang L, Lu Z (2007) Endofin, a FYVE domain protein, interacts with Smad4 and facilitates transforming growth factor-beta signaling. J Biol Chem 282(13):9688–9695. doi:10.1074/jbc.M611704200
13. Cicalese A, Bonizzi G, Pasi CE, Faretta M, Ronzoni S, Giulini B, Brisken C, Minucci S, Di Fiore PP, Pelicci PG (2009) The tumor suppressor p53 regulates polarity of self-renewing divisions in mammary stem cells. Cell 138(6):1083–1095. doi:10.1016/j.cell.2009.06.048
14. Collinet C, Stoter M, Bradshaw CR, Samusik N, Rink JC, Kenski D, Habermann B, Buchholz F, Henschel R, Mueller MS, Nagel WE, Fava E, Kalaidzidis Y, Zerial M (2010) Systems survey of endocytosis by multiparametric image analysis. Nature 464(7286):243–249. doi:10.1038/nature08779
15. Cosker KE, Courchesne SL, Segal RA (2008) Action in the axon: generation and transport of signaling endosomes. Curr Opin Neurobiol 18(3):270–275. doi:10.1016/j.conb.2008.08.005
16. Coumailleau F, Das V, Alcover A, Raposo G, Vandormael-Pournin S, Le Bras S, Baldacci P, Dautry-Varsat A, Babinet C, Cohen-Tannoudji M (2004) Over-expression of Rififylin, a new RING finger and FYVE-like domain-containing protein, inhibits recycling from the endocytic recycling compartment. Mol Biol Cell 15(10):4444–4456. doi:10.1091/mbc.E04-04-0274
17. Coumailleau F, Furthauer M, Knoblich JA, Gonzalez-Gaitan M (2009) Directional Delta and Notch trafficking in Sara endosomes during asymmetric cell division. Nature 458(7241):1051–1055. doi:10.1038/nature07854

18. Deepa SS, Dong LQ (2009) APPL1: role in adiponectin signaling and beyond. Am J Physiol Endocrinol Metab 296(1):E22–E36. doi:10.1152/ajpendo.90731.2008
19. Di Guglielmo GM, Baass PC, Ou WJ, Posner BI, Bergeron JJ (1994) Compartmentalization of SHC, GRB2 and mSOS, and hyperphosphorylation of Raf-1 by EGF but not insulin in liver parenchyma. EMBO J 13(18):4269–4277
20. Di Guglielmo GM, Le Roy C, Goodfellow AF, Wrana JL (2003) Distinct endocytic pathways regulate TGF-beta receptor signalling and turnover. Nat Cell Biol 5(5):410–421. doi:10.1038/ncb975
21. Dikic I (2006) Endosomes. Molecular Biology Intelligence Unit. Landes Bioscience/Springer, Austin, TX
22. Disanza A, Frittoli E, Palamidessi A, Scita G (2009) Endocytosis and spatial restriction of cell signaling. Mol Oncol 3(4):280–296. doi:10.1016/j.molonc.2009.05.008
23. Dozynkiewicz MA, Jamieson NB, Macpherson I, Grindlay J, van den Berghe PV, von Thun A, Morton JP, Gourley C, Timpson P, Nixon C, McKay CJ, Carter R, Strachan D, Anderson K, Sansom OJ, Caswell PT, Norman JC (2012) Rab25 and CLIC3 collaborate to promote integrin recycling from late endosomes/lysosomes and drive cancer progression. Dev Cell 22(1):131–145. doi:10.1016/j.devcel.2011.11.008
24. Dugani CB, Klip A (2005) Glucose transporter 4: cycling, compartments and controversies. EMBO Rep 6(12):1137–1142. doi:10.1038/sj.embor.7400584
25. Evin G, Barakat A, Masters CL (2010) BACE: therapeutic target and potential biomarker for Alzheimer's disease. Int J Biochem Cell Biol 42(12):1923–1926. doi:10.1016/j.biocel.2010.08.017
26. Ferrandon S, Feinstein TN, Castro M, Wang B, Bouley R, Potts JT, Gardella TJ, Vilardaga JP (2009) Sustained cyclic AMP production by parathyroid hormone receptor endocytosis. Nat Chem Biol 5(10):734–742. doi:10.1038/nchembio.206
27. Fevrier B, Raposo G (2004) Exosomes: endosomal-derived vesicles shipping extracellular messages. Curr Opin Cell Biol 16(4):415–421. doi:10.1016/j.ceb.2004.06.003
28. Futter CE, Hopkins CR (1989) Subfractionation of the endocytic pathway: isolation of compartments involved in the processing of internalised epidermal growth factor-receptor complexes. J Cell Sci 94(pt 4):685–694
29. Gagliardi M, Piddini E, Vincent JP (2008) Endocytosis: a positive or a negative influence on Wnt signalling? Traffic 9(1):1–9. doi:10.1111/j.1600-0854.2007.00662.x
30. Grimes ML, Zhou J, Beattie EC, Yuen EC, Hall DE, Valletta JS, Topp KS, LaVail JH, Bunnett NW, Mobley WC (1996) Endocytosis of activated TrkA: evidence that nerve growth factor induces formation of signaling endosomes. J Neurosci 16(24):7950–7964
31. Gruenberg J (2003) Lipids in endocytic membrane transport and sorting. Curr Opin Cell Biol 15(4):382–388
32. Gruenberg J, Stenmark H (2004) The biogenesis of multivesicular endosomes. Nat Rev Mol Cell Biol 5(4):317–323. doi:10.1038/nrm1360
33. Harrington AW, St Hillaire C, Zweifel LS, Glebova NO, Philippidou P, Halegoua S, Ginty DD (2011) Recruitment of actin modifiers to TrkA endosomes governs retrograde NGF signaling and survival. Cell 146(3):421–434. doi:10.1016/j.cell.2011.07.008
34. Henne WM, Buchkovich NJ, Emr SD (2011) The ESCRT pathway. Dev Cell 21(1):77–91. doi:10.1016/j.devcel.2011.05.015
35. Hisata S, Sakisaka T, Baba T, Yamada T, Aoki K, Matsuda M, Takai Y (2007) Rap1-PDZ-GEF1 interacts with a neurotrophin receptor at late endosomes, leading to sustained activation of Rap1 and ERK and neurite outgrowth. J Cell Biol 178(5):843–860. doi:10.1083/jcb.200610073
36. Hoepfner S, Severin F, Cabezas A, Habermann B, Runge A, Gillooly D, Stenmark H, Zerial M (2005) Modulation of receptor recycling and degradation by the endosomal kinesin KIF16B. Cell 121(3):437–450. doi:10.1016/j.cell.2005.02.017
37. Howe CL (2005) Modeling the signaling endosome hypothesis: why a drive to the nucleus is better than a (random) walk. Theor Biol Med Model 2:43. doi:10.1186/1742-4682-2-43
38. Huotari J, Helenius A (2011) Endosome maturation. EMBO J 30(17):3481–3500. doi:10.1038/emboj.2011.286

39. Israel MA, Yuan SH, Bardy C, Reyna SM, Mu Y, Herrera C, Hefferan MP, Van Gorp S, Nazor KL, Boscolo FS, Carson CT, Laurent LC, Marsala M, Gage FH, Remes AM, Koo EH, Goldstein LS (2012) Probing sporadic and familial Alzheimer's disease using induced pluripotent stem cells. Nature 482(7384):216–220. doi:10.1038/nature10821
40. Jiang Y, Mullaney KA, Peterhoff CM, Che S, Schmidt SD, Boyer-Boiteau A, Ginsberg SD, Cataldo AM, Mathews PM, Nixon RA (2010) Alzheimer's-related endosome dysfunction in Down syndrome is Abeta-independent but requires APP and is reversed by BACE-1 inhibition. Proc Natl Acad Sci U S A 107(4):1630–1635. doi:10.1073/pnas.0908953107
41. Jovic M, Sharma M, Rahajeng J, Caplan S (2010) The early endosome: a busy sorting station for proteins at the crossroads. Histol Histopathol 25(1):99–112
42. Jung CH, Ro SH, Cao J, Otto NM, Kim DH (2010) mTOR regulation of autophagy. FEBS Lett 584(7):1287–1295. doi:10.1016/j.febslet.2010.01.017
43. Kermorgant S, Parker PJ (2008) Receptor trafficking controls weak signal delivery: a strategy used by c-Met for STAT3 nuclear accumulation. J Cell Biol 182(5):855–863. doi:10.1083/jcb.200806076
44. Korolchuk VI, Saiki S, Lichtenberg M, Siddiqi FH, Roberts EA, Imarisio S, Jahreiss L, Sarkar S, Futter M, Menzies FM, O'Kane CJ, Deretic V, Rubinsztein DC (2011) Lysosomal positioning coordinates cellular nutrient responses. Nat Cell Biol 13(4):453–460. doi:10.1038/ncb2204
45. Lanzetti L, Di Fiore PP (2008) Endocytosis and cancer: an 'insider' network with dangerous liaisons. Traffic 9(12):2011–2021. doi:10.1111/j.1600-0854.2008.00816.x
46. Lemmon MA (2008) Membrane recognition by phospholipid-binding domains. Nat Rev Mol Cell Biol 9(2):99–111. doi:10.1038/nrm2328
47. Liao W, Xiao Q, Tchikov V, Fujita K, Yang W, Wincovitch S, Garfield S, Conze D, El-Deiry WS, Schutze S, Srinivasula SM (2008) CARP-2 is an endosome-associated ubiquitin ligase for RIP and regulates TNF-induced NF-kappaB activation. Curr Biol 18(9):641–649. doi:10.1016/j.cub.2008.04.017
48. Lu Z, Murray JT, Luo W, Li H, Wu X, Xu H, Backer JM, Chen YG (2002) Transforming growth factor beta activates Smad2 in the absence of receptor endocytosis. J Biol Chem 277(33):29363–29368. doi:10.1074/jbc.M203495200
49. Mao X, Kikani CK, Riojas RA, Langlais P, Wang L, Ramos FJ, Fang Q, Christ-Roberts CY, Hong JY, Kim RY, Liu F, Dong LQ (2006) APPL1 binds to adiponectin receptors and mediates adiponectin signalling and function. Nat Cell Biol 8(5):516–523. doi:10.1038/ncb1404
50. Matsuo H, Chevallier J, Mayran N, Le Blanc I, Ferguson C, Faure J, Blanc NS, Matile S, Dubochet J, Sadoul R, Parton RG, Vilbois F, Gruenberg J (2004) Role of LBPA and Alix in multivesicular liposome formation and endosome organization. Science 303(5657):531–534. doi:10.1126/science.1092425
51. Mayinger P (2012) Phosphoinositides and vesicular membrane traffic. Biochim Biophys Acta 1821(8):1104–1113. doi:10.1016/j.bbalip. 2012.01.002
52. Miaczynska M, Bar-Sagi D (2010) Signaling endosomes: seeing is believing. Curr Opin Cell Biol 22(4):535–540. doi:10.1016/j.ceb.2010.05.007
53. Miaczynska M, Christoforidis S, Giner A, Shevchenko A, Uttenweiler-Joseph S, Habermann B, Wilm M, Parton RG, Zerial M (2004) APPL proteins link Rab5 to nuclear signal transduction via an endosomal compartment. Cell 116(3):445–456
54. Miaczynska M, Pelkmans L, Zerial M (2004) Not just a sink: endosomes in control of signal transduction. Curr Opin Cell Biol 16(4):400–406. doi:10.1016/j.ceb.2004.06.005
55. Miura S, Takeshita T, Asao H, Kimura Y, Murata K, Sasaki Y, Hanai JI, Beppu H, Tsukazaki T, Wrana JL, Miyazono K, Sugamura K (2000) Hgs (Hrs), a FYVE domain protein, is involved in Smad signaling through cooperation with SARA. Mol Cell Biol 20(24):9346–9355
56. Mosesson Y, Mills GB, Yarden Y (2008) Derailed endocytosis: an emerging feature of cancer. Nat Rev Cancer 8(11):835–850. doi:nrc2521 [pii] 10.1038/nrc2521
57. Mullock BM, Hinton RH, Peppard JV, Slot JW, Luzio JP (1987) The preparative isolation of endosome fractions: a review. Cell Biochem Funct 5(4):235–243. doi:10.1002/cbf.290050402
58. Murphy JE, Padilla BE, Hasdemir B, Cottrell GS, Bunnett NW (2009) Endosomes: a legitimate platform for the signaling train. Proc Natl Acad Sci U S A 106(42):17615–17622. doi:10.1073/pnas.0906541106

59. Nada S, Hondo A, Kasai A, Koike M, Saito K, Uchiyama Y, Okada M (2009) The novel lipid raft adaptor p18 controls endosome dynamics by anchoring the MEK-ERK pathway to late endosomes. EMBO J 28(5):477–489. doi:10.1038/emboj.2008.308
60. Ohkawara B, Glinka A, Niehrs C (2011) Rspo3 binds syndecan 4 and induces Wnt/PCP signaling via clathrin-mediated endocytosis to promote morphogenesis. Dev Cell 20(3):303–314. doi:10.1016/j.devcel.2011.01.006
61. Oksvold MP, Skarpen E, Wierod L, Paulsen RE, Huitfeldt HS (2001) Re-localization of activated EGF receptor and its signal transducers to multivesicular compartments downstream of early endosomes in response to EGF. Eur J Cell Biol 80(4):285–294
62. Palamidessi A, Frittoli E, Garre M, Faretta M, Mione M, Testa I, Diaspro A, Lanzetti L, Scita G, Di Fiore PP (2008) Endocytic trafficking of Rac is required for the spatial restriction of signaling in cell migration. Cell 134(1):135–147. doi:10.1016/j.cell.2008.05.034
63. Panopoulou E, Gillooly DJ, Wrana JL, Zerial M, Stenmark H, Murphy C, Fotsis T (2002) Early endosomal regulation of Smad-dependent signaling in endothelial cells. J Biol Chem 277(20):18046–18052. doi:10.1074/jbc.M107983200
64. Pilecka I, Banach-Orlowska M, Miaczynska M (2007) Nuclear functions of endocytic proteins. Eur J Cell Biol 86(9):533–547. doi:10.1016/j.ejcb.2007.04.004
65. Platta HW, Stenmark H (2011) Endocytosis and signaling. Curr Opin Cell Biol 23(4):393–403. doi:10.1016/j.ceb.2011.03.008
66. Pyrzynska B, Pilecka I, Miaczynska M (2009) Endocytic proteins in the regulation of nuclear signaling, transcription and tumorigenesis. Mol Oncol 3(4):321–338. doi:10.1016/j.molonc.2009.06.001
67. Raiborg C, Stenmark H (2009) The ESCRT machinery in endosomal sorting of ubiquitylated membrane proteins. Nature 458(7237):445–452. doi:10.1038/nature07961
68. Rink J, Ghigo E, Kalaidzidis Y, Zerial M (2005) Rab conversion as a mechanism of progression from early to late endosomes. Cell 122(5):735–749. doi:10.1016/j.cell.2005.06.043
69. Roepstorff K, Grandal MV, Henriksen L, Knudsen SL, Lerdrup M, Grovdal L, Willumsen BM, van Deurs B (2009) Differential effects of EGFR ligands on endocytic sorting of the receptor. Traffic 10(8):1115–1127. doi:10.1111/j.1600-0854.2009.00943.x
70. Rutherford AC, Traer C, Wassmer T, Pattni K, Bujny MV, Carlton JG, Stenmark H, Cullen PJ (2006) The mammalian phosphatidylinositol 3-phosphate 5-kinase (PIKfyve) regulates endosome-to-TGN retrograde transport. J Cell Sci 119(pt 19):3944–3957. doi:10.1242/jcs.03153
71. Sadowski L, Pilecka I, Miaczynska M (2009) Signaling from endosomes: location makes a difference. Exp Cell Res 315(9):1601–1609. doi:10.1016/j.yexcr.2008.09.021
72. Saito T, Jones CC, Huang S, Czech MP, Pilch PF (2007) The interaction of Akt with APPL1 is required for insulin-stimulated Glut4 translocation. J Biol Chem 282(44):32280–32287. doi:10.1074/jbc.M704150200
73. Sancak Y, Bar-Peled L, Zoncu R, Markhard AL, Nada S, Sabatini DM (2010) Ragulator-Rag complex targets mTORC1 to the lysosomal surface and is necessary for its activation by amino acids. Cell 141(2):290–303. doi:10.1016/j.cell.2010.02.024
74. Schenck A, Goto-Silva L, Collinet C, Rhinn M, Giner A, Habermann B, Brand M, Zerial M (2008) The endosomal protein App l1 mediates Akt substrate specificity and cell survival in vertebrate development. Cell 133(3):486–497. doi:10.1016/j.cell.2008.02.044
75. Scott CC, Gruenberg J (2011) Ion flux and the function of endosomes and lysosomes: pH is just the start: the flux of ions across endosomal membranes influences endosome function not only through regulation of the luminal pH. Bioessays 33(2):103–110. doi:10.1002/bies.201000108
76. Shah M, Patel K, Mukhopadhyay S, Xu F, Guo G, Sehgal PB (2006) Membrane-associated STAT3 and PY-STAT3 in the cytoplasm. J Biol Chem 281(11):7302–7308. doi:10.1074/jbc.M508527200
77. Shin HW, Hayashi M, Christoforidis S, Lacas-Gervais S, Hoepfner S, Wenk MR, Modregger J, Uttenweiler-Joseph S, Wilm M, Nystuen A, Frankel WN, Solimena M, De Camilli P, Zerial M (2005) An enzymatic cascade of Rab5 effectors regulates phosphoinositide turnover in the endocytic pathway. J Cell Biol 170(4):607–618. doi:10.1083/jcb.200505128

78. Sigismund S, Argenzio E, Tosoni D, Cavallaro E, Polo S, Di Fiore PP (2008) Clathrin-mediated internalization is essential for sustained EGFR signaling but dispensable for degradation. Dev Cell 15(2):209–219. doi:10.1016/j.devcel.2008.06.012
79. Simons M, Raposo G (2009) Exosomes—vesicular carriers for intercellular communication. Curr Opin Cell Biol 21(4):575–581. doi:10.1016/j.ceb.2009.03.007
80. Sorkin A, von Zastrow M (2009) Endocytosis and signalling: intertwining molecular networks. Nat Rev Mol Cell Biol 10(9):609–622. doi:10.1038/nrm2748
81. Sousa LP, Lax I, Shen H, Ferguson SM, Camilli PD, Schlessinger J (2012) Suppression of EGFR endocytosis by dynamin depletion reveals that EGFR signaling occurs primarily at the plasma membrane. Proc Natl Acad Sci USA 109(12):4419–4424. doi:10.1073/pnas.1200164109
82. Stenmark H (2009) Rab GTPases as coordinators of vesicle traffic. Nat Rev Mol Cell Biol 10(8):513–525. doi:10.1038/nrm2728
83. Taelman VF, Dobrowolski R, Plouhinec JL, Fuentealba LC, Vorwald PP, Gumper I, Sabatini DD, De Robertis EM (2010) Wnt signaling requires sequestration of glycogen synthase kinase 3 inside multivesicular endosomes. Cell 143(7):1136–1148. doi:10.1016/j.cell.2010.11.034
84. Taub N, Teis D, Ebner HL, Hess MW, Huber LA (2007) Late endosomal traffic of the epidermal growth factor receptor ensures spatial and temporal fidelity of mitogen-activated protein kinase signaling. Mol Biol Cell 18(12):4698–4710. doi:10.1091/mbc.E07-02-0098
85. Teis D, Taub N, Kurzbauer R, Hilber D, de Araujo ME, Erlacher M, Offterdinger M, Villunger A, Geley S, Bohn G, Klein C, Hess MW, Huber LA (2006) p14-MP1-MEK1 signaling regulates endosomal traffic and cellular proliferation during tissue homeostasis. J Cell Biol 175(6):861–868. doi:10.1083/jcb.200607025
86. Teis D, Wunderlich W, Huber LA (2002) Localization of the MP1-MAPK scaffold complex to endosomes is mediated by p14 and required for signal transduction. Dev Cell 3(6):803–814
87. Tsukazaki T, Chiang TA, Davison AF, Attisano L, Wrana JL (1998) SARA, a FYVE domain protein that recruits Smad2 to the TGFbeta receptor. Cell 95(6):779–791
88. Valadi H, Ekstrom K, Bossios A, Sjostrand M, Lee JJ, Lotvall JO (2007) Exosome-mediated transfer of mRNAs and microRNAs is a novel mechanism of genetic exchange between cells. Nat Cell Biol 9(6):654–659. doi:10.1038/ncb1596
89. Wu J, Petralia RS, Kurushima H, Patel H, Jung MY, Volk L, Chowdhury S, Shepherd JD, Dehoff M, Li Y, Kuhl D, Huganir RL, Price DL, Scannevin R, Troncoso JC, Wong PC, Worley PF (2011) Arc/Arg3.1 regulates an endosomal pathway essential for activity-dependent beta-amyloid generation. Cell 147(3):615–628. doi:10.1016/j.cell.2011.09.036
90. Wunderlich W, Fialka I, Teis D, Alpi A, Pfeifer A, Parton RG, Lottspeich F, Huber LA (2001) A novel 14-kilodalton protein interacts with the mitogen-activated protein kinase scaffold mp1 on a late endosomal/lysosomal compartment. J Cell Biol 152(4):765–776
91. Yamamoto H, Sakane H, Michiue T, Kikuchi A (2008) Wnt3a and Dkk1 regulate distinct internalization pathways of LRP6 to tune the activation of beta-catenin signaling. Dev Cell 15(1):37–48. doi:10.1016/j.devcel.2008.04.015
92. Yang W, Rozan LM, McDonald ER III, Navaraj A, Liu JJ, Matthew EM, Wang W, Dicker DT, El-Deiry WS (2007) CARPs are ubiquitin ligases that promote MDM2-independent p53 and phospho-p53ser20 degradation. J Biol Chem 282(5):3273–3281. doi:10.1074/jbc.M610793200
93. Zhang Y, Moheban DB, Conway BR, Bhattacharyya A, Segal RA (2000) Cell surface Trk receptors mediate NGF-induced survival while internalized receptors regulate NGF-induced differentiation. J Neurosci 20(15):5671–5678
94. Zimmerman B, Simaan M, Akoume MY, Houri N, Chevallier S, Seguela P, Laporte SA (2011) Role of ssarrestins in bradykinin B2 receptor-mediated signalling. Cell Signal 23(4):648–659. doi:10.1016/j.cellsig.2010.11.016
95. Zoncu R, Bar-Peled L, Efeyan A, Wang S, Sancak Y, Sabatini DM (2011) mTORC1 senses lysosomal amino acids through an inside-out mechanism that requires the vacuolar H(+)-ATPase. Science 334(6056):678–683. doi:10.1126/science.1207056
96. Zoncu R, Perera RM, Balkin DM, Pirruccello M, Toomre D, De Camilli P (2009) A phosphoinositide switch controls the maturation and signaling properties of APPL endosomes. Cell 136(6):1110–1121. doi:10.1016/j.cell.2009.01.032

Chapter 8
Nuclear Functions and Trafficking of Receptor Tyrosine Kinases

Ying-Nai Wang, Jennifer L. Hsu, and Mien-Chie Hung

Abstract Accumulating evidence suggests that alteration of subcellular protein localization and compartmentalization results in various types of cancer. Since the receptor tyrosine kinases (RTKs) are highly expressed and activated in human malignancies and frequently correlated with poor prognosis, it is critical to understand subcellular trafficking of the RTKs, such as the epidermal growth factor receptor (EGFR) family. A novel nuclear mode of EGFR signaling has been gradually deciphered, in which EGFR is shuttled from the cell surface to the nucleus after endocytosis. Nuclear EGFR acts as a transcriptional regulator, transmits signals, and is involved in multiple biological functions, including cell proliferation, DNA repair and replication, and chemo- and radio-resistance. In this chapter, we summarize the functions of nuclear EGFR family and the potential pathways by which cell surface EGFR is trafficked to a variety of intracellular organelles such as the Golgi apparatus, the endoplasmic reticulum, the mitochondria, as well as the nucleus. Understanding the molecular mechanism of EGFR trafficking will contribute to both the receptor biology and potential therapeutic targets of anti-EGFR therapies for clinical application.

8.1 Introduction to Receptor Tyrosine Kinases in the Nucleus

Receptor tyrosine kinases (RTKs) mediate crucial cellular signal transduction by extracellular ligand binding. Each RTK contains an extracellular ligand-binding domain, a transmembrane domain, and an intracellular tyrosine kinase domain.

Y.-N. Wang • J.L. Hsu • M.-C. Hung (✉)
Department of Molecular and Cellular Oncology, The University of Texas
MD Anderson Cancer Center, Houston, TX 77030, USA

Center for Molecular Medicine and Graduate Institute of Cancer Biology,
China Medical University, Taichung, Taiwan, R.O.C
e-mail: mhung@mdanderson.org

Y. Yarden and G. Tarcic (eds.), *Vesicle Trafficking in Cancer*,
DOI 10.1007/978-1-4614-6528-7_8,

Multiple cell surface receptors have been reported to localize in the nucleus. In this chapter, we will focus on the epidermal growth factor receptor (EGFR) family, as it is the best characterized family in terms of nuclear translocation among the RTKs. There are four members in the EGFR family: EGFR/ErbB-1/HER-1, ErbB-2/HER-2/neu, ErbB-3/HER-3, and ErbB-4/HER-4. With the exception of ErbB-3, all ErbB family members are associated with tyrosine kinase activity. Upon ligand binding, EGFR family proteins dimerize by either receptor homo- or hetero-dimerization to activate tyrosine kinase activity and trigger a myriad of downstream signaling pathways. These, including phosphatidylinositol-3 kinase, mitogen-activated protein kinase, signal transducer and activator of transcription (STAT), phospholipase C, and the modulation of calcium channels, regulate proliferation, mobility, and differentiation in many different cell types [1–4].

The EGFR family proteins, such as EGFR and ErbB-2, are overexpressed and/or constitutively activated in human tumors of epithelial origin, leading to aggressive tumor behaviors, including cancer initiation, increased tumor growth/progression, poor patient outcome, metastasis, and chemo-resistance [5–8]. Therefore, EGFR and ErbB-2 have been considered as effective targets for anticancer therapies. Many of them have been developed and approved by the US Food and Drug Administration, including ectodomain-binding monoclonal antibodies, e.g., cetuximab and trastuzumab, and small-molecule tyrosine kinase inhibitors (TKIs) targeting EGFR and ErbB-2, e.g., erlotinib, gefitinib, and lapatinib. The role of ErbB-4 as an oncogene or a tumor suppressor is less clear, and controversial results have been reported [9, 10]. Further systematic investigation is required to clarify this issue.

In addition to their roles in traditional signaling cascades, several lines of evidence to date from various groups have indicated unique biological functions associated with nuclear translocation of the EGFR family receptors in which they are shuttled from the cell surface to the nucleus [11–35], also known as membrane receptors in the nucleus (MRIN) [36]. More recently, studies have unraveled the molecular mechanism underlying the cell surface membrane-to-nucleus trafficking of EGFR [37, 38]. In this chapter, recent discoveries in the MRIN field and the subcellular trafficking pathways of the EGFR family proteins from the cell surface to a variety of cellular organelles, including the Golgi apparatus, the endoplasmic reticulum (ER), the mitochondria, as well as the nucleus, will be discussed.

8.2 The Discovery of EGFR in the Nucleus

8.2.1 Detection of Nuclear EGFR and its Clinical Relevance

The expression of EGFR in the nucleus was first detected in hepatocytes during regeneration [39–42]. The first function of nuclear EGFR was then shown to be involved in transcriptional regulation [43]. Nuclear EGFR was further shown to be involved in cell proliferation, DNA replication, DNA repair, and chemo- and radio-resistance [44–48]. The translocation of EGFR into the nucleus is also associated with DNA damage events, including cisplatin treatment, oxidative stress, heat treatment, ionizing radiation, ultraviolet irradiation, and cetuximab/C225 treatment, a monoclonal anti-EGFR antibody, in a ligand-independent manner [49–52]. In diverse cancer types, such as breast cancer, ovarian cancer, and oropharyngeal and esophageal squamous cell carcinomas, nuclear EGFR is associated with poor clinical prognosis [53–58].

8.2.2 The Role of Nuclear EGFR in Therapeutic Resistance

The presence of nuclear EGFR has been shown to contribute to therapeutic resistance, e.g., radiation, cisplatin, and cetuximab [59–61]. In head and neck squamous cell carcinoma, cetuximab- and radiation-induced EGFR nuclear translocation can be blocked by dasatinib, a potent Src inhibitor [28]. Lapatinib, a dual TKI of EGFR and ErbB-2, has also been reported to inhibit the nuclear transport of EGFR and ErbB-2 and sensitizes cancer cells to fluoropyrimidine by downregulating thymidylate synthase, a commonly overexpressed protein in fluoropyrimidine-resistant cancer cells [62]. Two studies have recently demonstrated that nuclear EGFR confers acquired resistance to gefitinib, an EGFR-TKI, through increased expression of breast cancer-resistant protein (BCRP), an ATP-binding cassette transporter that pumps anticancer drugs out of cells [17, 63]. Collectively, evidence to date suggests that nuclear EGFR contributes to the resistance to EGFR-targeting treatments.

8.2.3 EGFR Variant Is Also Present in the Nucleus

Nuclear EGFRvIII, a constitutively activated EGFR type III variant, was first reported in hormone-refractory prostate cancer and associated with poor patient survival [64]. EGFRvIII can be found in the nucleus of normal glial cells and primary glioblastomas and forms an oncogenic complex with STAT3 to mediate EGFRvIII-dependent glial transformation [26]. Moreover, the nuclear EGFRvIII-STAT3 complex can also activate *cyclooxygenase-2* (*COX-2*) gene expression in glioblastoma cells [30]. Interestingly, a splicing variant of EGFR that lacks multiple domains, called mLEEK, functions as a transcription factor in the nucleus [16].

8.3 Nuclear Translocation of Other EGFR Family Members

8.3.1 ErbB-2/HER-2/Neu

Heregulin, a glycoprotein that binds to EGFR and thus elevates tyrosine phosphorylation of ErbB-2/neu receptor, has been shown to increase the levels of ErbB-2/neu in the nucleus [65]. The rat version of human ErbB-2, p185neu, was reported to be located in the nucleus and was the first membrane-associated RTK shown to associate with transcriptional activation [66]. In breast cancer cells, a full-length form of nuclear ErbB-2 plays a role in transactivation of the *COX2* gene promoter [67]. More recently, nuclear ErbB-2 was also demonstrated to activate transcription of ribosomal RNA genes by its association with RNA polymerase-I and β-actin to ribosomal DNA, which results in increased protein synthesis and cell growth [29]. In addition to full-length nuclear ErbB-2 ($p185^{ErbB-2}$), a truncated form of ErbB-2 (p95), which is generated by an alternative translation initiation site and lacks the N-terminal extracellular domain, has also been found in the nucleus [68, 69]. It has been reported that p95 contributes to acquired therapeutic resistance to ErbB-2 TKIs [70].

8.3.2 ErbB-3/HER-3 and ErbB-4/HER-4

Similar to EGFR and ErbB-2, full-length ErbB-3 has also been identified in the nucleus [24]. Interestingly, in patients with prostate cancer, low expression of nuclear ErbB-3 is a predictor of a higher risk of biochemical recurrence (rising levels of serum prostate-specific antigen found in patients with localized prostate cancer after radical prostatectomy) [71, 72]. While full-length ErbB-4 can be detected in the nuclei of some normal cells [73, 74], a truncated form of ErbB-4 in which its intracellular domain (ICD) is cleaved by γ-secretase has been reported in the nucleus of cancer cells [20, 75]. The role of nuclear ErbB-4 ICD remains ambiguous as it has been shown to be involved in both shorter patient survival [76] yet improves patient response to tamoxifen therapy [77] for estrogen receptor-α positive cancers. Specifically, these reports showed that nuclear cleavable ErbB-4 is associated with shorter survival than cell surface ErbB-4 in the estrogen receptor-positive subset of breast cancer patients [76]. On the other hand, researchers have demonstrated that nuclear ErbB-4 ICD acts as a co-activator of estrogen receptor-α and improves patient response to tamoxifen therapy [77]. In addition, ErbB-4 ICD has been shown to enhance the ubiquitination and degradation of the oncogene Hdm2, following increased expression of p53 [78], while blocking transcriptional repression mediated by Ero2, which is involved in cell differentiation [79]. Taken together, the controversial role of ErbB-4 ICD as an oncogene or a tumor suppressor requires further investigation.

8.4 Nuclear Translocation of Other Cell Surface Receptors

Accumulating evidence indicates that several full-length RTKs and cell surface receptors other than EGFR family proteins also translocate to the nucleus, for example, fibroblast growth factor receptor (FGFR), vascular endothelial growth factor receptor, insulin-like growth factor-1 receptor, interleukin receptors, interferon-γ receptor, growth hormone receptors, cMet, and TrkA [45, 46, 80]. In addition, receptor tyrosine kinase-like orphan receptor 1 (Ror1), which belongs to the ROR RTK family, also translocates to the nucleus via its juxtamembrane domain [81]. Nuclear translocation of cell surface prolactin receptor mediated by prolactin was shown to recruit a chromatin-modifying protein to activate Stat5a-driven gene expression [82]. Taken together, aside from members of the EGFR family, other cell surface receptors are also found in the nucleus, supporting that MRIN is likely a general phenomenon.

8.5 EGFR Family Ligands in the Nucleus

Studies have shown that ligands of the EGFR family, e.g., EGF, pro-transforming growth factor-α, pro-heparin-binding EGF-like growth factor, and Schwannoma-derived growth factor, are also found in the nucleus [83–88]. In addition, the ICD fragment of neuregulin-1, a ligand of ErbB-3/ErbB-4, also translocates into the nucleus and increases neuronal survival by repressing the cell death response [89]. Hence, in addition to RTKs, their ligands can also translocate into the nucleus. These findings suggest that the association of receptor and ligand might also take place in the nucleus. Indeed, a study has demonstrated the presence of the EGF/EGFR complex in the nucleus by a cross-linking experiment between EGF and EGFR [43].

8.6 Functions of Nuclear EGFR Family Proteins

8.6.1 Transcriptional Co-activator

EGFR family members, including EGFR, ErbB-2, and ErbB-4, contain an intrinsic transactivation activity at the acidic C-terminal region [20, 43, 66, 67] that can regulate transcription by activating transcriptional factors to enhance target gene expression. Promoters targeted by the nuclear EGFR family receptors include *cyclin D1* [43], *B-Myb* [90], *iNOS* [91], *Aurora-A* [92], *COX-2* [30], *c-Myc* [14], *thymidylate synthase* [62], and *BCRP* [17], all of which have been shown to play a role in tumorigenesis, chromosome instability, and chemo-resistance (Fig. 8.1). After ligand stimulation, EGFR acts as a transcriptional co-activator by binding to an AT-rich response sequence (ATRS) of the *cyclin D1* promoter and stimulating *cyclin D1* expression [43]. Using an unbiased mass spectrometry approach, Huo et al. identified RNA helicase A (RHA) as a nuclear EGFR-interacting protein and showed the EGFR-RHA complex bind to the ATRS through RHA to activate the transcription of *cyclin D1* and *iNOS* [93]. Nuclear EGFR also interacts with MUC1 to mediate *cyclin D1* gene expression [11]. Through interaction with transcription factors, such as E2F1, STAT3, and STAT5A, activated nuclear EGFR can bind to the ATRS motif and

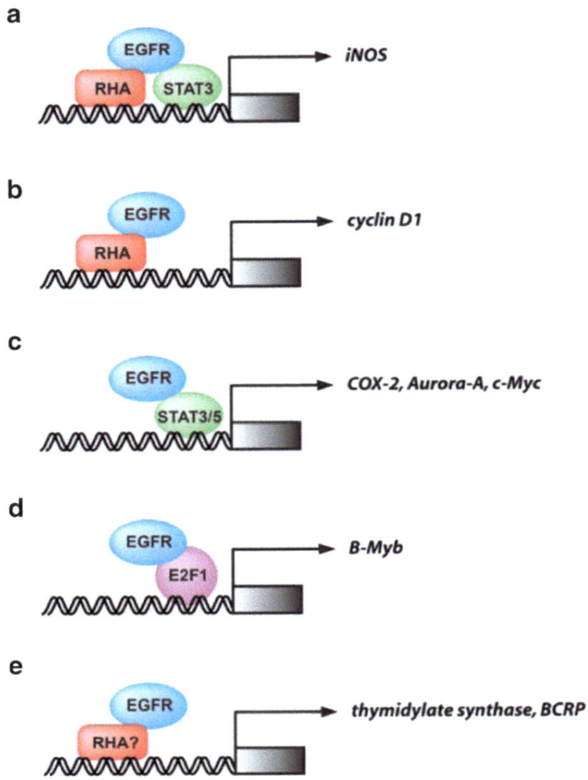

Fig. 8.1 A summary of nuclear function of EGFR as a transcriptional co-activator. Nuclear EGFR can function in transcriptional regulation to enhance expression levels of target genes, including iNOS (**a**), *cyclin D1* (**b**), *COX-2* (**c**), *Aurora-A* (**c**), *c-Myc* (**c**), *B-Myb* (**d**), *thymidylate synthase* (**e**), and *BCRP* (**e**), through activation of transcriptional factors, such as STAT3/5 and E2F1. Researchers further identified that EGFR can associate with RHA in the nucleus, where the EGFR/RHA complex binds to the target gene promoter, including *iNOS* (**a**) and *cyclinD1* (**b**), through the recruitment of RHA to the ATRS of the target gene promoter to mediate its transcriptional activation. In addition to RHA, EGFR is also recruited to the *iNOS* gene promoter through STAT3 to the STAT3-binding site (**a**). Furthermore, it is unexplored yet whether RHA is involved in the nuclear EGFR-mediated activation of *thymidylate synthase* and *BCRP* (**e**) (adapted from Wang et al., Cell and Bioscience, 2012)

promote the expression of *B-Myb*, *iNOS*, *COX-2*, and *Aurora-A* genes [30, 90–92]. Serine phosphorylation by Akt was shown to promote the transport of EGFR to the nucleus to target multiple ATRSs on the *BCRP* promoter in gefitinib-resistant cells, which is involved in chemo-resistance [17, 94]. In addition, both nuclear EGFR and ErbB-2 can activate transcription of the *thymidylate synthase* gene by binding to its promoter, and this can be blocked by the dual EGFR/HER2-TKI, lapatinib [62]. EGFR also forms a novel heteromeric nuclear complex with c-Src kinase and STAT3 that associates with the *c-Myc* promoter in pancreatic cancer, suggesting that this complex plays a role in *c-Myc* gene regulation [14]. Consistent with the studies of nuclear EGFR, nuclear ErbB-2 can also transactivate *COX-2* gene expression by binding to the HER2-associated sequence, a specific DNA element within the *COX-2* promoter. Currently, the transcriptional factors that are involved in the nuclear ErbB-2-mediated *COX2* activation have yet to be identified [67]. Moreover, in glioblastoma cells, nuclear EGFRvIII cooperates with STAT3 to activate *COX-2* gene expression that results in glioma tumorigenesis [26, 30]. Upon ligand stimulation and cleavage by the γ-secretase, the ICD of ErbB-4 translocates to the nucleus and associates with STAT5A to transactivate the promoter of β-*casein* [95]. Nuclear ErbB-4 ICD also interacts with the nuclear corepressor, Eto-2, and block Eto-2-dependent transcriptional repression [79]. A recent study that utilized unbiased approaches to profile the human protein-DNA interactome further demonstrated EGFR as one of the DNA-binding proteins [96], supporting the concept that EGFR functions in transcriptional regulation in the nucleus.

8.6.2 Protein Kinase and Protein-Protein Interaction of EGFR Family Proteins

Nuclear EGFR associates with and phosphorylates the chromatin-bound form of proliferative cell nuclear antigen (PCNA) to stabilize it for DNA replication and DNA damage repair [19]. Several studies have shown that the interaction between nuclear EGFR and DNA-dependent protein kinase (DNA-PK) can be induced by DNA damage cascades, including those activated by ultraviolet irradiation or cisplatin treatment [15, 50, 60]. DNA-PK is a central enzyme of the nonhomologous end-joining repair system of DNA double-strand breaks and has been shown to contribute to DNA repair and chemo- and radio-resistance. It is not yet determined whether EGFR phosphorylates DNA-PK to regulate its activity. ErbB-2 co-localizes both in the cytoplasm and the nucleus with the cyclin-dependent kinase $p34^{Cdc2}$ and phosphorylates it, which has been shown to contribute to taxol resistance in breast cancer [97], suggesting that the kinase activity of ErbB-2 remains while it is in the nucleus. The ICD fragment of ErbB-4 interacts with and phosphorylates the nuclear protein Hdm2 to enhance Hdm2 ubiquitination, leading to increased *p53* transcriptional activity and p53 and p21 expression [78].

8.7 Trafficking of EGFR from the Cell Surface to Subcellular Organelles

Membrane and vesicular trafficking processes regulated by receptor endocytosis determine the duration and intensity of transmembrane RTK signaling stimulated by extracellular ligands. After ligand-induced endocytosis, cargo proteins are carried in budding vesicles and delivered from donor membranes to acceptor subcellular organelles by membrane fusion pathways. In this process, both RTKs and their cognate ligands are internalized into cytoplasmic vesicles and sequentially detached from the cell surface [98, 99]. A wide array of evidence points to several potential destinations of the internalized EGFR embedded within the early endosomes through endosomal sorting: (1) EGFR is recycled back to the cell surface by either the recycling endosomes or a direct recycling pathway; (2) EGFR is sorted into the late endosomes destined for degradation by lysosomes. In addition to the above-mentioned trafficking routes, EGFR after endocytosis can also be transported from the cell surface to various cellular compartments, including the Golgi apparatus, the endoplasmic reticulum (ER), and the mitochondria, as well as the nucleus [25, 27, 45]. The subcellular trafficking mechanisms of EGFR are described in the following sections.

8.7.1 Golgi Apparatus and Endoplasmic Reticulum

After endocytosis, cargo proteins carried in budding vesicles are sorted to the biosynthetic/secretory compartments, such as the Golgi apparatus and the ER. This is also known as a retrograde transport pathway that is crucial to cargo protein transport. Mammalian cargo proteins utilize the retrograde route from the early endosomes to the Golgi apparatus [100] while exogenous viruses and toxins are commonly routed from the early endosomes to the ER [101]. The first example of retrograde trafficking in regulating the nuclear transport of cell surface RTKs was recently identified in which EGFR was transported to the ER from the Golgi, which is part of the regulatory mechanism of EGFR nuclear transport [37]. After EGF stimulation, the full-length EGFR anchors to the membranes of the Golgi and the ER, with the N-terminus masked inside the Golgi and ER lumen and the C-terminus exposed to the cytoplasm [37]. Other studies have reported that EGFR is trafficked from the cell surface to the ER in response to EGF [32]. It has also been shown that coat proteins, such as coat protein complex I (COPI), which plays a role in Golgi-to-ER retrograde transport and COPII in ER-to-Golgi anterograde transport, are crucial for vesicular transport of targets to reach their intracellular destination [102–104]. It was reported by Wang et al. [37] that γ-COP, which is one of the subunits of the COPI coatomer, associates with EGFR and mediates EGF-dependent EGFR nuclear transport. Together, these findings indicate that cell surface EGFR embedded in endosomal membranes remain in a membrane-associated environment as it travels from the cell surface to the nucleus along the Golgi and ER membranes by COPI-mediated vesicle trafficking. More recently, Yang et al. showed that COPI is not only involved in vesicular retrograde transport but in tubular formation for anterograde intra-Golgi transport [105]. Thus, it would be worthwhile to determine the physiological roles of COPI tubules to further unravel the mechanisms of how the COPI complex regulates subcellular cargo sorting.

8.7.2 Mitochondria

Full-length EGFR can also anchor to the mitochondria after EGF stimulation and associate with cytochrome *c* oxidase subunit II, which is a key component of the oxidative phosphorylation cascade in regulating apoptosis [106, 107]. A putative mitochondrial localization signal found within the juxtamembrane domain of EGFR and c-Src kinase activity is required for EGFR translocation to the mitochondria by clathrin-mediated endocytosis [107]. Subcellular localization of EGFR to the mitochondria may be an alternative survival pathway to modulate cytochrome *c* oxidase subunit II-dependent mitochondrial functions. A recent study showed that both EGFR and EGFRvIII can be transported to the mitochondria upon treatment of EGFR-TKI and apoptosis inducers, suggesting that mitochondrial targeting of these receptors contributes to drug resistance [108]. Furthermore, cetuximab has also been shown to enhance mitochondrial accumulation of EGFRvIII [109], implying that mitochondrial EGFR/EGFRvIII are responsible for therapeutic response to EGFR-targeting drugs. How EGFR is transported to the mitochondria remains unclear, and more investigations would be required to determine if the integration of EGFR into the mitochondrial membrane occurs through endosomal membrane fusion or other pathways.

8.7.3 Nucleus

EGFR family receptors are known to transport to the nucleus, where they exist as full-length or truncated forms to carry out a number of important biological functions. However, the mechanism of trafficking of endosome-embedded EGFR family proteins from cell surface to the nucleus has been unnoticed for many years. In addition to the recent report showing nuclear translocation of EGFR is regulated by COPI-mediated retrograde trafficking [37], there are also those that have identified the putative nuclear localization signals (NLSs) within all of the EGFR family members [24, 30, 67, 91, 95]. Unlike the traditional mono- and bipartite NLSs, EGFR contains a tripartite NLS, which consists of three clusters of basic amino acids and is conserved within the juxtamembrane regions among the EGFR family members [110]. Moreover, studies have demonstrated the role of the NLS and importin-β in the nuclear translocation of EGFR and ErbB-2 [111, 112]. Specifically, importin-β forms a complex with NLS-harboring proteins and binds to the nucleoporins of nuclear pore complexes during nuclear translocation (Fig. 8.2). In addition, studies have shown that nuclear translocation of EGFR and ErbB-2 requires receptor endocytosis and endosomal sorting by associating with early endosomal markers in the nucleus [111, 112].

Interestingly, ErbB-2 has also been observed to localize in the nucleolus, where it associates with RNA polymerase-I [29]. The trafficking mechanism of ErbB-2 to the nucleolus as well as other EGFR family receptors warrants further investigations. Moreover, the exportin CRM1 is also shown to play a role in the nuclear export of

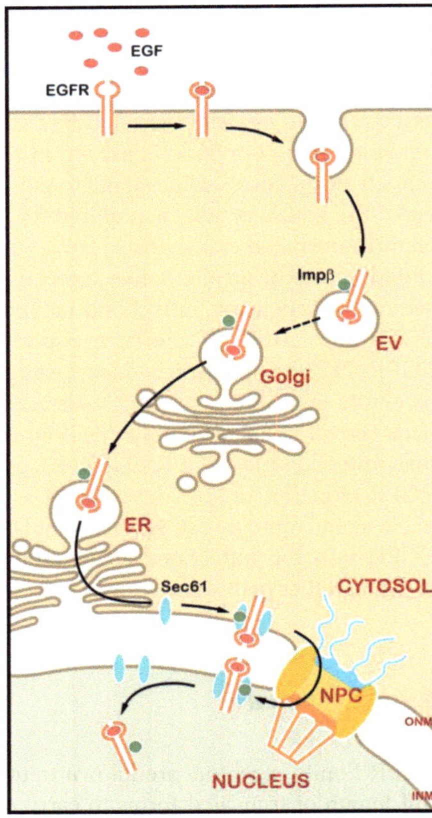

Fig. 8.2 Proposed model of EGFR trafficking from the cell surface to the nucleus. Diagram of integral trafficking of EGFR from the Golgi/ER/NE to the nucleus by EGF treatment. The scale of the diagram does not reflect the relative sizes of different molecules or subcellular structures. *EV* endocytic vesicle; *Impβ* importin β; *ER* endoplasmic reticulum; *NPC* nuclear pore complex (adapted from Wang et al., J. Biol. Chem, 2010)

cell surface RTKs, such as EGFR, ErbB-2, and ErbB-3, albeit unidentified nuclear export signals [24, 111, 112].

8.7.4 Inner Nuclear Membrane to the Nucleoplasm

EGFR is also present in the inner nuclear membrane (INM) or nuclear matrix [113, 114]; however, the exact trafficking mechanism in INM translocation is not well understood. A new pathway, namely, integral trafficking from the ER to the nuclear envelope transport (INTERNET) was identified to explain the mechanism of cell surface-to-IMM translocation of EGFR under EGF stimulation [38, 45]. In these studies, the INM-localized Sec61β translocon associates with EGFR that is anchored to the INM to mediate the release of EGFR into the nucleoplasm within the nucleus. This newly identified ER-associated translocon Sec61β demonstrates a previously unrecognized location and role in regulating EGFR nuclear transport and explains how the membrane-bound cell surface EGFR remains membrane-embedded while it travels from the lipid bilayer of the INM to the nucleus [38] (Fig. 8.2). However, other than these findings, the trafficking mechanism remains largely unexplored. It is possible that an endocytosis-like mechanism in the nuclear envelope is involved in the transport of EGFR from the INM to the nucleoplasm. More in-depth and systematic studies are needed to further address this notion. Traditionally thought to be localized in the ER, Sec61β has been proposed to extract EGFR from lipid layers of the ER membrane for delivery to the cytoplasm by an ER-associated degradation pathway, which then allows the cytoplasmic EGFR to be transported to the nucleus through the association of importin-β [32]. More investigations will be required to show that Sec61β-mediated ER-associated degradation model is involved in EGFR nuclear trafficking as there is no report of EGFR detected in the cytoplasm under EGF stimulation [32].

8.8 Exosome Secretion of EGFR from Cells

In addition to subcellular trafficking of cell surface EGFR within the cells, secretion of EGFR from cells into the exosomes has also been reported [115–118]. Exosomes are small membrane-bound vesicles (30–100 nm in diameter) derived from the multivesicular bodies and released into the extracellular environment via fusion of the multivesicular bodies with the cell surface membrane [119]. Exosomes that carry proteins, mRNAs, and microRNAs can function in intercellular communication and are involved in promoting angiogenesis, cell proliferation, and invasion, supporting the tumor microenvironment and suppressing the immune surveillance [120–123]. The existence of full-length and C-terminal remnant fragment of EGFR in the exosomes has been identified in human keratinocyte and pancreatic cancer cells [115, 116]. The secretion of EGFR into the exosomes can be stimulated by EGF treatment [115]. The exosomal forms of EGFR and EGFRvIII have also been detected in brain tumors [124]. Furthermore, a soluble extracellular domain of EGFR in the conditional culture media generated via metalloprotease-dependent ectodomain shedding has been found in different cancer cell lines [115–117, 125]. In addition to EGFR, its ligands, such as HB-EGF, TGF-α, and amphiregulin, have been found in the exosomes, and exosomal amphiregulin can contribute to increased invasiveness [126]. The recently discovered mechanism of exosomes has become more common in cancer; therefore, it is worthwhile to further investigate the potential roles and trafficking mechanism of EGFR in the exosomes.

8.9 Summary

Many integral membrane proteins, including all members of EGFR family, have been reported to function in the nucleus. Recently, researchers have delineated a logical pathway for the transport of EGFR from the cell surface to the nucleus in response to EGF. Specifically, studies have demonstrated that EGFR travels in form of a membrane-bound protein from the cell surface through the Golgi-to-ER retrograde pathway and INTERNET model to the INM in the nucleus [37, 38]. The entire process of EGFR traveling from the cell surface to the INM has been shown to be membrane-bound (Fig. 8.2). Nonetheless, there are many questions that remain unanswered: (1) How is membrane-embedded EGFR in the endosomal membrane shuttled to the Golgi apparatus? A possibility would be the involvement of the small GTPase protein Rab7b, which is essential for retrograde trafficking from the endosomes to the Golgi [127]. (2) Is membrane-bound trafficking a general mechanism for nuclear transport for all RTKs and cell surface receptors? For example, unlike EGFR family proteins, FGFR-1, which has an atypical transmembrane domain, can function as a membrane-embedded RTK or a soluble cytoplasmic protein [128–130]. It might be useful to compare the nuclear translocation trafficking mechanism between FGFR-1 and EGFR. (3) How does subcellular trafficking of EGFR to different compartments impact its role? Identifying the mechanisms underlying how cell surface RTKs travels to various destinations will advance our understanding of their unique functions and may have important clinical implication as RTKs are implicated in many cancers. The membrane-bound trafficking mechanism provides a new understanding of how EGFR as well as other cell surface receptors travel from the cell surface to the nucleus. Finally, it would be of great interest to determine if the same membrane-bound trafficking is reversible, i.e., if nuclear membrane-bound proteins also utilize this pathway to travel to the cell surface.

Acknowledgements The authors would like to acknowledge the National Breast Cancer Foundation for their generous support.

References

1. Avraham R, Yarden Y (2011) Feedback regulation of EGFR signalling: decision making by early and delayed loops. Nat Rev Mol Cell Biol 12(2):104–117
2. Citri A, Yarden Y (2006) EGF-ERBB signalling: towards the systems level. Nat Rev Mol Cell Biol 7(7):505–516
3. Lo HW, Hsu SC, Hung MC (2006) EGFR signaling pathway in breast cancers: from traditional signal transduction to direct nuclear translocalization. Breast Cancer Res Treat 95(3):211–218
4. Huang PH, Xu AM, White FM (2009) Oncogenic EGFR signaling networks in glioma. Sci Signal 2(87):re6
5. Irmer D, Funk JO, Blaukat A (2007) EGFR kinase domain mutations—functional impact and relevance for lung cancer therapy. Oncogene 26(39):5693–5701
6. Gazdar AF (2009) Activating and resistance mutations of EGFR in non-small-cell lung cancer: role in clinical response to EGFR tyrosine kinase inhibitors. Oncogene 28(suppl 1): S24–S31
7. Hynes NE, MacDonald G (2009) ErbB receptors and signaling pathways in cancer. Curr Opin Cell Biol 21(2):177–184

8. Linardou H, Dahabreh IJ, Bafaloukos D, Kosmidis P, Murray S (2009) Somatic EGFR mutations and efficacy of tyrosine kinase inhibitors in NSCLC. Nat Rev Clin Oncol 6(6):352–366
9. Jones FE (2008) HER4 intracellular domain (4ICD) activity in the developing mammary gland and breast cancer. J Mammary Gland Biol Neoplasia 13(2):247–258
10. Muraoka-Cook RS, Feng SM, Strunk KE, Earp HS III (2008) ErbB4/HER4: role in mammary gland development, differentiation and growth inhibition. J Mammary Gland Biol Neoplasia 13(2):235–246
11. Bitler BG, Goverdhan A, Schroeder JA (2010) MUC1 regulates nuclear localization and function of the epidermal growth factor receptor. J Cell Sci 123(pt 10):1716–1723
12. Merlin J, Stechly L, de Beauce S, Monte D, Leteurtre E, van Seuningen I et al (2011) Galectin-3 regulates MUC1 and EGFR cellular distribution and EGFR downstream pathways in pancreatic cancer cells. Oncogene 30(22):2514–2525
13. De Angelis Campos AC, Rodrigues MA, de Andrade C, de Goes AM, Nathanson MH, Gomes DA (2011) Epidermal growth factor receptors destined for the nucleus are internalized via a clathrin-dependent pathway. Biochem Biophys Res Commun 412(2):341–346
14. Jaganathan S, Yue P, Paladino DC, Bogdanovic J, Huo Q, Turkson J (2011) A functional nuclear epidermal growth factor receptor, SRC and Stat3 heteromeric complex in pancreatic cancer cells. PLoS One 6(5):e19605
15. Liccardi G, Hartley JA, Hochhauser D (2011) EGFR nuclear translocation modulates DNA repair following cisplatin and ionizing radiation treatment. Cancer Res 71(3):1103–1114
16. Piccione EC, Lieu TJ, Gentile CF, Williams TR, Connolly AJ, Godwin AK et al (2012) A novel epidermal growth factor receptor variant lacking multiple domains directly activates transcription and is overexpressed in tumors. Oncogene 31(24):2953–2967
17. Huang WC, Chen YJ, Li LY, Wei YL, Hsu SC, Tsai SL et al (2011) Nuclear translocation of epidermal growth factor receptor by Akt-dependent phosphorylation enhances breast cancer-resistant protein expression in gefitinib-resistant cells. J Biol Chem 286(23):20558–20568
18. Cornelissen B, Darbar S, Hernandez R, Kersemans V, Tullis I, Barber PR et al (2011) ErbB-2 blockade and prenyltransferase inhibition alter epidermal growth factor and epidermal growth factor receptor trafficking and enhance (111)In-DTPA-hEGF Auger electron radiation therapy. J Nucl Med 52(5):776–783
19. Wang SC, Nakajima Y, Yu YL, Xia W, Chen CT, Yang CC et al (2006) Tyrosine phosphorylation controls PCNA function through protein stability. Nat Cell Biol 8(12):1359–1368
20. Ni CY, Murphy MP, Golde TE, Carpenter G (2001) Gamma-secretase cleavage and nuclear localization of ErbB-4 receptor tyrosine kinase. Science 294(5549):2179–2181
21. Wanner G, Mayer C, Kehlbach R, Rodemann HP, Dittmann K (2008) Activation of protein kinase Cepsilon stimulates DNA-repair via epidermal growth factor receptor nuclear accumulation. Radiother Oncol 86(3):383–390
22. Das AK, Chen BP, Story MD, Sato M, Minna JD, Chen DJ et al (2007) Somatic mutations in the tyrosine kinase domain of epidermal growth factor receptor (EGFR) abrogate EGFR-mediated radioprotection in non-small cell lung carcinoma. Cancer Res 67(11):5267–5274
23. Kim J, Jahng WJ, Di Vizio D, Lee JS, Jhaveri R, Rubin MA et al (2007) The phosphoinositide kinase PIKfyve mediates epidermal growth factor receptor trafficking to the nucleus. Cancer Res 67(19):9229–9237
24. Offterdinger M, Schofer C, Weipoltshammer K, Grunt TW (2002) c-erbB-3: a nuclear protein in mammary epithelial cells. J Cell Biol 157(6):929–939
25. Massie C, Mills IG (2006) The developing role of receptors and adaptors. Nat Rev Cancer 6(5):403–409
26. de la Iglesia N, Konopka G, Puram SV, Chan JA, Bachoo RM, You MJ et al (2008) Identification of a PTEN-regulated STAT3 brain tumor suppressor pathway. Genes Dev 22(4):449–462
27. Mosesson Y, Mills GB, Yarden Y (2008) Derailed endocytosis: an emerging feature of cancer. Nat Rev Cancer 8(11):835–850

28. Li C, Iida M, Dunn EF, Wheeler DL (2010) Dasatinib blocks cetuximab- and radiation-induced nuclear translocation of the epidermal growth factor receptor in head and neck squamous cell carcinoma. Radiother Oncol 97(2):330–337
29. Li LY, Chen H, Hsieh YH, Wang YN, Chu HJ, Chen YH et al (2011) Nuclear ErbB2 enhances translation and cell growth by activating transcription of ribosomal RNA genes. Cancer Res 71(12):4269–4279
30. Lo HW, Cao X, Zhu H, Ali-Osman F (2010) Cyclooxygenase-2 is a novel transcriptional target of the nuclear EGFR-STAT3 and EGFRvIII-STAT3 signaling axes. Mol Cancer Res 8(2):232–245
31. Dittmann K, Mayer C, Fehrenbacher B, Schaller M, Kehlbach R, Rodemann HP (2010) Nuclear EGFR shuttling induced by ionizing radiation is regulated by phosphorylation at residue Thr654. FEBS Lett 584(18):3878–3884
32. Liao HJ, Carpenter G (2007) Role of the Sec61 translocon in EGF receptor trafficking to the nucleus and gene expression. Mol Biol Cell 18(3):1064–1072
33. Li CF, Fang FM, Wang JM, Tzeng CC, Tai HC, Wei YC et al (2012) EGFR nuclear import in gallbladder carcinoma: nuclear phosphorylated EGFR upregulates iNOS expression and confers independent prognostic impact. Ann Surg Oncol 19(2):443–454
34. Dekanic A, Dobrila Dintinjan R, Budisavljevic I, Pecanic S, Zuvic Butorac M, Jonjic N (2011) Strong nuclear EGFR expression in colorectal carcinomas is associated with cyclin-D1 but not with gene EGFR amplification. Diagn Pathol 6:108
35. Nowsheen S, Bonner JA, Yang ES (2011) The poly(ADP-Ribose) polymerase inhibitor ABT-888 reduces radiation-induced nuclear EGFR and augments head and neck tumor response to radiotherapy. Radiother Oncol 99(3):331–338
36. Wang SC, Hung MC (2009) Nuclear translocation of the epidermal growth factor receptor family membrane tyrosine kinase receptors. Clin Cancer Res 15(21):6484–6489
37. Wang YN, Wang H, Yamaguchi H, Lee HJ, Lee HH, Hung MC (2010) COPI-mediated retrograde trafficking from the Golgi to the ER regulates EGFR nuclear transport. Biochem Biophys Res Commun 399(4):498–504
38. Wang YN, Yamaguchi H, Huo L, Du Y, Lee HJ, Lee HH et al (2010) The translocon Sec61beta localized in the inner nuclear membrane transports membrane-embedded EGF receptor to the nucleus. J Biol Chem 285(49):38720–38729
39. Marti U, Burwen SJ, Wells A, Barker ME, Huling S, Feren AM et al (1991) Localization of epidermal growth factor receptor in hepatocyte nuclei. Hepatology (Baltimore, MD) 13(1):15–20
40. Marti U, Hug M (1995) Acinar and cellular distribution and mRNA expression of the epidermal growth factor receptor are changed during liver regeneration. J Hepatol 23(3):318–327
41. Marti U, Ruchti C, Kampf J, Thomas GA, Williams ED, Peter HJ et al (2001) Nuclear localization of epidermal growth factor and epidermal growth factor receptors in human thyroid tissues. Thyroid 11(2):137–145
42. Marti U, Wells A (2000) The nuclear accumulation of a variant epidermal growth factor receptor (EGFR) lacking the transmembrane domain requires coexpression of a full-length EGFR. Mol Cell Biol Res Commun 3(1):8–14
43. Lin SY, Makino K, Xia W, Matin A, Wen Y, Kwong KY et al (2001) Nuclear localization of EGF receptor and its potential new role as a transcription factor. Nat Cell Biol 3(9):802–808
44. Dittmann K, Mayer C, Rodemann HP (2010) Nuclear EGFR as novel therapeutic target: insights into nuclear translocation and function. Strahlenther Onkol 186(1):1–6
45. Wang YN, Yamaguchi H, Hsu JM, Hung MC (2010) Nuclear trafficking of the epidermal growth factor receptor family membrane proteins. Oncogene 29(28):3997–4006
46. Carpenter G, Liao HJ (2009) Trafficking of receptor tyrosine kinases to the nucleus. Exp Cell Res 315(9):1556–1566
47. Brand TM, Iida M, Li C, Wheeler DL (2011) The nuclear epidermal growth factor receptor signaling network and its role in cancer. Discov Med 12(66):419–432
48. Lo HW (2010) Nuclear mode of the EGFR signaling network: biology, prognostic value, and therapeutic implications. Discov Med 10(50):44–51

49. Xu Y, Shao Y, Zhou J, Voorhees JJ, Fisher GJ (2009) Ultraviolet irradiation-induces epidermal growth factor receptor (EGFR) nuclear translocation in human keratinocytes. J Cell Biochem 107(5):873–880
50. Dittmann K, Mayer C, Fehrenbacher B, Schaller M, Raju U, Milas L et al (2005) Radiation-induced epidermal growth factor receptor nuclear import is linked to activation of DNA-dependent protein kinase. J Biol Chem 280(35):31182–31189
51. Liao HJ, Carpenter G (2009) Cetuximab/C225-induced intracellular trafficking of epidermal growth factor receptor. Cancer Res 69(15):6179–6183
52. Dittmann K, Mayer C, Fehrenbacher B, Schaller M, Kehlbach R, Rodemann HP (2011) Nuclear epidermal growth factor receptor modulates cellular radio-sensitivity by regulation of chromatin access. Radiother Oncol 99(3):317–322
53. Xia W, Wei Y, Du Y, Liu J, Chang B, Yu YL et al (2009) Nuclear expression of epidermal growth factor receptor is a novel prognostic value in patients with ovarian cancer. Mol Carcinog 48(7):610–617
54. Hadzisejdic I, Mustac E, Jonjic N, Petkovic M, Grahovac B (2010) Nuclear EGFR in ductal invasive breast cancer: correlation with cyclin-D1 and prognosis. Mod Pathol 23(3): 392–403
55. Hoshino M, Fukui H, Ono Y, Sekikawa A, Ichikawa K, Tomita S et al (2007) Nuclear expression of phosphorylated EGFR is associated with poor prognosis of patients with esophageal squamous cell carcinoma. Pathobiology 74(1):15–21
56. Lo HW, Xia W, Wei Y, Ali-Seyed M, Huang SF, Hung MC (2005) Novel prognostic value of nuclear epidermal growth factor receptor in breast cancer. Cancer Res 65(1):338–348
57. Psyrri A, Egleston B, Weinberger P, Yu Z, Kowalski D, Sasaki C et al (2008) Correlates and determinants of nuclear epidermal growth factor receptor content in an oropharyngeal cancer tissue microarray. Cancer Epidemiol Biomarkers Prev 17(6):1486–1492
58. Psyrri A, Yu Z, Weinberger PM, Sasaki C, Haffty B, Camp R et al (2005) Quantitative determination of nuclear and cytoplasmic epidermal growth factor receptor expression in oropharyngeal squamous cell cancer by using automated quantitative analysis. Clin Cancer Res 11(16):5856–5862
59. Dittmann KH, Mayer C, Ohneseit PA, Raju U, Andratschke NH, Milas L et al (2008) Celecoxib induced tumor cell radiosensitization by inhibiting radiation induced nuclear EGFR transport and DNA-repair: a COX-2 independent mechanism. Int J Radiat Oncol Biol Phys 70(1):203–212
60. Hsu SC, Miller SA, Wang Y, Hung MC (2009) Nuclear EGFR is required for cisplatin resistance and DNA repair. Am J Transl Res 1(3):249–258
61. Li C, Iida M, Dunn EF, Ghia AJ, Wheeler DL (2009) Nuclear EGFR contributes to acquired resistance to cetuximab. Oncogene 28(43):3801–3813
62. Kim HP, Yoon YK, Kim JW, Han SW, Hur HS, Park J et al (2009) Lapatinib, a dual EGFR and HER2 tyrosine kinase inhibitor, downregulates thymidylate synthase by inhibiting the nuclear translocation of EGFR and HER2. PLoS One 4(6):e5933
63. Chen YJ, Huang WC, Wei YL, Hsu SC, Yuan P, Lin HY et al (2011) Elevated BCRP/ABCG2 expression confers acquired resistance to gefitinib in wild-type EGFR-expressing cells. PLoS One 6(6):e21428
64. Edwards J, Traynor P, Munro AF, Pirret CF, Dunne B, Bartlett JM (2006) The role of HER1-HER4 and EGFRvIII in hormone-refractory prostate cancer. Clin Cancer Res 12(1): 123–130
65. Bacus SS, Gudkov AV, Zelnick CR, Chin D, Stern R, Stancovski I et al (1993) Neu differentiation factor (heregulin) induces expression of intercellular adhesion molecule 1: implications for mammary tumors. Cancer Res 53(21):5251–5261
66. Xie Y, Hung MC (1994) Nuclear localization of p185neu tyrosine kinase and its association with transcriptional transactivation. Biochem Biophys Res Commun 203(3):1589–1598
67. Wang SC, Lien HC, Xia W, Chen IF, Lo HW, Wang Z et al (2004) Binding at and transactivation of the COX-2 promoter by nuclear tyrosine kinase receptor ErbB-2. Cancer Cell 6(3):251–261

68. Scaltriti M, Rojo F, Ocana A, Anido J, Guzman M, Cortes J et al (2007) Expression of p95HER2, a truncated form of the HER2 receptor, and response to anti-HER2 therapies in breast cancer. J Natl Cancer Inst 99(8):628–638
69. Anido J, Scaltriti M, Bech Serra JJ, Santiago Josefat B, Todo FR, Baselga J et al (2006) Biosynthesis of tumorigenic HER2 C-terminal fragments by alternative initiation of translation. EMBO J 25(13):3234–3244
70. Xia W, Liu Z, Zong R, Liu L, Zhao S, Bacus SS et al (2011) Truncated ErbB2 expressed in tumor cell nuclei contributes to acquired therapeutic resistance to ErbB2 kinase inhibitors. Mol Cancer Ther 10(8):1367–1374
71. Koumakpayi IH, Diallo JS, Le Page C, Lessard L, Filali-Mouhim A, Begin LR et al (2007) Low nuclear ErbB3 predicts biochemical recurrence in patients with prostate cancer. BJU Int 100(2):303–309
72. Koumakpayi IH, Diallo JS, Le Page C, Lessard L, Gleave M, Begin LR et al (2006) Expression and nuclear localization of ErbB3 in prostate cancer. Clin Cancer Res 12(9):2730–2737
73. Thompson M, Lauderdale S, Webster MJ, Chong VZ, McClintock B, Saunders R et al (2007) Widespread expression of ErbB2, ErbB3 and ErbB4 in non-human primate brain. Brain Res 1139:95–109
74. Bueter W, Dammann O, Zscheppang K, Korenbaum E, Dammann CE (2006) ErbB receptors in fetal endothelium—a potential linkage point for inflammation-associated neonatal disorders. Cytokine 36(5–6):267–275
75. Sardi SP, Murtie J, Koirala S, Patten BA, Corfas G (2006) Presenilin-dependent ErbB4 nuclear signaling regulates the timing of astrogenesis in the developing brain. Cell 127(1):185–197
76. Junttila TT, Sundvall M, Lundin M, Lundin J, Tanner M, Harkonen P et al (2005) Cleavable ErbB4 isoform in estrogen receptor-regulated growth of breast cancer cells. Cancer Res 65(4):1384–1393
77. Naresh A, Thor AD, Edgerton SM, Torkko KC, Kumar R, Jones FE (2008) The HER4/4ICD estrogen receptor coactivator and BH3-only protein is an effector of tamoxifen-induced apoptosis. Cancer Res 68(15):6387–6395
78. Arasada RR, Carpenter G (2005) Secretase-dependent tyrosine phosphorylation of Mdm2 by the ErbB-4 intracellular domain fragment. J Biol Chem 280(35):30783–30787
79. Linggi B, Carpenter G (2006) ErbB-4 s80 intracellular domain abrogates ETO2-dependent transcriptional repression. J Biol Chem 281(35):25373–25380
80. Lo HW, Hung MC (2006) Nuclear EGFR signalling network in cancers: linking EGFR pathway to cell cycle progression, nitric oxide pathway and patient survival. Br J Cancer 94(2):184–188
81. Tseng HC, Lyu PC, Lin WC (2010) Nuclear localization of orphan receptor protein kinase (Ror1) is mediated through the juxtamembrane domain. BMC Cell Biol 11:48
82. Fiorillo AA, Medler TR, Feeney YB, Liu Y, Tommerdahl KL, Clevenger CV (2011) HMGN2 inducibly binds a novel transactivation domain in nuclear PRLr to coordinate Stat5a-mediated transcription. Mol Endocrinol 25(9):1550–1564
83. Raper SE, Burwen SJ, Barker ME, Jones AL (1987) Translocation of epidermal growth factor to the hepatocyte nucleus during rat liver regeneration. Gastroenterology 92(5 pt 1): 1243–1250
84. Schausberger E, Eferl R, Parzefall W, Chabicovsky M, Breit P, Wagner EF et al (2003) Induction of DNA synthesis in primary mouse hepatocytes is associated with nuclear pro-transforming growth factor alpha and erbb-1 and is independent of c-jun. Carcinogenesis 24(5):835–841
85. Adam RM, Danciu T, McLellan DL, Borer JG, Lin J, Zurakowski D et al (2003) A nuclear form of the heparin-binding epidermal growth factor-like growth factor precursor is a feature of aggressive transitional cell carcinoma. Cancer Res 63(2):484–490
86. Clevenger CV (2003) Nuclear localization and function of polypeptide ligands and their receptors: a new paradigm for hormone specificity within the mammary gland? Breast Cancer Res 5(4):181–187

87. Kimura H (1993) Schwannoma-derived growth factor must be transported into the nucleus to exert its mitogenic activity. Proc Natl Acad Sci U S A 90(6):2165–2169
88. Hieda M, Isokane M, Koizumi M, Higashi C, Tachibana T, Shudou M et al (2008) Membrane-anchored growth factor, HB-EGF, on the cell surface targeted to the inner nuclear membrane. J Cell Biol 180(4):763–769
89. Bao J, Wolpowitz D, Role LW, Talmage DA (2003) Back signaling by the Nrg-1 intracellular domain. J Cell Biol 161(6):1133–1141
90. Hanada N, Lo HW, Day CP, Pan Y, Nakajima Y, Hung MC (2006) Co-regulation of B-Myb expression by E2F1 and EGF receptor. Mol Carcinog 45(1):10–17
91. Lo HW, Hsu SC, Ali-Seyed M, Gunduz M, Xia W, Wei Y et al (2005) Nuclear interaction of EGFR and STAT3 in the activation of the iNOS/NO pathway. Cancer Cell 7(6):575–589
92. Hung LY, Tseng JT, Lee YC, Xia W, Wang YN, Wu ML et al (2008) Nuclear epidermal growth factor receptor (EGFR) interacts with signal transducer and activator of transcription 5 (STAT5) in activating Aurora-A gene expression. Nucleic Acids Res 36(13):4337–4351
93. Huo L, Wang YN, Xia W, Hsu SC, Lai CC, Li LY et al (2010) RNA helicase A is a DNA-binding partner for EGFR-mediated transcriptional activation in the nucleus. Proc Natl Acad Sci U S A 107(37):16125–16130
94. Huang WC, Hung MC (2009) Induction of Akt activity by chemotherapy confers acquired resistance. J Formos Med Assoc 108(3):180–194
95. Williams CC, Allison JG, Vidal GA, Burow ME, Beckman BS, Marrero L et al (2004) The ERBB4/HER4 receptor tyrosine kinase regulates gene expression by functioning as a STAT5A nuclear chaperone. J Cell Biol 167(3):469–478
96. Hu S, Xie Z, Onishi A, Yu X, Jiang L, Lin J et al (2009) Profiling the human protein-DNA interactome reveals ERK2 as a transcriptional repressor of interferon signaling. Cell 139(3):610–622
97. Tan M, Jing T, Lan KH, Neal CL, Li P, Lee S et al (2002) Phosphorylation on tyrosine-15 of p34(Cdc2) by ErbB2 inhibits p34(Cdc2) activation and is involved in resistance to taxol-induced apoptosis. Mol Cell 9(5):993–1004
98. Sorkin A, von Zastrow M (2009) Endocytosis and signalling: intertwining molecular networks. Nat Rev Mol Cell Biol 10(9):609–622
99. Doherty GJ, McMahon HT (2009) Mechanisms of endocytosis. Annu Rev Biochem 78:857–902
100. Johannes L, Popoff V (2008) Tracing the retrograde route in protein trafficking. Cell 135(7):1175–1187
101. Spooner RA, Smith DC, Easton AJ, Roberts LM, Lord JM (2006) Retrograde transport pathways utilised by viruses and protein toxins. Virol J 3:26
102. Hsu VW, Lee SY, Yang JS (2009) The evolving understanding of COPI vesicle formation. Nat Rev Mol Cell Biol 10(5):360–364
103. Lee MC, Miller EA, Goldberg J, Orci L, Schekman R (2004) Bi-directional protein transport between the ER and Golgi. Annu Rev Cell Dev Biol 20:87–123
104. Cai H, Reinisch K, Ferro-Novick S (2007) Coats, tethers, Rabs, and SNAREs work together to mediate the intracellular destination of a transport vesicle. Dev Cell 12(5):671–682
105. Yang JS, Valente C, Polishchuk RS, Turacchio G, Layre E, Moody DB et al (2011) COPI acts in both vesicular and tubular transport. Nat Cell Biol 13(8):996–1003
106. Boerner JL, Demory ML, Silva C, Parsons SJ (2004) Phosphorylation of Y845 on the epidermal growth factor receptor mediates binding to the mitochondrial protein cytochrome c oxidase subunit II. Mol Cell Biol 24(16):7059–7071
107. Demory ML, Boerner JL, Davidson R, Faust W, Miyake T, Lee I et al (2009) Epidermal growth factor receptor translocation to the mitochondria: regulation and effect. J Biol Chem 284(52):36592–36604
108. Cao X, Zhu H, Ali-Osman F, Lo HW (2011) EGFR and EGFRvIII undergo stress- and EGFR kinase inhibitor-induced mitochondrial translocalization: a potential mechanism of EGFR-driven antagonism of apoptosis. Mol Cancer 10:26

109. Dreier A, Barth S, Goswami A, Weis J (2012) Cetuximab induces mitochondrial translocalization of EGFRvIII, but not EGFR: involvement of mitochondria in tumor drug resistance? Tumour Biol 33(1):85–94
110. Hsu SC, Hung MC (2007) Characterization of a novel tripartite nuclear localization sequence in the EGFR family. J Biol Chem 282(14):10432–10440
111. Giri DK, Ali-Seyed M, Li LY, Lee DF, Ling P, Bartholomeusz G et al (2005) Endosomal transport of ErbB-2: mechanism for nuclear entry of the cell surface receptor. Mol Cell Biol 25(24):11005–11018
112. Lo HW, Ali-Seyed M, Wu Y, Bartholomeusz G, Hsu SC, Hung MC (2006) Nuclear-cytoplasmic transport of EGFR involves receptor endocytosis, importin beta1 and CRM1. J Cell Biochem 98(6):1570–1583
113. Wang ZH, Tian XX, Cheng Y, Yam GH, Ng HK, Ding MX et al (1998) Association of EGFR gene fragments with nuclear matrices in glioblastoma cell lines. Anticancer Res 18(6A):4329–4332
114. Klein C, Gensburger C, Freyermuth S, Nair BC, Labourdette G, Malviya AN (2004) A 120 kDa nuclear phospholipase Cgamma1 protein fragment is stimulated in vivo by EGF signal phosphorylating nuclear membrane EGFR. Biochemistry 43(50):15873–15883
115. Sanderson MP, Keller S, Alonso A, Riedle S, Dempsey PJ, Altevogt P (2008) Generation of novel, secreted epidermal growth factor receptor (EGFR/ErbB1) isoforms via metalloprotease-dependent ectodomain shedding and exosome secretion. J Cell Biochem 103(6):1783–1797
116. Adamczyk KA, Klein-Scory S, Tehrani MM, Warnken U, Schmiegel W, Schnolzer M et al (2011) Characterization of soluble and exosomal forms of the EGFR released from pancreatic cancer cells. Life Sci 89(9–10):304–312
117. Perez-Torres M, Valle BL, Maihle NJ, Negron-Vega L, Nieves-Alicea R, Cora EM (2008) Shedding of epidermal growth factor receptor is a regulated process that occurs with overexpression in malignant cells. Exp Cell Res 314(16):2907–2918
118. Arscott WT, Camphausen KA (2011) EGFR isoforms in exosomes as a novel method for biomarker discovery in pancreatic cancer. Biomark Med 5(6):821
119. Thery C, Ostrowski M, Segura E (2009) Membrane vesicles as conveyors of immune responses. Nat Rev Immunol 9(8):581–593
120. Zhang HG, Grizzle WE (2011) Exosomes and cancer: a newly described pathway of immune suppression. Clin Cancer Res 17(5):959–964
121. Mathivanan S, Ji H, Simpson RJ (2010) Exosomes: extracellular organelles important in intercellular communication. J Proteomics 73(10):1907–1920
122. Simons M, Raposo G (2009) Exosomes—vesicular carriers for intercellular communication. Curr Opin Cell Biol 21(4):575–581
123. Schorey JS, Bhatnagar S (2008) Exosome function: from tumor immunology to pathogen biology. Traffic 9(6):871–881
124. Graner MW, Alzate O, Dechkovskaia AM, Keene JD, Sampson JH, Mitchell DA et al (2009) Proteomic and immunologic analyses of brain tumor exosomes. FASEB J 23(5):1541–1557
125. Zhen Y, Caprioli RM, Staros JV (2003) Characterization of glycosylation sites of the epidermal growth factor receptor. Biochemistry 42(18):5478–5492
126. Higginbotham JN, Demory Beckler M, Gephart JD, Franklin JL, Bogatcheva G, Kremers GJ et al (2011) Amphiregulin exosomes increase cancer cell invasion. Curr Biol 21(9):779–786
127. Progida C, Cogli L, Piro F, De Luca A, Bakke O, Bucci C (2010) Rab7b controls trafficking from endosomes to the TGN. J Cell Sci 123(pt 9):1480–1491
128. Dunham-Ems SM, Pudavar HE, Myers JM, Maher PA, Prasad PN, Stachowiak MK (2006) Factors controlling fibroblast growth factor receptor-1's cytoplasmic trafficking and its regulation as revealed by FRAP analysis. Mol Biol Cell 17(5):2223–2235
129. Stachowiak MK, Maher PA, Stachowiak EK (2007) Integrative nuclear signaling in cell development—a role for FGF receptor-1. DNA Cell Biol 26(12):811–826
130. Reilly JF, Maher PA (2001) Importin beta-mediated nuclear import of fibroblast growth factor receptor: role in cell proliferation. J Cell Biol 152(6):1307–1312

Chapter 9
Efficient Enhancement of Signalling Capacity: The Ubiquitin System

Daniela Hoeller and Ivan Dikic

Abstract Ubiquitin (Ub) has emerged as a potent cellular signal that regulates a wide variety of processes in normal as well as malignant cells. Its potency is based on an elaborate system of "writing" and "reading" a diverse Ub code on substrate proteins. In this chapter we give an overview on the principles of Ub-based signalling and explain how efficiency and specificity is ensured in the cell. Moreover, we discuss how the capacity of Ub signalling is further enhanced by crosstalk with other post-translational modifications such as phosphorylation and acetylation.

9.1 Ubiquitin: A Signalling Module

9.1.1 Biochemical and Biophysical Properties of Ubiquitin

Ubiquitin is a highly conserved protein of 76 amino acids that fold into a compact, globular structure. The flexible C-terminal tail pokes out from this structure and exposes the characteristic RGG motif that can be covalently linked to lysine residues of substrate proteins. This event requires the subsequent action of three enzymes (Fig. 9.1): the Ub-activating enzyme (E1), the Ub-conjugating enzyme (E2) and the Ub-ligating enzyme (E3). In the first step of this enzymatic cascade, E1 activates the C-terminal glycine (G) by forming an E1~Ub thiolester in an ATP-dependent manner. The activated Ub is then transferred from E1 to the active cysteine of E2 forming an E2~Ub thiolester. In the last step, E2~Ub cooperates with the E3 Ub ligase to covalently couple Ub to the substrate protein by forming an isopeptide bond between the ε-amino group of a lysine and the C-terminal G of Ub (reviewed

D. Hoeller (✉) • I. Dikic (✉)
Buchmann Institute for Molecular Life Sciences and Institute of Biochemistry II,
Goethe University School of Medicine, Theodor-Stern-Kai 7, 60590 Frankfurt, Germany
e-mail: hoeller.d@gmail.com; Ivan.Dikic@biochem2.de

Y. Yarden and G. Tarcic (eds.), *Vesicle Trafficking in Cancer*,
DOI 10.1007/978-1-4614-6528-7_9,

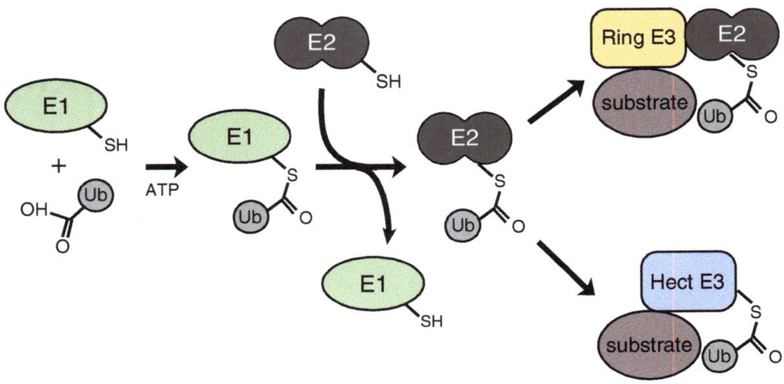

Fig. 9.1 Ub activation and conjugation by E1, E2 and E3. The covalent attachment of ubiquitin (Ub) to a substrate requires the subsequent action of three enzymes. In the first step, the C-terminus of Ub is activated in an ATP-dependent manner by forming a thiolester with the E1 Ub-activating enzyme. It is then passed on to the E2 conjugating enzyme. Ubiquitination occurs when an E3 ligase enzyme binds to both substrate and E2 ~ Ub, bringing them in proximity so that the ubiquitin is transferred from the E2 to the ε-amino group of a lysine in the substrate. Whereas RING-type ligases mediate the direct transfer of Ub from E2 to the substrate, HECT-type ligases form themselves a thioester intermediate with Ub to achieve ubiquitination

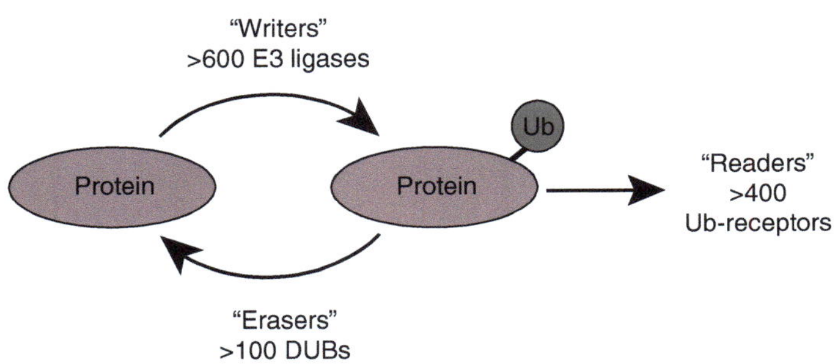

Fig. 9.2 Concept of Ub-based signalling. E3 enzymes (in cooperation with E2 enzymes) function as "writers" of Ub signals. They identify the substrate and assemble the Ub code. Up to now, around 600 E3 ligases have been identified. The message of the "writers" can be "erased" by de-ubiquitinating enzymes (DUBs). Currently around 100 "erasers" are known. Ub receptors are the "readers" of the Ub code. They are equipped with one or more Ub-binding domains (UBDs) that can distinguish different types of Ub modifications. The list of Ub receptors (and UBDs) is growing. To date around 1,000 proteins are known to function as "readers"

in [18, 34, 42]). This process is spatiotemporally regulated inside the cell, thus providing the compartment specific dynamics in the ubiquitination events [14]. To a lesser extent, also serine, threonine, cysteine [5, 37, 44] as well as the N-terminal amino group of the substrate protein [8] serve as acceptor sites for Ub. The ubiquitinated substrate can be relieved of its tag by the action of de-ubiquitinating enzymes (DUBs) that cleave Ub from the modified protein and recycle it to the cellular pool (Fig. 9.2) (reviewed in [2]). Five different subfamilies of DUBs are known:

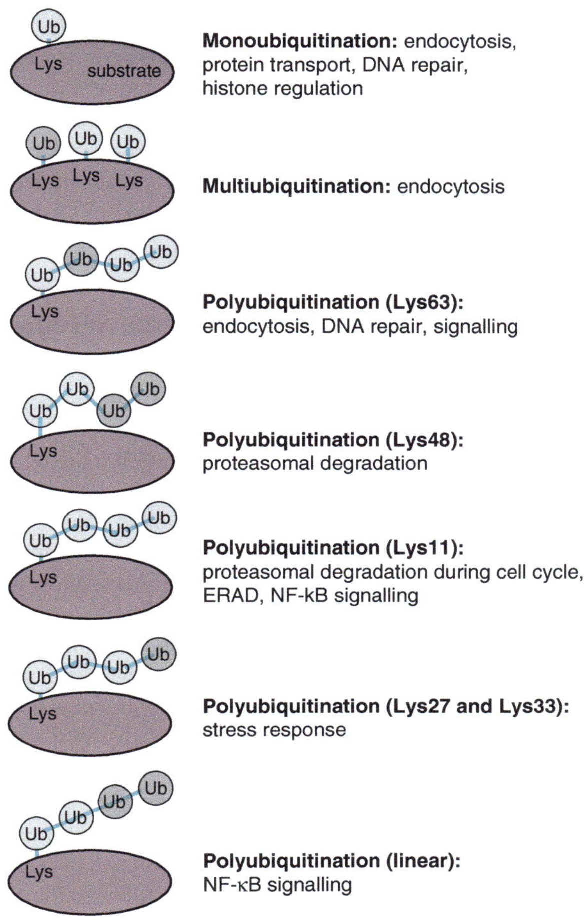

Fig. 9.3 Different types of Ub modification and their cellular function. Various types of ubiquitination have been detected in cells and linked to the indicated cellular processes. The attachment of a single Ub molecule is called "monoubiquitination". The monoubiquitination of multiple lysines within a substrate molecule results in "multiubiquitination". Polyubiquitination occurs when a lysine in ubiquitin serves as acceptor side in iterative rounds of ubiquitination. Dependent on the utilized lysine chains of different topologies is assembled. There is evidence that each possible type of Ub chains exists in the cell. However, the respective function has been studied for only few linkage types

ubiquitin-specific proteases (USP) (forming the largest family), ubiquitin carboxy-terminal hydrolases (UCH), Machado-Joseph disease protein domain proteases (MJD), ovarian tumour (otubain) proteases (OTU) and JAB1/MPN/Mov34 metalloenzyme motif proteases (JAMM) [29].

Ub itself contains seven lysines (Lys6, Lys11, Lys27, Lys29, Lys33, Lys48 and Lys63) that allow homotypic linkage types as well as a great variety of heterotypic linkages. In this way ubiquitination provides an enormous spectrum of specific and reversible signals that affect almost all aspects of cellular physiology. Indeed, each type of ubiquitination seems to trigger a specific cellular answer [24, 43] (Fig. 9.3). The attachment of a single Ub molecule (monoubiquitination) serves as an important signal during endocytic trafficking and lysosomal degradation of activated cell surface receptors. Lys48-linked chains label proteins for proteasomal degradation, Lys63-linked chains act as sorting signals and Lys11-linked chains promote proteasomal degradation of targets during cell cycle progression [25] and play an important role in endoplasmic reticulum-associated degradation (ERAD) [49]. The cellular roles of many other types of ubiquitination have not yet been studied in detail. However, mass spectrometry of cellular Ub conjugates has proven that each possible linkage type exists in the cell [49]. In order to exploit this signalling potential, the cell needs to ensure that the correct substrates are identified and modified at the correct time points with the correct type of Ub tag.

9.1.2 What Determines Substrate Specificity?

The attachment of ubiquitin to another protein is supposed to change its biochemical and biophysical properties [23]. It is therefore very important to ensure that the correct substrate is selected and modified. Within the ubiquitination machinery, it is the E3 ligase that is endowed with substrate specificity because it is the only component that directly binds to the substrate whereas neither E1 nor E2 touch it. Hence, the functions of Ub ligases are tightly regulated by signal-induced mechanisms such as compartmentalization, degradation, oligomerization and post-translational modifications (PTMs). The human proteome contains app. 600 E3 enzymes that can be divided in three groups, RING type [9], homologous to E6-AP carboxyl terminus (HECT) type [39] and U-box type. The most prevalent E3 ligases contain a zinc-finger domain called RING (really interesting new gene) that has an essential role in Ub transfer but does not possess catalytic activity itself. The RING-type E3 ligases rather function to bridge E2~Ub and the substrate. In contrast, the HECT-type ligases take over Ub from the E2-conjugating enzyme and catalyze the transfer to the substrate without further help of E2. The U-box E3 ligases are characterized by a modified RING motif and act in a similar way as RING E3s. Regardless of their way of Ub conjugation, E3 ligases have to first identify their substrates within a huge pool of proteins. In addition, efficient signal transduction usually requires that proteins are modified at a certain time point, e.g. during cell cycle or receptor stimulation, whereas before and/or after this event they are not supposed to be ubiquitinated. How can this be accomplished? In the first instance, specificity is provided by distinct substrate binding sites on the E3 and is further enhanced by additional layers of regulation.

The most common strategy to regulate ubiquitination events is to induce phosphorylation of the substrate, thereby creating (or destroying) a recognition signal for binding of an E3 ligase. This happens, for example, during downregulation of ligand-activated EGFR. Transphosphorylation at position Tyr1045 allows docking of c-Cbl that mediates multiubiquitination and thus ensures lysosomal sorting of the receptor [15]. Besides the substrate, the E3 ligase can be regulated by phosphorylation: for example, phosphorylation of c-Cbl at the linker region releases the autoinhibited conformation and allows E2 binding. This activation is required for EGFR ubiquitination [11]. Another example is the activation of Itch through JNK Ser/Thr phosphorylation [12] or inhibition through Fyn-mediated Tyr phosphorylation in the WW domain region [50].

The spatial separation of substrate and ligase (subcellular compartmentalization) is another mechanism by which the ubiquitination of proteins can be regulated. Specific signals induce cellular translocation of either substrate or ligase and allow the ligase to encounter its substrate [14].

A special way to achieve substrate selectivity is employed by multimeric cullin-RING ligases (CRLs) [33]. The best-studied example of CRLs are the SCF (Skp1-Cdc53/cullin F-box protein) complexes that contain four subunits: a scaffold (Cdc53/cullin) that binds the RING-H2 finger protein and the adaptor protein (Skp1) and finally, a substrate-recognition subunit (F-box protein). The adaptor Skp1 binds the F-box protein, which acts as the substrate-recognition subunit that positions the substrate for ubiquitination by the E2. The dynamic exchange of F-box proteins creates unique SCF complexes that bind and ubiquitinate distinct sets of substrates. To be fully active, SCF ligases need to be post-translationally modified with Nedd8 (neddylation), which promotes the recruitment of the E2 through direct interaction between Nedd8 and the E2 [27].

9.1.3 What Determines Linkage Topology?

Ubiquitin modifications come in many flavours: monoubiquitination (attachment of a single ubiquitin molecule), multiubiquitination (multiple single Ub molecules are attached) and polyubiquitination (Ub chains of specific linkage types are formed on the substrate) (Fig. 9.3). Importantly, ubiquitin tags are not generated randomly on a substrate but in a highly specific way and like molecular signatures each type of modification has specific consequences. It is not yet fully understood how the ubiquitination machinery accomplishes the generation of distinct tags based on a single building block. Why does a ligase stop after adding monoubiquitin to a given substrate while on another substrate it assembles polyubiquitin chains? And how is linkage specificity achieved?

Cellular strategies to create monoubiquitin signals include the use of monoubiquitination-dedicated E2-conjugating enzymes (e.g. Ube2w) or the trimming of a growing chain by DUBs. In the simplest scenario for the assembly of a specific type of chain, the E3 ligase exhibits a preference for one type of linkage. This is, for example, the case for some HECT ligases. Rsp5 is specialized on K63 chains; E6AP catalyzes mostly the assembly of K48 chains, while KIAA10 maintains some flexibility by conjugating both K48 and K29. It appears that conjugational specificity is primarily encoded in its C-terminal HECT lobe domain and is independent of the type of E2 used [28]. Unlike HECT ligases, RING-type ligases do not possess enzymatic activity themselves but operate by bridging the Ub-loaded E2 enzyme and the substrate. In this constellation the E2 enzymes play a major role in specifying the chain type. Ube2w supports monoubiquitination, Ubc13-Mms2 E2 conjugates K63 chains and Ube2S is specific for K11 chains [46], while Ube2k, CDC34 and E2-25K have been shown to conjugate K48 chains. By cooperating with different E2 enzymes, RING ligases can promote the generation of various types of Ub chains. BRCA1, for example, associates with at least eight diverse E2 enzymes, which govern its conjugational specificity [4, 7]. BRCA1 is also a classic example of a RING-type E3 ligase that can augment its conjugational ability by dimerization with another ligase (BARD1), allowing K6 polyubiquitin chain assembly [47]. Another example is LUBAC (linear ubiquitin chain assembly complex), a 600 kDa multisubunit ligase that utilizes another strategy of achieving conjugational specificity (reviewed in [43]). The ligase plays a major role in NF-κB signalling and is special because it assembles head-to-tail Ub chains. Instead of linking the Ub molecules via a lysine of the proximal Ub (the Ub molecule attached to the target protein), it utilizes the free N-terminal amino group. The resulting Ub chains exhibit an extended conformation that can be recognized in a specific manner with the ubiquitin-binding domain of NEMO adaptor molecule [35]. However, linear ubiquitin can also form a closed conformation [38]. Moreover, Lys-linked ubiquitin chains are dynamically changed by shifting from closed to open conformations (reviewed in [10]).

9.1.4 Diversity of Ub Signals: Different Types of Ub Modification and Their Functions

Like other PTMs, the attachment of Ub has a profound impact on the targeted protein. Allosteric effects may regulate binding specificity and/or enzymatic activity or expose modification sites or localization signals. The importance of Ub-dependent protein localization has been recently shown for Ras proteins (NRas, HRas and KRas) that play a central role in transducing signals that control cell proliferation, differentiation, motility and survival. Ras proteins can dynamically partition to different cellular membranes, and their localization dictates the signalling outcome. Rabex-5 is an E3 ligase that mono- and diubiquitinates Ras proteins and thereby promotes their endosomal association [48] as well as endosomal-specific signalling (or signal attenuation as is the case for Erk2). Yet, the factor that traps ubiquitinated Ras proteins at endosomal membranes needs to be identified.

In addition of having a direct impact on the functionality or the localization of the modified protein, new interaction surfaces are created that spur the assembly or disassembly of protein complexes with specific functions. Moreover, Ub signals are propagated by Ub receptors that detect Ub modifications on proteins. These effector proteins are equipped with Ub-binding domains (UBDs) that discriminate between different types of Ub modification [10].

9.2 Ub-Binding Proteins: Readers of Ub Signals

9.2.1 Different Types of UBDs and How They Manage to Discriminate Ub Signatures

There are currently around 20 known families of UBDs that can be divided into several subfamilies based on their structure [10]: UBDs containing single or multiple alpha helices (UBA, UIM, DUIM, MIU, CUE, GAT, VHS, UBAN), zinc fingers

(NZF, ZnF-UBP, ZnF-A20, UBZ), pleckstrin-homology fold (GLUE, PRU), UBC-like (UEV, UBC) or other structures (SH3, ubiquitin-binding motif [UBM], PFU, Jab1/MPN). Regardless of this apparent structural diversity, almost all of these UBDs bind to the same hydrophobic patch centred on Ile44 of the Ub molecule. There are only some exceptions to this rule, including the UBM of some Y-family translesion synthesis (TLS) polymerases (binds to the hydrophobic patch centred on Leu8), ZnF A20 domain of Rabex-5 (interacts with Asp58) and the ZnF-UBP domain of isopeptidase T (USP5) (contacts the C-terminal part of ubiquitin as well as a surface centred on Ile36).

Given the common mode of binding to Ub, the question arises on how UBDs distinguish different Ub modifications. In fact, in vitro most UBDs interact with monoubiquitin and do not show a clear preference for one linkage type either. Only a small subset of UBDs is known to recognize the linker region between two ubiquitin moieties and is thus able to directly discriminate between different linkage types. All other UBDs rely on linkage-specific avidity; in other words, they are able to detect the spatial distribution and positioning of the individual units of the chain without interacting with the linkage [10]. In order to do so, Ub receptors are either equipped with multiple UBDs or dimerize/oligomerize. Different types of chains have specific conformations in solution. K48-linked di- and tetraubiquitin chains adopt a closed conformation, with the hydrophobic patch surfaces around Ile44 buried at the inter-domain interface (interactions with UBD-containing proteins still occur since this conformation is dynamic, oscillating between the open and packed structure) [40]. Many UBA domains bind K48-linked chains in strong preference to monoubiquitin and other linkage types. K63-linked diubiquitin chains form an extended conformation without any contacts between hydrophobic surfaces and are fully exposed. UIM domains such as those present in endocytic adaptor proteins or NZF domains of TAB2 and TAB3 (subunits of the TAK1 kinase complex) show a high preference for K63-linked chains [10]. It should be noted however that the situation in vivo seems to be complex because the same type of domain can have another linkage specificity when present in a different Ub receptor. It is becoming clear that linkage-specific binding of UBDs is greatly influenced by the microenvironment in which the interaction takes place. Moreover, recent data indicate that the affinity and/or specificity of UBDs can be modulated by phosphorylation [31].

9.2.2 How Can Low-Affinity Ub/UBD Interactions Support a Specific and Efficient Signalling Network?

Another notable feature of UBD-Ub interactions is the low binding affinity, with Kd values in the micromolar range. However, interactions studies are mostly performed in vitro using the isolated UBD rather than the integral Ub receptor. In vivo avidity due to oligomerization of ubiquitin receptors and multi- or polyubiquitinated substrates, the presence of multiple UBDs in the Ub receptor (different UBDs in the same protein or multiplication of the same UBD within protein) plays a central role. Moreover, in comparison to recognition of phosphorylated proteins by phospho-specific-binding domains, it strikes that the interaction of monoUb with UBDs generally does not depend on the sequence of amino acids immediately around the ubiquitinated residue, whereas phospho-binding domains are influenced by the primary sequence embedding the pSer/pThr/pY. This raises the question of how specificity is determined. One likely possibility is that proteins harbouring a UBD have weak but direct interactions with proteins that are ubiquitinated or both proteins are part of the same large protein complex (signalosome). Such a bipartite interaction provides increased affinity and specificity (e.g. binding of Y-polymerases to proliferating cell nuclear antigen (PCNA) during TLS is mediated via Ub/UBZ and PCNA/PIP box interactions, see below).On the other hand, it has been proposed that the endolysosomal transport of ubiquitinated cell surface proteins indeed benefits from a low-affinity interaction network because this facilitates the efficient passage of the cargo along multiple sorting complexes [17].

9.2.3 UBD Proteins Regulate Specific Cellular Responses

The first described function of protein ubiquitination was the degradation of the modified protein by the proteasome. It is now clear that ubiquitination serves as an important non-proteolytic signal in numerous cellular processes including endocytosis, DNA repair, NF-κB signalling, virus budding and trafficking. The key elements in Ub-dependent processes are Ub receptors (UBD-containing proteins) that provide the link between Ub signals and a specific cellular pathway. Examples are endocytic adaptor proteins such as Epsin, Eps15, Hrs and STAM2. They are responsible for the delivery of plasma membrane proteins (e.g. activated RTKs) to the lysosome [22]. All of them contain one or more UBDs (mostly UIM domains) that detected monoUb as well as K63-linked diubiquitin on the endocytic cargo and link it with the sorting machinery. Proteins that are not labelled with ubiquitin are not recognized and recycled back to the plasma membrane instead of being degraded in the lysosome [22].

TLS is another pathway that depends on an ubiquitin signal. Upon DNA damage, monoubiquitination of PCNA induces bypass of the lesion by directing the replication machinery into the TLS pathway. The crucial step here is the recognition of the monoubiquitin by the UBZ domains of Y-polymerases. Y-polymerases also contain a PIP box that binds directly to PCNA. In order to induce TLS, both interaction surfaces must be engaged in binding [3].

9.3 Crosstalk Between Ub Signals and Other Post-translational Modifications

An additional degree of signalling complexity is generated by the crosstalk of ubiquitination with other PTMs such as phosphorylation or acetylation. The crosstalk usually functions in both directions and can be positive (one modification triggers the addition or removal of a second PTM) or negative (direct competition for modification of a single residue in a protein).

9.3.1 *Phosphorylation and Ubiquitylation: Inseparable Duo*

The intersection between phosphorylation and ubiquitination is probably the best-studied example of PTM crosstalk [21]. Both types of modifications are very abundant and implicated in the regulation of virtually any cellular process. Since they do not compete for the same residues, the nature of their crosstalk is positive. They share a number of common features (inducibility, reversibility, large number of transferases, i.e. kinases/Ub ligases, huge number of cellular targets and ATP dependence), and there are several cross–points of interactions between these systems. As indicated above phosphorylation events are fundamental for the specificity as well as the right timing of ubiquitination. Phosphorylation can regulate ubiquitination at multiple levels: (1) activity of the E3 ligases, (2) recognition of the substrate by E3 ligase and (3) substrate availability (subcellular compartmentalization) [14]. There is accumulating evidence that also the reading of Ub signals by Ub receptors can be regulated by phosphorylation [31]. On the other side, ubiquitination can affect kinase activity via degradation or allosteric activation/inactivation as well as Ub-dependent shuttling of kinase or substrate [21].

9.3.1.1 Regulation of Ubiquitination by Phosphorylation-Mediated Compartmentalization

Subcellular compartmentalization is an important and widely used strategy of cells to separate cellular processes such as DNA synthesis (nucleus), energy production (mitochondria) and protein/organelle digestion (lysosomes) for most efficient performance in a specialized environment. The same concept is also used for signal transduction, and the regulated trafficking of effector proteins between different compartments provides a means to control signalling pathways [32, 45]. Very often ubiquitination and phosphorylation team up to orchestrate these events. For example, the E3 ligase Parkin and the PINK1 kinase co-operate during autophagic clearance of depolarized mitochondria [13, 30]. While PINK1 is a mitochondrial protein, Parkin is evenly distributed in the cytosol. Parkin is recruited to depolarized mitochondria to promote their autophagic removal. Subcellular redistribution of Parkin requires mitochondrial localization and kinase activity of PINK1, while phosphorylation targets remain unknown. Once present on mitochondria Parkin ubiquitinates certain mitochondrial surface proteins such as VDAC1 and Mfn1/2 and initiates a cascade of events that govern the mitophagic clearance of damaged mitochondria [13, 30].

Another example is provided by the tumour suppressor p27 that is present in the nucleus where it acts as a cell cycle inhibitor. Phosphorylation on certain residues promotes the cytoplasmic translocation of p27 in G1 phase where it becomes a substrate of the cytosolic KPC ligase complex [26].

9.3.1.2 Regulation of Ub Receptors by Phosphorylation

Not only the ubiquitination machinery can be subjected to phosphorylation-dependent regulation. There is accumulating evidence that the modulation of Ub receptors by phosphorylation of critical residues within the UBD is a common way to affect their ability to read Ub signals. In principle, this modification can enhance, reduce or shift specificity of Ub binding. A recent example is provided by Syntenin-1, a PDZ domain-containing adaptor that controls trafficking of transmembrane proteins. Syntenin-1 contains a novel type of UBD characterized by a conserved LYPSL motif. It was shown that the Ser within this motif is targeted by the kinase Ulk1, thereby abolishing Ub binding of Syntenin-1 [36].

Another intriguing case of UBD regulation by phosphorylation is the UBA of the autophagic adaptor p62. Casein kinase 2 (CK2) mediates specific phosphorylation of p62 within its UBA domain. This modification increases the affinity between UBA domain and ubiquitin chains and enhances the ability of p62 to act as an autophagy receptor for ubiquitinated protein aggregates [31].

9.3.1.3 Regulation of Kinase Activity by Ubiquitination

Ubiquitination can affect kinase activity in two different ways: (1) permanent inactivation by degradation and (2) activation or inactivation by non-proteolytic ubiquitination [21]. The degradation of soluble protein kinases is mediated by K48-linked polyubiquitination and subsequent proteasomal cleavage. Receptor-tyrosine kinases, on the other hand, are degraded in the lysosome. Ligand-induced K63-linked diubiquitination and multiple monoubiquitination trigger internalization and endolysosomal sorting of the activated receptor. There is evidence that the activated receptor continues signalling until it reaches the lysosome. Importantly, the emitted signals (i.e. activated downstream effectors) differ from those emitted at the plasma membrane, and thus, ubiquitination influences not only the quantity but also the quality of kinase activity [20].

Protein kinases can also be negatively regulated by ubiquitination without being degraded. For example, polyubiquitination of AMPK (AMP-activated protein kinase)-related kinases, NUAK1 and MARK4, with K29- and K33-linked chains was shown to reduce their activity without affecting protein stability [1]. Ubiquitination can also support the activation of a kinase. For example, the TAK1 kinase complex is activated downstream of cytokine receptors and Toll-like receptors through TRAF6-mediated K63-linked polyubiquitination of TRAF6 or IRAK1. This leads to the recruitment of TAK1, which allows transphosphorylation and activation of TAK1 probably via a dimer proximity mechanism [6]. Another example of Ub-dependent kinase activation is provided by the activation of inhibitor-kappaB kinase (IKK) complex, which is composed of two kinases alpha and beta and one non-catalytic subunit called NEMO. NEMO ubiquitination and the ability to bind to linear and other Lys-linked ubiquitin chains are critical for the conformational change of the kinase complex, activation of kinase domains and the stimulation of the NF-κB pathway [35].

9.3.2 Acetylation Modulates Ubiquitin Networks

As acetylation and ubiquitination target lysine residues, both modifications in some cases directly compete with each other for the substrate. Thus proteins can be stabilized by acetylation of lysines that would otherwise serve as ubiquitination sites leading to protein degradation. A prominent example is the N-end rule pathway [41]. Proteins with acetylated amino-terminal group escape degradation via N-terminal ubiquitination.

In the case of Notch signalling, which is essential for cell specification and tissue patterning, it was shown reversible acetylation of the Notch1 intracellular domain (NICD) on conserved lysine residues adjusts the amplitude and duration of Notch responses by altering ubiquitination and protein turnover by blocking ubiquitination of the lysines. The NAD^+-dependent deacetylase SIRT1 associates with NICD and opposes the acetylation-induced NICD stabilization [16]. This is relevant for promoting a non-sprouting, stalk cell-like phenotype and impaired growth, sprout elongation and enhanced Notch target gene expression in response to DLL4 stimulation.

9.4 Conclusion

Rapid scientific progress in the last decade has revealed that cells utilize ubiquitin not only for regulation of protein stability but also for the establishment and enhancement of signalling capacity. In such signalling networks, ubiquitin, either as a monomer or diverse ubiquitin chains, functions as a versatile cellular signal that is recognized by more than 1,000 ubiquitin-binding proteins. Development of tools to identify and decode ubiquitin signals in vivo as well as visualization of their interplay in situ remains an important technological advance. In particular, recent progress in mass spectroscopy and proteomics, coupled with the development of ubiquitin chain-specific antibodies and ubiquitin sensors, has provided deeper insight into the complexity of ubiquitin signalling pathways. The remaining challenges are associated with the delineation of the spatio-temporal clues controlling the ubiquitin networks and the dissection of pathways by which deregulation of ubiquitin signalling results in the initiation and progression of various human diseases including neurodegeneration, inflammation and cancer [19].

Acknowledgement We would like to thank Tobias Kensche for his help with references and comments on the manuscript.

References

1. Al-Hakim AK, Zagorska A, Chapman L, Deak M, Peggie M, Alessi DR (2008) Control of AMPK-related kinases by USP9X and atypical Lys(29)/Lys(33)-linked polyubiquitin chains. Biochem J 411(2):249–260
2. Amerik AY, Hochstrasser M (2004) Mechanism and function of deubiquitinating enzymes. Biochim Biophys Acta 1695(1–3):189–207
3. Bienko M, Green CM, Crosetto N, Rudolf F, Zapart G, Coull B, Kannouche P, Wider G, Peter M, Lehmann AR, Hofmann K, Dikic I (2005) Ubiquitin-binding domains in Y-family polymerases regulate translesion synthesis. Science 310(5755):1821–1824
4. Brzovic PS, Keeffe JR, Nishikawa H, Miyamoto K, Fox D III, Ohta T, Fukuda M, Klevit R (2003) Binding and recognition in the assembly of an active BRCA1/BARD1 ubiquitin-ligase complex. Proc Natl Acad Sci U S A 100(10):5646–5651
5. Cadwell K, Coscoy L (2005) Ubiquitination on nonlysine residues by a viral E3 ubiquitin ligase. Science 309(5731):127–130
6. Chen ZJ (2012) Ubiquitination in signaling to and activation of IKK. Immunol Rev 246(1):95–106
7. Christensen DE, Brzovic PS, Klevit RE (2007) E2-BRCA1 RING interactions dictate synthesis of mono- or specific polyubiquitin chain linkages. Nat Struct Mol Biol 14(10):941–948
8. Ciechanover A, Ben-Saadon R (2004) N-terminal ubiquitination: more protein substrates join in. Trends Cell Biol 14(3):103–106
9. Deshaies RJ, Joazeiro CA (2009) RING domain E3 ubiquitin ligases. Annu Rev Biochem 78:399–434
10. Dikic I, Wakatsuki S, Walters KJ (2009) Ubiquitin-binding domains—from structures to functions. Nat Rev Mol Cell Biol 10(10):659–671
11. Dou H, Buetow L, Hock A, Sibbet GJ, Vousden KH, Huang DT (2012) Structural basis for autoinhibition and phosphorylation-dependent activation of c-Cbl. Nat Struct Mol Biol 19(2):184–192
12. Gallagher E, Gao M, Liu YC, Karin M (2006) Activation of the E3 ubiquitin ligase Itch through a phosphorylation-induced conformational change. Proc Natl Acad Sci USA 103(6):1717–1722
13. Geisler S, Holmstrom KM, Skujat D, Fiesel FC, Rothfuss OC, Kahle PJ, Springer W (2010) PINK1/Parkin-mediated mitophagy is dependent on VDAC1 and p62/SQSTM1. Nat Cell Biol 12(2):119–131
14. Grabbe C, Husnjak K, Dikic I (2011) The spatial and temporal organization of ubiquitin networks. Nat Rev Mol Cell Biol 12(5):295–307
15. Grovdal LM, Stang E, Sorkin A, Madshus IH (2004) Direct interaction of Cbl with pTyr 1045 of the EGF receptor (EGFR) is required to sort the EGFR to lysosomes for degradation. Exp Cell Res 300(2):388–395
16. Guarani V, Deflorian G, Franco CA, Kruger M, Phng LK, Bentley K, Toussaint L, Dequiedt F, Mostoslavsky R, Schmidt MH, Zimmermann B, Brandes RP, Mione M, Westphal CH, Braun T, Zeiher AM, Gerhardt H, Dimmeler S, Potente M (2011) Acetylation-dependent regulation of endothelial Notch signalling by the SIRT1 deacetylase. Nature 473(7346):234–238
17. Haglund K, Di Fiore PP, Dikic I (2003) Distinct monoubiquitin signals in receptor endocytosis. Trends Biochem Sci 28(11):598–603
18. Hershko A, Ciechanover A (1998) The ubiquitin system. Annu Rev Biochem 67:425–479
19. Hoeller D, Dikic I (2009) Targeting the ubiquitin system in cancer therapy. Nature 458(7237):438–444
20. Hoeller D, Volarevic S, Dikic I (2005) Compartmentalization of growth factor receptor signalling. Curr Opin Cell Biol 17(2):107–111
21. Hunter T (2007) The age of crosstalk: phosphorylation, ubiquitination, and beyond. Mol Cell 28(5):730–738
22. Hurley JH, Stenmark H (2011) Molecular mechanisms of ubiquitin-dependent membrane traffic. Annu Rev Biophys 40:119–142

23. Ikeda F, Crosetto N, Dikic I (2010) What determines the specificity and outcomes of ubiquitin signaling? Cell 143(5):677–681
24. Ikeda F, Dikic I (2008) Atypical ubiquitin chains: new molecular signals. 'Protein Modifications: Beyond the Usual Suspects' review series. EMBO Rep 9(6):536–542
25. Jin L, Williamson A, Banerjee S, Philipp I, Rape M (2008) Mechanism of ubiquitin-chain formation by the human anaphase-promoting complex. Cell 133(4):653–665
26. Kamura T, Hara T, Matsumoto M, Ishida N, Okumura F, Hatakeyama S, Yoshida M, Nakayama K, Nakayama KI (2004) Cytoplasmic ubiquitin ligase KPC regulates proteolysis of p27(Kip1) at G1 phase. Nat Cell Biol 6(12):1229–1235
27. Kawakami T, Chiba T, Suzuki T, Iwai K, Yamanaka K, Minato N, Suzuki H, Shimbara N, Hidaka Y, Osaka F, Omata M, Tanaka K (2001) NEDD8 recruits E2-ubiquitin to SCF E3 ligase. EMBO J 20(15):4003–4012
28. Kim HC, Huibregtse JM (2009) Polyubiquitination by HECT E3s and the determinants of chain type specificity. Mol Cell Biol 29(12):3307–3318
29. Komander D, Clague MJ, Urbe S (2009) Breaking the chains: structure and function of the deubiquitinases. Nat Rev Mol Cell Biol 10(8):550–563
30. Matsuda N, Sato S, Shiba K, Okatsu K, Saisho K, Gautier CA, Sou YS, Saiki S, Kawajiri S, Sato F, Kimura M, Komatsu M, Hattori N, Tanaka K (2010) PINK1 stabilized by mitochondrial depolarization recruits Parkin to damaged mitochondria and activates latent Parkin for mitophagy. J Cell Biol 189(2):211–221
31. Matsumoto G, Wada K, Okuno M, Kurosawa M, Nukina N (2011) Serine 403 phosphorylation of p62/SQSTM1 regulates selective autophagic clearance of ubiquitinated proteins. Mol Cell 44(2):279–289
32. Miaczynska M, Bar-Sagi D (2010) Signaling endosomes: seeing is believing. Curr Opin Cell Biol 22(4):535–540
33. Nakayama KI, Nakayama K (2006) Ubiquitin ligases: cell-cycle control and cancer. Nat Rev Cancer 6(5):369–381
34. Pickart CM (2004) Back to the future with ubiquitin. Cell 116(2):181–190
35. Rahighi S, Ikeda F, Kawasaki M, Akutsu M, Suzuki N, Kato R, Kensche T, Uejima T, Bloor S, Komander D, Randow F, Wakatsuki S, Dikic I (2009) Specific recognition of linear ubiquitin chains by NEMO is important for NF-kappaB activation. Cell 136(6):1098–1109
36. Rajesh S, Bago R, Odintsova E, Muratov G, Baldwin G, Sridhar P, Overduin M, Berditchevski F (2011) Binding to syntenin-1 protein defines a new mode of ubiquitin-based interactions regulated by phosphorylation. J Biol Chem 286(45):39606–39614
37. Ravid T, Hochstrasser M (2007) Autoregulation of an E2 enzyme by ubiquitin-chain assembly on its catalytic residue. Nat Cell Biol 9(4):422–427
38. Rohaim A, Kawasaki M, Kato R, Dikic I, Wakatsuki S (2012) Structure of a compact conformation of linear diubiquitin. Acta Crystallogr D Biol Crystallogr 68(Pt 2):102–108
39. Rotin D, Kumar S (2009) Physiological functions of the HECT family of ubiquitin ligases. Nat Rev Mol Cell Biol 10(6):398–409
40. Ryabov Y, Fushman D (2006) Interdomain mobility in di-ubiquitin revealed by NMR. Proteins 63(4):787–796
41. Sriram SM, Kim BY, Kwon YT (2011) The N-end rule pathway: emerging functions and molecular principles of substrate recognition. Nat Rev Mol Cell Biol 12(11):735–747
42. Varshavsky A (2005) Regulated protein degradation. Trends Biochem Sci 30(6):283–286
43. Walczak H, Iwai K, Dikic I (2012) Generation and physiological roles of linear ubiquitin chains. BMC Biol 10:23
44. Wang X, Herr RA, Chua WJ, Lybarger L, Wiertz EJ, Hansen TH (2007) Ubiquitination of serine, threonine, or lysine residues on the cytoplasmic tail can induce ERAD of MHC-I by viral E3 ligase mK3. J Cell Biol 177(4):613–624
45. Wortzel I, Seger R (2011) The ERK cascade: distinct functions within various subcellular organelles. Genes Cancer 2(3):195–209
46. Wu T, Merbl Y, Huo Y, Gallop JL, Tzur A, Kirschner MW (2010) UBE2S drives elongation of K11-linked ubiquitin chains by the anaphase-promoting complex. Proc Natl Acad Sci USA 107(4):1355–1360

47. Wu-Baer F, Lagrazon K, Yuan W, Baer R (2003) The BRCA1/BARD1 heterodimer assembles polyubiquitin chains through an unconventional linkage involving lysine residue K6 of ubiquitin. J Biol Chem 278(37):34743–34746
48. Xu L, Lubkov V, Taylor LJ, Bar-Sagi D (2010) Feedback regulation of Ras signaling by Rabex-5-mediated ubiquitination. Curr Biol 20(15):1372–1377
49. Xu P, Duong DM, Seyfried NT, Cheng D, Xie Y, Robert J, Rush J, Hochstrasser M, Finley D, Peng J (2009) Quantitative proteomics reveals the function of unconventional ubiquitin chains in proteasomal degradation. Cell 137(1):133–145
50. Yang C, Zhou W, Jeon MS, Demydenko D, Harada Y, Zhou H, Liu YC (2006) Negative regulation of the E3 ubiquitin ligase itch via Fyn-mediated tyrosine phosphorylation. Mol Cell 21(1):135–141

Chapter 10
Molecular Mechanism of Ubiquitin-Dependent Traffic

Elena Maspero, Hans-Peter Wollscheid, and Simona Polo

Abstract Posttranslational modification (PTM) of signaling receptors by the covalent attachment of one, or often more, ubiquitin (Ub) moieties has emerged as the major regulatory mechanism responsible for receptor "downregulation." Pioneering work in yeast has demonstrated that Ub is required for the first step in cargo internalization as well as for targeting cargos to vacuoles (the yeast equivalent of lysosomes) [1, 2]. Following these initial observations, there are now numerous reports of ubiquitination of a vast array of mammalian signaling receptors, such as RTKs, GPCRs, MHC-I, NOTCH, various channels and transporters, cytokine, and interferon receptors (reviewed in [3–8]). The molecular basis of Ub-dependent regulation of receptor endocytosis is being clarified. In this chapter, we will give a general overview of the mammalian system.

10.1 The Ubiquitin System

Ubiquitin is a small protein of 76 amino acids (aa) that is highly conserved from yeast to mammals. Ubiquitination of substrate proteins requires the sequential action of three classes of enzymes,ultimately resulting in the covalent attachment of

E. Maspero
IFOM, Fondazione Istituto FIRC di Oncologia Molecolare, Via Adamello 16, 20139 Milan, Italy

H.-P. Wollscheid
Dipartimento di Medicina, Chirurgia ed Odontoiatria, Universita' degli Studi di Milano, Via di Rudini' 8, 20122 Milan, Italy

S. Polo (✉)
IFOM, Fondazione Istituto FIRC di Oncologia Molecolare, Via Adamello 16, 20139 Milan, Italy

Dipartimento di Medicina, Chirurgia ed Odontoiatria, Universita' degli Studi di Milano, Via di Rudini' 8, 20122 Milan, Italy
e-mail: simona.polo@ifom-ieo-campus.it

Y. Yarden and G. Tarcic (eds.), *Vesicle Trafficking in Cancer*,
DOI 10.1007/978-1-4614-6528-7_10, © Springer Science+Business Media New York 2013

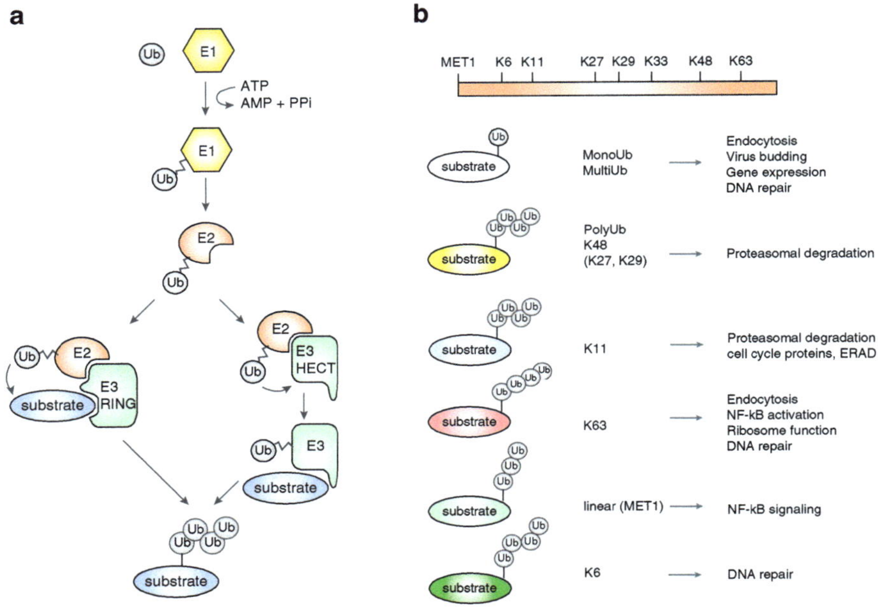

Fig. 10.1 The ubiquitin pathway. (**a**) Schematic representation of the ubiquitination process. A hierarchical set of three types of enzymes is required for substrate ubiquitination: ubiquitin-activating (E1), ubiquitin-conjugating, (E2) and ubiquitin-protein ligase (E3) enzymes. The two major classes of E3 ligases are depicted. (**b**) Schematic representation of the different Ub modifications with their functional roles

Ub to the ε-amino group of a Lys residue of the target protein (Fig. 10.1a) (reviewed in [9]). The E1 activating enzyme catalyzes the formation of a thioester bond with its active site Cys and the carboxy terminus of Ub through an adenylation intermediate that requires ATP. To date two E1 enzymes for ubiquitin were identified in mammals. In a second step ubiquitin is transferred to the active site Cys of one of approximately 40 E2 ubiquitin-conjugating enzymes. The final ubiquitin transfer to a Lys residue within the substrate protein is achieved by the action of ubiquitin E3 ligases. E3s have the ability to bind both to an E2 (via a RING finger, U box, or HECT "catalytic" domain) and to the substrate—in many E3s, those two binding sites reside in the same polypeptide. The number of ubiquitin E3 ligases in vertebrates is estimated at around 600 proteins, reflecting the high substrate specificity provided by these enzymes. From the structure and the function point of view we mainly discriminate between two types of ubiquitin E3s [10] (Fig. 10.1a). Members of the HECT (homologous to E6AP C-terminus)-type ligases receive ubiquitin from the E2 and form a thioester conjugate with Ub and their catalytic Cys from where it is then transferred to the substrate protein. Contrarily, RING (Really Interesting

New Gene)-type ligases do not form any kind of intermediate with Ub but facilitate its transfer from the E2 to the substrate. This is achieved not only by bringing E2 and substrate in close proximity to each other but also through an allosteric activation mechanism [11, 12].

An increase in complexity of the ubiquitin pathway is given by the fact that ubiquitin itself can serve as a substrate for the next steps of conjugation. Polymers of Ub can be linked via one of the seven Lys resides (Lys6, Lys11, Lys27, Lys33, Lys48, Lys63) or through the N-terminus (Met, forming linear chains) [13] (Fig. 10.1b). All eight Ub linkage types coexist in eukaryotic cells [14, 15] giving rise to a plethora of different signals with possibly diverse functions [13].

Lys48-linked ubiquitin chains are a bona fide targeting signal for substrate degradation via the 26S proteasome. For a long time this was believed to be the only function of protein ubiquitination. During the last two decades, it became clear that Lys63-linked ubiquitin chains and also monoubiquitination serve as non-proteolytic signaling tags. Proteins modified with these signals are involved in a variety of different cellular pathways, including DNA repair, kinase activation as well as in many steps of various endocytic routes, as we will discuss below. The signaling outcomes of chains other than Lys48- and Lys63-linked are yet poorly studied (for further reading, please see Chap. 9). Recently, tools to investigate the role of linear and Lys11-linked ubiquitin chains were generated, connecting these signals to immune response [16] and cell cycle-regulated degradation [17], respectively.

To deal with this relatively high number of different signals provided by the "ubiquitin code" [13, 18], the cell is equipped with a multitude of Ub receptors that are able to read and translate the signal into a specific outcome. These protein modules, which do not display strict sequence conservation, interact with Ub via diverse three-dimensional folds and direct the flow of information to specific signaling pathways [19]. For instance, once ubiquitinated the epidermal growth factor receptor (EGFR) is passed through a series of ubiquitin-binding domain (UBD)-containing proteins, responsible for the correct sorting of the receptor (described later in details).

In analogy to phosphorylation, ubiquitination is a very dynamic and reversible process. Removal of ubiquitin from the substrate protein serves as an additional layer of regulation and is accomplished by deubiquitinating enzymes (DUBs) [20]. The human genome encodes around 100 DUBs, which can be classified into five distinct families based on homologies within the catalytic domains. DUBs from the USP (ubiquitin-specific proteases), UCH (C-terminal hydrolases), OTU (ovarian tumor proteases), and Josephin families are cysteine proteases. JAB1/MPN/MOV34 metalloenzyme (JAMM, also known as MPN^+) domain DUBs form the fifth family belonging to the Zn^{2+}-dependent metalloproteases. In recent years several RNAi-based screenings have indeed demonstrated the critical role of DUBs during ubiquitin-dependent endocytic signaling. For instance USP15 and DUBA are found to act directly on the plasma membrane where they deubiquitinate TGF-beta receptor [21] and TRAF6, the Toll-like receptors associated E3 ligase [22], while Cezanne, counteracting EGFR ubiquitination, is able to enhance EGFR signaling[23]. Acting at different biological levels, DUBs balance the fate of the specific receptors to regulate the signaling outcome (for further reading, please see Chap. 12).

10.2 Ubiquitination and Internalization of Cargos

One of the first non-proteasomal functions discovered for Ub came from studies performed in yeast, in which Ub was shown to be a critical determinant for intracellular protein trafficking. Several groups have demonstrated that monoubiquitination alone of several transmembrane receptors (α-factor receptors, permeases, and transporters) is sufficient to trigger their internalization in yeast, although modification with Lys63-linked Ub chains speeds up this process [8].

In mammalian cells, the picture is complex, as Ub is one of the multiple internalization signals that might be present in various receptors. In addition not only the receptor but also the endocytic adaptors are often ubiquitinated in response to extracellular stimuli [24, 25]. The situation is even more complicated by the existence of various entry portals and by the fact that only some of them appear to be regulated by Ub(reviewed in [8]).

Here we will describe how Ub can regulate endocytosis, fate, and signaling of plasma membrane receptors using representative models.

10.2.1 GPCR

The first evidence for a role of Ub in receptor endocytosis was obtained studying the G-protein-coupled receptor (GPCR) Ste2p in *Saccharomyces cerevisiae*. Hicke and Riezman showed that Ste2p undergoes significant ubiquitination following addition of its physiological agonist, α-mating factor and that single lysine residue within the C-terminal internalization signal was required for rapid endocytosis of Ste2p [1]. In mammalian cells, the transmission of information through GPCR-initiated signaling pathways is modulated in several ways [5, 6]. It is generally thought that direct ubiquitination is not required for efficient internalization of GPCRs via clathrin-mediated endocytosis (CME) but the details of how GPCR endocytosis is regulated can vary substantially (see [26], Table 10.1 and references therein).

Early studies on the growth hormone receptor (GHR) have shown that although ubiquitination of the receptor itself is not essential for internalization, intact ubiquitination machinery is required [27]. Similar results were obtained for GPCRs that are ubiquitinated upon agonist stimulation. In the case of the β2-adrenergic receptor and the chemokine receptor, CXCR4, mutation of all the lysine residues in the cytoplasmic tail did not affect the initial internalization event, but severely impaired the downstream endocytic sorting step that targets receptors to the degradative pathway [28, 29]. In this case, ubiquitination of GPCR-associated β-arrestins (ARR), mediated by MDM2, has been proposed to be necessary for the internalization of receptors in clathrin-coated pits ([29], see also below). Intriguingly, the interaction between β-arrestin2 and MDM2 was enhanced by GPCR activation, leading to sequestration of MDM2 at the plasma membrane and to a reduction of MDM2-mediated p53 ubiquitination and degradation [30]. Moreover, β-arrestin2 has the additional role of

Table 10.1 GPCRs regulated by ubiquitination (adapted from [26])

Receptor	E3 ligase	Agonist-induced ubiquitination	ESCRT required?	References
Beta1AR	?	No	?	[129]
Beta2AR	Nedd4	Yes	?	[29, 31, 130]
CLR	?	No	Yes	[131, 132]
CB1	?	?	?	[133]
CXCR4	ITCH/AIP4	Yes	Yes	[28, 32, 33, 134, 135]
DOR	ITCH/AIP4	Yes	Yes	[136, 137]
D2R	?	?	?	[138]
KOR	?	Yes	?	[139, 140]
NK1	?	Yes	?	[141, 142]
PAF	Cbl	?	?	[143]
PAR1	?	No	No	[35, 144]
PAR2	Cbl	Yes	Yes	[131, 145]
V2R	?	Yes	?	[146]

recruiting Nedd4, the E3 responsible for β2-adrenergic receptor ubiquitination and degradation [31]. The detailed mechanism for ubiquitin-dependent lysosomal degradation of the β2-adrenergic receptor has yet to be determined.

The best understood model for GPCR ubiquitin-regulated endocytosis is provided by the chemokine receptor CXCR4. The HECT E3 ligase ITCH is responsible for the agonist-induced ubiquitination of CXCR4 [32]. ITCH can interact directly with CXCR4 via a noncanonical WW domain-mediated interaction involving serine residues within the carboxyl-terminal tail of CXCR4. These serine residues are phosphorylated upon agonist activation and are critical for mediating agonist-promoted binding of ITCH and the subsequent ubiquitination and degradation of CXCR4 [33]. Ubiquitination of CXCR4 serves as a sorting signal on endosomes for entry into the degradative pathway and long-term attenuation of signaling [33].

An alternative function for Ub-mediated regulation has been described recently for the protease-activated receptor PAR1 [34]. PAR1 displays two modes of internalization, which are clathrin- and dynamin-dependent but are independent of arrestins. Constitutive internalization of unactivated PAR1 is mediated by the clathrin adaptor protein complex-2 (AP-2) that binds directly to a tyrosine-based motif localized within the receptor C-tail domain. Agonist-induced internalization of PAR1 requires AP-2 and is negatively regulated by ubiquitination. Activated PAR1 is phosphorylated, rapidly internalized, and sorted from endosomes to lysosomes through an SNX1-dependent pathway [35]. Upon agonist stimulation PAR1 basal ubiquitination is reduced rather than increased, pointing to a negative regulation of PAR1 internalization by receptor ubiquitination. Indeed, a Lys mutant, which is defective in ubiquitination, exhibits increased internalization, whereas a C-terminal PAR1 fusion with ubiquitin inhibited PAR1 constitutive internalization [34].

Table 10.2 Tyrosine kinase receptors regulated by ubiquitination (adapted from [3]).

RTK	E3 ligase	Type of Ub modification	References
EGFR	Cbl, Cbl-b	MultimonoUb	[38, 39]
		PolyUb Lys63	
VEGFR-1	Cbl	?	[147]
VEGFR-2	Cbl	?	[148, 149]
	Nedd4		
PDGFR	Cbl, Cbl-b	PolyUb	[37, 38, 150–152]
		MultimonoUb	
FGFR	Cbl	MultimonoUb	[153–155]
IGF-1R	Nedd4	?	[156–158]
	MDM2	PolyUb Lys63	
	Cbl	PolyUb Lys48	
MET/HGFR	Cbl, Cbl-b	MultimonoUb	[82, 159]
		PolyUb Lys48	
TrkA	Cbl	PolyUb Lys63	[160–162]
	TRAF6	MultimonoUb	
	Nedd4L		
ErbB3	NRDP1	PolyUb	[163, 164]
ErbB4	NRDP1	PolyUb	[164–167]
	Nedd4	?	
	WWP1	?	
	ITCH	MonoUb	
c-KIT	Cbl, Cbl-b	?	[118, 168, 169]
		MonoUb	
TGFβ	Smad7/Smurf2	PolyUb	[170–173]
	WWP1		
	Nedd4L		
RET	Cbl	MonoUb/PolyUb	[174, 175]
CSF-1R	Cbl	?	[176–178]

10.2.2 EGFR

The role of ubiquitination in receptor tyrosine kinases (RTKs) internalization has been extensively studied. Indeed, platelet-derived growth factor receptor (PDGFR) and EGFR were the first mammalian receptors found to be ubiquitinated [36, 37]. Since these studies, several other RTKs have been shown to be ubiquitinated in a ligand-dependent manner by the E3 ligase Cbl (see Table 10.2 and references therein).

Ligand-induced EGFR trafficking is one of the best-characterized examples of how the regulation of receptor turnover is modulated by the Ub signal. Initial studies based on the use of a chimeric protein composed of EGFR and a Ub mutant that could not be extended by polyubiquitination showed that a single Ub was sufficient to drive internalization, although at a lower rate compared to wild-type receptor [38]. Later mass-spectrometry analysis has revealed that EGFR is both mono- and polyubiquitinated through Lys63-linked chains [39]. Whether these two PTMs may have different impact on EGFR, fate is not known. Although monoUb is sufficient for internalization, it might be that Lys63 polyubiquitination generates a more efficient internalization signal, possibly by increasing the binding avidity of the endocytic Ub receptors [19, 40]. Another possibility, which needs to be experimentally verified, is that monoUb and Lys63-linked chains represent distinct signals acting at different steps along the endocytic route. Recent data obtained again in yeast indeed suggest that Lys63 chains are required for cargo sorting into the multivesicular body (MVB) pathway [41, 42].

EGFR ubiquitination is not essential for internalization through clathrin-coated pits. Indeed, EGFR mutants defective in ubiquitination, either mutagenized in the E3 ligase binding site or in the Ub acceptor lysines, show no major internalization defects [43–45]. This apparent discrepancy between ubiquitination being sufficient, but not required, for EGFR internalization can be explained by the existence of alternative internalization pathways possibly regulated by different signals beside Ub, the AP-2 recognition motif and multiple phosphorylated tyrosines present in receptor tail [46]. Interestingly, the level of EGFR ubiquitination, regulated by ligand concentration, correlates with the differential recruitment of the EGFR into distinct endocytic pathways [18, 44]. At low EGF doses, receptor ubiquitination is not detected and the EGFR is internalized through CME only, while at high EGF doses, both clathrin-dependent and clathrin-independent internalization routes come into play as the receptor becomes ubiquitinated. Indeed, ubiquitination appears to be required for clathrin-independent endocytosis while it is dispensable for clathrin-dependent internalization [44]. Recent observations suggest the existence of a cooperative mechanism controlling receptor ubiquitination, which displays sigmoidal behavior as a function of EGF dose (EM and SP, unpublished). One possible explanation for the threshold in EGFR ubiquitination could reside in the mechanism of Cbl recruitment to the receptor. Indeed multiple direct and indirect (through Grb2) binding sites for Cbl are present in the receptor tail of EGFR and these sites might cooperate in E3 binding (Fig. 10.2).

The existence of distinct endocytic pathways raises a number of questions. Are all endocytic routes equivalent from a functional point of view? Or are they associated with different receptor functions? Finally, if endocytosis is required for both receptor attenuation and signaling, how are these two opposing outcomes coordinated? Interestingly, the majority of EGFRs internalized via CME is not targeted to degradation, but rather are recycled to the cell surface [47]. Conversely, clathrin-independent internalization preferentially commits the receptor to degradation [47]. This has profound implications for signaling, as by skewing EGFR fate towards recycling rather than degradation, CME prolongs the duration of the signal [47]. Thus, Ub might play a critical role in deciding between receptor signaling and downmodulation, already at the internalization step.

10.2.3 SLCs

The solute carrier (SLC) superfamily is a major group of membrane transport proteins present in mammalian cells and is conserved from yeast to humans [48, 49]. These molecules are gatekeepers for cells and organelles and control the uptake and

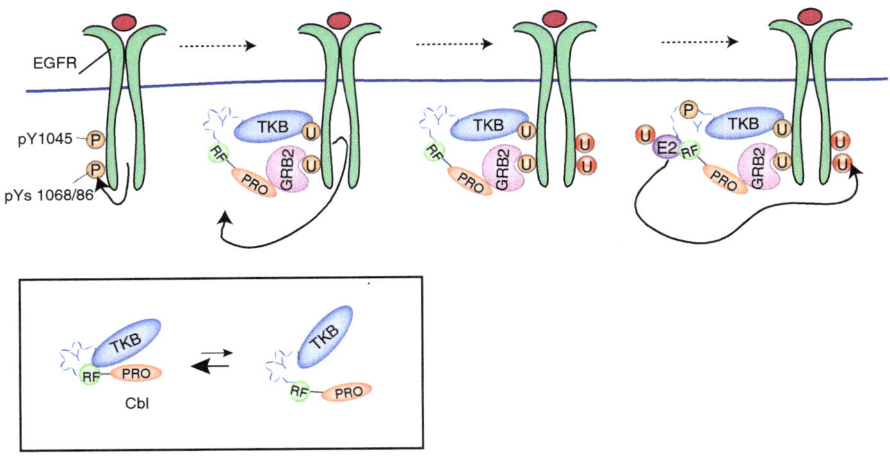

Fig. 10.2 Activation of Cbl and recruitment to the EGFR. *Bottom panel*, unphosphorylated c-Cbl exists in an equilibrium between an open, catalytically competent conformation and a closed, auto-inhibited conformation, in which the E2-binding surface of the RING associates with the TKBD. Upon EGF-dependent receptor activation, the GRB2-Cbl complex binds to the receptor through interactions of (1) the SH2 domain of GRB2 with pY1045 of EGFR and (2) the TKB domain of Cbl (either c-Cbl or Cbl-b) with pY1068 or pY1086. This substrate interaction may either stabilize or select for a partially open Cbl conformation (see *bottom panel* and main text). EGFR-bound Cbl becomes phosphorylated on a critical tyrosine, leading to full rotation of the linker region. This, in turn, exposes the RING domain for ubiquitin-charged E2 binding, resulting in the allosteric activation of the E2 by Cbl and ubiquitination of the EGFR. Note that, to simplify the picture, Cbl bound to one receptor molecule is depicted to ubiquitinate the other molecule of the dimer. No available data suggest that this is indeed the case. For simplicity, the EGF receptor is depicted as monoubiquitinated: in reality, it is both multimono- and polyubiquitinated (see main text)

the efflux of important metabolites such as glucose and other sugars, amino acids, vitamins, neurotransmitters, nucleotides, and inorganic and organic ions. Important insights into the regulation of transporter endocytosis by extracellular signaling events have been obtained from studies performed in yeast. In this model system most transporters are endocytosed in an inducible manner, through sudden changes in the availability of the cognate substrate [50].

Interestingly, transporter ubiquitination is among the first events that occur after stimulation and is a prerequisite for transporter internalization [51–53]. A mutation in the critical E3 enzyme responsible for this ubiquitination, Rsp5, leads to strong defects in endocytosis [54].

In a striking parallelism with the yeast model the few SLCs that have been characterized at the molecular level (the dopamine transporter DAT/SLC6A3 and the cationic amino acid transporter CAT-1/SLC7A1) are regulated by the human homologues of Rsp5, the E3 ligases Nedd4 and Nedd4L [7]. In the case of DAT, Nedd4L-specific KD leads to impaired transporter ubiquitination and degradation, whereas overexpression of Nedd4L resulted in the opposite effect [55, 56].

Once ubiquitinated, DAT is internalized via the CME pathway thanks to the endocytic Ub receptors Epsin, Eps15, and Eps15R that co-localize with internalized DAT in clathrin-coated pits [55]. PKC activation leads to increased transporter Lys63-linked ubiquitination and mutagenesis of three Lys residues mapped within the DAT N-terminus resulted in a strong decrease of PKC-induced ubiquitination and consequently, in a significantly impaired CME [56]. A very similar mechanism has been described for the cationic amino acid transporter CAT-1. Upon PKC activation, CAT-1 is ubiquitinated and internalized via CME. Notably, in contrast to DAT that is ubiquitinated by Nedd4L, CAT-1 was shown to be ubiquitinated by Nedd4 and Nedd4L in a cell type-dependent manner [57].

Due to the remarkable analogy between the above-described SLCs, it is tempting to speculate a generalization of the model in which Ub controls the function of various SLCs by regulating their internalization and endocytic trafficking. Interestingly, a recent proteomic approach demonstrated for the first time EGF-mediated ubiquitination of SLC proteins [58]. Although the molecular mechanism behind deserves further investigations, the noticeable number of SLC proteins ubiquitinated upon EGF stimulation suggests an intriguing connection between trafficking of EGFR and regulation of SLCs/transporters.

10.2.4 NOTCH

Signaling can also be modulated by the regulation of ligand accessibility through endocytosis. NOTCH receptor signaling provides a specialized example of this kind, since not only the receptor but also the ligand is regulated via endocytic mechanisms (reviewed in [59, 60]).

The NOTCH pathway is an evolutionary conserved signaling pathway that controls key developmental processes and numerous cell-fate specialization events in higher organisms [61]. In order to fulfill its multitude of different functions, NOTCH signaling is tightly controlled in time and space.

NOTCH signaling is induced through the direct engagement of the receptor by ligands of the DSL (Delta, Serrate, Lag-2) family that are present, in a membrane-anchored form, on an adjacent signal-sending cell. Following binding, a series of events ensues in the NOTCH-containing (signal-receiving) cell leading to two proteolytic cuts in NOTCH (the so-called S2 and S3 cuts). The S3 cut, executed by γ-secretases, occurs in the transmembrane region of NOTCH, leading to the release of a soluble cytoplasmic fragment of NOTCH, which translocates to the nucleus and activates the expression of target genes. Genetic evidences provided mainly by studies in *Drosophila melanogaster* revealed the importance of the endocytic machinery for the activation of the NOTCH receptor. Key-endocytic genes like dynamin, rab5, and the endocytic proteins syntaxin and avalanche are necessary for the ligand-dependent receptor activation [26]. Studies performed in vivo and in cell culture show a correlation between the expression of the E3 ligase Deltex and NOTCH ubiquitination [60]. Monoubiquitination of NOTCH is reported to be a sorting signal for the NOTCH/γ-secretase complex into endocytic compartments allowing an efficient S3 cleavage.

NOTCH activation in the signal-receiving cell requires specific endocytosis to occur also in the signal-sending cell. Extensive studies of the ubiquitin ligases Mindbomb (Mib) and Neuralized (Neur) revealed their importance in the DSL-ligand endocytosis [60]. Both Neur and Mib1 bind and ubiquitinate Delta and Serrate inducing their endocytosis, a process required to activate the NOTCH-ligand complex through a not fully understood mechanism. Several possibilities are envisioned. One model contemplates a mechanotransduction mechanism in which pulling forces exerted by the internalizing DSL ligand "strip" the extracellular domain of NOTCH from the intracellular membrane-anchored moiety thereby allowing proteolytic cleavage of NOTCH, as recently demonstrated in *Drosophila* germ lines [62]. In another, not mutually exclusive model, endocytosis and recycling of Delta to restricted regions of the plasma membrane may be required to maintain loc.lly high levels of ligand in order to obtain robust NOTCH activation [63]. Consistent with this idea, clustering of DSL ligands can potentiate their signaling effects in mammalian cell culture assays [64]. Monoubiquitination of DSL intracellular domains might also activate the ligands through possible conformational changes not yet identified. In sum the exact mechanism of Ub-dependent activation in the signal-sending cell remains to be clarified.

10.3 Endocytic Adaptors and Their Ubiquitination

Ub-mediated internalization/sorting of membrane receptors requires accurate recognition of the ubiquitinated cargo by endocytic Ub receptors, proteins containing one or more UBDs. Such Ub-binding "route controllers" inexorably ferry the internalized receptor towards a degradative fate in lysosomes and away from a recycling pathway (reviewed in [3, 7, 65]).

These endocytic adaptors have the ability to interact on one hand with the ubiquitinated cargos and on the other hand with different components of the endocytic machinery like clathrin and AP-2. Clathrin adaptor proteins are collectively called CLASPs (clathrin-associated sorting proteins) to distinguish them from AP-2 [66]. CLASPs can be classified on the basis of the motif recognized on the cargo and among them Epsins, Eps15, and Eps15R are the ones able to recognize ubiquitinated cargos. For these adaptors, cargo selectivity comes from tandemly arrayed Ub-interacting motifs (UIMs) [25]. Epsins associate with Eps15 through the binding of Epsin NPF motifs to Eps15 EH domain; moreover, Epsins are able to bind to clathrin [67] and to bend the membrane [68]. These properties of Epsins allow assembly of clathrin coat in the absence of AP-2.

Due to their overlapping binding abilities, CLASP adaptors often play redundant roles. Depending on cell context, cargo, and signal, the same machinery might be utilized by cells to trigger different intracellular routes. Indeed, although both Eps15s and Epsins have traditionally been linked to CME [68], they are absolutely required for clathrin-independent endocytosis of the EGFR [44]. This latter function is exerted through the binding of adaptor UIMs to the ubiquitinated EGFR [44].

Another class of CLASPs is the previously mentioned ARR family of proteins that is able to direct internalization of the GPCR cargo. Signaling from activated GPCRs is terminated when GPCRs are phosphorylated by G-protein-coupled receptor kinases (GRKs), leading to the recruitment of ARR that binds to AP-2 and clathrin, causing the whole complex to be internalized [69]. These endocytic adaptors apparently do not possess any UBD.

In parallel with receptor ubiquitination, ubiquitination of endocytic adaptors plays a critical role in endocytosis. Indeed, agonist-stimulated ubiquitination of ARR mediated by the E3 Ub ligase MDM2 is critical for rapid receptor internalization [70]. MDM2-ARR binding occurs constitutively and does not persist after receptor activation, suggesting that Ub modification might cause a conformational change on ARR required to promote internalization. Several components of the downstream endocytic machinery are modified by monoubiquitination upon EGFR activation [25, 71–73]. In most cases, these adaptors are Ub receptors that are ubiquitinated by the E3 ligase Nedd4. The presence of a UBD is required for monoubiquitination of the UBD-harboring adaptor, in a process termed "coupled monoubiquitination" whose molecular workings have been elucidated using the endocytic proteins Eps15s and Epsins as model systems [25, 74, 75]. On the contrary, the mechanism by which the upstream signal induced by the activated EGFR causes Nedd4 recruitment remains to be clarified.

What is the role of adaptor ubiquitination? Monoubiquitination might permit the formation of several tiers of ubiquitination-dependent interactions in the endosome, by allowing binding of ubiquitinated cargo (through UBDs) and recruiting another layer of Ub receptors through a monoUb signal. The result would be signal amplification and progression of ubiquitinated cargos along the endocytic pathway [18, 40]. Monoubiquitination of Ub receptors may also result in an intramolecular interaction between their UBDs and monoubiquitinated residues, with resulting dissociation from the ubiquitinated cargo [76, 77]. These two possibilities are not mutually exclusive and both mechanisms may be involved in the regulation of endocytic processes, possibly by acting at distinct trafficking steps and/or regulating different endocytic adaptors.

Summarizing all these data, we can conclude that, acting in parallel with the "phosphorylation code" on the receptor carboxyl tail, Ub modifications on both adaptors and receptors result in a "ubiquitination code" that fine-tunes signal strength, localization, and cellular functions of signaling receptors.

10.4 E3 Ligases Involved in the Endocytic Process

In the endocytic pathway distinct E3 enzymes may catalyze the ubiquitination of cell-surface receptors and of the endosomal sorting proteins [3, 78]. Therefore the endocytic sorting of a given target generally involves more than one E3 ubiquitin ligase. Below, we discuss the E3 ligases that have a prominent role in the endocytic process.

10.4.1 The Cbl Family

Elucidation of tyrosine-based signaling pathways has led to the discovery of the E3 ligase Cbl, which is responsible for the ubiquitination of several RTKs [79, 80]. The mammalian Cbl protein family consists of the three homologues c-Cbl, Cbl-b, and Cbl-3, all of which associate with a wide variety of signaling proteins [81]. Two highly conserved amino-terminal domains (NTDs) contribute strongly to E3 regulatory function. First, the N-terminal tyrosine kinase binding (TKB) domain of Cbl recognizes phosphotyrosine residues and allows Cbl to interact directly with activated RTKs at the plasma membrane (Fig. 10.2). Second, the RING finger domain recruits Ub-loaded E2s, whose interaction with Cbl results in the ubiquitination, and subsequent degradation of the associated RTK. In the case of the EGFR and the hepatocyte growth factor receptor Met, the molecular mechanism of receptor ubiquitination has been investigated in detail. In both cases, Cbl binds directly to phosphotyrosine (pY)-sites on the activated receptor through its TKB [80, 82], as well as indirectly through its constitutive partner GRB2, which is recruited to receptors via other pY sites [7, 83–85]. Both direct and indirect interactions of Cbl with the EGFR or Met are required for full ubiquitination of these receptors (Fig. 10.2). Once bound, the ligase is phosphorylated and consequently activated [86]. Two structural studies have now shed light on the mechanism of phosphorylation-induced activation of c-Cbl and Cbl-b [87, 88]. In the absence of substrate binding, the TKB and RING domains form a compact structure that masks the E2 binding site. Binding of the TKB to the substrate induces a first rotation of the linker region, allowing phosphorylation of tyrosine 371 (363 in Cbl-b). This phosphorylation event induces a complete rotation of the linker region that unmasks the RING E2 binding surface and activates the ligase ([87, 88] and Fig. 10.2).

10.4.2 The Nedd4 Family

Another class of E3 ligases, the HECT Nedd4 family [89] whose regulation has been extensively studied, also regulates endocytosis and sorting of numerous signaling receptors [3, 7, 65]. These enzymes present a conserved modular organization with an N-terminal C2 domain that is crucial for membrane localization, between two and four WW domains capable of recognizing substrates and adaptor proteins through PY motifs, and a C-terminal catalytic HECT domain. In contrast to RING-based ligases in which the RING is an allosteric activator of the E2, HECT-containing E3s have intrinsic catalytic activity and directly ubiquitinate their targets. In humans, there are nine members of this family: Nedd4 (also known as Nedd4-1), Nedd4L (also known as Nedd4-2), ITCH (also known as AIP4), WWP1, WWP2, SMURF1, SMURF2, NEDL1 (also known as HECW1), and NEDL2 (also known as HECW2). Rsp5 is the unique, essential member of the Nedd4 family in *S. cerevisiae*. In normal conditions most of them appear to be

in an inactive state due to an intramolecular inhibitory interaction between the C-terminal HECT and the N-terminal C2 domain (in the case of SMURF2, Nedd4, and WWP2 [90]) or the WW domains (in the case of ITCH [91]). Activation of this class of enzyme can occur in various ways that will be briefly described below (Fig. 10.3).

As previously mentioned, ITCH is the E3 ligase for the chemokine receptor CXCR4 [32]. The Ub moiety on CXCR4 serves as a signal on endosomes for entry into the degradative pathway and long-term attenuation or downregulation of signaling [32]. Also in this case, the ligase appears to be regulated by phosphorylation. ITCH phosphorylation is activated by JNK1 [91], which presumably leads to conformational changes that disrupt the inhibitory intramolecular interactions between its WW and the HECT domains (Fig. 10.3a).

In the case of SMURF2, autoinhibition of the HECT domain by the C2 domain helps in maintaining the steady-state levels of this E3 ligase and can be relieved by adaptor-mediated substrate targeting [90]. SMURF1 and SMURF2 bind to TGF-β family receptors via the inhibitory Smads, SMAD6, and SMAD7, to induce their ubiquitin-dependent degradation. Wiesner et al. demonstrated that intramolecular interactions between the C2 and HECT inhibit SMURF2 catalytic activity interfering with ubiquitin thioester formation [90]. This in cisautoinhibition can be relieved by binding of the NTD of the adaptor protein SMAD7 to the E3 HECT domain. In addition, the SMAD7 NTD further enhances the catalytic activity of the SMURF2 ligase by recruiting the E2 UbcH7 to the HECT domain [92]. By releasing C2-mediated autoinhibition, stimulating E2 binding, and recruiting SMURF targets, SMAD7 functions at multiple levels to control E3 activity and ensure specificity in SMURF-catalyzed ubiquitination (Fig. 10.3b).

Recently, a role for a UBD present on the N-lobe of the HECT domain of Nedd4 and Rsp5 has been identified [93, 94]. The ability of the HECT domain to bind noncovalently to the distal Ub at the growing end of the polyubiquitin chain on the substrate allows enzyme processivity [93]. It is tempting to attribute an inhibitory role of the C2 binding for this critical feature of these enzymes. Accessibility of the UBD may be restored in response to upstream signaling events capable of inducing phosphorylation and/or ubiquitination of critical sites in the C2 or in the HECT domain, leading to full ligase activation (Fig. 10.3c). While this hypothesis needs to be experimentally verified, we notice that ubiquitination of Nedd4 is a critical event for the coupled monoubiquitination of Eps15 [74].

In some cases, such as for the epithelial Na^+ channel (ENaC), receptor-ligase interaction—and consequent receptor ubiquitination—is the default pathway, with phosphorylation negatively regulating ligase activity. Nedd4L binds constitutively to ENaCPPxY-containing motifs and catalyzes its ubiquitination, internalization, and lysosomal targeting. This prevents Na^+ overload in epithelial cells and is necessary for the maintenance of salt and fluid balance in the body. To increase ENaC abundance at the surface and enhance epithelial Na^+ absorption, Nedd4L is phosphorylated by various kinases, including PKA, SGK, and IKKβ. Phosphorylation induces binding of 14-3-3, which prevents Nedd4L from binding to ENaC [95, 96].

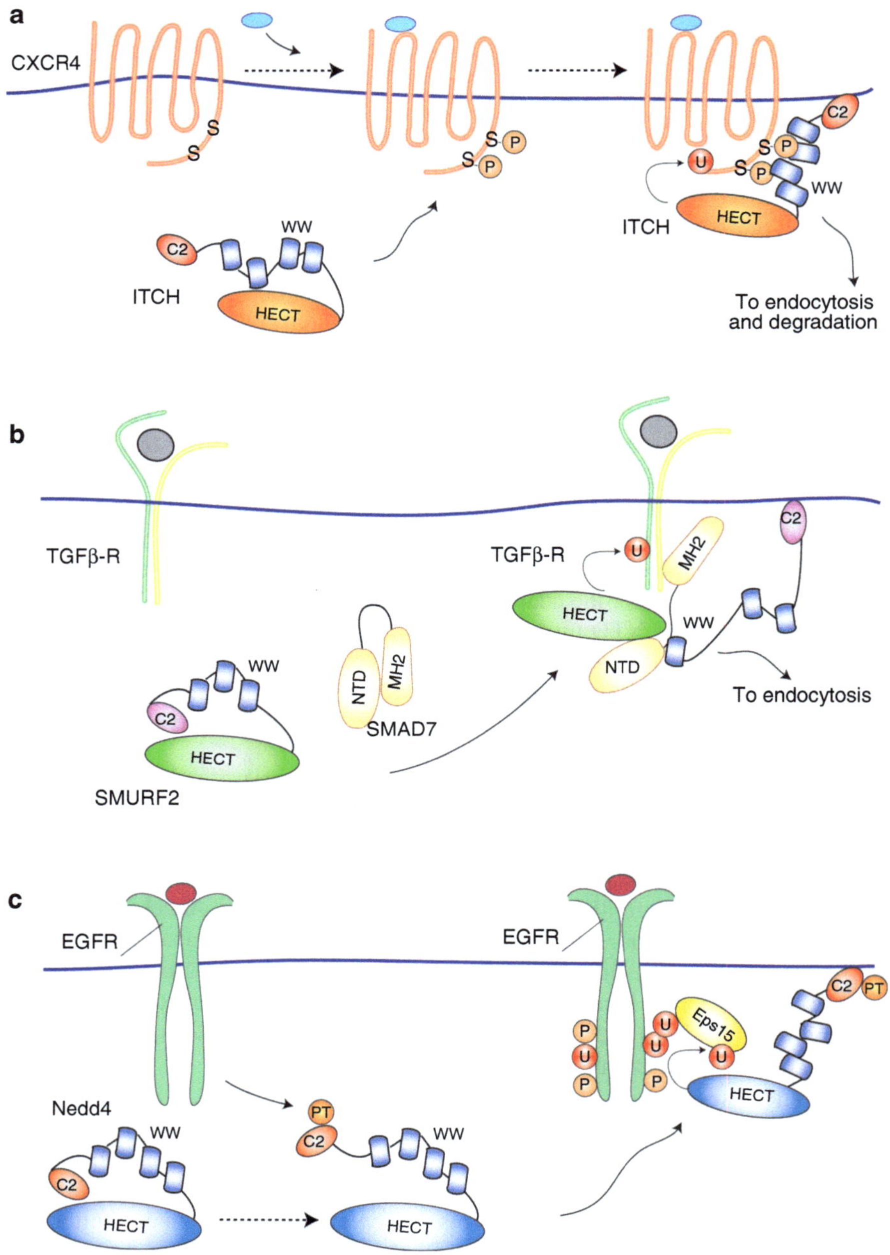

Fig. 10.3 Activation of HECT E3 ubiquitin ligases and recruitment to their receports. (**a**) Ubiquitination of CXCR4 by ITCH. ITCH activity is inhibited as a result of the intramolecular interaction between the WW domain and the carboxy-terminal catalytic HECT domain. Upon agonist-mediated activation, CXCR4 becomes phosphorylated at Ser324 and Ser325 by an unknown kinase. This leads to the recruitment of ITCH, through its WW domain, and consequent release of the inhibitory intramolecular interaction, allowing ubiquitination of the receptor. (**b**) Ubiquitination of the TGF-β receptor

10.4.3 E3 Ligases Involved in NOTCH Signaling

Endocytosis both in the signal-generating (Delta-expressing) cell and in the signal-receiving (NOTCH-expressing) cell is required for appropriate NOTCH activation [60]. During these processes, ubiquitination plays a crucial role in both cellular contexts. The activity of DSL ligands is tightly controlled by two distinct RING-containing E3 ligases, Neuralized (Neur 1 and Neur2 in mammals) and Mind bomb (Mib) 1 and 2 that directly promote DSL ubiquitination and endocytosis [97]. Various E3 ligases (Fig. 10.4) have been shown to ubiquitinate NOTCH, regulating its constitutive internalization, sorting, and ultimately its signal activity [60]. These E3s belong both to the RING family (Deltex (Dx) and Cbl) and the HECT family (Suppressor of Deltex [Su(Dx)] and Nedd4) and are implicated in the ubiquitination of the inactivated NOTCH in order to promote its recognition by the ESCRT (endosomal sorting complex required for transport) machinery. The detailed functional role of these E3s in NOTCH ubiquitination has yet to be determined.

10.5 Impact of Ubiquitination on the Subsequent Steps of Endocytosis

Following internalization, cargos are routed to early endosomes, where they are subjected to distinct trafficking paths that ultimately lead to their fate: either degradation into lysosomes or recycling to the plasma membrane. Early endosomes therefore represent an initial common sorting station, where cargos destined for degradation are separated from those destined for recycling. The delivery of transmembrane receptors into the lumen of the lysosome requires the sorting of these proteins into intralumenal vesicles (ILV) in an endosomal compartment, giving rise to the MVB.

Ligand-induced ubiquitination plays a key and essential role in the MVB targeting. Ub-directed sorting into MVBs is mediated by the ESCRT multiprotein complexes harboring UBDs ([98–101], also reviewed in [65]). This conserved machinery performs three distinct but connected functions: first, it recognizes ubiquitinated cargos and prevents their recycling and retrograde trafficking; second, it deforms the

Fig. 10.3 (continued) by the SMURF2-SMAD7 complex. SMURF2 activity is inhibited as a result of the intramolecular interaction between the amino-terminal C2 and the carboxy-terminal catalytic HECT domain. The interaction with SMAD7 NTD displaces the C2 domain of SMURF2 from the HECT domain and activates the ligase. The activated SMURF2-SMAD7 complex associates with activated TGF-β receptor complexes at the membrane via the displaced C2 domain, causing receptor ubiquitination. (**c**) Similarly to SMURF2, Nedd4 activity is inhibited as a result of the intramolecular interaction between the amino-terminal C2 and the carboxy-terminal catalytic HECT domain. Release of the C2, promoted by EGF-induced PTMs (PT in the figure), may activate the enzyme leading to its re-localization at the membrane where it interacts with its substrates (Eps15, depicted)

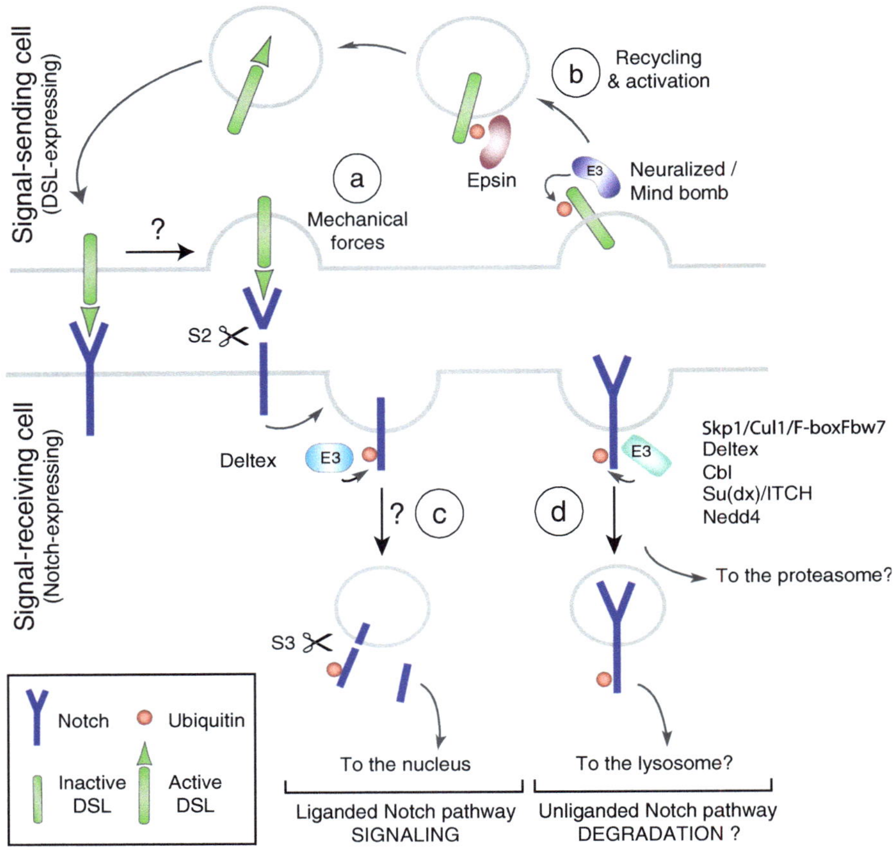

Fig. 10.4 Dual regulation of Notch signaling by ubiquitin. Notch is processed and activated through a series of proteolytic cleavages. The endocytosis of Notch and its ligands is required for Notch activity. Internalization of DSL depends on ubiquitination by the E3 ligases Neuralized and Mind bomb. Downstream events require the Ub-binding protein Epsin/lqf. Models to explain why DSL endocytosis is required for Notch activation are shown. (**a**) Endocytosis of DSLs generates conditions (by mechanical "pulling" forces) that unmask the Notch S2 site. (**b**) Inactive DSLs are endocytosed, "activated" in endosomes, and recycled to the surface. (**c**) Ligand-engaged Notch requires endocytosis (possibly dependent on ubiquitination by Deltex) for its activation. (**d**) Unliganded Notch is continuously endocytosed, through ubiquitination by the E3 ligases depicted, to prevent sporadic activation. The *question marks* indicate the processes still unclear

endosomal membrane, allowing cargo to be sorted into endosomal invaginations; third, it catalyzes the final abscission (breaking off) of the endosomal invaginations, forming intraluminal vesicles that contain the sorted cargo.

Today we have the complete inventory of the proteins that make up the ESCRT complexes that act sequentially at the MVBs. Fourteen genes have been first identified

in yeast and quickly it became clear that every component as well as their function is conserved in mammals: (1) ESCRT-0 is composed of two interacting proteins HRS and STAM [65, 102]. (2) ESCRT-I is a heterotetramer of Vps23, Vps28, Vps37, and Mvb12 [65, 102]. (3) ESCRT-II is a heterotetramer composed of one molecule of Vps22, one molecule of Vps36, and two molecules of Vps25 [65, 102]. (4) ESCRT-III, unlike other ESCRTs which are stable complexes, is a dynamic polymer of ESCRT-III proteins (CHMPs in mammals) that does not have a clearly defined or unique composition [65, 102]. The ESCRT apparatus is filled with a plethora of UBDs that ensure recognition and delivery of the ubiquitinated cargos [40]. The ESCRT-0 heterodimer, for example, has no less than five UBDs, housed in both α-helical UIMs and the N-terminal VHS domains [103].

ESCRT complexes are also involved in the recruitment of the enzymes that deubiquitinate receptors before they are packaged into ILVs. Recent data have established crucial, but distinct, roles for two DUBs in the fine regulation of EGFR degradation. AMSH (Associated Molecule with the SH3-domain of STAM) rescues EGFR from sorting to MVBs and consequently degradation, by removing Ub from the receptor at early stages of endosomal sorting, thus promoting its recycling [104]. AMSH might also act at a later step of MVB sorting, by competing with Vps4 for ESCRT-III binding, possibly inhibiting vesicles budding [105]. In contrast, USP8 (also called UBPY) activity seems to be required for the final commitment of the EGFR to MVBs and degradation, similarly to Doa4 in yeast. To exert this function, USP8/UBPY acts on multiple substrates along the sorting pathway, targeting the receptor itself, as well as endocytic adaptors and components of the ESCRT machinery, e.g., STAM [106]. Acting at different steps along the EGFR route to lysosomes, these DUBs balance the fate of EGFR between downregulation and recycling.

Another DUB recently implicated in the EGFR pathway is Usp18 [107]. Usp18 regulates EGFR at the transcriptional level possibly through a micro-RNA-dependent mechanism [107, 108]. Finally, through an RNAi-based screening, Cezanne was found as negative regulator of EGFR degradation and enhancer of EGFR signaling [23], although its exact mechanism of action remains to be established. Considering the fine-tuning regulation of the EGFR signaling and downregulation, it is easy to predict that more DUBs able to directly or indirectly impact on EGFR fate will be identified in the future.

10.6 Impact of Ubiquitination on Receptor Fate, Relevance for Cancer

Since the rate of receptor downregulation and MVB targeting typically correlates with the extent of receptor ubiquitination in endosomes, interference with this post-translational processing enhances signaling, such as for mutants in EGFR ubiquitination sites [39]. Similarly, RNA or genetic interference with the Ub adaptor HRS in mammalian cells or in *Drosophila* in vivo results in enhanced signaling by various RTKs, including EGFR and VEGFR (vascular endothelial growth factor receptor) [109, 110]. The opposite effect (signal impairment of various RTKs) is observed in conditional mouse knockouts of the deubiquitinating enzyme USP8/UBPY [111]. Furthermore, genetic disruption of members of the ESCRT complexes, which are required for membrane fission events, including those that lead to endosomal intraluminal vesicle formation, leads to sustained EGFR signaling in mice [112] and, in *Drosophila*, NOTCH hyperactivation and neoplastic transformation [113]. This latter observation underscores the emerging involvement of endosomal sorting, and endocytosis in general, in tumorigenesis [114]. PM cargos are frequently mutated in human cancers, in specific determinants that alter their vesicular traffic (reviewed in [114, 115]). This is the case for several RTKs, like EGFR, MET, and KIT [116–118]. The most frequent genetic alterations, in these occurrences, consist of deletions that affect the region encoding portions of the intracellular domains of RTKs, usually encompassing the binding region for Cbl, the major E3 ligase involved in RTK ubiquitination [114].

In addition to this, other mechanisms are exploited by cancer cells in order to evade endocytosis-mediated desensitization. For instance, somatic mutations in the kinase domain of the EGFR have been reported in non-small cell lung cancers, and they have been shown in vitro to cause reduced receptor phosphorylation at Y1045, the major Cbl binding site and, consequently, defective receptor downregulation [119]. Similarly, EGFRvIII—an oncogenic deletion mutant of the EGFR, frequently observed in glioblastoma—shows hypo-phosphorylation of Y1045 and reduced degradation ([120, 121], see also [114] and Chap. 14 for more detailed explanations).

An endocytic-dependent mechanism has been proposed to contribute to the transforming effects of ERBB-2 overexpression in breast cancer. ERBB-2 belongs to the EGFR family of RTKs; at variance with EGFR, however, ERBB-2 is internalization-impaired [122]. Heterodimerization of ERBB-2 with ligand-occupied EGFRs seems to influence the endocytic trafficking of both ERBB-2 and EGFR. Indeed, it has been shown that EGFR-ERBB-2 heterodimers display delayed endocytosis, are not efficiently sorted to lysosomes, and are preferentially recycled back to the cell surface, causing aberrant signaling [123–125]. One possibility is that EGFR and ERBB-2 are not fully ubiquitinated in the heterodimers. Indeed, while activated ERBB-2 can recruit Cbl, this recruitment is less efficient compared to EGFR [79]. An alternative possibility is that heterodimers display reduced affinity for EGF and dissociate from the ligand in endosomes, due to the release of the ligand in the acidic environment of endosomes, being recycled back to the surface [125]. However, computational modeling of the trafficking of EGFR-ERBB-2 heterodimers predicted that elevated dissociation of ligand in endosomes could not explain the observed trafficking patterns of the heterodimers [124]. Rather, the reduced degradation of EGFR might be explained by a mechanism through which ERBB-2 directly competes with EGFR for a stoichiometrically limited quantity of endosomal retention components, thereby reducing endosomal retention and degradation of EGFR [124]. Whatever the case, it appears that altered trafficking of EGFR might be one mechanism through which ERBB-2 exerts its oncogenic potential.

In addition to RTKs, many GPCRs are overexpressed in human cancers and contribute to tumor progression (reviewed in [126]). Recent work has revealed that deregulated trafficking of CXCR4 and PAR1 through the endosomal-lysosomal station leads to increased surface expression of these cargos in breast cancer cells, contributing to cancer progression [5, 6]. Interplay with ERBB-2 seems to have a role in breast cancers that display elevated CXCR4 surface levels. Indeed, ERBB-2 overexpression seems to enhance CXCR4 levels both by increasing protein synthesis and by impairing CXCR4 ubiquitination and lysosomal degradation mediated by AIP4, the E3 ligase involved in this process [127]. It has been proposed that in this case, the mechanism may involve CISK, an Ser/Thr kinase downstream of PI3K signaling, which phosphorylates and inactivates AIP4, thereby contributing to the increased CXCR4 levels [6, 128].

In conclusion, while we have had to necessarily limit ourselves to the description of a few paradigmatic cases, it is evident that subversion of endocytosis might be involved in cancer in multiple ways. Given this, a deeper an.lysis of the endocytic process is predicted not only to advance our understanding of cell regulation and how it connects to the pathogenetic mechanisms of cancer but should also help to identify novel targets for molecular therapies and clinically relevant biomarkers for prognostic, diagnostic, and therapeutic purposes.

Acknowledgements Research in the Polo laboratory is supported by grants from the Associazione Italiana per la Ricerca sul Cancro, the Italian University Research Program for the Development of Research of National Interest and of Health (PRIN), the Association of International Cancer Research, the CARIPLO Foundation, and the EMBO Young Investigator Program. HPW is recipient of a Marie Curie/AIRC Cofund International Fellowship.

References

1. Hicke L, Riezman H (1996) Ubiquitination of a yeast plasma membrane receptor signals its ligand-stimulated endocytosis. Cell 84(2):277–87
2. Kolling R, Hollenberg CP (1994) The ABC-transporter Ste6 accumulates in the plasma membrane in a ubiquitinated form in endocytosis mutants. EMBO J 13(14):3261–71
3. Acconcia F, Sigismund S, Polo S (2009) Ubiquitin in trafficking: the network at work. Exp Cell Res 315(9):1610–8
4. Bianchi K, Meier P (2009) A tangled web of ubiquitin chains: breaking news in TNF-R1 signaling. Mol Cell 36(5):736–42
5. Hanyaloglu AC, von Zastrow M (2008) Regulation of GPCRs by endocytic membrane trafficking and its potential implications. Annu Rev Pharmacol Toxicol 48:537–68
6. Marchese A, Paing MM, Temple BR, Trejo J (2008) G protein-coupled receptor sorting to endosomes and lysosomes. Annu Rev Pharmacol Toxicol 48:601–29
7. Miranda M, Sorkin A (2007) Regulation of receptors and transporters by ubiquitination: new insights into surprisingly similar mechanisms. Mol Interv 7(3):157–67
8. Lauwers E, Erpapazoglou Z, Haguenauer-Tsapis R, Andre B (2010) The ubiquitin code of yeast permease trafficking. Trends Cell Biol 20(4):196–204
9. Pickart CM (2001) Mechanisms underlying ubiquitination. Annu Rev Biochem 70:503–33
10. Metzger MB, Hristova VA, Weissman AM (2012) HECT and RING finger families of E3 ubiquitin ligases at a glance. J Cell Sci 125(pt 3):531–7
11. Deshaies RJ, Joazeiro CA (2009) RING domain E3 ubiquitin ligases. Annu Rev Biochem 78:399–434

12. Ozkan E, Yu H, Deisenhofer J (2005) Mechanistic insight into the allosteric activation of a ubiquitin-conjugating enzyme by RING-type ubiquitin ligases. Proc Natl Acad Sci U S A 102(52):18890–5
13. Komander D, Rape M (2012) The ubiquitin code. Annu Rev Biochem 81:203–29
14. Xu P, Duong DM, Seyfried NT, Cheng D, Xie Y, Robert J, Rush J, Hochstrasser M, Finley D, Peng J (2009) Quantitative proteomics reveals the function of unconventional ubiquitin chains in proteasomal degradation. Cell 137(1):133–45
15. Peng J, Schwartz D, Elias JE, Thoreen CC, Cheng D, Marsischky G, Roelofs J, Finley D, Gygi SP (2003) A proteomics approach to understanding protein ubiquitination. Nat Biotechnol 21(8):921–6
16. Tokunaga F, Nakagawa T, Nakahara M, Saeki Y, Taniguchi M, Sakata S, Tanaka K, Nakano H, Iwai K (2011) SHARPIN is a component of the NF-kappaB-activating linear ubiquitin chain assembly complex. Nature 471(7340):633–6
17. Wickliffe KE, Williamson A, Meyer HJ, Kelly A, Rape M (2011) K11-linked ubiquitin chains as novel regulators of cell division. Trends Cell Biol 21(11):656–63
18. Woelk T, Sigismund S, Penengo L, Polo S (2007) The ubiquitination code: a signalling problem. Cell Div 2:11
19. Husnjak K, Dikic I (2012) Ubiquitin-binding proteins: decoders of ubiquitin-mediated cellular functions. Annu Rev Biochem 81:291–322
20. Reyes-Turcu FE, Ventii KH, Wilkinson KD (2009) Regulation and cellular roles of ubiquitin-specific deubiquitinating enzymes. Annu Rev Biochem 78:363–97
21. Eichhorn PJ, Rodon L, Gonzalez-Junca A, Dirac A, Gili M, Martinez-Saez E, Aura C, Barba I, Peg V, Prat A et al (2012) USP15 stabilizes TGF-beta receptor I and promotes oncogenesis through the activation of TGF-beta signaling in glioblastoma. Nat Med 18(3):429–35
22. Kayagaki N, Phung Q, Chan S, Chaudhari R, Quan C, O'Rourke KM, Eby M, Pietras E, Cheng G, Bazan JF et al (2007) DUBA: a deubiquitinase that regulates type I interferon production. Science 318(5856):1628–32
23. Pareja F, Ferraro DA, Rubin C, Cohen-Dvashi H, Zhang F, Aulmann S, Ben-Chetrit N, Pines G, Navon R, Crosetto N et al (2012) Deubiquitination of EGFR by Cezanne-1 contributes to cancer progression. Oncogene 31:4599–608
24. Di Fiore PP, Polo S, Hofmann K (2003) When ubiquitin meets ubiquitin receptors: a signalling connection. Nat Rev Mol Cell Biol 4(6):491–7
25. Polo S, Sigismund S, Faretta M, Guidi M, Capua MR, Bossi G, Chen H, De Camilli P, Di Fiore PP (2002) A single motif responsible for ubiquitin recognition and monoubiquitination in endocytic proteins. Nature 416(6879):451–5
26. Hislop JN, von Zastrow M (2011) Role of ubiquitination in endocytic trafficking of G-protein-coupled receptors. Traffic 12(2):137–48
27. Govers R, ten Broeke T, van Kerkhof P, Schwartz AL, Strous GJ (1999) Identification of a novel ubiquitin conjugation motif, required for ligand-induced internalization of the growth hormone receptor. EMBO J 18(1):28–36
28. Marchese A, Benovic JL (2001) Agonist-promoted ubiquitination of the G protein-coupled receptor CXCR4 mediates lysosomal sorting. J Biol Chem 276(49):45509–12
29. Shenoy SK, McDonald PH, Kohout TA, Lefkowitz RJ (2001) Regulation of receptor fate by ubiquitination of activated beta 2-adrenergic receptor and beta-arrestin. Science 294(5545):1307–13
30. Wang P, Wu Y, Ge X, Ma L, Pei G (2003) Subcellular localization of beta-arrestins is determined by their intact N domain and the nuclear export signal at the C terminus. J Biol Chem 278(13):11648–53
31. Shenoy SK, Xiao K, Venkataramanan V, Snyder PM, Freedman NJ, Weissman AM (2008) Nedd4 mediates agonist-dependent ubiquitination, lysosomal targeting, and degradation of the beta2-adrenergic receptor. J Biol Chem 283(32):22166–76
32. Marchese A, Raiborg C, Santini F, Keen JH, Stenmark H, Benovic JL (2003) The E3 ubiquitin ligase AIP4 mediates ubiquitination and sorting of the G protein-coupled receptor CXCR4. Dev Cell 5(5):709–22

33. Bhandari D, Robia SL, Marchese A (2009) The E3 ubiquitin ligase atrophin interacting protein 4 binds directly to the chemokine receptor CXCR4 via a novel WW domain-mediated interaction. Mol Biol Cell 20(5):1324–39
34. Wolfe BL, Marchese A, Trejo J (2007) Ubiquitination differentially regulates clathrin-dependent internalization of protease-activated receptor-1. J Cell Biol 177(5):905–16
35. Gullapalli A, Wolfe BL, Griffin CT, Magnuson T, Trejo J (2006) An essential role for SNX1 in lysosomal sorting of protease-activated receptor-1: evidence for retromer-, Hrs-, and Tsg101-independent functions of sorting nexins. Mol Biol Cell 17(3):1228–38
36. Galcheva-Gargova Z, Theroux SJ, Davis RJ (1995) The epidermal growth factor receptor is covalently linked to ubiquitin. Oncogene 11(12):2649–55
37. Mori S, Heldin CH, Claesson-Welsh L (1993) Ligand-induced ubiquitination of the platelet-derived growth factor beta-receptor plays a negative regulatory role in its mitogenic signaling. J Biol Chem 268(1):577–83
38. Haglund K, Sigismund S, Polo S, Szymkiewicz I, Di Fiore PP, Dikic I (2003) Multiple monoubiquitination of RTKs is sufficient for their endocytosis and degradation. Nat Cell Biol 5(5):461–6
39. Huang F, Kirkpatrick D, Jiang X, Gygi S, Sorkin A (2006) Differential regulation of EGF receptor internalization and degradation by multiubiquitination within the kinase domain. Mol Cell 21(6):737–48
40. Macgurn JA, Hsu PC, Emr SD (2012) Ubiquitin and membrane protein turnover: from cradle to grave. Annu Rev Biochem 81:231–59
41. Lauwers E, Jacob C, Andre B (2009) K63-linked ubiquitin chains as a specific signal for protein sorting into the multivesicular body pathway. J Cell Biol 185(3):493–502
42. Erpapazoglou Z, Dhaoui M, Pantazopoulou M, Giordano F, Mari M, Leon S, Raposo G, Reggiori F, Haguenauer-Tsapis R (2012) A dual role for K63-linked ubiquitin chains in multivesicular bodies biogenesis and cargo sorting. Mol Biol Cell 23:2170–83
43. Huang F, Goh LK, Sorkin A (2007) EGF receptor ubiquitination is not necessary for its internalization. Proc Natl Acad Sci U S A 104(43):16904–9
44. Sigismund S, Woelk T, Puri C, Maspero E, Tacchetti C, Transidico P, Di Fiore PP, Polo S (2005) Clathrin-independent endocytosis of ubiquitinated cargos. Proc Natl Acad Sci U S A 102(8):2760–5
45. Duan L, Miura Y, Dimri M, Majumder B, Dodge IL, Reddi AL, Ghosh A, Fernandes N, Zhou P, Mullane-Robinson K et al (2003) Cbl-mediated ubiquitinylation is required for lysosomal sorting of epidermal growth factor receptor but is dispensable for endocytosis. J Biol Chem 278(31):28950–60
46. Sorkin A (2004) Cargo recognition during clathrin-mediated endocytosis: a team effort. Curr Opin Cell Biol 16(4):392–9
47. Sigismund S, Argenzio E, Tosoni D, Cavallaro E, Polo S, Di Fiore PP (2008) Clathrin-mediated internalization is essential for sustained EGFR signaling but dispensable for degradation. Dev Cell 15(2):209–19
48. He L, Vasiliou K, Nebert DW (2009) Analysis and update of the human solute carrier (SLC) gene superfamily. Hum Genomics 3(2):195–206
49. Hediger MA, Romero MF, Peng JB, Rolfs A, Takanaga H, Bruford EA (2004) The ABCs of solute carriers: physiological, pathological and therapeutic implications of human membrane transport proteinsIntroduction. Pflugers Arch 447(5):465–8
50. Edinger AL (2007) Controlling cell growth and survival through regulated nutrient transporter expression. Biochem J 406(1):1–12
51. Galan JM, Moreau V, Andre B, Volland C, Haguenauer-Tsapis R (1996) Ubiquitination mediated by the Npi1p/Rsp5p ubiquitin-protein ligase is required for endocytosis of the yeast uracil permease. J Biol Chem 271(18):10946–52
52. Springael JY, Galan JM, Haguenauer-Tsapis R, Andre B (1999) NH4+-induced down-regulation of the Saccharomyces cerevisiae Gap1p permease involves its ubiquitination with lysine-63-linked chains. J Cell Sci 112(pt 9):1375–83

53. Springael JY, Andre B (1998) Nitrogen-regulated ubiquitination of the Gap1 permease of Saccharomyces cerevisiae. Mol Biol Cell 9(6):1253–63
54. Rotin D, Staub O, Haguenauer-Tsapis R (2000) Ubiquitination and endocytosis of plasma membrane proteins: role of Nedd4/Rsp5p family of ubiquitin-protein ligases. J Membr Biol 176(1):1–17
55. Sorkina T, Miranda M, Dionne KR, Hoover BR, Zahniser NR, Sorkin A (2006) RNA interference screen reveals an essential role of Nedd4-2 in dopamine transporter ubiquitination and endocytosis. J Neurosci 26(31):8195–205
56. Miranda M, Dionne KR, Sorkina T, Sorkin A (2007) Three ubiquitin conjugation sites in the amino terminus of the dopamine transporter mediate protein kinase C-dependent endocytosis of the transporter. Mol Biol Cell 18(1):313–23
57. Vina-Vilaseca A, Bender-Sigel J, Sorkina T, Closs EI, Sorkin A (2011) Protein kinase C-dependent ubiquitination and clathrin-mediated endocytosis of the cationic amino acid transporter CAT-1. J Biol Chem 286(10):8697–706
58. Argenzio E, Bange T, Oldrini B, Bianchi F, Peesari R, Mari S, Di Fiore PP, Mann M, Polo S (2011) Proteomic snapshot of the EGF-induced ubiquitin network. Mol Syst Biol 7:462
59. Fortini ME (2009) Notch signaling: the core pathway and its posttranslational regulation. Dev Cell 16(5):633–47
60. Le Bras S, Loyer N, Le Borgne R (2011) The multiple facets of ubiquitination in the regulation of notch signaling pathway. Traffic 12(2):149–61
61. Kopan R, Ilagan MX (2009) The canonical Notch signaling pathway: unfolding the activation mechanism. Cell 137(2):216–33
62. Windler SL, Bilder D (2010) Endocytic internalization routes required for delta/notch signaling. Curr Biol 20(6):538–43
63. Rajan A, Tien AC, Haueter CM, Schulze KL, Bellen HJ (2009) The Arp2/3 complex and WASp are required for apical trafficking of Delta into microvilli during cell fate specification of sensory organ precursors. Nat Cell Biol 11(7):815–24
64. Shimizu K, Chiba S, Saito T, Takahashi T, Kumano K, Hamada Y, Hirai H (2002) Integrity of intracellular domain of Notch ligand is indispensable for cleavage required for release of the Notch2 intracellular domain. EMBO J 21(3):294–302
65. Raiborg C, Stenmark H (2009) The ESCRT machinery in endosomal sorting of ubiquitylated membrane proteins. Nature 458(7237):445–52
66. Owen DJ, Collins BM, Evans PR (2004) Adaptors for clathrin coats: structure and function. Annu Rev Cell Dev Biol 20:153–91
67. Wendland B (2002) Epsins: adaptors in endocytosis? Nat Rev Mol Cell Biol 3(12):971–7
68. Horvath CA, Vanden Broeck D, Boulet GA, Bogers J, De Wolf MJ (2007) Epsin: inducing membrane curvature. Int J Biochem Cell Biol 39(10):1765–70
69. Santini F, Gaidarov I, Keen JH (2002) G protein-coupled receptor/arrestin3 modulation of the endocytic machinery. J Cell Biol 156(4):665–76
70. Shenoy SK, Modi AS, Shukla AK, Xiao K, Berthouze M, Ahn S, Wilkinson KD, Miller WE, Lefkowitz RJ (2009) Beta-arrestin-dependent signaling and trafficking of 7-transmembrane receptors is reciprocally regulated by the deubiquitinase USP33 and the E3 ligase Mdm2. Proc Natl Acad Sci U S A 106(16):6650–5
71. Haglund K, Shimokawa N, Szymkiewicz I, Dikic I (2002) Cbl-directed monoubiquitination of CIN85 is involved in regulation of ligand-induced degradation of EGF receptors. Proc Natl Acad Sci U S A 99(19):12191–6
72. Shih SC, Katzmann DJ, Schnell JD, Sutanto M, Emr SD, Hicke L (2002) Epsins and Vps27p/Hrs contain ubiquitin-binding domains that function in receptor endocytosis. Nat Cell Biol 4(5):389–93
73. Katz M, Shtiegman K, Tal-Or P, Yakir L, Mosesson Y, Harari D, Machluf Y, Asao H, Jovin T, Sugamura K et al (2002) Ligand-independent degradation of epidermal growth factor receptor involves receptor ubiquitylation and Hgs, an adaptor whose ubiquitin-interacting motif targets ubiquitylation by Nedd4. Traffic 3(10):740–51
74. Woelk T, Oldrini B, Maspero E, Confalonieri S, Cavallaro E, Di Fiore PP, Polo S (2006) Molecular mechanisms of coupled monoubiquitination. Nat Cell Biol 8(11):1246–54

75. Fallon L, Belanger CM, Corera AT, Kontogiannea M, Regan-Klapisz E, Moreau F, Voortman J, Haber M, Rouleau G, Thorarinsdottir T et al (2006) A regulated interaction with the UIM protein Eps15 implicates parkin in EGF receptor trafficking and PI(3)K-Akt signalling. Nat Cell Biol 8(8):834–42
76. Hoeller D, Crosetto N, Blagoev B, Raiborg C, Tikkanen R, Wagner S, Kowanetz K, Breitling R, Mann M, Stenmark H et al (2006) Regulation of ubiquitin-binding proteins by monoubiquitination. Nat Cell Biol 8(2):163–9
77. Mattera R, Bonifacino JS (2008) Ubiquitin binding and conjugation regulate the recruitment of Rabex-5 to early endosomes. EMBO J 27(19):2484–94
78. Polo S (2012) Signaling-mediated control of ubiquitin ligases in endocytosis. BMC Biol 10:25
79. Levkowitz G, Klapper LN, Tzahar E, Freywald A, Sela M, Yarden Y (1996) Coupling of the c-Cbl protooncogene product to ErbB-1/EGF-receptor but not to other ErbB proteins. Oncogene 12(5):1117–25
80. Levkowitz G, Waterman H, Ettenberg SA, Katz M, Tsygankov AY, Alroy I, Lavi S, Iwai K, Reiss Y, Ciechanover A et al (1999) Ubiquitin ligase activity and tyrosine phosphorylation underlie suppression of growth factor signaling by c-Cbl/Sli-1. Mol Cell 4(6):1029–40
81. Schmidt MH, Dikic I (2005) The Cbl interactome and its functions. Nat Rev Mol Cell Biol 6(12):907–18
82. Peschard P, Fournier TM, Lamorte L, Naujokas MA, Band H, Langdon WY, Park M (2001) Mutation of the c-Cbl TKB domain binding site on the Met receptor tyrosine kinase converts it into a transforming protein. Mol Cell 8(5):995–1004
83. Waterman H, Katz M, Rubin C, Shtiegman K, Lavi S, Elson A, Jovin T, Yarden Y (2002) A mutant EGF-receptor defective in ubiquitylation and endocytosis unveils a role for Grb2 in negative signaling. EMBO J 21(3):303–13
84. Jiang X, Huang F, Marusyk A, Sorkin A (2003) Grb2 Regulates Internalization of EGF Receptors through Clathrin-coated pits. Mol Biol Cell 14(3):858–70
85. Huang F, Sorkin A (2005) Growth factor receptor binding protein 2-mediated recruitment of the RING domain of Cbl to the epidermal growth factor receptor is essential and sufficient to support receptor endocytosis. Mol Biol Cell 16(3):1268–81
86. Kassenbrock CK, Hunter S, Garl P, Johnson GL, Anderson SM (2002) Inhibition of Src family kinases blocks epidermal growth factor (EGF)-induced activation of Akt, phosphorylation of c-Cbl, and ubiquitination of the EGF receptor. J Biol Chem 277(28):24967–75
87. Kobashigawa Y, Tomitaka A, Kumeta H, Noda NN, Yamaguchi M, Inagaki F (2011) Autoinhibition and phosphorylation-induced activation mechanisms of human cancer and autoimmune disease-related E3 protein Cbl-b. Proc Natl Acad Sci U S A 108(51):20579–84
88. Dou H, Buetow L, Hock A, Sibbet GJ, Vousden KH, Huang DT (2012) Structural basis for autoinhibition and phosphorylation-dependent activation of c-Cbl. Nat Struct Mol Biol 19(2):184–92
89. Rotin D, Kumar S (2009) Physiological functions of the HECT family of ubiquitin ligases. Nat Rev Mol Cell Biol 10(6):398–409
90. Wiesner S, Ogunjimi AA, Wang HR, Rotin D, Sicheri F, Wrana JL, Forman-Kay JD (2007) Autoinhibition of the HECT-type ubiquitin ligase Smurf2 through its C2 domain. Cell 130(4):651–62
91. Gallagher E, Gao M, Liu YC, Karin M (2006) Activation of the E3 ubiquitin ligase Itch through a phosphorylation-induced conformational change. Proc Natl Acad Sci U S A 103(6):1717–22
92. Ogunjimi AA, Briant DJ, Pece-Barbara N, Le Roy C, Di Guglielmo GM, Kavsak P, Rasmussen RK, Seet BT, Sicheri F, Wrana JL (2005) Regulation of Smurf2 ubiquitin ligase activity by anchoring the E2 to the HECT domain. Mol Cell 19(3):297–308
93. Maspero E, Mari S, Valentini E, Musacchio A, Fish A, Pasqualato S, Polo S (2011) Structure of the HECT:ubiquitin complex and its role in ubiquitin chain elongation. EMBO Rep 12(4):342–9
94. Kim HC, Steffen AM, Oldham ML, Chen J, Huibregtse JM (2011) Structure and function of a HECT domain ubiquitin-binding site. EMBO Rep 12(4):334–41

95. Ichimura T, Yamamura H, Sasamoto K, Tominaga Y, Taoka M, Kakiuchi K, Shinkawa T, Takahashi N, Shimada S, Isobe T (2005) 14-3-3 proteins modulate the expression of epithelial Na+ channels by phosphorylation-dependent interaction with Nedd4-2 ubiquitin ligase. J Biol Chem 280(13):13187–94
96. Snyder PM (2009) Down-regulating destruction: phosphorylation regulates the E3 ubiquitin ligase Nedd4-2. Sci Signal 2(79):pe41
97. Le Borgne R (2006) Regulation of Notch signalling by endocytosis and endosomal sorting. Curr Opin Cell Biol 18(2):213–22
98. Babst M, Katzmann DJ, Estepa-Sabal EJ, Meerloo T, Emr SD (2002) Escrt-III: an endosome-associated heterooligomeric protein complex required for mvb sorting. Dev Cell 3(2):271–82
99. Bache KG, Brech A, Mehlum A, Stenmark H (2003) Hrs regulates multivesicular body formation via ESCRT recruitment to endosomes. J Cell Biol 162(3):435–42
100. Katzmann DJ, Babst M, Emr SD (2001) Ubiquitin-dependent sorting into the multivesicular body pathway requires the function of a conserved endosomal protein sorting complex, ESCRT-I. Cell 106(2):145–55
101. Lu Q, Hope LW, Brasch M, Reinhard C, Cohen SN (2003) TSG101 interaction with HRS mediates endosomal trafficking and receptor down-regulation. Proc Natl Acad Sci U S A 100(13):7626–31
102. Hurley JH, Hanson PI (2010) Membrane budding and scission by the ESCRT machinery: it's all in the neck. Nat Rev Mol Cell Biol 11(8):556–66
103. Henne WM, Buchkovich NJ, Emr SD (2011) The ESCRT pathway. Dev Cell 21(1):77–91
104. McCullough J, Clague MJ, Urbe S (2004) AMSH is an endosome-associated ubiquitin isopeptidase. J Cell Biol 166(4):487–92
105. Agromayor M, Martin-Serrano J (2006) Interaction of AMSH with ESCRT-III and deubiquitination of endosomal cargo. J Biol Chem 281(32):23083–91
106. Row PE, Prior IA, McCullough J, Clague MJ, Urbe S (2006) The ubiquitin isopeptidase UBPY regulates endosomal ubiquitin dynamics and is essential for receptor down-regulation. J Biol Chem 281(18):12618–24
107. Duex JE, Sorkin A (2009) RNA interference screen identifies Usp18 as a regulator of epidermal growth factor receptor synthesis. Mol Biol Cell 20(6):1833–44
108. Duex JE, Comeau L, Sorkin A, Purow B, Kefas B (2011) Usp18 regulates epidermal growth factor (EGF) receptor expression and cancer cell survival via microRNA-7. J Biol Chem 286(28):25377–86
109. Lloyd TE, Atkinson R, Wu MN, Zhou Y, Pennetta G, Bellen HJ (2002) Hrs regulates endosome membrane invagination and tyrosine kinase receptor signaling in Drosophila. Cell 108(2):261–9
110. Hammond DE, Carter S, McCullough J, Urbe S, Vande Woude G, Clague MJ (2003) Endosomal dynamics of Met determine signaling output. Mol Biol Cell 14(4):1346–54
111. Niendorf S, Oksche A, Kisser A, Lohler J, Prinz M, Schorle H, Feller S, Lewitzky M, Horak I, Knobeloch KP (2007) Essential role of ubiquitin-specific protease 8 for receptor tyrosine kinase stability and endocytic trafficking in vivo. Mol Cell Biol 27(13):5029–39
112. Bache KG, Stuffers S, Malerod L, Slagsvold T, Raiborg C, Lechardeur D, Walchli S, Lukacs GL, Brech A, Stenmark H (2006) The ESCRT-III subunit hVps24 is required for degradation but not silencing of the epidermal growth factor receptor. Mol Biol Cell 17(6):2513–23
113. Vaccari T, Lu H, Kanwar R, Fortini ME, Bilder D (2008) Endosomal entry regulates Notch receptor activation in Drosophila melanogaster. J Cell Biol 180(4):755–62
114. Mosesson Y, Mills GB, Yarden Y (2008) Derailed endocytosis: an emerging feature of cancer. Nat Rev Cancer 8(11):835–50
115. Sorkin A, Goh LK (2009) Endocytosis and intracellular trafficking of ErbBs. Exp Cell Res 315(4):683–96
116. Lai AZ, Abella JV, Park M (2009) Crosstalk in Met receptor oncogenesis. Trends Cell Biol 19(10):542–51
117. Pines G, Kostler WJ, Yarden Y (2010) Oncogenic mutant forms of EGFR: lessons in signal transduction and targets for cancer therapy. FEBS Lett 584(12):2699–706

118. Zeng S, Xu Z, Lipkowitz S, Longley JB (2005) Regulation of stem cell factor receptor signaling by Cbl family proteins (Cbl-b/c-Cbl). Blood 105(1):226–32
119. Shtiegman K, Kochupurakkal BS, Zwang Y, Pines G, Starr A, Vexler A, Citri A, Katz M, Lavi S, Ben-Basat Y et al (2007) Defective ubiquitinylation of EGFR mutants of lung cancer confers prolonged signaling. Oncogene 26(49):6968–78
120. Grandal MV, Zandi R, Pedersen MW, Willumsen BM, van Deurs B, Poulsen HS (2007) EGFRvIII escapes down-regulation due to impaired internalization and sorting to lysosomes. Carcinogenesis 28(7):1408–17
121. Han W, Zhang T, Yu H, Foulke JG, Tang CK (2006) Hypophosphorylation of residue Y1045 leads to defective downregulation of EGFRvIII. Cancer Biol Ther 5(10):1361–8
122. Baulida J, Kraus MH, Alimandi M, Di Fiore PP, Carpenter G (1996) All ErbB receptors other than the epidermal growth factor receptor are endocytosis impaired. J Biol Chem 271(9):5251–7
123. Austin CD, De Maziere AM, Pisacane PI, van Dijk SM, Eigenbrot C, Sliwkowski MX, Klumperman J, Scheller RH (2004) Endocytosis and sorting of ErbB2 and the site of action of cancer therapeutics trastuzumab and geldanamycin. Mol Biol Cell 15(12):5268–82
124. Hendriks BS, Opresko LK, Wiley HS, Lauffenburger D (2003) Coregulation of epidermal growth factor receptor/human epidermal growth factor receptor 2 (HER2) levels and locations: quantitative analysis of HER2 overexpression effects. Cancer Res 63(5):1130–7
125. Lenferink AE, Pinkas-Kramarski R, van de Poll ML, van Vugt MJ, Klapper LN, Tzahar E, Waterman H, Sela M, van Zoelen EJ, Yarden Y (1998) Differential endocytic routing of homo- and hetero-dimeric ErbB tyrosine kinases confers signaling superiority to receptor heterodimers. EMBO J 17(12):3385–97
126. Dorsam RT, Gutkind JS (2007) G-protein-coupled receptors and cancer. Nat Rev Cancer 7(2):79–94
127. Li YM, Pan Y, Wei Y, Cheng X, Zhou BP, Tan M, Zhou X, Xia W, Hortobagyi GN, Yu D et al (2004) Upregulation of CXCR4 is essential for HER2-mediated tumor metastasis. Cancer Cell 6(5):459–69
128. Slagsvold T, Marchese A, Brech A, Stenmark H (2006) CISK attenuates degradation of the chemokine receptor CXCR4 via the ubiquitin ligase AIP4. EMBO J 25(16):3738–49
129. Liang W, Fishman PH (2004) Resistance of the human beta1-adrenergic receptor to agonist-induced ubiquitination: a mechanism for impaired receptor degradation. J Biol Chem 279(45):46882–9
130. Nabhan JF, Pan H, Lu Q (2010) Arrestin domain-containing protein 3 recruits the NEDD4 E3 ligase to mediate ubiquitination of the beta2-adrenergic receptor. EMBO Rep 11(8):605–11
131. Hasdemir B, Bunnett NW, Cottrell GS (2007) Hepatocyte growth factor-regulated tyrosine kinase substrate (HRS) mediates post-endocytic trafficking of protease-activated receptor 2 and calcitonin receptor-like receptor. J Biol Chem 282(40):29646–57
132. Padilla BE, Cottrell GS, Roosterman D, Pikios S, Muller L, Steinhoff M, Bunnett NW (2007) Endothelin-converting enzyme-1 regulates endosomal sorting of calcitonin receptor-like receptor and beta-arrestins. J Cell Biol 179(5):981–97
133. Martini L, Waldhoer M, Pusch M, Kharazia V, Fong J, Lee JH, Freissmuth C, Whistler JL (2007) Ligand-induced down-regulation of the cannabinoid 1 receptor is mediated by the G-protein-coupled receptor-associated sorting protein GASP1. FASEB J 21(3):802–11
134. Bhandari D, Trejo J, Benovic JL, Marchese A (2007) Arrestin-2 interacts with the ubiquitin-protein isopeptide ligase atrophin-interacting protein 4 and mediates endosomal sorting of the chemokine receptor CXCR4. J Biol Chem 282(51):36971–9
135. Malik R, Marchese A (2010) Arrestin-2 interacts with the endosomal sorting complex required for transport machinery to modulate endosomal sorting of CXCR4. Mol Biol Cell 21(14):2529–41
136. Tanowitz M, Von Zastrow M (2002) Ubiquitination-independent trafficking of G protein-coupled receptors to lysosomes. J Biol Chem 277(52):50219–22
137. Hislop JN, Henry AG, Marchese A, von Zastrow M (2009) Ubiquitination regulates proteolytic processing of G protein-coupled receptors after their sorting to lysosomes. J Biol Chem 284(29):19361–70

138. Bartlett SE, Enquist J, Hopf FW, Lee JH, Gladher F, Kharazia V, Waldhoer M, Mailliard WS, Armstrong R, Bonci A et al (2005) Dopamine responsiveness is regulated by targeted sorting of D2 receptors. Proc Natl Acad Sci U S A 102(32):11521–6
139. Li JG, Haines DS, Liu-Chen LY (2008) Agonist-promoted Lys63-linked polyubiquitination of the human kappa-opioid receptor is involved in receptor down-regulation. Mol Pharmacol 73(4):1319–30
140. Li JG, Chen C, Liu-Chen LY (2002) Ezrin-radixin-moesin-binding phosphoprotein-50/Na+/H+ exchanger regulatory factor (EBP50/NHERF) blocks U50,488H-induced down-regulation of the human kappa opioid receptor by enhancing its recycling rate. J Biol Chem 277(30): 27545–52
141. Cottrell GS, Padilla B, Pikios S, Roosterman D, Steinhoff M, Gehringer D, Grady EF, Bunnett NW (2006) Ubiquitin-dependent down-regulation of the neurokinin-1 receptor. J Biol Chem 281(38):27773–83
142. Roosterman D, Cottrell GS, Padilla BE, Muller L, Eckman CB, Bunnett NW, Steinhoff M (2007) Endothelin-converting enzyme 1 degrades neuropeptides in endosomes to control receptor recycling. Proc Natl Acad Sci U S A 104(28):11838–43
143. Dupre DJ, Chen Z, Le Gouill C, Theriault C, Parent JL, Rola-Pleszczynski M, Stankova J (2003) Trafficking, ubiquitination, and down-regulation of the human platelet-activating factor receptor. J Biol Chem 278(48):48228–35
144. Wang Y, Zhou Y, Szabo K, Haft CR, Trejo J (2002) Down-regulation of protease-activated receptor-1 is regulated by sorting nexin 1. Mol Biol Cell 13(6):1965–76
145. Jacob C, Cottrell GS, Gehringer D, Schmidlin F, Grady EF (2005) Bunnett NW: c-Cbl mediates ubiquitination, degradation, and down-regulation of human protease-activated receptor 2. J Biol Chem 280(16):16076–87
146. Martin NP, Lefkowitz RJ, Shenoy SK (2003) Regulation of V2 vasopressin receptor degradation by agonist-promoted ubiquitination. J Biol Chem 278(46):45954–9
147. Kobayashi S, Sawano A, Nojima Y, Shibuya M, Maru Y (2004) The c-Cbl/CD2AP complex regulates VEGF-induced endocytosis and degradation of Flt-1 (VEGFR-1). FASEB J 18(7): 929–31
148. Duval M, Bedard-Goulet S, Delisle C, Gratton JP (2003) Vascular endothelial growth factor-dependent down-regulation of Flk-1/KDR involves Cbl-mediated ubiquitination. Consequences on nitric oxide production from endothelial cells. J Biol Chem 278(22): 20091–7
149. Murdaca J, Treins C, Monthouel-Kartmann MN, Pontier-Bres R, Kumar S, Van Obberghen E, Giorgetti-Peraldi S (2004) Grb10 prevents Nedd4-mediated vascular endothelial growth factor receptor-2 degradation. J Biol Chem 279(25):26754–61
150. Reddi AL, Ying G, Duan L, Chen G, Dimri M, Douillard P, Druker BJ, Naramura M, Band V, Band H (2007) Binding of Cbl to a phospholipase Cgamma1-docking site on platelet-derived growth factor receptor beta provides a dual mechanism of negative regulation. J Biol Chem 282(40):29336–47
151. Joazeiro CA, Wing SS, Huang H, Leverson JD, Hunter T, Liu YC (1999) The tyrosine kinase negative regulator c-Cbl as a RING-type, E2-dependent ubiquitin-protein ligase. Science 286(5438):309–12
152. Miyake S, Lupher ML Jr, Druker B, Band H (1998) The tyrosine kinase regulator Cbl enhances the ubiquitination and degradation of the platelet-derived growth factor receptor alpha. Proc Natl Acad Sci U S A 95(14):7927–32
153. Haugsten EM, Malecki J, Bjorklund SM, Olsnes S, Wesche J (2008) Ubiquitination of fibroblast growth factor receptor 1 is required for its intracellular sorting but not for its endocytosis. Mol Biol Cell 19(8):3390–403
154. Wong A, Lamothe B, Lee A, Schlessinger J, Lax I (2002) FRS2 alpha attenuates FGF receptor signaling by Grb2-mediated recruitment of the ubiquitin ligase Cbl. Proc Natl Acad Sci U S A 99(10):6684–9
155. Cho JY, Guo C, Torello M, Lunstrum GP, Iwata T, Deng C, Horton WA (2004) Defective lysosomal targeting of activated fibroblast growth factor receptor 3 in achondroplasia. Proc Natl Acad Sci U S A 101(2):609–14

156. Vecchione A, Marchese A, Henry P, Rotin D, Morrione A (2003) The Grb10/Nedd4 complex regulates ligand-induced ubiquitination and stability of the insulin-like growth factor I receptor. Mol Cell Biol 23(9):3363–72
157. Girnita L, Girnita A, Larsson O (2003) Mdm2-dependent ubiquitination and degradation of the insulin-like growth factor 1 receptor. Proc Natl Acad Sci U S A 100(14):8247–52
158. Sehat B, Andersson S, Girnita L, Larsson O (2008) Identification of c-Cbl as a new ligase for insulin-like growth factor-I receptor with distinct roles from Mdm2 in receptor ubiquitination and endocytosis. Cancer Res 68(14):5669–77
159. Carter S, Urbe S, Clague MJ (2004) The met receptor degradation pathway: requirement for Lys48-linked polyubiquitin independent of proteasome activity. J Biol Chem 279(51): 52835–9
160. Takahashi Y, Shimokawa N, Esmaeili-Mahani S, Morita A, Masuda H, Iwasaki T, Tamura J, Haglund K, Koibuchi N (2011) Ligand-induced downregulation of TrkA is partly regulated through ubiquitination by Cbl. FEBS Lett 585(12):1741–7
161. Geetha T, Wooten MW (2008) TrkA receptor endolysosomal degradation is both ubiquitin and proteasome dependent. Traffic 9(7):1146–56
162. Geetha T, Jiang J, Wooten MW (2005) Lysine 63 polyubiquitination of the nerve growth factor receptor TrkA directs internalization and signaling. Mol Cell 20(2):301–12
163. Qiu XB, Goldberg AL (2002) Nrdp1/FLRF is a ubiquitin ligase promoting ubiquitination and degradation of the epidermal growth factor receptor family member, ErbB3. Proc Natl Acad Sci U S A 99(23):14843–8
164. Cao Z, Wu X, Yen L, Sweeney C, Carraway KL III (2007) Neuregulin-induced ErbB3 downregulation is mediated by a protein stability cascade involving the E3 ubiquitin ligase Nrdp1. Mol Cell Biol 27(6):2180–8
165. Yen L, Cao Z, Wu X, Ingalla ER, Baron C, Young LJ, Gregg JP, Cardiff RD, Borowsky AD, Sweeney C et al (2006) Loss of Nrdp1 enhances ErbB2/ErbB3-dependent breast tumor cell growth. Cancer Res 66(23):11279–86
166. Zeng F, Xu J, Harris RC (2009) Nedd4 mediates ErbB4 JM-a/CYT-1 ICD ubiquitination and degradation in MDCK II cells. FASEB J 23(6):1935–45
167. Li Y, Zhou Z, Alimandi M, Chen C (2009) WW domain containing E3 ubiquitin protein ligase 1 targets the full-length ErbB4 for ubiquitin-mediated degradation in breast cancer. Oncogene 28(33):2948–58
168. Masson K, Heiss E, Band H, Ronnstrand L (2006) Direct binding of Cbl to Tyr568 and Tyr936 of the stem cell factor receptor/c-Kit is required for ligand-induced ubiquitination, internalization and degradation. Biochem J 399(1):59–67
169. Sun J, Pedersen M, Bengtsson S, Ronnstrand L (2007) Grb2 mediates negative regulation of stem cell factor receptor/c-Kit signaling by recruitment of Cbl. Exp Cell Res 313(18): 3935–42
170. Kuratomi G, Komuro A, Goto K, Shinozaki M, Miyazawa K, Miyazono K, Imamura T (2005) NEDD4-2 (neural precursor cell expressed, developmentally down-regulated 4-2) negatively regulates TGF-beta (transforming growth factor-beta) signalling by inducing ubiquitin-mediated degradation of Smad2 and TGF-beta type I receptor. Biochem J 386(pt 3):461–70
171. Komuro A, Imamura T, Saitoh M, Yoshida Y, Yamori T, Miyazono K, Miyazawa K (2004) Negative regulation of transforming growth factor-beta (TGF-beta) signaling by WW domain-containing protein 1 (WWP1). Oncogene 23(41):6914–23
172. Kavsak P, Rasmussen RK, Causing CG, Bonni S, Zhu H, Thomsen GH, Wrana JL (2000) Smad7 binds to Smurf2 to form an E3 ubiquitin ligase that targets the TGF beta receptor for degradation. Mol Cell 6(6):1365–75
173. Ebisawa T, Fukuchi M, Murakami G, Chiba T, Tanaka K, Imamura T, Miyazono K (2001) Smurf1 interacts with transforming growth factor-beta type I receptor through Smad7 and induces receptor degradation. J Biol Chem 276(16):12477–80
174. Scott RP, Eketjall S, Aineskog H, Ibanez CF (2005) Distinct turnover of alternatively spliced isoforms of the RET kinase receptor mediated by differential recruitment of the Cbl ubiquitin ligase. J Biol Chem 280(14):13442–9

175. Pierchala BA, Tsui CC, Milbrandt J, Johnson EM (2007) NGF augments the autophosphorylation of Ret via inhibition of ubiquitin-dependent degradation. J Neurochem 100(5): 1169–76
176. Wang Y, Yeung YG, Stanley ER (1999) CSF-1 stimulated multiubiquitination of the CSF-1 receptor and of Cbl follows their tyrosine phosphorylation and association with other signaling proteins. J Cell Biochem 72(1):119–34
177. Pakuts B, Debonneville C, Liontos LM, Loreto MP, McGlade CJ (2007) The Src-like adaptor protein 2 regulates colony-stimulating factor-1 receptor signaling and down-regulation. J Biol Chem 282(25):17953–63
178. Lee PS, Wang Y, Dominguez MG, Yeung YG, Murphy MA, Bowtell DD, Stanley ER (1999) The Cbl protooncoprotein stimulates CSF-1 receptor multiubiquitination and endocytosis, and attenuates macrophage proliferation. EMBO J 18(13):3616–28

Chapter 11
Cbl as a Master Regulator of Receptor Tyrosine Kinase Trafficking

Ke Ma, Stephen C. Kales, Marion M. Nau, and Stanley Lipkowitz

Abstract The Cbl proteins are a family of RING finger ubiquitin ligases which are found throughout metazoans. In mammalian cells there are three Cbl proteins, Cbl, Cbl-b, and Cbl-c. The RING finger domain, responsible for ubiquitin ligase activity, is surrounded by protein interaction motifs that allow Cbl proteins to interact with a large number of signaling proteins and thus function in many signaling pathways. Receptor tyrosine kinases (RTKs) are rapidly internalized upon activation and can either be recycled to the cell surface or degraded in the lysosome (a process known as downregulation). The Cbl proteins ubiquitinate the activated RTKs and mediate their trafficking to the lysosome for degradation. Thus, they are critical regulators of RTK downregulation. This process is tightly regulated by RTK-mediated phosphorylation of the Cbl proteins that activates the ubiquitin ligase activity of the Cbl protein. In addition, multiple proteins can attenuate Cbl-mediated ubiquitination and downregulation of the RTK. Mutations which disrupt the ubiquitin ligase activity of the Cbl proteins result in oncogenic forms, and such mutations have been described in human myeloid neoplasms. In addition mutations in the RTK or overexpression of negative regulators of Cbl proteins can result in aberrant RTK downregulation and transformation. Thus, the Cbl proteins are critical regulators of RTK trafficking and serve to tune the level of RTK activity.

Ke Ma and Stephen C. Kales contributed equally to this work.

K. Ma • S.C. Kales • M.M. Nau • S. Lipkowitz (✉)
Laboratory of Cellular and Molecular Biology, Center for Cancer Research, National Cancer Institute, National Institutes of Health, Building 37, Room 2066, 37 Convent Drive, Bethesda, MD 20892, USA
e-mail: lipkowis@mail.nih.gov

Y. Yarden and G. Tarcic (eds.), *Vesicle Trafficking in Cancer*,
DOI 10.1007/978-1-4614-6528-7_11, © Springer Science+Business Media New York 2013

11.1 The Cbl Proteins

The Cbl proteins comprise a conserved family of RING finger ubiquitin ligase (E3) proteins found in metazoans (multicellular animals) that regulates signal transduction in a wide variety of pathways [1, 2]. The covalent modification of proteins by ubiquitin occurs via the sequential activation and conjugation of ubiquitin to target proteins by the ubiquitin-activating enzyme (E1), a ubiquitin-conjugating enzyme (E2), and a ubiquitin protein ligase (E3) [3, 4]. The E3 directs the conjugation of ubiquitin to one or more lysines of the specific target protein and thus confers specificity to the process. The majority of E3s contain a RING finger domain or the related U box domains [5]. RING finger E3s can mediate the addition of a single ubiquitin to the substrate (monoubiquitination), single ubiquitin molecules to multiple different sites on the substrate (multi-monoubiquitination), or chains of ubiquitin on one or more lysines of the substrate (polyubiquitination) [3, 4, 6]. These ubiquitin chains are formed between the C-terminal glycine of the ubiquitin molecule and an internal lysine on the preceding ubiquitin, and chains can potentially be formed on all of the seven internal lysines of ubiquitin (lysine 6, 11, 27, 29, 33, 48 or 63). Ubiquitination with chains of four or more ubiquitin molecules linked via lysine 48 on ubiquitin was first demonstrated to target proteins for degradation by the 26S proteasome [3, 4]. Also, ubiquitination regulates diverse processes such as endocytosis, activation of signal transduction cascades, and DNA repair [4, 6]. The outcome of ubiquitination is specified by the type of ubiquitin chain formed on the substrate, cellular location and context, interactions with proteins containing ubiquitin interacting motifs, and the activity of deubiquitinating enzymes [4, 6]. As will be discussed below, Cbl proteins are E3s which ubiquitinate activated receptor tyrosine kinases (RTKs) and regulate their endocytic trafficking to and degradation in the lysosome.

v-Cbl, the first Cbl gene identified, is the transforming gene of the Cas NS-1 murine retrovirus which causes leukemia and lymphoma in mice [7]. Subsequent cloning of the cellular proto-oncogene c-Cbl (herein called Cbl) revealed that v-Cbl is a gag-Cbl fusion protein containing only the N-terminal 355 amino acids of Cbl (Fig. 11.1) [8]. There are three mammalian Cbl proteins encoded by separate genes: Cbl (a.k.a. c-Cbl, CBL2, RNF55), Cbl-b (a.k.a. RNF56), and Cbl-c (a.k.a. Cbl-3, Cbl-SL, RNF57) (Fig. 11.1) [8–13]. Sequences for Cbl, Cbl-b, and Cbl-c are found in numerous mammalian species [1, 2]. Thus, it appears that all mammals have three Cbl genes encoding Cbl, Cbl-b, and Cbl-c.

Cbl genes are found also in nonmammalian chordates [1, 2]. Orthologues of Cbl and Cbl-b are present in birds, lizards, amphibians, and fish. A third Cbl gene is found in lizards and several species of fish and encodes a predicted protein which lacks an ubiquitin-associated (UBA) domain. Thus, the predicted protein appears to structurally resemble Cbl-c although sequence comparisons of this protein to the other Cbl proteins suggest that it has greater similarity to Cbl and Cbl-b than to Cbl-c [1, 2]. Sea squirt, the most primitive chordate, has a single Cbl gene whose predicted protein contains tyrosine kinase binding (TKB), RING finger, proline-rich, and UBA domains (see structural features below) [1, 2]. This protein shares similar

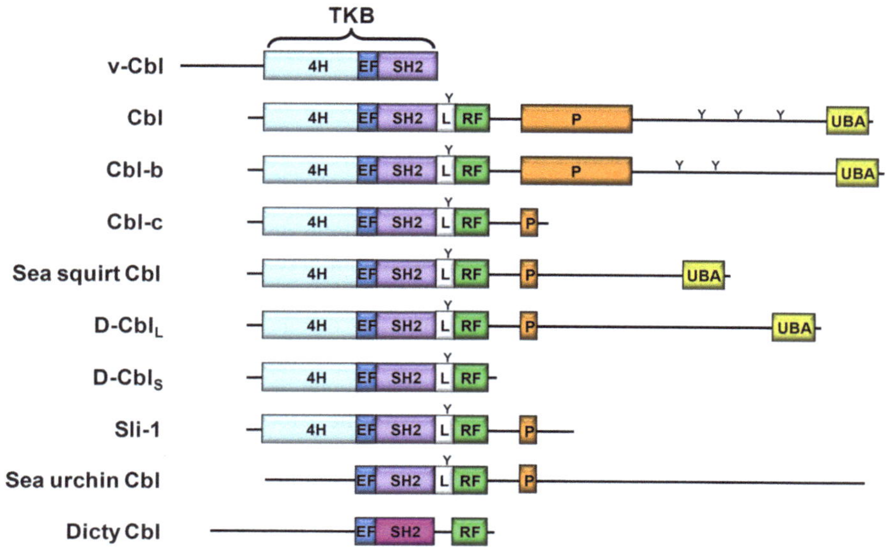

Fig. 11.1 Structure of the Cbl proteins. v-Cbl: the Gag-Cbl fusion protein of the Cas NS-1 murine retrovirus; Cbl: the human proto-oncogene of v-Cbl; Cbl-b: the second human Cbl protein; Cbl-c: the third human Cbl protein; sea squirt Cbl is the Cbl protein from primitive chordate *C. intestinalis*; D-Cbl$_L$ and D-Cbl$_S$: the long and short spliced isoforms of the *D. melanogaster* Cbl protein, respectively; Sli-1: the *C. elegans* Cbl protein; sea urchin Cbl is the putative Cbl protein from *S. purpuratus*; Dicty Cbl: the putative *D. discoideum* Cbl protein. The tyrosine kinase binding domain (TKB) is comprised of a 4-helix bundle (4H), an EF hand (EF), and a variant SH2 domain (SH2). The SH2 domain of the *D. discoideum* Cbl is indicated in a different color as it is a true SH2 domain. The linker (L), RING finger (RF), proline-rich (P), and the ubiquitin-associated (UBA) domains are indicated on the diagram. The critical linker tyrosine and the tyrosines in the C-terminal of c-Cbl and Cbl-b that are phosphorylated are indicated (Y) (figure adapted with modification from [2])

homology to both Cbl and Cbl-b and thus is likely to represent a precursor to both of the long Cbl genes found in chordates.

Cbl genes are also found in invertebrates. A single Cbl gene is present in a variety of insects [1, 2]. The fruit fly *Drosophila melanogaster* (*D. melanogaster*) produces two alternatively spliced mRNAs encoding a short and long form of the Cbl proteins (D-Cbl$_S$ and D-Cbl$_L$, respectively) (Fig. 11.1) [1, 14–16]. Roundworms, including *Caenorhabditis elegans* (*C. elegans*), the trichinosis worm (*C. briggsae*), cotton root-knot worm, and the elephantiasis worm, also have a single Cbl orthologue [1, 2, 17]. The roundworm Cbl genes (known as *Sli-1* in *C. elegans*) are most similar in structure to the mammalian Cbl-c gene (Fig. 11.1) [2, 10, 12, 17]. The sea urchin, a member of the *Echinodermata*, contains a predicted Cbl protein containing a truncated TKB domain, a linker region, a RING finger, and a proline-rich region but lacks the UBA domain found in other long Cbl proteins (Fig. 11.1) [2].

The genome of the social amoeba, *Dictyostelium discoideum* (*D. discoideum*), contains a putative Cbl orthologue [2, 18]. This finding suggests that Cbl-like genes may have originated in unicellular eukaryotes. Database searches have not identified Cbl genes in other single cell organisms or in plants.

11.2 The Structure of Cbl Proteins

All Cbl proteins contain a RING finger, the catalytic domain responsible for E3 activity (Fig. 11.1) [19–23]. RING fingers were first described based on homology to the protein "Really Interesting New Gene 1" (RING1) and were predicted to form a zinc coordination complex with two molecules of zinc [24]. Unlike other zinc coordination complexes in proteins, the structure of the RING finger is a cross-braced coordination complex [25, 26]. The RING finger domains of the Cbl proteins are highly conserved and directly interact with the E2 enzymes [2, 27–29]. The region immediately downstream of the core RING finger (called the RING finger tail) is highly conserved in Cbl proteins [2, 30]. An arginine residue at the beginning of this domain makes contact with the E2 [28, 29]. This arginine is conserved in all Cbl proteins including the putative *D. discoideum* Cbl protein [2]. Additional residues within this domain make contact with the TKB domain [29]. This region appears to regulate the E3 activity of Cbl proteins [30].

The RING finger of Cbl proteins is surrounded by a number of domains which interact with the substrate proteins and/or regulate the activity of the RING finger (Fig. 11.1). All metazoan Cbl proteins contain a highly conserved N-terminal TKB domain. The TKB domain is a novel domain found only in Cbl proteins that mediates interactions between Cbl proteins and phosphorylated tyrosines on other proteins [31]. It is composed of a 4-helix bundle, a calcium binding EF hand, and a variant SH2 domain [32]. The SH2 domain is not recognizable from the primary amino acid sequence and was identified by the crystal structure of the N-terminal domain of the Cbl protein [32]. The putative Cbl protein in the sea urchin has a truncated TKB containing only the EF hand and the variant SH2 domain found in the other Cbl proteins but lacks the 4-helix bundle (Fig. 11.1) [2]. Similarly, the predicted putative Cbl protein from *D. discoideum* contains the EF hand and partial SH2 domain but lacks the 4-helix bundle (Fig. 11.1) [2, 18]. Furthermore, the SH2 domain in the *D. discoideum* Cbl is a partial SH2 domain with homology to classical SH2 domains rather than to the variant SH2 domains found in the other Cbl proteins [2].

The TKB domain was first shown to bind to phosphotyrosines within the amino acid sequence (D/N)XpYXX(D/E)Ψ (where Ψ represents a hydrophobic residue) in the ZAP-70 tyrosine kinase [31]. The crystal structure of the TKB domain complexed to its binding site in ZAP-70 revealed that the primary specificity-determining interactions are located at the hydrophobic residue at pY+4 and the acidic residue at pY+3 [32]. This consensus TKB binding sequence is found in other Cbl protein binding partners including the epidermal growth factor receptor (EGFR), Syk, and Sprouty proteins [21, 33–35]. Recently, the TKB domain of Cbl proteins has been shown to interact with the sequence DpYR in the Met receptor [36]. The DpYR

motif is conserved in the Met RTK family (i.e., Met, Ron, and Sea) and also is found in plexins [36]. A third binding motif, RA(V/I)XNQpY(S/T), has been identified in the adaptor protein with a PH and SH2 domain (APS) [37]. The ability of the TKB region to interact with several unrelated pY consensus sequences is not seen in other phosphotyrosine interacting domains. Also, the TKB domain of Cbl constitutively interacts with SLAP (Src-like adaptor protein) and binds to microtubules [38–40]. These interactions appear to be phosphotyrosine independent.

The RING finger is separated from the TKB domain by a short, conserved alpha-helical sequence known as the linker region [2]. The linker region and the RING finger domain directly interact with the ubiquitin-charged E2 and catalyze transfer of ubiquitin to the target substrate [29]. In addition, amino acid residues within the linker region make contact with the TKB domain [29]. The linker region is recognizable in all of the Cbl proteins, including the sea urchin Cbl protein, but not in the putative *D. discoideum* Cbl protein [2]. This region is critical for the regulation of the Cbl protein E3 activity (discussed below).

The Cbl proteins, except v-Cbl and d-Cbl_S, have proline-rich domains that mediate interactions with SH3-containing proteins [2]. The long forms of Cbl proteins (e.g., Cbl, Cbl-b, D-Cbl_L, and sea squirt Cbl) share additional areas of homology in the C-terminal half of the proteins [2]. The long Cbl proteins have more extensive proline-rich regions than either Cbl-c or Sli-1. There are multiple SH3-containing proteins which interact with the proline-rich region of the Cbl proteins [41]. Additional proline-based interactions occur outside of the proline-rich region. For example, Cin85 interacts via an SH3 domain with a proline motif near the UBA domains of Cbl and Cbl-b [42]. The long Cbl proteins (e.g., Cbl and Cbl-b) become heavily tyrosine phosphorylated upon activation of tyrosine kinase pathways, and many SH2 proteins can interact with the phosphotyrosine residues in the C-terminal region of these Cbl proteins [41].

UBA domains are found at the C-terminus of the long forms of the Cbl proteins. UBA domains were identified originally as a motif containing three alpha helices found in proteins involved in ubiquitin metabolism and were shown subsequently to mediate ubiquitin binding [43, 44]. Interestingly, the UBA domain of Cbl-b, but not the UBA domain of Cbl, mediates binding of Cbl-b to ubiquitin and ubiquitinated proteins [45, 46]. The UBA domains of Cbl and Cbl-b have been shown to mediate homodimerization via distinct mechanisms. The UBA domain of Cbl mediates dimerization of Cbl via an interaction between the second two alpha helices formed by the UBA domain [47–49]. The UBA domain of Cbl-b mediates dimerization via the same alpha helices as that of Cbl [50]. However, the interaction is weak and requires binding of the Cbl-b UBA domain to ubiquitin chains to stabilize the dimer [50]. Disruption of dimerization via the UBA domain of Cbl and Cbl-b has been shown to lead to reduced phosphorylation of the Cbl protein, decreased recruitment, and decreased ubiquitination of RTKs [47, 48, 50]. Thus, dimerization of Cbl proteins by the UBA domain is essential for their optimal functional interaction with RTKs.

The multiple binding motifs allow Cbl proteins to target many proteins for ubiquitination that interact via distinct mechanisms. For example, Cbl interacts with the EGFR directly via the TKB domain and indirectly via an SH3-mediated interaction with Grb2. This leads to EGFR ubiquitination and downregulation [21, 51, 52]. In contrast, Cbl proteins interact with phosphoinositide-3 kinase (PI3K) predominantly via the binding of the SH2 domain of the p85 subunit of PI3K to phosphotyrosine residues in the C-terminal region of Cbl proteins. In T cell receptor signaling, this results in ubiquitination of PI3K without degradation [53, 54]. In addition, the multiple protein-binding motifs in Cbl proteins allow the Cbl proteins to function as adaptor proteins in signaling pathways (reviewed in [41, 55]).

11.3 Cbl Proteins as Regulators of Endocytic Trafficking and Degradation of Activated RTKs

It has been known since the 1970s that ligand activation of many RTKs such as the EGFR results in a rapid decrease in the number of cell surface receptors due to internalization and to an eventual decrease in the cellular content of activated receptors due to degradation of the RTK—a process known as "downregulation" [56–58]. Upon ligand-mediated activation of the growth factor receptors, the receptors localize to specific membrane regions known as clathrin-coated pits. Clathrin-coated pits are membrane invaginations coated by a lattice of clathrin proteins, and these pits are the major site of endocytosis of activated membrane receptors under physiological conditions [58, 59]. These invaginations pinch off to form clathrin-coated vesicles—thus internalizing the membrane proteins localized within these clathrin-coated pits. Specific amino acid sequences within membrane protein cytoplasmic tails target proteins to the clathrin-coated pits [58, 59]. Mechanisms of non-clathrin-dependent endocytosis have been described for RTKs, and the relative contribution of each may be cell type dependent (reviewed in [58]). Subsequently, the internalized RTKs localize to a vesicular membrane compartment known as the early endosome [58, 59]. This compartment sorts the RTKs to either a recycling compartment which returns the protein to the plasma membrane or to an endocytic pathway which trafficks the RTK to the multivesicular body (MVB) and then to the lysosome for degradation [58, 59].

Ubiquitination of membrane proteins was first found when sequence analysis of purified proteins of the growth hormone receptor and the platelet-derived growth factor receptor (PDGFR) yielded two sequences, one for the receptors and one for ubiquitin [60, 61]. This suggested that the receptors were covalently modified by ubiquitin. Subsequently it was shown that several yeast and mammalian membrane proteins are modified by ubiquitin (reviewed in [62]). Also, it was demonstrated that ligand activation resulted in ubiquitination of the cytosolic tails of a number of RTKs and that ubiquitination was implicated in the downregulation of these receptors (e.g., CSF1R, EGFR, FGFR, KIT, MET, and PDGFR) [63–68]. Regulation of internalization and endocytic trafficking by ubiquitin was clearly demonstrated in studies of G-protein-coupled receptors and transporters in *Saccharomyces cerevisiae* [69, 70]. These receptors are mono- and di-ubiquitinated upon ligand activation, and the ubiquitinated membrane proteins are degraded in the yeast lysosome-like vacuole [69–71]. The di-ubiquitin chains are formed by linkage of the C-terminal glycine of the second ubiquitin molecule to lysine 63 of the first ubiquitin molecule. Subsequent work has demonstrated that ubiquitinated membrane proteins are recognized by proteins containing ubiquitin-binding motifs which form part of a multiprotein ESCRT complex which in turn mediates fusion of the vesicles with the MVB [72]. This leads to degradation of the proteins in the MVB and lysosome.

The initial identification of v-Cbl as a viral oncogene did not shed light on the mechanisms by which Cbl proteins work [7, 8]. Similarly, when Cbl was first cloned, the functions of the critical structural motifs (i.e., the TKB, RING finger, and UBA domain) were generally unknown, again providing scant clues to the function of Cbl proteins. Presence of a zinc finger (the RING finger) and a putative atypical leucine zipper (the UBA domain) led to speculation that Cbl was a nuclear transcription factor [73, 74]. The recognition that Cbl proteins were involved in signal transduction came from a series of studies that showed that Cbl and Cbl-b interacted with other signaling proteins and/or became tyrosine phosphorylated in response to activation of a number of tyrosine kinase dependent signaling pathways [11, 75–85]. Also, activation induced tyrosine phosphorylation and association of Cbl proteins with RTKs such as the EGFR and the CSF1R was reported [78, 85–88]. Simultaneously, genetic experiments in *C. elegans* demonstrated that the *C. elegans* Cbl protein Suppressor of Lineage-1 (Sli-1) is a negative regulator of EGFR signaling [17, 89]. Specifically they found that inactivating mutations of Sli-1 could rescue a vulvaless phenotype caused by a hypomorphic mutation of Let-23, the *C. elegans* orthologue of the EGFR [17, 89]. One of the mutations identified was a missense mutation (G315E) which disrupted what was subsequently identified as the variant SH2 domain [17]. This mutation identified a highly conserved site within the Cbl variant SH2 domain that is required for the interaction between the TKB domain of Cbl proteins and phosphotyrosine [90]. Subsequent work demonstrated that the TKB domain and the RING finger of Sli-1 were necessary and sufficient for the negative regulation of Let-23 and that an autophosphorylated tyrosine residue on Let-23 was essential for this interaction [91]. In parallel, similar results were reported for the human EGFR [21, 92]. Subsequent experiments in *D. melanogaster* demonstrated that overexpression of D-Cbl in the eye of *D. melanogaster* embryos inhibits EGFR-dependent photoreceptor cell development [15]. During embryogenesis dorsoventral patterning in *D. melanogaster* is determined by a gradient of EGFR activity, and inactivation of D-Cbl results in abnormal dorsoventral patterning as a consequence of EGFR hyperactivity [93, 94]. Overexpression of either D-Cbl$_L$ or D-Cbl$_S$ rescues the mutant phenotype [93, 94].

The findings in *C. elegans* stimulated functional studies of the role of mammalian Cbl proteins in EGFR signaling as well as in that of other RTKs. Several studies demonstrated that overexpression of Cbl proteins could inhibit EGFR signaling in mammalian cells [10, 19, 95]. In an early study, it was noted that CSF stimulation of the CSFR-1 RTK resulted in recruitment of Cbl to a membrane fraction and that the Cbl protein associated with membranes was tyrosine phosphorylated and ubiquitinated [88]. Subsequent studies found that Cbl proteins enhanced ubiquitination and degradation of the activated PDGFR and EGFR (Fig. 11.2) [96, 97]. In the latter study,

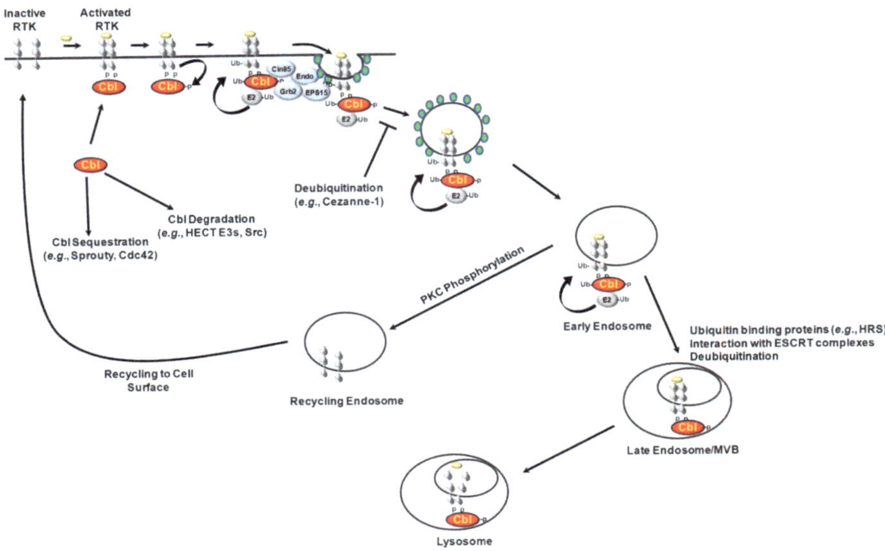

Fig. 11.2 Cbl and RTK endocytosis (see text for details). Cbl proteins are recruited to the activated RTK by binding to phosphotyrosines on the RTK. The Cbl proteins are in turn phosphorylated on a critical linker tyrosine, which activates the E3 function of the Cbl protein leading to ubiquitination of the RTK and autoubiquitination of the Cbl protein. The Cbl proteins also recruit other proteins involved in endocytosis (e.g., Grb2, Cin85, endophilins, and Eps15) to the activated RTK. The complex is internalized in clathrin-coated pits (clathrin shown in green) and moves to the early endosome. For simplicity, only clathrin-dependent mechanisms are shown (see text for discussion). At the early endosome, ubiquitination is critical for the trafficking of the RTK to the late endosome/multivesicular body (MVB) and eventual degradation in the lysosome. The complex is deubiquitinated at the MVB. Cbl-mediated RTK downregulation can be inhibited by proteins that sequester or degrade Cbl, by deubiquitination of the RTK, or by PKC-mediated phosphorylation of the RTK. The details of endocytic trafficking are covered in greater detail elsewhere in this book

it was shown that the Cbl protein is found in a diffuse localization in unstimulated cells, but upon EGF stimulation, Cbl colocalizes with the activated EGFR in an intracellular vesicular-like compartment [96]. These observations led to the demonstration that Cbl proteins are E3s that mediate the ubiquitination of activated RTKs [20, 21, 98]. The ubiquitination of the RTK requires kinase activity by the RTK, the TKB domain of Cbl which interacts with the tyrosine-phosphorylated RTK, phosphorylation of the linker domain, and the RING finger, which is the catalytic domain required for E3 activity [20, 21, 92, 98]. Cbl-mediated ubiquitination and downregulation of RTKs have been demonstrated for many RTKs (Table 11.1).

Mechanistic studies of Cbl protein-mediated downregulation of RTKs, primarily using EGFR as the model substrate, have demonstrated that Cbl proteins bind directly to the EGFR via an interaction between the TKB domain of Cbl and a specific phosphotyrosine on the EGFR (tyrosine 1045) leading to EGFR ubiquitination [21]. Also, Cbl can bind to the EGFR indirectly by binding to the SH3 domains

Table 11.1 RTKs and Cbl

RTK	Evidence of Cbl function	Ref
Alk	Activation induced Cbl phosphorylation and association with Alk	[167]
Axl	Activation induced Cbl phosphorylation, association with Axl and ubiquitination of Axl	[168]
CSFR1 (FMS)	Activation induced Cbl phosphorylation, association with CSFR1, ubiquitination, and downregulation	[163]
EphA1	Activation induced Cbl phosphorylation	[169]
EphA2	Activation induced downregulation of EphA2 by Cbl. TKB and RING finger dependent. No direct evidence of ubiquitination	[170]
EphA3	Activation induced Cbl phosphorylation and Cbl induced downregulation. No direct evidence of ubiquitination	[169]
EphA4	Activation induced Cbl phosphorylation, association with EphA4, and Cbl induced downregulation. No direct evidence of ubiquitination	[169]
EphB1	Activation does not induce Cbl phosphorylation or association. No evidence for Cbl-mediated downregulation	[170]
EphB2	Activation does not induce Cbl phosphorylation or association. No evidence for Cbl-mediated downregulation	[170]
EphB6	Kinase dead family member. Cross-linking of EphB6 induces association with Cbl. No evidence for ubiquitination or downregulation	[171–173]
ErbB1 (EGFR, HER1)	Activation induced Cbl phosphorylation, association with EGFR, ubiquitination, and downregulation (see text for discussion)	See text
ErbB2 (HER2, Neu)	Activation induced Cbl phosphorylation, association with ErbB2, ubiquitination, and downregulation when ErbB2 is homodimerized. Not recruited to ErbB2 when heterodimerized with other ErbB family members	[166, 174]
ErbB3 (HER3)	Cbl not phosphorylated upon activation of ErbB3	[87]
ErbB4 (HER4)	Cbl not phosphorylated upon activation of ErbB4	[87]
FGFR1	Activation induced Cbl phosphorylation, association with FGFR1, ubiquitination, and downregulation	[175]
FGFR2	Activation induced Cbl phosphorylation, association with FGFR2, ubiquitination, and downregulation	[176–179]
FGFR3	Activation induced Cbl phosphorylation, ubiquitination, and downregulation	[180]
Flt1(VEGFR1)	Activation induced Cbl phosphorylation, association with Flt1, Cbl mediated ubiquitination, and Cbl induced downregulation	[181]
Flt3	Activation induced Cbl phosphorylation, association with Flt3, ubiquitination, and downregulation	[139]
IGF1R	Activation induced association of Cbl with IGF1R, Cbl mediated ubiquitination, and downregulation of IGF1R	[182]
INSR	Activation induced Cbl phosphorylation. Indirect association of Cbl with insulin receptor via Cbl associated protein (CAP). No evidence for role of Cbl in ubiquitination or downregulation of insulin receptor	[183, 184]
KDR (VEGFR2)	Activation induced Cbl phosphorylation, association with KDR, ubiquitination, and downregulation of KDR	[185]

(continued)

Table 11.1 (continued)

RTK	Evidence of Cbl function	Ref
KIT	Activation induced Cbl phosphorylation, association with KIT, ubiquitination, and downregulation of KIT	[109]
LTK	Activation induced Cbl phosphorylation. Indirect association of Cbl via binding to Shc and Grb2	[186]
MET	Activation induced Cbl phosphorylation, association with MET, ubiquitination, and downregulation of MET	[36]
MUSK	Evidence of no association with MUSK upon activation	[187]
PDGFRα	Activation induced Cbl phosphorylation, association with PDGFRα, ubiquitination, and downregulation of PDGFRα	[97]
PDGFRβ	Activation induced Cbl phosphorylation, association with PDGFRβ, ubiquitination, and downregulation of PDGFRβ	[188]
RET	Activation induced Cbl phosphorylation, indirect association with RET via Shc or Grb2, ubiquitination, and downregulation of Ret. Evidence for CD2AP-dependent Cbl-c association with and ubiquitination of RET	[189, 190]
RON	Activation induced Cbl phosphorylation, association with RON, ubiquitination, and downregulation of RON	[191]
TRKA	Activation induced Cbl association with TRKA, ubiquitination, and downregulation of TRKA	[192]
TRKB	Activation induced Cbl phosphorylation. No evidence for ubiquitination or downregulation of TRKB by Cbl	[193]
TRKC	Cbl-mediated downregulation is not induced by oxidant induced activation of TRKC. No data on the effects of Cbl proteins on ligand activated TRKC	[194]

There are no references demonstrating any interaction for Cbl and the following RTKs: CCK4, DDR1, DDR2, EphA5, EphA8, EphB3, EphB4, FGFR4, FLT4(VEGFR-3), IRR, MER, ROR1, ROR2, ROS, RYK, TEC, TIE1, TIE2, TYRO3

of Grb2 which in turn binds via its SH2 domain to specific phosphotyrosines on the EGFR [51]. Cbl mutants which cannot bind to EGFR via the TKB domain depend on the indirect Grb2-mediated interaction to mediate ubiquitination and downregulation of the EGFR [51]. Interestingly, Grb2 has been shown to be critical for EGFR internalization and degradation (reviewed in [58]). Quantitative mass spectroscopy of the activated EGFR has demonstrated that approximately half of the ubiquitinated lysines are monoubiquitinated, and approximately 40% of the ubiquitinated lysines have chains formed via lysine 63 of ubiquitin [99]. Cbl-mediated monoubiquitination of the EGFR is sufficient for RTK downregulation [100, 101]. These results are consistent with data indicating a role for monoubiquitin or lysine 63 linked chains in endocytosis [62, 71]. A number of proteins which contain ubiquitin-binding domains (e.g., Epsin, EPS15, and HRS) are recruited to the ubiquitinated EGFR complex via their ubiquitin-binding domains and are implicated in the trafficking of the EGFR to the MVB for degradation. The cellular location where the EGFR becomes ubiquitinated by Cbl is not clear. Cbl proteins can associate with and ubiquitinate the activated EGFR at the plasma membrane when endocytosis is blocked [102]. However, Cbl proteins or ubiquitination may not be absolutely necessary for the initial internalization of the EGFR. In some studies internalization of activated EGFR is not impaired in cells lacking Cbl proteins [99, 103, 104]. Other studies have found that while Cbl and Grb2 are essential for internalization, ubiquitination of the EGFR by Cbl is not required [99]. Further, CHO cells with a temperature-sensitive mutation of the mammalian E1 had markedly decreased ubiquitination of the EGFR at the nonpermissive temperature, but internalization was not impaired [103]. In contrast, Cbl protein-mediated ubiquitination is required for sorting of the internalized EGFR from the early endosome to the late endosome/MVB [103, 105, 106]. Interestingly, both the RTK and the Cbl proteins are ubiquitinated, and the complex formed between Cbl proteins, the activated RTK, and other signaling molecules (e.g., Grb2, Cin85) appears to traffic en masse for degradation in the lysosome [102, 107–109].

Also, the Cbl proteins associate with CIN85, an adaptor protein containing three SH3 domains that interact with prolines near the UBA domain of Cbl and Cbl-b [42, 110, 111]. The interaction between Cbl and CIN85 is increased upon RTK-induced phosphorylation of the Cbl protein although the mechanism underlying this increase is not clear [42]. Cbl proteins also monoubiquitinate CIN85 [108]. CIN85 is constitutively associated with endophilins which in turn bind to proteins which regulate endocytosis such as dyamin, amphiphysin, and synaptojanin (Fig. 11.2) [110, 111]. These interactions potentially couple the binding of Cbl proteins to the EGFR and clathrin-mediated endocytosis [110, 111]. Together these data suggest multiple roles for Cbl in the internalization and sorting of RTKs although the indispensible role is likely to be the ubiquitin-mediated sorting step from the early endosome to the late endosome/MVB.

A number of mechanisms have been shown to regulate Cbl-mediated ubiquitination and downregulation of the EGFR (Fig. 11.2) (reviewed in [58, 112]). The Cbl proteins are inactive in the absence of their substrate. The N-terminal TKB region of Cbl proteins has been shown to regulate negatively the E3 activity of the Cbl protein [113, 114]. Based on structural studies, in the absence of substrate peptide binding to the TKB or phosphorylation of the linker tyrosine, the TKB domain forms a compact structure with the RING finger that masks the E2 binding sites [27, 28]. This effectively inhibits the E3 activity of the Cbl proteins. Phosphorylation of the conserved tyrosine in the linker preceding the RING finger has been shown to be essential for activation of the E3 activity of all Cbl proteins both in vivo and in vitro [21, 113, 114]. The linker tyrosine is inaccessible for phosphorylation in the compact, inactive state [27–29]. However, the inactive protein exists in an equilibrium between the tightly folded unphosphorylated protein and a partially unfolded structure, even in the absence of substrate binding that would make the linker tyrosine accessible for phosphorylation [27]. Upon phosphorylation of the linker tyrosine, there is a dramatic rotation of the linker domain which fully exposes the E2 binding surface of the RING finger, increasing the affinity of the E2 for the RING finger, and simultaneously positioning the E2 in close proximity to the substrate bound to the TKB domain [27, 28]. Phosphorylation results in increased E3 activity by the Cbl proteins [21, 27, 28, 113, 114]. Thus, autophosphorylation of the RTK creates a docking site for the Cbl proteins TKB, and the subsequent RTK-dependent phosphorylation of the Cbl proteins stimulates the E3 activity of the Cbl proteins.

Deubiquitination of the EGFR can also negatively regulate Cbl-mediated downregulation of the EGFR. Recent work has identified the deubiquitinating enzyme Cezanne-1 as a protein which binds to the EGFR and deubiqutinates the EGFR, and thereby attenuates downregulation [115]. Cezanne-1 is amplified and overexpressed in approximately one third of breast cancer tumors, and high expression is associated with poor prognosis [115].

Several reports have described degradation of Cbl proteins which can potentially dysregulate RTK activity. The HECT E3s Nedd4 and Itch bind to and ubiquitinate all of the mammalian Cbl proteins and target them for proteasomal degradation [116]. Consistent with these observations, Nedd4 prevents Cbl-mediated EGFR ubiquitination and downregulation and results in the persistence of downstream signaling by the EGFR [116]. Similar regulation of EGFR signaling has been described for Itch [117]. While a physiological role for HECT-mediated degradation of Cbl proteins has not been demonstrated in RTK signaling, mice that are null for Nedd4 have impaired T cell receptor signaling in part due to the loss of ubiquitination and degradation of Cbl-b [118]. Also, activated forms of Src induce ubiquitination and proteasomal degradation of c-Cbl thereby preventing downregulation of the activated EGFR by Cbl [23, 119]. Activated Src potentiates the transforming activity of the EGFR, and this may be due, at least in part, to degradation of the Cbl proteins by activated Src [119–121]. Src kinase activity is increased in a wide range of human epithelial malignancies compared to normal tissues (reviewed in [122]). Increased Src activity would be predicted to cause Cbl degradation, and this may contribute to the development or progression of these cancers by increasing RTK activity.

A number of proteins have been described which can sequester Cbl proteins and thereby increase RTK signaling. For example, upon EGFR stimulation the Ras-related GTPase Cdc42 becomes activated to the GTP-bound form which then forms a complex with the protein p85Cool-1 (a.k.a. β-Pix) and Cbl or Cbl-b [123]. This interaction sequesters the Cbl protein and prevents the binding of Cbl to the activated EGFR, resulting in decreased ubiquitination and downregulation of the activated EGFR and prolonged downstream signaling by the EGFR [123]. Hydrolysis of GTP to GDP by Cdc42 results in dissociation of the Cdc42/p85Cool-1/Cbl complex and allows Cbl proteins to bind to the EGFR mediating its downregulation. Under normal physiological conditions, the interaction with Cdc42 could provide a temporal switch for Cbl proteins to prevent too rapid a recruitment of the Cbl proteins to the activated EGFR. Interestingly, Cdc42 and p85Cool-1 are overexpressed in some human cancers [124–128]. Constitutively active, transforming mutants of Cdc42 sequester Cbl, resulting in enhanced mitogenic signaling by the EGFR, and mutations that disrupt the Cdc42/p85Cool-1/Cbl complex prevent transformation by activated Cdc42 [123]. As another example, Sprouty proteins bind constitutively to Cbl and Cbl-b via the RING finger domain in a weak interaction [129, 130]. Upon EGFR activation, Sprouty2 becomes phosphorylated on tyrosine 55 and interacts with Cbl proteins via the Cbl TKB domain in a higher affinity interaction [129–131]. The phosphorylation-dependent binding of Cbl to Sprouty2 results in ubiquitination and proteasomal degradation of Sprouty2 [129–131]. The binding site on Sprouty2 conforms to the (D/N)XpYXXXP motif found in the Cbl TKB binding site on the EGFR and ZAP70 proteins [21, 132]. Thus, phosphorylated Sprouty2 competes with the activated EGFR for binding to the Cbl TKB domain, and this results in decreased Cbl-mediated ubiquitination and downregulation of the EGFR [129–131, 133]. Consistent with this, Sprouty2 positively regulates transformation of fibroblasts by activated H-Ras, and loss of Sprouty2 leads to decreased EGFR activity due to enhanced EGFR downregulation [134].

Phosphorylation of a juxtamembrane threonine on the EGFR by protein kinase C results in decreased degradation of the activated EGFR [135]. Phosphorylation of the EGFR prevents Cbl-mediated ubiquitination of the EGFR. This leads to increased recycling of the EGFR from the early endosome and decreased trafficking of the EGFR to the late endosome/MVB. The mechanism remains to be completely elucidated.

11.4 Pathophysiological Functions of Cbl Proteins and Cancer

11.4.1 Cbl Mutations in Myeloid Neoplasms

The mechanistic studies of Cbl proteins suggest that as a negative regulator of RTK signaling, Cbl proteins might serve as tumor suppressor genes. However, v-Cbl was originally identified as an oncogene, causing leukemia in mice and transforming NIH3T3 cells [7, 73]. The v-Cbl protein contains only the TKB domain of Cbl and when expressed in cells prevents RTK ubiquitination and downregulation most likely by acting as a dominant negative protein and preventing the recruitment of endogenous Cbl proteins to the RTK [96, 136]. Two other transforming forms of the murine Cbl protein have been identified from a chemically induced murine pre-B cell lymphoma and a chemically induced histiocytic lymphoma (70Z Cbl and p95 Cbl, respectively) [8, 137]. These mutants both contain deletions within the linker and RING finger regions, and thus these proteins have lost E3 activity [8, 137]. Interestingly, the 70Z Cbl deletion arises due to a point mutation in the splice acceptor for the exon containing the distal portion of the linker region and the proximal portion of the RING finger (exon 8) [75]. This mutation results in the deletion of 17 amino acids including the critical linker tyrosine that must be phosphorylated for activation of the E3 ligase and the first cysteine involved in the first zinc coordination complex of the RING finger.

While mice deficient in Cbl, Cbl-b, or Cbl-c do not show evidence of leukemia, mice that have a RING finger mutant Cbl knockin develop myeloid leukemia [138]. Cbl has been shown to ubiquitinate and downregulate the Flt3 RTK [139]. The development of leukemia in these mice is dependent on Flt3 activity, thus identifying Flt3 as the RTK dysregulated by the expression of the mutant Cbl protein [138].

Further evidence that Cbl mutations may contribute to myeloid leukemogenesis comes from the identification of a Cbl mutant lacking the exon containing the distal portion of the linker region and the proximal portion of the RING finger (exon 8) in acute myelogenous leukemia (AML) cells arising in a transgenic murine model of the progression of myelodysplastic syndrome (MDS) to AML [140]. Like the 70Z Cbl, this mutation arises due to a deletion of the splice acceptor site at the beginning of exon 8. The absence of leukemia in Cbl knockout mice and the development of leukemia in mice with a Cbl RING finger mutant knockin are most readily explained by a dominant negative function of the mutant protein. Indeed, mice deficient in both Cbl and Cbl-b in hematopoietic stem cells develop early onset of myeloid leukemia [141]. However, the positive functions of Cbl proteins in signaling based on the adaptor function of Cbl suggest that the mutant proteins may have both loss of tumor suppressor function (i.e., the loss of the negative regulatory E3 function) and a gain of oncogene function (e.g., coupling the RTK to downstream signaling pathways such as PI3K). Consistent with this, the transforming 70Z form of Cbl activates the EGFR in the absence of ligand and enhances activity of the EGFR and downstream signaling upon ligand stimulation [142].

Over the past 5 years, Cbl mutations have been found in ~5% of a wide variety of myeloid neoplasms including myelodysplastic syndrome, myelofibrosis, refractory anemia with excess blasts, de novo and secondary acute myeloid leukemia (AML and sAML, respectively), atypical chronic myelogenous leukemia (aCML), CML in blast crisis, chronic myelomonocytic leukemia (CMML), and juvenile myelomonocytic leukemia (JMML) [139, 143–153]. However, the frequency of Cbl mutations appears to be highest in JMML (~15%), CMML (~13%), sAML (~10%), and aCML (8%) [147]. The majority of these mutations are missense mutations that cluster within the linker region and at or near the zinc coordinating amino acids within the RING finger domain. These mutations have been shown to disrupt E3 function. The critical linker tyrosine (Y371 in Cbl), whose phosphorylation is required for E3 activity (as described above), is frequently mutated in myeloid neoplasms accounting for ~ 15% of all missense mutations [147, 153]. These mutations at Y371 occur mostly in patients with JMML and CMML [146, 148, 151–153]. Deletions of all or portions of the Cbl exon containing the distal portion of the linker region and the proximal portion of the RING finger have been described [143, 144, 146, 148, 150–153]. These deletions of this exon result from missplicing due to mutation, insertions, or deletions in the splice donor and acceptor sites surrounding exon 8. Nonsense mutations, frame shift mutations, and insertions within the linker and RING finger regions have been found as well [147]. The missense mutations of Cbl are usually homozygous mutations (resulting from copy neutral loss of heterozygosity—also known as uniparental disomy) while the deletions that arise from splicing mutations are more commonly heterozygous [139, 143–146, 148–152, 154]. In vitro transformation assays in NIH 3T3 cells found that deletions of the linker domain were transforming, while point mutations in the linker or RF were not [155]. In addition, one group found that 70Z Cbl induces greater ligand-independent proliferation and survival than the R420Q mutation [156]. However, others found no difference in transformation efficiency between 70Z Cbl and a variety of point mutants found in patients [152]. Thus, it is unclear why most missense mutations are homozygous and the deletion mutations are heterozygous.

Mutations of Cbl-b and Cbl-c are much less common in myeloid neoplasms. Two studies have described a total of five mutations within Cbl-b, all of which are either frame shift or missense mutations within the RF domain [144, 149]. These Cbl-b mutants have not been characterized further, but the locations would be predicted to disrupt E3 function [144, 149]. Other studies have not found mutations of Cbl-b suggesting that the frequency of Cbl-b mutations is low in myeloid neoplasms relative to Cbl [139, 146, 150]. One report describes four samples with a frame shift polymorphism in the RF domain of Cbl-c [149]. However, expression of Cbl-c is restricted to epithelial cells, so the significance of these abnormalities remains to be determined [10, 12, 157].

Functional studies have demonstrated that the mutant Cbl proteins found in myeloid neoplasms lose E3 activity, increase activity of Flt3, increase downstream signaling of the PI3K and STAT pathways, and are transforming [139, 146, 151, 152]. The most straightforward explanation is that the mutant Cbl proteins act as dominant negative proteins to endogenous Cbl or Cbl-b. However, similar to the ligand-independent activation of EGFR by 70Z Cbl, the Cbl mutants found in myeloid neoplasms can cause ligand-independent activation of Flt3 and enhance ligand-dependent activity [139, 146, 151]. In addition, expression of mutant Cbl proteins in hematopoietic stem/progenitor cells (HSPCs) results in increased proliferative response to a variety of cytokines when introduced into Cbl null HSPCs but not in Cbl wt HSPCs [152]. This suggests that there is a gain of function or oncogene function of the mutant Cbl proteins that is unmasked when the normal allele is lost. The recent demonstration that mice with loss of both Cbl and Cbl-b in HSPCs develop leukemia suggests that the oncogenic function of the mutant Cbl proteins is most likely an inhibition (or dominant negative effect) of endogenous Cbl and Cbl-b [141].

11.4.2 Cbl Mutations in Other Tumor Types

While v-Cbl also caused B cell lymphomas in mice, mutation in human lymphoid malignancies appears rare. Sequencing of Cbl in more than 500 lymphoid malignancies found five somatic mutations, three of which represent splice site mutations resulting in the loss of RING finger containing exon 8 [151, 153, 158, 159].

Somatic mutations of Cbl have been found in 10 non-small cell lung tumors out of 452 samples [153, 160]. All but one of the mutations described are outside the linker and RING finger, and all are heterozygous. For those mutants analyzed, E3 activity was maintained, but overexpression of these mutants in lung cancer cells resulted in increased viability and motility [160]. This suggests that they may be impairing the association of Cbl with a critical substrate but the mechanism by which these mutants affected viability or motility is not known.

Mutations of Cbl in other cancers or mutations of Cbl-b or Cbl-c have not been further described.

11.4.3 RTK Mutations That Inhibit Cbl Function

Mutations of a tyrosine in the cytoplasmic domain of the human CSF1R that enhanced the transforming activity of this RTK were found in children with secondary myelodysplasia and secondary sAML [161, 162]. Cbl binds to this tyrosine and ubiquitinates the CSF1R upon ligand activation [163]. Mutation of this tyrosine abrogates the interaction with Cbl and the downregulation of the CSF1R [163]. Similarly, mutations of Met that abrogate Cbl binding enhance the transformation of NIH 3T3 cells by Met [36, 164].

While mutations of the Cbl binding site have not been described for other RTKs in cancer, several reports suggest that there may be disruption of Cbl protein function in human cancers. Activating mutations of the EGFR found in non-small cell lung cancer tumors may interfere with Cbl-mediated downregulation of the EGFR although the mechanism is not clear [165]. Also, overexpression of ErbB-2 can inhibit Cbl-mediated downregulation of EGFR by forming heterodimers with the activated EGFR and preventing Cbl binding to the activated EGFR [166].

11.4.4 Inhibition of Cbl Function by Overexpression of Negative Regulators

Evidence exists that proteins which negatively regulate Cbl-mediated downregulation of RTKs (e.g., Cdc42, Cortactin, Dub-2, Src, and Cezanne-1) are aberrantly active or overexpressed in human epithelial malignancies and thus may contribute to transformation by inhibiting downregulation of RTKs [112, 115].

11.5 Summary

Activated RTKs are rapidly internalized and move to a vesicular compartment known as the early endosome where they are sorted either to a recycling or to a degradation pathway. The Cbl proteins are RING finger E3 proteins that mediate RTK downregulation. Upon activation of the RTKs, Cbl proteins are recruited to the RTK, the E3 function is activated by RTK-dependent phosphorylation of the Cbl protein, and in turn the Cbl proteins ubiquitinate the RTK. The ubiquitination of the RTK leads to its sorting from the early endosome to the multivesicular body and eventually to the lysosome, where it is degraded. This process is regulated by many proteins that affect the stability of Cbl proteins, sequester the Cbl proteins away from the activated RTK, and prevent ubiquitination of the RTK by Cbl proteins or deubiquitinate the RTK. Inappropriate activity of RTKs is important in the pathogenesis of many cancers. Mutations that abrogate the E3 activity of Cbl proteins lead to unregulated RTK activity and thereby result in transformation. Such transforming mutations of Cbl, originally found in murine tumors, have now been identified in a subset of human myeloid neoplasms. Additionally, it is now appreciated that disruption of Cbl-mediated RTK downregulation by aberrations in the RTK or other regulatory mechanisms may also contribute to the pathogenesis of cancer. Overall, the Cbl proteins are key regulators of RTK downregulation, and the importance of Cbl-mediated downregulation is highlighted by the aberrations of this process that are associated with the development of cancer.

Acknowledgement Financial Support: This research was supported by the Intramural Research Program of the NIH, National Cancer Institute, Center for Cancer Research

References

1. Nau MM, Lipkowitz S (2003) Comparative genomic organization of the cbl genes. Gene 308:103–113
2. Nau MM, Lipkowitz S (2008) Welcome to the family: Cbl-family gene organization, overview of structure and functions of Cbl-related proteins in various taxonomical groups. In: Tsygankov AY (ed) Cbl proteins, 1st edn. Nova Science, New York, pp 3–25
3. Hershko A, Ciechanover A (1998) The ubiquitin system. Annu Rev Biochem 67:425–479
4. Weissman AM (2001) Themes and variations on ubiquitylation. Nat Rev Mol Cell Biol 2:169–178
5. Deshaies RJ, Joazeiro CA (2009) RING domain E3 ubiquitin ligases. Annu Rev Biochem 78:399–434
6. Lipkowitz S, Weissman AM (2011) RINGs of good and evil: RING finger ubiquitin ligases at the crossroads of tumour suppression and oncogenesis. Nat Rev Cancer 11:629–643
7. Langdon WY, Hartley JW, Klinken SP, Ruscetti SK, Morse HC III (1989) v-cbl, an oncogene from a dual-recombinant murine retrovirus that induces early B-lineage lymphomas. Proc Natl Acad Sci USA 86:1168–1172
8. Blake TJ, Shapiro M, Morse HC III, Langdon WY (1991) The sequences of the human and mouse c-cbl proto-oncogenes show v-cbl was generated by a large truncation encompassing a proline-rich domain and a leucine zipper-like motif. Oncogene 6:653–657
9. Bustelo XR, Crespo P, Lopez-Barahona M, Gutkind JS, Barbacid M (1997) Cbl-b, a member of the Sli-1/c-Cbl protein family, inhibits Vav- mediated c-Jun N-terminal kinase activation. Oncogene 15:2511–2520
10. Keane MM, Ettenberg SA, Nau MM et al (1999) cbl-3: a new mammalian cbl family protein. Oncogene 18:3365–3375
11. Keane MM, Rivero-Lezcano OM, Mitchell JA, Robbins KC, Lipkowitz S (1995) Cloning and characterization of cbl-b: a SH3 binding protein with homology to the c-cbl proto-oncogene. Oncogene 10:2367–2377
12. Kim M, Tezuka T, Suziki Y, Sugano S, Hirai M, Yamamoto T (1999) Molecular cloning and characterization of a novel cbl-family gene, cbl-c. Gene 239:145–154
13. Lupher ML Jr, Rao N, Eck MJ, Band H (1999) The Cbl protooncoprotein: a negative regulator of immune receptor signal transduction. Immunol Today 20:375–382
14. Hime GR, Dhungat MP, Ng A, Bowtell DD (1997) D-Cbl, the Drosophila homologue of the c-Cbl proto-oncogene, interacts with the Drosophila EGF receptor in vivo, despite lacking C-terminal adaptor binding sites. Oncogene 14:2709–2719
15. Meisner H, Daga A, Buxton J et al (1997) Interactions of Drosophila Cbl with epidermal growth factor receptors and role of Cbl in R7 photoreceptor cell development. Mol Cell Biol 17:2217–2225

16. Robertson H, Hime GR, Lada H, Bowtell DD (2000) A Drosophila analogue of v-Cbl is a dominant-negative oncoprotein in vivo. Oncogene 19:3299–3308
17. Yoon CH, Lee J, Jongeward GD, Sternberg PW (1995) Similarity of sli-1, a regulator of vulval development in C. elegans, to the mammalian proto-oncogene c-cbl. Science 269:1102–1105
18. Eichinger L, Pachebat JA, Glockner G et al (2005) The genome of the social amoeba Dictyostelium discoideum. Nature 435:43–57
19. Ettenberg SA, Rubinstein YR, Banerjee P, Nau MM, Keane MM, Lipkowitz S (1999) Cbl-b inhibits EGF-receptor-induced apoptosis by enhancing ubiquitination and degradation of activated receptors. Mol Cell Biol Res Commun 2:111–118
20. Joazeiro CA, Wing SS, Huang H, Leverson JD, Hunter T, Liu YC (1999) The tyrosine kinase negative regulator c-Cbl as a RING-type, E2-dependent ubiquitin-protein ligase. Science 286:309–312
21. Levkowitz G, Waterman H, Ettenberg SA et al (1999) Ubiquitin ligase activity and tyrosine phosphorylation underlie suppression of growth factor signaling by c-Cbl/Sli-1. Mol Cell 4:1029–1040
22. Lorick KL, Jensen JP, Fang S, Ong AM, Hatakeyama S, Weissman AM (1999) RING fingers mediate ubiquitin-conjugating enzyme (E2)-dependent ubiquitination. Proc Natl Acad Sci USA 96:11364–11369
23. Yokouchi M, Kondo T, Sanjay A et al (2001) Src-catalyzed phosphorylation of c-Cbl leads to the interdependent ubiquitination of both proteins. J Biol Chem 276:35185–35193
24. Lovering R, Hanson IM, Borden KL et al (1993) Identification and preliminary characterization of a protein motif related to the zinc finger. Proc Natl Acad Sci USA 90:2112–2116
25. Borden KL, Boddy MN, Lally J et al (1995) The solution structure of the RING finger domain from the acute promyelocytic leukaemia proto-oncoprotein PML. EMBO J 14:1532–1541
26. Borden KL, Freemont PS (1996) The RING finger domain: a recent example of a sequence-structure family. Curr Opin Struct Biol 6:395–401
27. Kobashigawa Y, Tomitaka A, Kumeta H, Noda NN, Yamaguchi M, Inagaki F (2011) Autoinhibition and phosphorylation-induced activation mechanisms of human cancer and autoimmune disease-related E3 protein Cbl-b. Proc Natl Acad Sci USA 108:20579–20584
28. Dou H, Buetow L, Hock A, Sibbet GJ, Vousden KH, Huang DT (2012) Structural basis for autoinhibition of and phosphorylation-dependent activation of c-Cbl. Nat Struct Mol Biol 19(2):184–192
29. Zheng N, Wang P, Jeffrey PD, Pavletich NP (2000) Structure of a c-Cbl-UbcH7 complex: RING domain function in ubiquitin- protein ligases. Cell 102:533–539
30. Visser GD, Lill NL (2005) The Cbl RING finger C-terminal flank controls epidermal growth factor receptor fate downstream of receptor ubiquitination. Exp Cell Res 311:281–293
31. Lupher ML Jr, Reedquist KA, Miyake S, Langdon WY, Band H (1996) A novel phosphotyrosine-binding domain in the N-terminal transforming region of Cbl interacts directly and selectively with ZAP-70 in T cells. J Biol Chem 271:24063–24068
32. Meng W, Sawasdikosol S, Burakoff SJ, Eck MJ (1999) Structure of the amino-terminal domain of Cbl complexed to its binding site on ZAP-70 kinase. Nature 398:84–90
33. Deckert M, Elly C, Altman A, Liu YC (1998) Coordinated regulation of the tyrosine phosphorylation of Cbl by Fyn and Syk tyrosine kinases. J Biol Chem 273:8867–8874
34. Lupher MJ, Rao N, Lill NL et al (1998) Cbl-mediated negative regulation of the Syk tyrosine kinase. A critical role for Cbl phosphotyrosine-binding domain binding to Syk phosphotyrosine 323. J Biol Chem 273:35273–35281
35. Rubin C, Litvak V, Medvedovsky H, Zwang Y, Lev S, Yarden Y (2003) Sprouty fine-tunes EGF signaling through interlinked positive and negative feedback loops. Curr Biol 13:297–307
36. Peschard P, Ishiyama N, Lin T, Lipkowitz S, Park M (2004) A conserved DpYR motif in the juxtamembrane domain of the Met receptor family forms an atypical c-Cbl/Cbl-b tyrosine kinase binding domain binding site required for suppression of oncogenic activation. J Biol Chem 279:29565–29571

37. Hu J, Hubbard SR (2005) Structural characterization of a novel Cbl phosphotyrosine recognition motif in the APS family of adapter proteins. J Biol Chem 280:18943–18949
38. Sosinowski T, Killeen N, Weiss A (2001) The Src-like adaptor protein downregulates the T cell receptor on CD4+CD8+ thymocytes and regulates positive selection. Immunity 15:457–466
39. Tang J, Sawasdikosol S, Chang JH, Burakoff SJ (1999) SLAP, a dimeric adapter protein, plays a functional role in T cell receptor signaling. Proc Natl Acad Sci USA 96:9775–9780
40. Teckchandani AM, Birukova AA, Tar K, Verin AD, Tsygankov AY (2005) The multidomain protooncogenic protein c-Cbl binds to tubulin and stabilizes microtubules. Exp Cell Res 306:114–127
41. Schmidt MH, Dikic I (2005) The Cbl interactome and its functions. Nat Rev Mol Cell Biol 6:907–919
42. Szymkiewicz I, Kowanetz K, Soubeyran P, Dinarina A, Lipkowitz S, Dikic I (2002) CIN85 participates in Cbl-b-mediated down-regulation of receptor tyrosine kinases. J Biol Chem 277:39666–39672
43. Buchberger A (2002) From UBA, to UBX: new words in the ubiquitin vocabulary. Trends Cell Biol 12:216–221
44. Hofmann K, Bucher P (1996) The UBA domain: a sequence motif present in multiple enzyme classes of the ubiquitination pathway. Trends Biochem Sci 21:172–173
45. Davies GC, Ettenberg SA, Coats AO et al (2004) Cbl-b interacts with ubiquitinated proteins; differential functions of the UBA domains of c-Cbl and Cbl-b. Oncogene 23:7104–7115
46. Raasi S, Varadan R, Fushman D, Pickart CM (2005) Diverse polyubiquitin interaction properties of ubiquitin-associated domains. Nat Struct Mol Biol 12:708–714
47. Bartkiewicz M, Houghton A, Baron R (1999) Leucine zipper-mediated homodimerization of the adaptor protein c-Cbl. A role in c-Cbl's tyrosine phosphorylation and its association with epidermal growth factor receptor. J Biol Chem 274:30887–30895
48. Kozlov G, Peschard P, Zimmerman B et al (2007) Structural basis for UBA-mediated dimerization of c-Cbl ubiquitin ligase. J Biol Chem 282:27547–27555
49. Liu J, DeYoung SM, Hwang JB, O'Leary EE, Saltiel AR (2003) The roles of Cbl-b and c-Cbl in insulin-stimulated glucose transport. J Biol Chem 278:36754–36762
50. Peschard P, Kozlov G, Lin T et al (2007) Structural basis for ubiquitin-mediated dimerization and activation of the ubiquitin protein ligase Cbl-b. Mol Cell 27:474–485
51. Waterman H, Katz M, Rubin C et al (2002) A mutant EGF-receptor defective in ubiquitylation and endocytosis unveils a role for Grb2 in negative signaling. EMBO J 21:303–313
52. Waterman H, Levkowitz G, Alroy I, Yarden Y (1999) The RING Finger of c-Cbl Mediates Desensitization of the Epidermal Growth Factor Receptor. J Biol Chem 274:22151–22154
53. Fang D, Liu YC (2001) Proteolysis-independent regulation of PI3K by Cbl-b-mediated ubiquitination in T cells. Nat Immunol 2:870–875
54. Fang D, Wang HY, Fang N, Altman Y, Elly C, Liu YC (2001) Cbl-b, a RING-type E3 ubiquitin ligase, targets phosphatidylinositol 3-kinase for ubiquitination in T cells. J Biol Chem 276:4872–4878
55. Tsygankov AY, Teckchandani AM, Feshchenko EA, Swaminathan G (2001) Beyond the RING: CBL proteins as multivalent adapters. Oncogene 20:6382–6402
56. Carpenter G, Cohen S (1976) 125I-labeled human epidermal growth factor. Binding, internalization, and degradation in human fibroblasts. J Cell Biol 71:159–171
57. Haigler H, Ash JF, Singer SJ, Cohen S (1978) Visualization by fluorescence of the binding and internalization of epidermal growth factor in human carcinoma cells A-431. Proc Natl Acad Sci USA 75:3317–3321
58. Sorkin A, Goh LK (2009) Endocytosis and intracellular trafficking of ErbBs. Exp Cell Res 315:683–696
59. Waterman H, Yarden Y (2001) Molecular mechanisms underlying endocytosis and sorting of ErbB receptor tyrosine kinases. FEBS Lett 490:142–152
60. Spencer SA, Hammonds RG, Henzel WJ, Rodriguez H, Waters MJ, Wood WI (1988) Rabbit liver growth hormone receptor and serum binding protein. Purification, characterization, and sequence. J Biol Chem 263:7862–7867

61. Yarden Y, Escobedo JA, Kuang WJ et al (1986) Structure of the receptor for platelet-derived growth factor helps define a family of closely related growth factor receptors. Nature 323:226–232
62. Hicke L (1999) Gettin' down with ubiquitin: turning off cell-surface receptors, transporters and channels. Trends Cell Biol 9:107–112
63. Galcheva-Gargova Z, Theroux SJ, Davis RJ (1995) The epidermal growth factor receptor is covalently linked to ubiquitin. Oncogene 11:2649–2655
64. Miyazawa K, Toyama K, Gotoh A, Hendrie PC, Mantel C, Broxmeyer HE (1994) Ligand-dependent polyubiquitination of c-kit gene product: a possible mechanism of receptor down modulation in M07e cells. Blood 83:137–145
65. Mori S, Claesson-Welsh L, Okuyama Y, Saito Y (1995) Ligand-induced polyubiquitination of receptor tyrosine kinases. Biochem Biophys Res Commun 213:32–39
66. Mori S, Heldin CH, Claesson-Welsh L (1992) Ligand-induced polyubiquitination of the platelet-derived growth factor beta-receptor. J Biol Chem 267:6429–6434
67. Mori S, Heldin CH, Claesson-Welsh L (1993) Ligand-induced ubiquitination of the platelet-derived growth factor beta-receptor plays a negative regulatory role in its mitogenic signaling. J Biol Chem 268:577–583
68. Yee NS, Hsiau CW, Serve H, Vosseller K, Besmer P (1994) Mechanism of down-regulation of c-kit receptor. Roles of receptor tyrosine kinase, phosphatidylinositol 3'-kinase, and protein kinase C. J Biol Chem 269:31991–31998
69. Hicke L, Riezman H (1996) Ubiquitination of a yeast plasma membrane receptor signals its ligand-stimulated endocytosis. Cell 84:277–287
70. Terrell J, Shih S, Dunn R, Hicke L (1998) A function for monoubiquitination in the internalization of a G protein-coupled receptor. Mol Cell 1:193–202
71. Hicke L, Dunn R (2003) Regulation of membrane protein transport by ubiquitin and ubiquitin-binding proteins. Annu Rev Cell Dev Biol 19:141–172
72. Katzmann DJ, Odorizzi G, Emr SD (2002) Receptor downregulation and multivesicular-body sorting. Nat Rev Mol Cell Biol 3:893–905
73. Blake TJ, Heath KG, Langdon WY (1993) The truncation that generated the v-cbl oncogene reveals an ability for nuclear transport, DNA binding and acute transformation. EMBO J 12:2017–2026
74. Langdon WY, Hyland CD, Grumont RJ, Morse HC 3rd (1989) The c-cbl proto-oncogene is preferentially expressed in thymus and testis tissue and encodes a nuclear protein. J Virol 63:5420–5424
75. Andoniou CE, Thien CBF, Langdon WY (1994) Tumour induction by activated abl involves tyrosine phosphorylation of the product of the cbl oncogene. EMBO J 13:4515–4523
76. Cory GO, Lovering RC, Hinshelwood S, MacCarthy-Morrogh L, Levinsky RJ, Kinnon C (1995) The protein product of the c-cbl protooncogene is phosphorylated after B cell receptor stimulation and binds the SH3 domain of Bruton's tyrosine kinase. J Exp Med 182:611–615
77. Donovan JA, Wange RL, Langdon WY, Samelson LE (1994) The protein product of the c-cbl protooncogene is the 120-kDa tyrosine-phosphorylated protein in Jurkat cells activated via the T cell antigen receptor. J Biol Chem 269:22921–22924
78. Fukazawa T, Miyake S, Band V, Band H (1996) Tyrosine phosphorylation of Cbl upon epidermal growth factor (EGF) stimulation and its association with EGF receptor and downstream signaling proteins. J Biol Chem 271:14554–14559
79. Fukazawa T, Reedquist KA, Trub T et al (1995) The SH3 domain-binding T cell tyrosyl phosphoprotein p120. Demonstration of its identity with the c-cbl protooncogene product and in vivo complexes with Fyn, Grb2, and phosphatidylinositol 3-kinase. J Biol Chem 270:19141–19150
80. Hartley D, Meisner H, Corvera S (1995) Specific association of the beta isoform of the p85 subunit of phosphatidylinositol-3 kinase with the proto-oncogene c-cbl. J Biol Chem 270:18260–18263
81. Meisner H, Conway BR, Hartley D, Czech MP (1995) Interactions of Cbl with Grb2 and phosphatidylinositol 3'-kinase in activated Jurkat cells. Mol Cell Biol 15:3571–3578

82. Meisner H, Czech MP (1995) Coupling of the proto-oncogene product c-Cbl to the epidermal growth factor receptor. J Biol Chem 270:25332–25335
83. Odai H, Sasaki K, Iwamatsu A et al (1995) The proto-oncogene product c-Cbl becomes tyrosine phosphorylated by stimulation with GM-CSF or Epo and constitutively binds to the SH3 domain of Grb2/Ash in human hematopoietic cells. J Biol Chem 270: 10800–10805
84. Rivero-Lezcano OM, Sameshima JH, Marcilla A, Robbins KC (1994) Physical association between Src homology 3 elements and the protein product of the c-cbl proto-oncogene. J Biol Chem 269:17363–17366
85. Tanaka S, Neff L, Baron R, Levy JB (1995) Tyrosine phosphorylation and translocation of the c-cbl protein after activation of tyrosine kinase signaling pathways. J Biol Chem 270:14347–14351
86. Galisteo ML, Dikic I, Batzer AG, Langdon WY, Schlessinger J (1995) Tyrosine phosphorylation of the c-cbl proto-oncogene protein product and association with epidermal growth factor (EGF) receptor upon EGF stimulation. J Biol Chem 270:20242–20245
87. Levkowitz G, Klapper LN, Tzahar E, Freywald A, Sela M, Yarden Y (1996) Coupling of the c-cbl protooncogene product to ErbB-1/EGF-receptor but not to other ErbB proteins. Oncogene 12:1117–1125
88. Wang Y, Yeung YG, Langdon WY, Stanley ER (1996) c-Cbl is transiently tyrosine-phosphorylated, ubiquitinated, and membrane-targeted following Csf-1 stimulation of macrophages. J Biol Chem 271:17–20
89. Jongeward GD, Clandinin TR, Sternberg PW (1995) sli-1, a negative regulator of let-23-mediated signaling in C. elegans. Genetics 139:1553–1566
90. Thien CB, Langdon WY (1997) EGF receptor binding and transformation by v-cbl is ablated by the introduction of a loss-of-function mutation from the Caenorhabditis elegans sli-1 gene. Oncogene 14:2239–2249
91. Yoon CH, Chang C, Hopper NA, Lesa GM, Sternberg PW (2000) Requirements of multiple domains of SLI-1, a Caenorhabditis elegans homologue of c-Cbl, and an inhibitory tyrosine in LET-23 in regulating vulval differentiation. Mol Biol Cell 11:4019–4031
92. Lill NL, Douillard P, Awwad RA et al (2000) The evolutionarily conserved N-terminal region of Cbl is sufficient to enhance down-regulation of the epidermal growth factor receptor. J Biol Chem 275:367–377
93. Pai LM, Barcelo G, Schupbach T (2000) D-cbl, a negative regulator of the Egfr pathway, is required for dorsoventral patterning in Drosophila oogenesis. Cell 103:51–61
94. Pai LM, Wang PY, Chen SR et al (2006) Differential effects of Cbl isoforms on Egfr signaling in Drosophila. Mech Dev 123:450–462
95. Ettenberg SA, Keane MM, Nau MM et al (1999) cbl-b inhibits epidermal growth factor receptor signaling. Oncogene 18:1855–1866
96. Levkowitz G, Waterman H, Zamir E et al (1998) c-Cbl/Sli-1 regulates endocytic sorting and ubiquitination of the epidermal growth factor receptor. Genes Dev 12:3663–3674
97. Miyake S, Lupher ML Jr, Druker B, Band H (1998) The tyrosine kinase regulator Cbl enhances the ubiquitination and degradation of the platelet-derived growth factor receptor alpha. Proc Natl Acad Sci USA 95:7927–7932
98. Yokouchi M, Kondo T, Houghton A et al (1999) Ligand-induced ubiquitination of the epidermal growth factor receptor involves the interaction of the c-Cbl RING finger and UbcH7. J Biol Chem 274:31707–31712
99. Huang F, Kirkpatrick D, Jiang X, Gygi S, Sorkin A (2006) Differential regulation of EGF receptor internalization and degradation by multiubiquitination within the kinase domain. Mol Cell 21:737–748
100. Haglund K, Sigismund S, Polo S, Szymkiewicz I, Di Fiore PP, Dikic I (2003) Multiple monoubiquitination of RTKs is sufficient for their endocytosis and degradation. Nat Cell Biol 5:461–466
101. Mosesson Y, Shtiegman K, Katz M et al (2003) Endocytosis of receptor tyrosine kinases is driven by monoubiquitylation, not polyubiquitylation. J Biol Chem 278:21323–21326

102. de Melker AA, van der Horst G, Calafat J, Jansen H, Borst J (2001) c-Cbl ubiquitinates the EGF receptor at the plasma membrane and remains receptor associated throughout the endocytic route. J Cell Sci 114:2167–2178
103. Duan L, Miura Y, Dimri M et al (2003) Cbl-mediated ubiquitinylation is required for lysosomal sorting of epidermal growth factor receptor but is dispensable for endocytosis. J Biol Chem 278:28950–28960
104. Huang F, Goh LK, Sorkin A (2007) EGF receptor ubiquitination is not necessary for its internalization. Proc Natl Acad Sci USA 104:16904–16909
105. Longva KE, Blystad FD, Stang E, Larsen AM, Johannessen LE, Madshus IH (2002) Ubiquitination and proteasomal activity is required for transport of the EGF receptor to inner membranes of multivesicular bodies. J Cell Biol 156:843–854
106. Visser Smit GD, Place TL, Cole SL et al (2009) Cbl controls EGFR fate by regulating early endosome fusion. Sci Signal 2:ra86
107. Ettenberg SA, Magnifico A, Cuello M et al (2001) Cbl-b-dependent coordinated degradation of the epidermal growth factor receptor signaling complex. J Biol Chem 276:27677–27684
108. Haglund K, Shimokawa N, Szymkiewicz I, Dikic I (2002) Cbl-directed monoubiquitination of CIN85 is involved in regulation of ligand-induced degradation of EGF receptors. Proc Natl Acad Sci USA 99:12191–12196
109. Zeng S, Xu Z, Lipkowitz S, Longley JB (2005) Regulation of stem cell factor receptor signaling by Cbl family proteins (Cbl-b/c-Cbl). Blood 105:226–232
110. Petrelli A, Gilestro GF, Lanzardo S, Comoglio PM, Migone N, Giordano S (2002) The endophilin-CIN85-Cbl complex mediates ligand-dependent downregulation of c-Met. Nature 416:187–190
111. Soubeyran P, Kowanetz K, Szymkiewicz I, Langdon WY, Dikic I (2002) Cbl-CIN85-endophilin complex mediates ligand-induced downregulation of EGF receptors. Nature 416:183–187
112. Ryan PE, Davies GC, Nau MM, Lipkowitz S (2006) Regulating the regulator: negative regulation of Cbl ubiquitin ligases. Trends Biochem Sci 31:79–88
113. Kassenbrock CK, Anderson SM (2004) Regulation of ubiquitin protein ligase activity in c-Cbl by phosphorylation-induced conformational change and constitutive activation by tyrosine to glutamate point mutations. J Biol Chem 279:28017–28027
114. Ryan PE, Sivadasan-Nair N, Nau MM, Nicholas S, Lipkowitz S (2010) The N-terminus of Cbl-c regulates ubiquitin ligase activity by modulating affinity for the ubiquitin conjugating enzyme. J Biol Chem 285:23687–23698
115. Pareja F, Ferraro DA, Rubin C et al (2012) Deubiquitination of EGFR by Cezanne-1 contributes to cancer progression. Oncogene 31(43):4599–4608
116. Magnifico A, Ettenberg S, Yang C et al (2003) WW domain HECT E3s target Cbl RING finger E3s for proteasomal degradation. J Biol Chem 278:43169–43177
117. Azakir BA, Angers A (2009) Reciprocal regulation of the ubiquitin ligase Itch and the epidermal growth factor receptor signaling. Cell Signal 21:1326–1336
118. Yang B, Gay DL, MacLeod MK et al (2008) Nedd4 augments the adaptive immune response by promoting ubiquitin-mediated degradation of Cbl-b in activated T cells. Nat Immunol 9:1356–1363
119. Bao J, Gur G, Yarden Y (2003) Src promotes destruction of c-Cbl: implications for oncogenic synergy between Src and growth factor receptors. Proc Natl Acad Sci USA 100:2438–2443
120. Luttrell DK, Luttrell LM, Parsons SJ (1988) Augmented mitogenic responsiveness to epidermal growth factor in murine fibroblasts that overexpress pp 60c-src. Mol Cell Biol 8:497–501
121. Maa MC, Leu TH, McCarley DJ, Schatzman RC, Parsons SJ (1995) Potentiation of epidermal growth factor receptor-mediated oncogenesis by c-Src: implications for the etiology of multiple human cancers. Proc Natl Acad Sci USA 92:6981–6985
122. Irby RB, Yeatman TJ (2000) Role of Src expression and activation in human cancer. Oncogene 19:5636–5642
123. Wu WJ, Tu S, Cerione RA (2003) Activated Cdc42 sequesters c-Cbl and prevents EGF receptor degradation. Cell 114:715–725

124. Ahn SJ, Chung KW, Lee RA et al (2003) Overexpression of betaPix-a in human breast cancer tissues. Cancer Lett 193:99–107
125. Fritz G, Brachetti C, Bahlmann F, Schmidt M, Kaina B (2002) Rho GTPases in human breast tumours: expression and mutation analyses and correlation with clinical parameters. Br J Cancer 87:635–644
126. Fritz G, Just I, Kaina B (1999) Rho GTPases are over-expressed in human tumors. Int J Cancer 81:682–687
127. Kamai T, Yamanishi T, Shirataki H et al (2004) Overexpression of RhoA, Rac1, and Cdc42 GTPases is associated with progression in testicular cancer. Clin Cancer Res 10:4799–4805
128. Zhang X, Machii T, Matsumura I et al (2003) Constitutively activated Rho guanosine triphosphatases regulate the growth and morphology of hairy cell leukemia cells. Int J Hematol 77:263–273
129. Fong CW, Leong HF, Wong ES, Lim J, Yusoff P, Guy GR (2003) Tyrosine phosphorylation of Sprouty2 enhances its interaction with c-Cbl and is crucial for its function. J Biol Chem 278:33456–33464
130. Guy GR, Wong ES, Yusoff P et al (2003) Sprouty: how does the branch manager work? J Cell Sci 116:3061–3068
131. Hall AB, Jura N, DaSilva J, Jang YJ, Gong D, Bar-Sagi D (2003) hSpry2 is targeted to the ubiquitin-dependent proteasome pathway by c-Cbl. Curr Biol 13:308–314
132. Lupher ML Jr, Songyang Z, Shoelson SE, Cantley LC, Band H (1997) The Cbl phosphotyrosine-binding domain selects a D(N/D)XpY motif and binds to the Tyr292 negative regulatory phosphorylation site of ZAP-70. J Biol Chem 272:33140–33144
133. Wong ES, Lim J, Low BC, Chen Q, Guy GR (2001) Evidence for direct interaction between Sprouty and Cbl. J Biol Chem 276:5866–5875
134. Lito P, Mets BD, Kleff S, O'Reilly S, Maher VM, McCormick JJ (2008) Evidence that sprouty 2 is necessary for sarcoma formation by H-Ras oncogene-transformed human fibroblasts. J Biol Chem 283:2002–2009
135. Bao J, Alroy I, Waterman H et al (2000) Threonine phosphorylation diverts internalized EGF-receptors from a degradative pathway to the recycling endosome. J Biol Chem 275:26178–26186
136. Bonita DP, Miyake S, Lupher MLJ, Langdon WY, Band H (1997) Phosphotyrosine binding domain-dependent upregulation of the platelet-derived growth factor receptor alpha signaling cascade by transforming mutants of Cbl: implications for Cbl's function and oncogenicity. Mol Cell Biol 17:4597–4610
137. Bisson SA, Ujack EE, Robbins SM (2002) Isolation and characterization of a novel, transforming allele of the c-Cbl proto-oncogene from a murine macrophage cell line. Oncogene 21:3677–3687
138. Rathinam C, Thien CB, Flavell RA, Langdon WY (2010) Myeloid leukemia development in c-Cbl RING finger mutant mice is dependent on FLT3 signaling. Cancer Cell 18:341–352
139. Sargin B, Choudhary C, Crosetto N et al (2007) Flt3-dependent transformation by inactivating c-Cbl mutations in AML. Blood 110(3):1004–1012
140. Slape C, Liu LY, Beachy S, Aplan PD (2008) Leukemic transformation in mice expressing a NUP98-HOXD13 transgene is accompanied by spontaneous mutations in Nras, Kras, and Cbl. Blood 112:2017–2019
141. Naramura M, Nandwani N, Gu H, Band V, Band H (2010) Rapidly fatal myeloproliferative disorders in mice with deletion of Casitas B-cell lymphoma (Cbl) and Cbl-b in hematopoietic stem cells. Proc Natl Acad Sci USA 107:16274–16279
142. Thien CB, Langdon WY (1997) Tyrosine kinase activity of the EGF receptor is enhanced by the expression of oncogenic 70Z-Cbl. Oncogene 15:2909–2919
143. Abbas S, Rotmans G, Lowenberg B, Valk PJ (2008) Exon 8 splice site mutations in the gene encoding the E3-ligase CBL are associated with core binding factor acute myeloid leukemias. Haematologica 93:1595–1597
144. Caligiuri MA, Briesewitz R, Yu J et al (2007) Novel c-CBL and CBL-B ubiquitin ligase mutations in human acute myeloid leukemia. Blood 110(3):1022–1024

145. Dunbar AJ, Gondek LP, O'Keefe CL et al (2008) 250 K single nucleotide polymorphism array karyotyping identifies acquired uniparental disomy and homozygous mutations, including novel missense substitutions of c-Cbl, in myeloid malignancies. Cancer Res 68: 10349–10357
146. Grand FH, Hidalgo-Curtis CE, Ernst T et al (2009) Frequent CBL mutations associated with 11q acquired uniparental disomy in myeloproliferative neoplasms. Blood 113:6182–6192
147. Kales SC, Ryan PE, Nau MM, Lipkowitz S (2010) Cbl and human myeloid neoplasms: the Cbl oncogene comes of age. Cancer Res 70:4789–4794
148. Loh ML, Sakai DS, Flotho C et al (2009) Mutations in CBL occur frequently in juvenile myelomonocytic leukemia. Blood 114:1859–1863
149. Makishima H, Cazzolli H, Szpurka H et al (2009) Mutations of e3 ubiquitin ligase cbl family members constitute a novel common pathogenic lesion in myeloid malignancies. J Clin Oncol 27:6109–6116
150. Muramatsu H, Makishima H, Jankowska AM et al (2010) Mutations of E3 ubiquitin ligase Cbl family members but not TET2 mutations are pathogenic in juvenile myelomonocytic leukemia. Blood 115(10):1969–1975
151. Reindl C, Quentmeier H, Petropoulos K et al (2009) CBL exon 8/9 mutants activate the FLT3 pathway and cluster in core binding factor/11q deletion acute myeloid leukemia/myelodysplastic syndrome subtypes. Clin Cancer Res 15:2238–2247
152. Sanada M, Suzuki T, Shih LY et al (2009) Gain-of-function of mutated C-CBL tumour suppressor in myeloid neoplasms. Nature 460:904–908
153. Cosmic. Catalogue of Somatic Mutations in Cancer. (2012). Available from: http://www.sanger.ac.uk/genetics/CGP/cosmic/
154. Beer PA, Delhommeau F, Lecouedic JP et al (2010) Two routes to leukemic transformation following a JAK2 mutation-positive myeloproliferative neoplasm. Blood 115(14):2891–2900
155. Thien CB, Walker F, Langdon WY (2001) RING finger mutations that abolish c-Cbl-directed polyubiquitination and downregulation of the EGF receptor are insufficient for cell transformation. Mol Cell 7:355–365
156. Bandi SR, Brandts C, Rensinghoff M et al (2009) E3 ligase-defective Cbl mutants lead to a generalized mastocytosis and myeloproliferative disease. Blood 114:4197–4208
157. Griffiths EK, Sanchez O, Mill P et al (2003) Cbl-3-deficient mice exhibit normal epithelial development. Mol Cell Biol 23:7708–7718
158. Nicholson L, Knight T, Matheson E et al (2012) Casitas B lymphoma mutations in childhood acute lymphoblastic leukemia. Genes Chromosomes Cancer 51:250–256
159. McKeller MR, Robetorye RS, Dahia PL, Aguiar RC (2009) Integrity of the CBL gene in mature B-cell malignancies. Blood 114:4321–4322
160. Tan YH, Krishnaswamy S, Nandi S et al (2010) CBL Is Frequently Altered in Lung Cancers: its Relationship to Mutations in MET and EGFR Tyrosine Kinases. PLoS One 5:e8972
161. Baker A, Cachia P, Ridge S et al (1995) FMS mutations in patients following cytotoxic therapy for lymphoma. Leuk Res 19:309–318
162. Ridge SA, Worwood M, Oscier D, Jacobs A, Padua RA (1990) FMS mutations in myelodysplastic, leukemic, and normal subjects. Proc Natl Acad Sci USA 87:1377–1380
163. Mancini A, Koch A, Wilms R, Tamura T (2002) c-Cbl associates directly with the C-terminal tail of the receptor for the macrophage colony-stimulating factor, c-Fms, and down-modulates this receptor but not the viral oncogene v-Fms. J Biol Chem 277:14635–14640
164. Peschard P, Fournier TM, Lamorte L et al (2001) Mutation of the c-Cbl TKB domain binding site on the Met receptor tyrosine kinase converts it into a transforming protein. Mol Cell 8:995–1004
165. Mosesson Y, Mills GB, Yarden Y (2008) Derailed endocytosis: an emerging feature of cancer. Nat Rev Cancer 8:835–850
166. Muthuswamy SK, Gilman M, Brugge JS (1999) Controlled dimerization of ErbB receptors provides evidence for differential signaling by homo- and heterodimers. Mol Cell Biol 19:6845–6857

167. Motegi A, Fujimoto J, Kotani M, Sakuraba H, Yamamoto T (2004) ALK receptor tyrosine kinase promotes cell growth and neurite outgrowth. J Cell Sci 117:3319–3329
168. Valverde P (2005) Effects of Gas6 and hydrogen peroxide in Axl ubiquitination and down-regulation. Biochem Biophys Res Commun 333:180–185
169. Sharfe N, Freywald A, Toro A, Roifman CM (2003) Ephrin-A1 induces c-Cbl phosphorylation and EphA receptor down-regulation in T cells. J Immunol 170:6024–6032
170. Wang Y, Ota S, Kataoka H et al (2002) Negative regulation of EphA2 receptor by Cbl. Biochem Biophys Res Commun 296:214–220
171. Freywald A, Sharfe N, Roifman CM (2002) The kinase-null EphB6 receptor undergoes transphosphorylation in a complex with EphB1. J Biol Chem 277:3823–3828
172. Luo H, Wan X, Wu Y, Wu J (2001) Cross-linking of EphB6 resulting in signal transduction and apoptosis in Jurkat cells. J Immunol 167:1362–1370
173. Truitt L, Freywald T, DeCoteau J, Sharfe N, Freywald A (2010) The EphB6 receptor cooperates with c-Cbl to regulate the behavior of breast cancer cells. Cancer Res 70:1141–1153
174. Levkowitz G, Oved S, Klapper LN et al (2000) c-Cbl is a suppressor of the neu oncogene. J Biol Chem 275:35532–35539
175. Wong A, Lamothe B, Lee A, Schlessinger J, Lax I (2002) FRS2 alpha attenuates FGF receptor signaling by Grb2-mediated recruitment of the ubiquitin ligase Cbl. Proc Natl Acad Sci USA 99:6684–6689
176. Dufour C, Guenou H, Kaabeche K, Bouvard D, Sanjay A, Marie PJ (2008) FGFR2-Cbl interaction in lipid rafts triggers attenuation of PI3K/Akt signaling and osteoblast survival. Bone 42:1032–1039
177. Kaabeche K, Lemonnier J, Le Mee S, Caverzasio J, Marie PJ (2004) Cbl-mediated degradation of Lyn and Fyn induced by constitutive fibroblast growth factor receptor-2 activation supports osteoblast differentiation. J Biol Chem 279:36259–36267
178. Severe N, Miraoui H, Marie PJ (2011) The Casitas B lineage lymphoma (Cbl) mutant G306E enhances osteogenic differentiation in human mesenchymal stromal cells in part by decreased Cbl-mediated platelet-derived growth factor receptor alpha and fibroblast growth factor receptor 2 ubiquitination. J Biol Chem 286:24443–24450
179. Xian W, Schwertfeger KL, Rosen JM (2007) Distinct roles of fibroblast growth factor receptor 1 and 2 in regulating cell survival and epithelial-mesenchymal transition. Mol Endocrinol 21:987–1000
180. Cho JY, Guo C, Torello M et al (2004) Defective lysosomal targeting of activated fibroblast growth factor receptor 3 in achondroplasia. Proc Natl Acad Sci USA 101:609–614
181. Kobayashi S, Sawano A, Nojima Y, Shibuya M, Maru Y (2004) The c-Cbl/CD2AP complex regulates VEGF-induced endocytosis and degradation of Flt-1 (VEGFR-1). FASEB J 18:929–931
182. Sehat B, Andersson S, Girnita L, Larsson O (2008) Identification of c-Cbl as a new ligase for insulin-like growth factor-I receptor with distinct roles from Mdm2 in receptor ubiquitination and endocytosis. Cancer Res 68:5669–5677
183. Baumann CA, Ribon V, Kanzaki M et al (2000) CAP defines a second signalling pathway required for insulin-stimulated glucose transport. Nature 407:202–207
184. Ribon V, Printen JA, Hoffman NG, Kay BK, Saltiel AR (1998) A novel, multifunctional c-Cbl binding protein in insulin receptor signaling in 3 T3-L1 adipocytes. Mol Cell Biol 18:872–879
185. Duval M, Bedard-Goulet S, Delisle C, Gratton JP (2003) Vascular endothelial growth factor-dependent down-regulation of Flk-1/KDR involves Cbl-mediated ubiquitination. Consequences on nitric oxide production from endothelial cells. J Biol Chem 278:20091–20097
186. Ueno H, Honda H, Nakamoto T et al (1997) The phosphatidylinositol 3' kinase pathway is required for the survival signal of leukocyte tyrosine kinase. Oncogene 14:3067–3072
187. Lu Z, Je HS, Young P, Gross J, Lu B, Feng G (2007) Regulation of synaptic growth and maturation by a synapse-associated E3 ubiquitin ligase at the neuromuscular junction. J Cell Biol 177:1077–1089

188. Mullane-Robinson KP, Lill NL, Douillard P, Band H (1999) Cbl-mediated negative regulation of platelet-derived growth factor receptor-dependent cell proliferation. A critical role for Cbl tyrosine kinase-binding domain. J Biol Chem 274:16619–16628
189. Scott RP, Eketjall S, Aineskog H, Ibanez CF (2005) Distinct turnover of alternatively spliced isoforms of the RET kinase receptor mediated by differential recruitment of the Cbl ubiquitin ligase. J Biol Chem 280:13442–13449
190. Tsui CC, Pierchala BA (2008) CD2AP and Cbl-3/Cbl-c constitute a critical checkpoint in the regulation of ret signal transduction. J Neurosci 28:8789–8800
191. Penengo L, Rubin C, Yarden Y, Gaudino G (2003) c-Cbl is a critical modulator of the Ron tyrosine kinase receptor. Oncogene 22:3669–3679
192. Takahashi Y, Shimokawa N, Esmaeili-Mahani S et al (2011) Ligand-induced downregulation of TrkA is partly regulated through ubiquitination by Cbl. FEBS Lett 585:1741–1747
193. McCarty JH, Feinstein SC (1999) The TrkB receptor tyrosine kinase regulates cellular proliferation via signal transduction pathways involving SHC, PLCgamma, and CBL. J Recept Signal Transduct Res 19:953–974
194. Li Z, Dong T, Proschel C, Noble M (2007) Chemically diverse toxicants converge on Fyn and c-Cbl to disrupt precursor cell function. PLoS Biol 5:e35

Chapter 12
Regulation of Endocytic Trafficking and Signalling by Deubiquitylating Enzymes

Han Liu, Sylvie Urbé, and Michael J. Clague

Abstract As the major route by which activated Receptor Tyrosine Kinases are degraded, the endolysosomal pathway may be seen as a tumour suppressor pathway. The appendage of ubiquitin chains to activated receptors provides a sorting signal for sorting into multivesicular bodies which go on to fuse directly with lysosomes. Deubiquitylating (DUB) activities, such as the endosome-localised AMSH and USP8, can favour recycling of receptors by reducing this active sorting into MVBs. These enzymes have an overlapping set of binding partners at the endosome, which include both early- and late-acting components of the Endosomal Sorting Complex Required for Transport (ESCRT) machinery. The exact interplay between these enzymes is still under debate. The consequences of depletion can be complex and need to be interpreted with care. Generically endosomal DUBs can influence receptor trafficking by direct deubiquitylation of receptors or associated proteins, by stabilisation of sorting factors or by contributing to free ubiquitin homeostasis by recycling ubiquitin once a MVB cargo molecule has been committed to degradation. We propose that a single endosomal DUB may carry out multiple functions depending on the suite of interactions being employed. Recent studies have provided further examples of DUBs which may associate with endosomes in a transient manner to influence the sorting of RTKs but also other types of receptors, such as GPCRs and various channels.

H. Liu • S. Urbé • M.J. Clague (✉)
Cellular and Molecular Physiology, Institute of Translational Medicine,
The University of Liverpool, Liverpool L69 3BX, UK
e-mail: clague@liv.ac.uk

Y. Yarden and G. Tarcic (eds.), *Vesicle Trafficking in Cancer*,
DOI 10.1007/978-1-4614-6528-7_12,

12.1 Introduction

Ubiquitylation is a reversible post-translational modification, implicated in most complex cellular activities. These include each of three major cellular protein degradation pathways: proteasomal, endolysosomal and autophagolysosomal degradation [1]. Ubiquitylation requires activity of the E1-E2-E3 cascade in which more than 300 E3 ligases provide substrate specificity, whilst the attached ubiquitin moiety can be removed by cognate deubiquitylating enzymes (DUBs) [2]. Hence, a major function of DUBs is to promote protein stability by rescuing ubiquitylated proteins from degradation. Herein, we shall focus on their roles within the endolysosomal degradative pathway.

This pathway represents the major means by which Receptor Tyrosine Kinases (RTKs) are turned over, in a manner which is coupled to receptor activation. A key regulator of this pathway is the E3-ligase, c-CBL, which ubiquitylates many of the activated RTKs that have been studied so far (see Chap. 10). In some circumstances, ubiquitylation can provide an internalisation signal, but more crucial is its role in the sorting of cargo to lumenal vesicles of the multivesicular body (MVB), governed by the Endosomal Sorting Complex Required for Transport (ESCRT) machinery [3] (see chapter by Ma, Kales, Nau and Lipkowitz). Once fully mature, MVBs fuse directly with lysosomes [4]. Failure to downregulate activated receptors in this manner may lead to cellular transformation [5]. Major lines of evidence supporting this view come from the transforming properties of v-Cbl, transforming mutations in RTKs associated with loss of Cbl binding [6, 7] and transformation associated with loss of ESCRT function in fly models [8]. Other key signalling pathways such as Wnt and Hedgehog may also be regulated through the determination of receptor availability by the endocytic pathway.

As receptors traffic through the endocytic pathway, they may continue to signal, up until the point when they are sequestered away from the cytosol into MVBs. In fact it has been estimated that up to 80 % of the signalling lifetime of an EGFR may be manifested at intracellular locations [9]. As receptors transit intracellular compartments, they are exposed to a changing palette of substrates and effectors, such that changes to their itinerary may affect signalling output [10]. Furthermore, ubiquitylation provides for more than a simple degradation signal and is frequently used as a means to assemble networks of protein–protein interactions, much like phosphorylation [11]. Hence, endosomal DUBs may influence signalling through regulation of receptor trafficking or by directly reversing ubiquitin-based signalling outputs.

Several resident endosomal proteins, including components of the ESCRT machinery, are themselves monoubiquitylated although the functional consequences of this remain murky [12]. Appendage of a single ubiquitin is sufficient to provide an MVB sorting signal, but many receptors are modified by multiple types of polyubiquitin chains [13–17]. Pre-eminent amongst these are chains which are extended through isopeptide linkages at Lys63 within ubiquitin which are proposed to increase the efficiency of endosomal sorting [18–21].

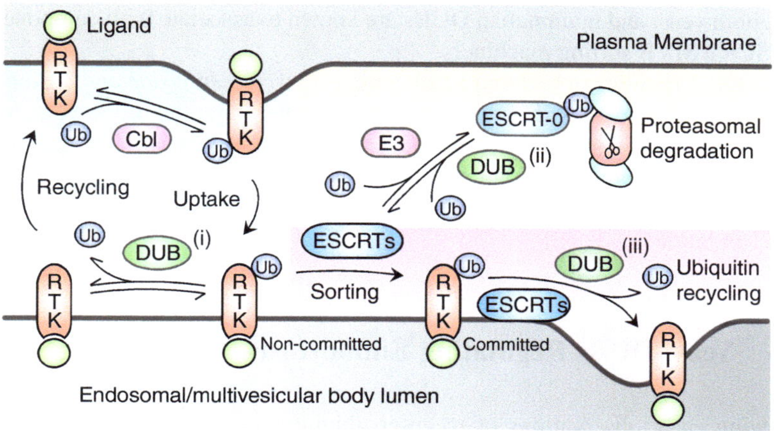

Fig. 12.1 Generic functions of deubiquitylases at the sorting endosome. Activated Receptor Tyrosine Kinases (RTKs) are internalised from the plasma membrane and then actively sorted into lumenal vesicles of multivesicular bodies. Cbl-dependent ubiquitylation of the RTK provides a sorting signal which is recognised by the ESCRT machinery. (**i**) DUBs can negatively regulate this process by deubiquitylating receptors at an early stage of the process, (**ii**) endosomal DUBs can regulate the stability of the sorting machinery itself by rescuing from proteasomal degradation or (**iii**) recycle ubiquitin from receptors which are irreversibly committed to the MVB/lysosomal pathway, thereby maintaining free ubiquitin levels

Both the proteasome and the sorting endosome have a complement of associated DUBs [22, 23]. Although the choreography and specificities of individual DUB activities are complex, one function is clear in that they ensure the recycling of ubiquitin once a substrate is committed to degradation. Without these activities free ubiquitin levels run down and globally impact upon the ubiquitin economy [24, 25]. In this review we will highlight three major functions of endosomal DUBs: (i) maintenance of ubiquitin homeostasis; (ii) direct deubiquitylation of receptors in transit, favouring recycling; and (iii) control of stability of the endosomal sorting machinery (Fig. 12.1).

12.2 DUBs in the Human and Yeast Genomes

The human genome encodes ~100 DUB family members, of which 79 are predicted to possess enzyme activity. Based on sequence similarity within the catalytic domain, DUBs are sorted into five subfamilies: ubiquitin-specific proteases (USP); ubiquitin C-terminal hydrolases (UCH); ovarian tumour (OTU) and Machado-Joseph domain proteins (MJD), all of which are cysteine proteases, as well as a distinct group of zinc-dependent metalloproteases, Jab1/MPN/Mov34 metalloproteases (JAMM) [26, 27]. Many DUBs harbour further protein domains which mediate localisation and incorporation into protein complexes [27, 28]. Central to the current topic, both yeast and mammalian DUBs are known to associate with components of the ESCRT MVB sorting machinery.

The DUB families in budding yeast *Saccharomyces cerevisiae* and fission yeast *Schizosaccharomyces pombe* are considerably smaller. They are proposed to include 17 and 20 members, respectively, and lack MJD proteins [29, 30]. There appears to be a high level of redundancy amongst the yeast DUBs. None of them are indispensable for viability with the exception of the proteasome-associated DUB, Rpn11 (POH1 in human) [29–32].

12.3 Yeast DUBs Regulating Endocytosis

Following early observations of receptor ubiquitylation in mammalian cells [33, 34], a definitive role for ubiquitin in receptor endocytosis was first established in yeast cells [35, 36]. Similarly functional studies of DUBs on the endocytic pathway were pioneered in yeast systems.

The USP family members, Doa4, Ubp7 and Ubp2, are the critical endosomal regulators in *S. cerevisiae*. The endosomal recruitment of Doa4 is mediated by Bro1, a class E Vps protein that binds the ESCRT-III component Snf7 (CHMP4B in human) [37–39]. In addition to recruiting Doa4 to the endosome, Bro1 also stimulates its deubiquitylase activity [40]. Early studies discovered that depletion of Doa4 rendered yeast cells with multiple defects [41]. Endocytosis and degradation of various membrane proteins such as maltose transporter, uracil permease Fur4p, the a-factor transporter Ste6p and general amino acid permease Gap1p were found to be impaired [42–45]. Using GFP-Doa4, Amerik et al. observed its localisation to the late endosome/pre-vacuolar compartment in a yeast strain deficient for the vacuolar protein sorting-associated protein 4 (Vps4), an AAA ATPase required for normal endosome function [46]. In *Doa4Δ* yeast that also lack vacuolar protease activity, large amounts of ubiquitin conjugates accumulate in membrane-enriched fractions and free ubiquitin levels decline [24, 45, 47]. Some of the effects of Doa4 loss may hence be ascribed to depletion of the free ubiquitin pool, but Gap1 permease down-regulation still relies upon the catalytic activity of Doa4 even if the free ubiquitin pool is restored. This suggests a direct role of Doa4 as a positive regulator of the MVB pathway [48].

The E3-ligase Rsp5 generates Lys63-linked ubiquitin chains on substrate proteins, which accumulate if the endosomal DUB Ubp2 is deleted [49, 50]. Based on depletion experiments, an active role for Ubp2 in MVB sorting has also been proposed [51, 52]. Both these antagonistic activities (Rsp5 and Ubp2 in complex with Rup1) are recruited to the ESCRT-0 protein Hse1 (STAM in mammals) through an adaptor protein Hua1 [50]. Hse1 can also bind directly to a second DUB, Ubp7, the deletion of which can restore MVB sorting made defective by compromising Rsp5 [51].

In *S. pombe*, multiple redundant DUBs are implicated in endocytosis, as severe defects in endocytosis and cell polarity require deletion of 5 DUBs: Ubp4, Ubp5, Ubp9, Ubp15 and Sst2 (AMSH in human, not present in catalytic form in

S. cerevisiae) [23, 29]. How these DUBs cooperate to regulate endocytosis awaits further investigation. Interestingly a suppressor screen identified Sst2 as a suppressor of the sterility defect due to loss of Ste12, a PtdIns kinase which has been linked to endosomal trafficking (Fab1 and PIKfyve in *S. cerevisiae* and human, respectively) [53, 54]. Sst2 deletion has been shown to result in a class E Vps phenotype typical of components of the endosomal sorting machinery [55].

12.4 Human DUBs Regulating Endocytosis

We have recently mapped the localisation of 66 human GFP-DUBs in HeLa cells [56]. The most convincing examples of endosome-localised DUBs are the JAMM domain proteins AMSH and AMSH-LP, as well as USP8 which translocates to endosomes following EGF stimulation [57]. Of course, this does not preclude the dynamic association of other DUBs with endosomal compartments, merely reflecting the steady-state distribution of over-expressed protein.

12.4.1 AMSH

AMSH was identified as a protein interacting with the SH3 (Src Homology 3) domain of the ESCRT-0 component, STAM (signal transducing adapter molecule). It has been linked to interleukin 2 (IL-2) and granulocyte-macrophage colony-stimulating factor (GM-CSF)-mediated signalling as well as the bone morphogenetic protein (BMP)-Smad pathway, prior to the recognition of its deubiquitylating activity [58, 59]. Expressing GFP-AMSH, McCullough et al. observed a prominent nuclear fraction, together with cytosolic and distinct endosomal pools in HeLa cells [60]. The nuclear function of AMSH remains unexplored, whilst most attention has been paid to its endosomal engagements.

Three motifs in AMSH contribute to its endosomal localisation: a clathrin binding site, a non-canonical SH3 binding motif PX(V/I)(D/N)RXXKP and a MIT domain (Fig. 12.2) [23, 61, 62]. AMSH interacts with the terminal domain of clathrin heavy chain, which is recruited via the ESCRT-0 component Hrs, to a specialised "bilayered coat" region on the vacuolar aspect of the sorting endosome [63–65]. Clathrin binding appears to be the essential determinant regarding endosomal localisation of AMSH or the related AMSH-LP, judged by immunofluorescence localisation experiments [66]. Deletion of the clathrin binding site in AMSH or short-inference RNA (siRNA)-mediated depletion of clathrin abrogates AMSH accumulation on endosomes in distinction to the deletion of the SH3 binding site or MIT domain [66–68]. Two interactions, which can be engaged simultaneously, link AMSH (but not AMSH-LP) directly to the ESCRT machinery [23, 69]. Firstly it binds the SH3 domain of the ESCRT-0 component, STAM. Secondly, the MIT domain binds to MIT domain-interacting motifs (MIMs) within the C-terminus of

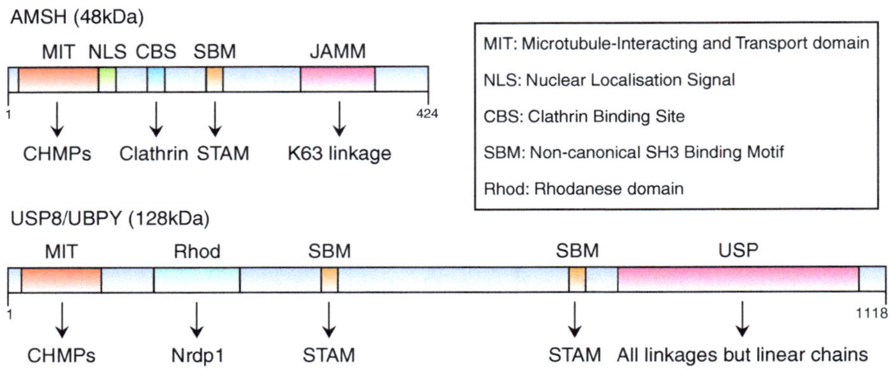

Fig. 12.2 Architecture of the major endosomal DUBs, AMSH and USP8. Catalytic domains are indicated with attendant ubiquitin chain-linkage specificities as JAMM and USP for AMSH and USP8, respectively, reflecting their different familial associations. Common elements specify binding to ESCRT machinery components. In each case non-canonical SH3 binding motifs (SBM) mediate interaction with the ESCRT-0 component STAM, and a MIT domain mediate interactions with partially overlapping sets of CHMP proteins belonging to ESCRT-III. AMSH also contains a clathrin binding site which is critical for its endosomal localisation. UBPY directly interacts with Nrdp1 an E3-ligase implicated in the control of surface levels of ErbB3 by control of the secretory pathway [110]

CHMPs (charged multivesicular body proteins), components of the ESCRT-III complex, which function at a late stage of the lumenal vesicle budding process [3, 70]. AMSH binds to a subset of CHMPs including CHMP1A, 1B, 2A, 2B, 3 and 4C [71–73]. Amongst MIT-MIM interactions tested to date, binding between AMSH MIT and CHMP3 MIM4 exhibits a significantly higher affinity, with a Kd of ~60nM [74]. Although the SH3 binding site and MIT domain are not required for endosomal localisation of AMSH per se, they play important functional roles at the endosome. STAM association via its SH3 domain is required for cargo deubiquitylation as well as stimulating DUB activity. HeLa cells expressing an AMSH mutant lacking the MIT domain accumulate large amounts of ubiquitylated proteins on endosomes [67, 69].

AMSH provided the first example of chain-linkage selectivity amongst the DUB enzymes and remains one of those showing the highest stringency of selection [60, 69, 75]. It is proposed to exclusively process Lys63-linked chains, due to two characteristic insertions (shared with AMSH-LP) in its catalytic domain [76]. As Lys63 linkages are of pre-eminence with regard to receptor trafficking [18, 20, 21, 77], the specificity matches up well with proposed function.

The involvement of AMSH in the regulation of endocytic receptor trafficking was first demonstrated with epidermal growth factor receptor (EGFR) as a model endosomal cargo [60]. Following siRNA-mediated depletion of AMSH in HeLa cells, acutely stimulated EGFR and its ligand are degraded at an increased rate [60, 78, 79], implying a negative role of AMSH in regulating EGF-EGFR lysosomal trafficking. As a working model, it is proposed that this effect is exerted through direct counteraction of c-CBL-dependent ubiquitylation of EGFR at the sorting endosome, thereby tipping the balance in favour of recycling. However, one should note that a requirement for deubiquitylating activity has not been formally demonstrated, and inconsistent siRNA depletion effects on EGFR trafficking have been reported by others [68]. AMSH has also been reported to positively regulate the MVB/lysosomal routing of various G protein-coupled receptors, such as chemokine receptor CXCR4 [80], calcium-sensing receptor (CaR) [81] and protease-activated receptor 2 (PAR2) [82].

12.4.2 USP8

USP8, also called UBPY, was originally identified as a growth-regulated DUB [83]. Mice lacking USP8 exhibit embryonic lethality, whilst induced USP8 loss in adult mice results in fatal liver failure, suggesting essential and non-redundant roles of USP8 during development and adult life [84].

USP8 was shown to interact with STAM (component of ESCRT-0), through two non-canonical SH3 domain binding motifs PX(V/I)(D/N)RXXKP also possessed by AMSH [61]. Thus, USP8 competes with AMSH for a common binding site on STAM. Its domain structure is shown in Fig. 12.2, encompassing an N-terminal MIT domain, a rhodanese domain, two non-canonical SH3 binding motifs and the catalytic USP domain [23, 73]. Compared with AMSH, endosomal association of USP8 is less prominent, possibly due to the absence of a clathrin binding site. However, endosomal accumulation is observed in cells expressing catalytically inactive USP8 and with wild-type USP8 following over-expression of Hrs or a dominant negative form of Vps4 [57, 85]. Furthermore, under conditions of acute EGF stimulation a clear translocation to endosomal structures is observed [57, 85]. The MIT domain of USP8 is required for endosomal localisation and shares an overlapping set of CHMP protein binding partners with AMSH (CHMP1A, 1B, 2A, 2B, 4C) but differs in not binding to CHMP3 [73]. The rhodanese domain interacts with Nrdp1, an E3 ligase stabilised by USP8 [86, 87].

As with AMSH, the DUB activity of USP8 is stimulated by STAM binding but in contrast to AMSH shows no chain-linkage specificity [73]. In a recent report, Sixma et al. performed a comprehensive kinetic analysis of 12 USPs, amongst which USP8 shows highest *kcat* and processes all seven types of ubiquitin isopeptide linkages but not linear chains [88]. Interestingly, the structure of the USP8 catalytic domain reveals an occluded catalytic site, suggesting conformational flexibility and the potential for allosteric regulation [86].

The regulation of endocytic trafficking by USP8 was first investigated with EGFR as a model cargo using siRNA-mediated depletion. Mizuno et al. reported accelerated degradation of EGFR following acute EGF treatment in USP8-depleted HeLa cells, thereby proposing a negative regulatory role of USP8 in lysosomal trafficking akin to that proposed for AMSH above [85]. However, following prolonged knockdown of USP8, Row et al. observed aberrant endosomal structures enriched with ubiquitin conjugates, as well as inhibited EGFR and c-Met degradation in HeLa cells, consistent with data from experiments by Bowers et al. with radioisotope-labelled EGF [57, 78]. Later work from Mizuno et al. also concurred with these observations following prolonged depletion of USP8 [89]. Crucially these effects of USP8 depletion on EGFR downregulation could be rescued by expression of siRNA-resistant USP8 but neither by a catalytically inactive mutant nor by a truncated form lacking the MIT domain which fails to localise to endosomes [73]. As USP8 depletion elicits a failure to deubiquitylate activated EGFR [57], one might have expected enhanced sorting into MVBs and an increased rate of degradation. However, the effects of USP8 depletion are pleiotropic. For unknown reasons it leads to a global increase in ubiquitin conjugation [57, 83]. It also leads to the accumulation of clustered MVBs "stitched" together by regularly spaced tethers of unknown origin, which we speculate could be related to the HOPs tethering complex [90]. Finally it leads to a depletion of the ESCRT-0 components Hrs and STAM, by rescuing them through its deubiquitylating activity from constitutive breakdown by proteasomal degradation [57].

There is now a substantial body of evidence suggesting a positive role of USP8 on endocytic protein sorting and degradation, similar to that of its yeast orthologue Doa4. However, experiments carried out in *Drosophila* or S2 cells suggest an opposite role for USP8 as a negative regulator of receptor downregulation. Deubiquitylation by USP8 promotes the recycling and the consequent plasma membrane accummulation of frizzled (Fz) and smoothened (Smo), key cell surface molecules which can determine signal strength associated with the Wingless and Hedgehog signalling pathways, respectively [91–93].

12.4.3 Different Outcomes of Deubiquitylating Activity at Different Stages of Lysosomal Sorting

There is clearly a complex interplay between AMSH and USP8 that includes competition for binding partners such as Hrs and CHMPs, possible redundancy and distinct substrate activities. This always needs to be born in mind when interpreting the effects of depletion or over-expression of one or the other. According to a recent estimate of copy numbers in NIH3T3 mouse fibroblasts, AMSH (9,054 copies per cell) exceeds USP8 (1,596 copies per cell) by several fold [94]. Both AMSH and USP8 are capable of deubiquitylating EGFR directly in vitro [60, 85] and positive and negative regulatory roles in receptor trafficking have been proposed for each. Here we argue that negative and positive regulatory effects could reflect the stage of the endocytic process where deubiquitylation occurs.

The first point of engagement of ubiquitylated receptor with the lysosomal sorting machinery is proposed to be through the interaction with ESCRT-0, which concentrates cargo in areas of the endosomes covered by clathrin containing coated structures, which differ from traditional clathrin coats [63, 95, 96]. Although both AMSH and USP8 compete for interaction with the ESCRT-0 component STAM, only AMSH binds to clathrin and requires this binding for endosomal localisation.

At this early juncture, we propose that receptor is not yet fully committed to the degradation pathway and can be diverted towards recycling by deubiquitylating activity counteracting any E3-ligase activity. The fraction of cargo recycled depends on many factors, but the key determinant is the balance between ubiquitylation and deubiquitylation that promote lysosomal sorting and membrane recycling, respectively. There may be a "proofreading" element to this process that is sensitive to chain-linkage type or particular substrates [1].

Both AMSH and USP8 bind to CHMP proteins. These make up the ESCRT-III complex that is believed to mediate the final steps of lumenal vesicle budding from the limiting membrane of the MVB [3, 97]. DUB interaction with CHMPs requires their respective MIT domains, which they partially share with the AAA ATPase, Vps4, an enzyme which couples vesicle scission with the disassembly of the ESCRT-III complex [98]. This suggests a deubiquitylation function late in the sorting process that may be used to recycle ubiquitin once an irreversible commitment to sorting has taken place. It is tempting to speculate that somehow the endosomal DUBs via MIT-MIM interactions block access to Vps4 until this ubiquitin-recycling function is accomplished. Coupling to vesicle scission would ensure maximal efficiency of recycling, just as activity of the proteasomal DUB POH1 is coupled to proteasomal degradation [23, 72].

12.4.4 Other Mammalian DUBs

To date, AMSH and USP8 are undoubtedly the most thoroughly studied DUBs on the endocytic pathway in mammalian cells, for which interactions with the endosomal sorting machinery are well characterised. However several other DUBs may be transiently recruited to endosomes or in fact the plasma membrane to perform similar functions to those described above. Recently, Cezanne-1, a member of the OTU family, was reported to promote oncogenesis, by negatively regulating EGFR endocytosis and degradation [79]. USP33 was shown to inhibit β-arrestin-dependent internalisation of G protein-coupled receptors by deubiquitylating β-arrestin [99]. In addition, USP33 and its paralogue USP20 constitutively bind β2-adrenergic receptor (β2AR) undergoing agonist-induced dissociation leading to receptor ubiquitylation and internalisation. They then reassociate with the receptor after prolonged agonist treatment to deubiquitylate and recycle it to the plasma membrane [100].

A specific splice variant of USP2, USP2-45 deubiquitylates the epithelial sodium channel (ENaC), therefore suppressing its ubiquitylation-mediated endocytic trafficking [101]. The vasopressin-induced USP10 can also regulate endocytosis of ENaC indirectly, by deubiquitylating and stabilising sorting nexin 3 (SNX3), which promotes recycling [102]. Moreover, USP10 was also reported to deubiquitylate cystic fibrosis transmembrane conductance regulator (CFTR) and promote its recycling from early endosomes [103, 104]. There is also some evidence that proteasomes bind to endosomes and influence RTK trafficking [105–107]. Allied to this the deubiquitylation activity associated with the proteasome is capable of trimming Lys63 ubiquitin chains from tropomyosin-regulated kinase A (TrkA) receptor [108].

12.5 Conclusions

As the physiological roles of DUBs are beginning to be unveiled, they are now implicated in a wide range of cellular activities. Amongst these the governance of endocytic trafficking ranks highly. Being either cysteine- or metalloproteases, DUBs can be considered to be druggable targets and are now undergoing intensive investigation as such [109]. Since the endocytic pathway can function as a tumour suppressor pathway and also represents a major route for targeted drug delivery, pharmaceutical manipulation of this pathway offers substantial therapeutic opportunities.

Bibliography

1. Clague MJ, Urbe S (2010) Ubiquitin: same molecule, different degradation pathways. Cell 143:682–685
2. Hershko A, Ciechanover A (1998) The ubiquitin system. Annu Rev Biochem 67:425–479
3. Henne WM, Buchkovich NJ, Emr SD (2011) The ESCRT Pathway. Dev Cell 21:77–91
4. Futter CE, Pearse A, Hewlett LJ, Hopkins CR (1996) Multivesicular endosomes containing internalized EGF-EGF receptor complexes mature and then fuse directly with lysosomes. J Cell Biol 132:1011–1023
5. Mosesson Y, Mills GB, Yarden Y (2008) Derailed endocytosis: an emerging feature of cancer. Nat Rev Cancer 8:835–850
6. Peschard P, Park M (2003) Escape from Cbl-mediated downregulation: a recurrent theme for oncogenic deregulation of receptor tyrosine kinases. Cancer Cell 3:519–523
7. Abella JV, Peschard P, Naujokas MA, Lin T, Saucier C, Urbe S, Park M (2005) Met/Hepatocyte growth factor receptor ubiquitination suppresses transformation and is required for Hrs phosphorylation. Mol Cell Biol 25:9632–9645
8. Thompson BJ, Mathieu J, Sung HH, Loeser E, Rorth P, Cohen SM (2005) Tumor suppressor properties of the ESCRT-II complex component Vps25 in Drosophila. Dev Cell 9:711–720
9. Sorkin A (1998) Endocytosis and intracellular sorting of receptor tyrosine kinases. Front Biosci 3:729–738
10. Sigismund S, Confalonieri S, Ciliberto A, Polo S, Scita G, Di Fiore PP (2012) Endocytosis and signaling: cell logistics shape the eukaryotic cell plan. Physiol Rev 92:273–366
11. Grabbe C, Husnjak K, Dikic I (2011) The spatial and temporal organization of ubiquitin networks. Nat Rev Mol Cell Biol 12:295–307
12. Woelk T, Oldrini B, Maspero E, Confalonieri S, Cavallaro E, Di Fiore PP, Polo S (2006) Molecular mechanisms of coupled monoubiquitination. Nat Cell Biol 8:1246–1254
13. Huang F, Kirkpatrick D, Jiang X, Gygi S, Sorkin A (2006) Differential regulation of EGF receptor internalization and degradation by multiubiquitination within the kinase domain. Mol Cell 21:737–748
14. Vina-Vilaseca A, Sorkin A (2010) Lysine 63-linked polyubiquitination of the dopamine transporter requires WW3 and WW4 domains of Nedd4-2 and UBE2D ubiquitin-conjugating enzymes. J Biol Chem 285:7645–7656
15. Reggiori F, Pelham HR (2001) Sorting of proteins into multivesicular bodies: ubiquitin-dependent and -independent targeting. EMBO J 20:5176–5186
16. Sigismund S, Woelk T, Puri C, Maspero E, Tacchetti C, Transidico P, Di Fiore PP, Polo S (2005) Clathrin-independent endocytosis of ubiquitinated cargos. Proc Natl Acad Sci U S A 102:2760–2765

17. Haglund K, Dikic I (2012) The role of ubiquitylation in receptor endocytosis and endosomal sorting. J Cell Sci 125:265–275
18. Lauwers E, Erpapazoglou Z, Haguenauer-Tsapis R, Andre B (2010) The ubiquitin code of yeast permease trafficking. Trends Cell Biol 20:196–204
19. Behrends C, Harper JW (2011) Constructing and decoding unconventional ubiquitin chains. Nat Struct Mol Biol 18:520–528
20. Li JG, Haines DS, Liu-Chen LY (2008) Agonist-promoted Lys63-linked polyubiquitination of the human kappa-opiod receptor is involved in receptor down-regulation. Mol Pharmacol 73:1319–1330
21. Duncan LM, Piper S, Dodd RB, Saville MK, Sanderson CM, Luzio JP, Lehner PJ (2006) Lysine-63-linked ubiquitination is required for endolysosomal degradation of class I molecules. EMBO J 25:1635–1645
22. Finley D (2009) Recognition and processing of ubiquitin-protein conjugates by the proteasome. Ann Rev Biochem 78:477–513
23. Clague MJ, Urbe S (2006) Endocytosis: the DUB version. Trends Cell Biol 16:551–559
24. Kimura Y, Yashiroda H, Kudo T, Koitabashi S, Murata S, Kakizuka A, Tanaka K (2009) An inhibitor of a deubiquitinating enzyme regulates ubiquitin homeostasis. Cell 137:549–559
25. Carter S, Urbe S, Clague MJ (2004) The met receptor degradation pathway: requirement for Lys48-linked polyubiquitin independent of proteasome activity. J Biol Chem 279:52835–52839
26. Nijman SM, Luna-Vargas MP, Velds A, Brummelkamp TR, Dirac AM, Sixma TK, Bernards R (2005) A genomic and functional inventory of deubiquitinating enzymes. Cell 123:773–786
27. Komander D, Clague MJ, Urbe S (2009) Breaking the chains: structure and function of the deubiquitinases. Nat Rev Mol Cell Biol 10:550–563
28. Ventii KH, Wilkinson KD (2008) Protein partners of deubiquitinating enzymes. Biochem J 414:161–175
29. Kouranti I, McLean JR, Feoktistova A, Liang P, Johnson AE, Roberts-Galbraith RH, Gould KL (2010) A global census of fission yeast deubiquitinating enzyme localization and interaction networks reveals distinct compartmentalization profiles and overlapping functions in endocytosis and polarity. PLoS Biol 8(9)
30. Amerik AY, Li SJ, Hochstrasser M (2000) Analysis of the deubiquitinating enzymes of the yeast Saccharomyces cerevisiae. Biol Chem 381:981–992
31. Amerik AY, Hochstrasser M (2004) Mechanism and function of deubiquitinating enzymes. Biochim Biophys Acta 1695:189–207
32. Shimanuki M, Saka Y, Yanagida M, Toda T (1995) A novel essential fission yeast gene pad1+ positively regulates pap1(+)-dependent transcription and is implicated in the maintenance of chromosome structure. J Cell Sci 108(Pt 2):569–579
33. Strous GJ, Van Kerhhof P, Govers R, Ciechanover A, Schwartz AL (1996) The ubiquitin conjugation system is required for ligand-induced endocytosis and degradation of the growth-hormone receptor. EMBO J 15:3806–3812
34. Cenciarelli C, Hou D, Hsu KC, Rellahan BL, Wiest DL, Smith HT, Fried VA, Weissman AM (1992) Activation-induced ubiquitination of the T cell antigen receptor. Science 257:795–797
35. Hicke L, Riezman H (1996) Ubiquitination of a yeast plasma membrane receptor signals its ligand-stimulated endocytosis. Cell 84:277–287
36. Galan JM, Moreau V, Andre B, Volland C, Haguenauer-Tsapis R (1996) Ubiquitination mediated by the Npi1p/Rsp5p ubiquitin-protein ligase is required for endocytosis of the yeast uracil permease. J Biol Chem 271:10946–10952
37. Odorizzi G, Katzmann DJ, Babst M, Audhya A, Emr SD (2003) Bro1 is an endosome-associated protein that functions in the MVB pathway in Saccharomyces cerevisiae. J Cell Sci 116:1893–1903
38. Nikko E, Marini AM, Andre B (2003) Permease recycling and ubiquitination status reveal a particular role for Bro1 in the multivesicular body pathway. J Biol Chem 278:50732–50743

39. Luhtala N, Odorizzi G (2004) Bro1 coordinates deubiquitination in the multivesicular body pathway by recruiting Doa4 to endosomes. J Cell Biol 166:717–729
40. Richter C, West M, Odorizzi G (2007) Dual mechanisms specify Doa4-mediated deubiquitination at multivesicular bodies. EMBO J 26:2454–2464
41. Papa FR, Hochstrasser M (1993) The yeast DOA4 gene encodes a deubiquitinating enzyme related to a product of the human tre-2 oncogene. Nature 366:313–319
42. Galan JM, Haguenauer-Tsapis R (1997) Ubiquitin lys63 is involved in ubiquitination of a yeast plasma membrane protein. EMBO J 16:5847–5854
43. Springael JY, Galan JM, Haguenauer-Tsapis R, Andre B (1999) NH4+–induced down-regulation of the Saccharomyces cerevisiae Gap1p permease involves its ubiquitination with lysine-63-linked chains. J Cell Sci 112(Pt 9):1375–1383
44. Loayza D, Michaelis S (1998) Role for the ubiquitin-proteasome system in the vacuolar degradation of Ste6p, the a-factor transporter in Saccharomyces cerevisiae. Mol Cell Biol 18:779–789
45. Lucero P, Lagunas R (1997) Catabolite inactivation of the yeast maltose transporter requires ubiquitin-ligase npi1/rsp5 and ubiquitin-hydrolase npi2/doa4. FEMS Microbiol Lett 147:273–277
46. Amerik AY, Nowak J, Swaminathan S, Hochstrasser M (2000) The DoA4 deubiquitinating enzyme is functionally linked to the vacuolar protein-sorting and endocytic pathways. Mol Biol Cell 11:3365–3380
47. Dupre S, Haguenauer-Tsapis R (2001) Deubiquitination step in the endocytic pathway of yeast plasma membrane proteins: crucial role of Doa4p ubiquitin isopeptidase. Mol Cell Biol 21:4482–4494
48. Nikko E, Andre B (2007) Evidence for a direct role of the Doa4 deubiquitinating enzyme in protein sorting into the MVB pathway. Traffic 8:566–581
49. Kee Y, Lyon N, Huibregtse JM (2005) The Rsp5 ubiquitin ligase is coupled to and antagonized by the Ubp2 deubiquitinating enzyme. EMBO J 24:2414–2424
50. Kee Y, Munoz W, Lyon N, Huibregtse JM (2006) The deubiquitinating enzyme Ubp2 modulates Rsp5-dependent Lys63-linked polyubiquitin conjugates in Saccharomyces cerevisiae. J Biol Chem 281:36724–36731
51. Ren J, Kee Y, Huibregtse JM, Piper RC (2007) Hse1, a component of the yeast Hrs-STAM ubiquitin-sorting complex, associates with ubiquitin peptidases and a ligase to control sorting efficiency into multivesicular bodies. Mol Biol Cell 18:324–335
52. Lam MH, Urban-Grimal D, Bugnicourt A, Greenblatt JF, Haguenauer-Tsapis R, Emili A (2009) Interaction of the deubiquitinating enzyme Ubp2 and the e3 ligase Rsp5 is required for transporter/receptor sorting in the multivesicular body pathway. PLoS One 4:e4259
53. Onishi M, Nakamura Y, Koga T, Takegawa K, Fukui Y (2003) Isolation of suppressor mutants of phosphatidylinositol 3-phosphate 5-kinase deficient cells in Schizosaccharomyces pombe. Biosci Biotechnol Biochem 67:1772–1779
54. de Lartigue J, Polson H, Feldman M, Shokat K, Tooze SA, Urbe S, Clague MJ (2009) PIKfyve regulation of endosome-linked pathways. Traffic 10:883–893
55. Iwaki T, Onishi M, Ikeuchi M, Kita A, Sugiura R, Giga-Hama Y, Fukui Y, Takegawa K (2007) Essential roles of class E Vps proteins for sorting into multivesicular bodies in Schizosaccharomyces pombe. Microbiology 153:2753–2764
56. Urbe S, Liu H, Hayes SD, Heride C, Rigden DJ, Clague MJ (2012) Systematic survey of deubiquitinase localisation identifies USP21 as a regulator of centrosome and microtubule associated functions. Mol Biol Cell 23(6):1095–1103
57. Row PE, Prior IA, McCullough J, Clague MJ, Urbe S (2006) The ubiquitin isopeptidase UBPY regulates endosomal ubiquitin dynamics and is essential for receptor down-regulation. J Biol Chem 281:12618–12624
58. Tanaka N, Kaneko K, Asao H, Kasai H, Endo Y, Fujita T, Takeshita T, Sugamura K (1999) Possible involvement of a novel STAM-associated molecule "AMSH" in intracellular signal transduction mediated by cytokines. J Biol Chem 274:19129–19135
59. Itoh F, Asao H, Sugamura K, Heldin CH, ten Dijke P, Itoh S (2001) Promoting bone morphogenetic protein signaling through negative regulation of inhibitory Smads. EMBO J 20:4132–4142

60. McCullough J, Clague MJ, Urbe S (2004) AMSH is an endosome-associated ubiquitin isopeptidase. J Cell Biol 166:487–492
61. Kato M, Miyazawa K, Kitamura N (2000) A de-ubiquitinating enzyme UBPY interacts with the SH3 domain of Hrs binding protein via a novel binding motif Px(V/I)(D/N)RxxKP. J Biol Chem 275:37481–37487
62. Hurley JH, Yang D (2008) MIT domainia. Dev Cell 14:6–8
63. Sachse M, Urbé S, Oorschot V, Strous GJ, Klumperman J (2002) Bilayered clathrin coats on endosomal vacuoles are involved in protein sorting toward lysosomes. Mol Biol Cell 13:1313–1328
64. Clague MJ (2002) Membrane transport: a coat for ubiquitin. Curr Biol 12:R529–R531
65. Raiborg C, Wesche J, Malerod L, Stenmark H (2006) Flat clathrin coats on endosomes mediate degradative protein sorting by scaffolding Hrs in dynamic microdomains. J Cell Sci 119:2414–2424
66. Nakamura M, Tanaka N, Kitamura N, Komada M (2006) Clathrin anchors deubiquitinating enzymes, AMSH and AMSH-like protein, on early endosomes. Genes Cells 11:593–606
67. Kyuuma M, Kikuchi K, Kojima K, Sugawara Y, Sato M, Mano N, Goto J, Takeshita T, Yamamoto A, Sugamura K et al (2007) AMSH, an ESCRT-III associated enzyme, deubiquitinates cargo on MVB/late endosomes. Cell Struct Funct 31:159–172
68. Ma YM, Boucrot E, Villen J, Affar el B, Gygi SP, Gottlinger HG, Kirchhausen T (2007) Targeting of AMSH to endosomes is required for epidermal growth factor receptor degradation. J Biol Chem 282:9805–9812
69. McCullough J, Row PE, Lorenzo O, Doherty M, Beynon R, Clague MJ, Urbe S (2006) Activation of the endosome-associated ubiquitin isopeptidase AMSH by STAM, a component of the multivesicular body-sorting machinery. Curr Biol 16:160–165
70. Hurley JH, Stenmark H (2011) Molecular mechanisms of ubiquitin-dependent membrane traffic. Annu Rev Biophys 40:119–142
71. Tsang HT, Connell JW, Brown SE, Thompson A, Reid E, Sanderson CM (2006) A systematic analysis of human CHMP protein interactions: additional MIT domain-containing proteins bind to multiple components of the human ESCRT III complex. Genomics 88(3):333–346
72. Agromayor M, Martin-Serrano J (2006) Interaction of AMSH with ESCRT-III and deubiquitination of endosomal cargo. J Biol Chem 281:1374–1387
73. Row PE, Liu H, Hayes S, Welchman R, Charalabous P, Hofmann K, Clague MJ, Sanderson CM, Urbe S (2007) The MIT domain of UBPY constitutes a CHMP binding and endosomal localization signal required for efficient epidermal growth factor receptor degradation. J Biol Chem 282:30929–30937
74. Solomons J, Sabin C, Poudevigne E, Usami Y, Hulsik DL, Macheboeuf P, Hartlieb B, Gottlinger H, Weissenhorn W (2011) Structural Basis for ESCRT-III CHMP3 Recruitment of AMSH. Structure 19:1149–1159
75. Komander D, Reyes-Turcu F, Licchesi JDF, Odenwaelder P, Wilkinson KD, Barford D (2009) Molecular discrimination of structurally equivalent Lys63-linked and linear polyubiquitin chains. EMBO Rep 10:466–473
76. Sato Y, Yoshikawa A, Yamagata A, Mimura H, Yamashita M, Ookata K, Nureki O, Iwai K, Komada M, Fukai S (2008) Structural basis for specific cleavage of Lys 63-linked polyubiquitin chains. Nature 455:358–362
77. Lauwers E, Jacob C, Andre B (2009) K63-linked ubiquitin chains as a specific signal for protein sorting into the multivesicular body pathway. J Cell Biol 185:493–502
78. Bowers K, Piper SC, Edeling MA, Gray SR, Owen DJ, Lehner PJ, Luzio JP (2006) Degradation of endocytosed epidermal growth factor and virally ubiquitinated major histocompatibility complex class I is independent of mammalian ESCRTII. J Biol Chem 281:5094–5105
79. Pareja F, Ferraro DA, Rubin C, Cohen-Dvashi H, Zhang F, Aulmann S, Ben-Chetrit N, Pines G, Navon R, Crosetto N et al (2012) Deubiquitination of EGFR by Cezanne-1 contributes to cancer progression. Oncogene 31(43):4599–4608
80. Sierra MI, Wright MH, Nash PD (2010) AMSH interacts with ESCRT-0 to regulate the stability and trafficking of CXCR4. J Biol Chem 285:13990–14004

81. Reyes-Ibarra AP, Garcia-Regalado A, Ramirez-Rangel I, Esparza-Silva AL, Valadez-Sanchez M, Vazquez-Prado J, Reyes-Cruz G (2007) Calcium-sensing receptor endocytosis links extracellular calcium signaling to parathyroid hormone-related peptide secretion via a Rab11a-dependent and AMSH-sensitive mechanism. Mol Endocrinol 21:1394–1407
82. Hasdemir B, Murphy JE, Cottrell GS, Bunnett NW (2009) Endosomal deubiquitinating enzymes control ubiquitination and down-regulation of protease-activated receptor 2. J Biol Chem 284:28453–28466
83. Naviglio S, Mattecucci C, Matoskova B, Nagase T, Nomura N, Di Fiore PP, Draetta GF (1998) UBPY: a growth-regulated human ubiquitin isopeptidase. EMBO J 17:3241–3250
84. Niendorf S, Oksche A, Kisser A, Lohler J, Prinz M, Schorle H, Feller S, Lewitzky M, Horak I, Knobeloch KP (2007) Essential role of ubiquitin-specific protease 8 for receptor tyrosine kinase stability and endocytic trafficking in vivo. Mol Cell Biol 27:5029–5039
85. Mizuno E, Iura T, Mukai A, Yoshimori T, Kitamura N, Komada M (2005) Regulation of epidermal growth factor receptor down-regulation by ubpy-mediated deubiquitination at endosomes. Mol Biol Cell 16:5163–5174
86. Avvakumov GV, Walker JR, Xue S, Finerty PJ Jr, Mackenzie F, Newman EM, Dhe-Paganon S (2006) Amino-terminal dimerization, NRDP1-rhodanese interaction, and inhibited catalytic domain conformation of the ubiquitin-specific protease 8 (USP8). J Biol Chem 281:38061–38070
87. Wu X, Yen L, Irwin L, Sweeney C, Carraway KL 3rd (2004) Stabilization of the E3 ubiquitin ligase Nrdp1 by the deubiquitinating enzyme USP8. Mol Cell Biol 24:7748–7757
88. Faesen AC, Luna-Vargas MPA, Geurink PP, Clerici M, Merkx R, van Dijk WJ, Hameed DS, El Oualid F, Ovaa H, Sixma TK (2011) The differential modulation of USP activity by internal regulatory domains, interactors and eight ubiquitin chain types. Chem Biol 18:1550–1561
89. Mizuno E, Kobayashi K, Yamamoto A, Kitamura N, Komada M (2006) A deubiquitinating enzyme UBPY regulates the level of protein ubiquitination on endosomes. Traffic 7:1017–10131
90. Brocker C, Kuhlee A, Gatsogiannis C, Balderhaar HJ, Honscher C, Engelbrecht-Vandre S, Ungermann C, Raunser S (2011) Molecular architecture of the multisubunit homotypic fusion and vacuole protein sorting (HOPS) tethering complex. Proc Natl Acad Sci U S A 109:1991–1996
91. Mukai A, Yamamoto-Hino M, Awano W, Watanabe W, Komada M, Goto S (2010) Balanced ubiquitylation and deubiquitylation of Frizzled regulate cellular responsiveness to Wg/Wnt. EMBO J 29:2114–2125
92. Xia R, Jia H, Fan J, Liu Y, Jia J (2012) USP8 promotes smoothened signaling by preventing its ubiquitination and changing its subcellular localization. PLoS Biol 10:e1001238
93. Li S, Chen Y, Shi Q, Yue T, Wang B, Jiang J (2012) Hedgehog-regulated ubiquitination controls smoothened trafficking and cell surface expression in Drosophila. PLoS Biol 10:e1001239
94. Schwanhausser B, Busse D, Li N, Dittmar G, Schuchhardt J, Wolf J, Chen W, Selbach M (2011) Global quantification of mammalian gene expression control. Nature 473:337–342
95. Raiborg C, Gronvold Bache K, Mehlum A, Stang E, Stenmark H (2001) Hrs recruits clathrin to early endosomes. EMBO J 20:5008–5021
96. Raiborg C, Bache KG, Gillooly DJ, Madshus IH, Stang E, Stenmark H (2002) Hrs sorts ubiquitinated proteins into clathrin -coated microdomains of early endosomes. Nat Cell Biol 4:394–398
97. Wollert T, Hurley JH (2010) Molecular mechanism of multivesicular body biogenesis by ESCRT complexes. Nature 464:864–869
98. Scott A, Gaspar J, Stuchell-Brereton MD, Alam SL, Skalicky JJ, Sundquist WI (2005) Structure and ESCRT-III protein interactions of the MIT domain of human VPS4A. Proc Natl Acad Sci U S A 102:13813–13818
99. Shenoy SK, Modi AS, Shukla AK, Xiao K, Berthouze M, Ahn S, Wilkinson KD, Miller WE, Lefkowitz RJ (2009) Beta-arrestin-dependent signaling and trafficking of 7-transmembrane

receptors is reciprocally regulated by the deubiquitinase USP33 and the E3 ligase Mdm2. Proc Natl Acad Sci U S A 106:6650–6655
100. Berthouze M, Venkataramanan V, Li Y, Shenoy SK (2009) The deubiquitinases USP33 and USP20 coordinate beta2 adrenergic receptor recycling and resensitization. EMBO J 28:1684–1696
101. Fakitsas P, Adam G, Daidie D, van Bemmelen MX, Fouladkou F, Patrignani A, Wagner U, Warth R, Camargo SM, Staub O et al (2007) Early aldosterone-induced gene product regulates the epithelial sodium channel by deubiquitylation. J Am Soc Nephrol 18:1084–1092
102. Boulkroun S, Ruffieux-Daidie D, Vitagliano JJ, Poirot O, Charles RP, Lagnaz D, Firsov D, Kellenberger S, Staub O (2008) Vasopressin-inducible ubiquitin-specific protease 10 increases ENaC cell surface expression by deubiquitylating and stabilizing sorting nexin 3. Am J Physiol Renal Physiol 295:F889–F900
103. Bomberger JM, Barnaby RL, Stanton BA (2010) The deubiquitinating enzyme USP10 regulates the endocytic recycling of CFTR in airway epithelial cells. Channels (Austin) 4:150–154
104. Bomberger JM, Barnaby RL, Stanton BA (2009) The deubiquitinating enzyme USP10 regulates the post-endocytic sorting of cystic fibrosis transmembrane conductance regulator in airway epithelial cells. J Biol Chem 284:18778–18789
105. Gorbea C, Pratt G, Ustrell V, Bell R, Sahasrabudhe S, Hughes RE, Rechsteiner M (2010) A protein interaction network for ECM29 links the 26S proteasome to molecular motors and endosomal components. J Biol Chem 285:31616–31633
106. Hammond DE, Urbe S, Vande Woude GF, Clague MJ (2001) Down-regulation of MET, the receptor for hepatocyte growth factor. Oncogene 20:2761–2770
107. Alwan HA, van Zoelen EJ, van Leeuwen JE (2003) Ligand-induced lysosomal epidermal growth factor receptor (EGFR) degradation is preceded by proteasome-dependent EGFR de-ubiquitination. J Biol Chem 278:35781–35790
108. Geetha T, Wooten MW (2008) TrkA receptor endolysosomal degradation is both ubiquitin and proteasome dependent. Traffic 9:1146–1156
109. Colland F (2010) The therapeutic potential of deubiquitinating enzyme inhibitors. Biochem Soc Trans 38:137–143
110. Fry WH, Simion C, Sweeney C, Carraway KL 3rd (2011) Quantity control of the ErbB3 receptor tyrosine kinase at the endoplasmic reticulum. Mol Cell Biol 31:3009–3018

Chapter 13
RTKs as Models for Trafficking Regulation: c-Met/HGF Receptor-c-Met Signalling in Cancer—Location Counts

Carine Joffre, Rachel Barrow, Ludovic Ménard, and Stéphanie Kermorgant

Abstract c-Met, the receptor tyrosine kinase of Hepatocyte Growth Factor (HGF), plays a major role in tumour progression and constitutes an attractive target for cancer therapy. Upon ligand binding at the plasma membrane, c-Met triggers various signalling pathways, which promote cell responses such as proliferation, migration and survival. Recent studies have shown that HGF binding to c-Met also triggers a rapid endocytosis of c-Met, leading to its intracellular trafficking and ultimately degradation. However, c-Met remains bound to HGF and activated on endosomes and in fact its endocytosis is required for optimal signalling. Moreover c-Met exploits both its endocytosis and its subsequent intracellular trafficking to alter its signalling capacity. Finally, recent studies on oncogenic mutants of c-Met, reported in human cancers, have shown that activated c-Met accumulation on endosomes can lead to cell transformation. Here, we review the present knowledge about c-Met endocytosis and trafficking, the interplay between c-Met trafficking and signalling and the consequences on tumorigenesis. We discuss the idea that c-Met localisation, in addition to its activation, plays a key role in its signalling and possibly in cancer progression.

C. Joffre
Spatial Signalling Team, Centre for Tumour Biology,
Barts Cancer Institute—A Cancer Research UK Centre of Excellence,
Queen Mary University of London, John Vane Science Centre,
Charterhouse Square, London EC1M 6BQ, UK

Institut Curie, INSERM U-830, 26 rue d'Ulm, 75248 Paris Cedex 05, France

R. Barrow • L. Ménard • S. Kermorgant (✉)
Spatial Signalling Team, Centre for Tumour Biology,
Barts Cancer Institute—A Cancer Research UK Centre of Excellence,
Queen Mary University of London, John Vane Science Centre,
Charterhouse Square, London EC1M 6BQ, UK
e-mail: s.kermorgant@qmul.ac.uk

Y. Yarden and G. Tarcic (eds.), *Vesicle Trafficking in Cancer*,
DOI 10.1007/978-1-4614-6528-7_13, © Springer Science+Business Media New York 2013

13.1 The c-Met Receptor Tyrosine Kinase

13.1.1 c-Met and HGF

c-Met is a receptor tyrosine kinase (RTK), belonging to a subfamily containing Met, Ron and the avian receptor Sea [14]. It is expressed on the surface of both epithelial and endothelial cells, where it binds specifically to its ligand, the Hepatocyte Growth Factor (HGF), also called scatter factor (SF). HGF is produced by mesenchymal cells, in close proximity to the epithelial/endothelial cells expressing the c-Met receptor. Thus HGF and c-Met signal in a paracrine manner [23,68].

13.1.2 Cellular Functions Regulated by HGF and c-Met

HGF and c-Met mediate numerous cell functions including cell survival, motility (such as scattering, migration and invasion), proliferation, tubulogenesis and angiogenesis [23,68]. Expression of both c-Met and HGF is necessary during development, thus mice lacking either die in *utero* [7,61,70]. In adults, c-Met normally is expressed at very low levels; however, expression of HGF and c-Met often is increased in injured tissues [23,68] and this concordant upregulation has been shown to be required for wound healing [12].

13.1.3 c-Met and Cancer

c-Met is a proto-oncogene and becomes oncogenic either through gene rearrangement, leading to a fusion protein (translocated promoter region (TPR)–MET) [16] or, more frequently, through gene amplification and/or overexpression, as reported in many cancer types including solid tumours in the lung, breast and gut [23,68]. The oncogenic c-Met is involved in cancer formation and progression. This usually occurs through paracrine, and sometimes autocrine, activation of c-Met [22]. In addition, activating mutations of c-Met, leading to its constitutive activation, were found in many cancers including papillary renal carcinomas [62] and ovarian, gastric, liver and head and neck cancer [6]. Thus c-Met currently is under investigation for the development of novel anti-cancer treatments [15,23].

13.1.4 c-Met Signalling

Binding of HGF to the c-Met receptor leads to the stable dimerisation of two molecules of c-Met. This enables trans-autophosphorylation of the tyrosine kinase domain at tyrosine residues Y1234 and Y1235, followed by trans-phosphorylation

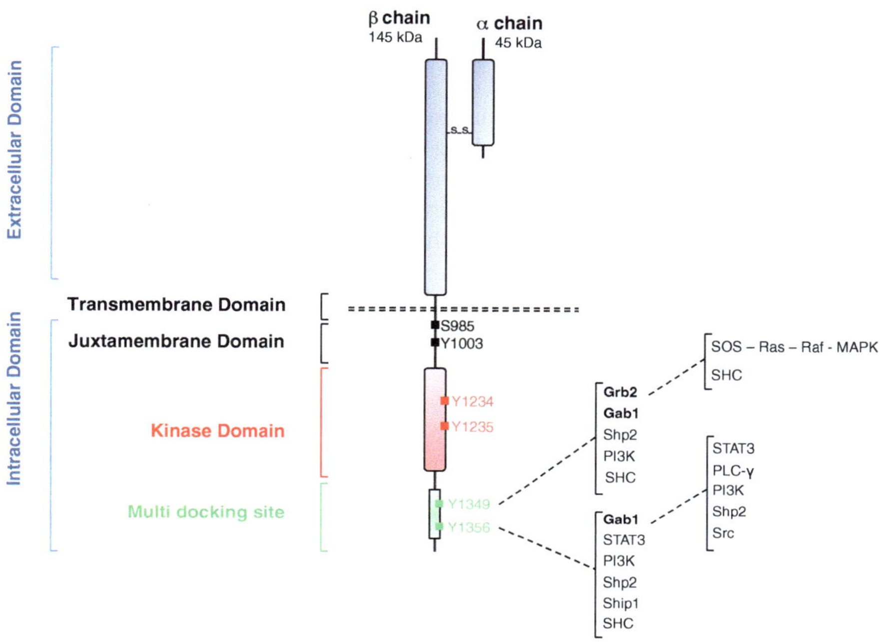

Fig. 13.1 c-Met receptor. c-Met receptor is composed of an extracellular α-chain linked by a disulphide bond to a β-chain, which contains an extracellular domain, a transmembrane domain and an intracellular domain. The intracellular domain is composed of a juxtamembrane sequence which contains the tyrosine 1003 (which, when phosphorylated, binds to c-Cbl leading to c-Met degradation) and the serine 985 (which, when phosphorylated, downregulates c-Met kinase activity); a kinase domain with the two major tyrosines 1234 and 1235, which transphosphorylate upon HGF binding and c-Met dimerisation; and a multifunctional docking site with the two docking tyrosines 1349 and 1356 which recruit the indicated adaptors and signalling molecules

of the two tyrosine residues Y1349 and Y1356 [23,68] in the C-terminal region. These two tyrosines form a multisubstrate docking site, which is unique to members of the c-Met subfamily [53] (Fig. 13.1), most RTKs having several tyrosines located in their intracellular domain, each of which will be responsible for binding to a specific signalling molecule. The Y1349/Y1356 site alone is responsible for c-Met binding to multiple substrates, either directly or through adaptors, leading to the activation of a variety of signalling pathways [53].

c-Met adaptor proteins include growth factor receptor-bound protein 2 (Grb2) and Grb2-associated binder 1 (Gab1). Gab1 can bind directly to c-Met on either Y1349 or Y1356 through its specific "c-Met binding site" consisting of 13 amino acids [59] or indirectly through Grb2 [5,47], which binds to c-Met at Y1356 [47], through an Src Homology 2 (SH2) binding site [19]. The signalling molecules bind to c-Met through their SH2 domain and include the small GTPase Ras, mitogen-activated protein kinase (MAPK), phosphatidylinositol 3-kinase (PI3K), Src, SH2 domain-containing inositol-5-phosphatase (Ship1), phospholipase C-γ (PLC-γ),

SH2-containing protein tyrosine phosphatase 2 (Shp2), the c-Jun N-terminal kinase (JNK) and the transcription factor signal transducer and activator of transcription 3 (STAT3) [8,19,21,26,58] (Fig. 13.1).

Due to the wide variety of cellular processes that are initiated following c-Met signalling, it is important that its signalling is tightly controlled. The phosphorylation of the serine residue S985, in the juxtamembrane region of c-Met, by either PKC or Ca^{2+}-dependent kinases, negatively regulates the tyrosine kinase activity of c-Met [20].

Several studies have aimed to identify which individual downstream signalling molecules are responsible for specific cellular responses. In fact, one pathway can regulate several functions while one function may be regulated by many pathways [68]. For example, activation of PI3K-AKT, either by direct binding to c-Met or through Ras or Gab1, triggers cell survival as well as cell growth and cell scattering [17,18,54]. The activation of the MAPK pathway by c-Met, leading to the activation of ERK1/2, triggers cell proliferation, survival and motility [17,54]. The activation of Shp2 via Gab1 is important for branching morphogenesis [59], while the activation of the pathway Grb2–son of sevenless (SOS)–Ras MAPK–paxillin–PI3K–focal adhesion kinase (FAK)–Rac1 regulates cell migration and tubulogenesis [29]. The transcription factor STAT3 plays an important role in c-Met-dependent tubulogenesis, anchorage-independent growth and tumorigenesis [8,74].

How are these many pathways interconnected and how are they integrated in space and time to produce cell responses? How the signalling molecules "compete with each other" to access the c-Met docking? How the decision to utilise specific pathways is made? Answers to these questions have remained elusive. However the emerging idea is that c-Met signalling is spatially and temporally regulated, such that different pathways are activated at specific subcellular locations following c-Met activation and not just at the plasma membrane as it was assumed (see Sect. 13.3).

13.2 c-Met Internalisation, Trafficking and Degradation

13.2.1 c-Met Internalisation

Ligand binding triggers rapid internalisation of the receptor, such that a large percentage of c-Met can be internalised from the cell surface within a few minutes [27,35]. This classically occurs via clathrin- and dynamin-mediated endocytosis. Indeed, c-Met endocytosis is impaired upon clathrin heavy chain knockdown by siRNA [36], dynamin mutant overexpression [27,36] or incubation with the pharmacological inhibitor of dynamin, dynasore [34]. The endocytic adaptors regulating the clathrin-mediated endocytosis of c-Met include eps15 [71] and AP180 [35].

The ubiquitin ligase activity of c-Cbl is critical for clathrin-dependent c-Met internalisation and occurs following the indirect binding of c-Cbl to c-Met via Grb2 [40]. c-Cbl also plays the role of an endocytic adaptor through connecting c-Met to the adaptor CIN85, which in turn binds to endophilin [52], involved in the invagination of the plasma membrane [9] (Fig. 13.2).

13.2.2 c-Met Trafficking

Internalised c-Met is recruited in peripheral endosomes which are positive for the early endosome antigen 1 (EEA1) [33,34] and thus correspond to the early/sorting endosome. This localisation of c-Met is predominant at 15 min of HGF stimulation [35]. c-Met then traffics along the microtubules and gets progressively accumulated in a perinuclear endosome containing Lamp1 and Rab7 and thus corresponds to the late endosome/multivesicular body (MVB). This localisation is predominant at 120 min of stimulation [35]. While PKC inhibition has no effect on c-Met internalisation, c-Met trafficking from peripheral to perinuclear endosomes is promoted by the activity of the classical PKC isotype, PKC-α (Fig. 13.2). However, the trafficking of c-Met to the perinuclear location is not required for its degradation [35].

13.2.3 c-Met Degradation

Internalised c-Met is sorted for degradation. This occurs progressively with time such that, for example, in HeLa cells, after 120 min post-HGF stimulation, 50% of c-Met has degraded [35]. It has recently been reported that a proportion of c-Met can recycle through interaction with Golgi-localised gamma ear-containing Arf-binding protein 3(GGA3) within Rab4-positive endosomes [50].

c-Met is targeted for degradation through ubiquitination by c-Cbl following a direct binding of c-Cbl to the phosphorylated tyrosine Y1003 in the juxtamembrane domain of c-Met [51]. This interaction between Y1003 and c-Cbl, however, is not required for c-Met endocytosis [2]. Ubiquitinated c-Met interacts with and phosphorylates Hrs (HGF-regulated tyrosine kinase substrate) [2,27], a molecule that recognises ubiquitinated receptors within endosomes and consequently prevents them from recycling back to the plasma membrane, targeting them for degradation instead [56]. A requirement for K48 polyubiquitination in the c-Met degradation pathway has been shown [10]. A transcytosis of internalised c-Met to Gab1-positive dorsal ruffles, followed by the collapse of these ruffles and the subsequent delivery of c-Met to Rab5- and Hrs-positive endosomes, has been reported to lead to an efficient c-Met degradation [1]. Additionally, the transmembrane protein LRIG1 (leucine-rich repeats and immunoglobulin-like domains 1) can interact with c-Met and this, independently of HGF stimulation, leads to c-Met destabilisation and degradation in a c-Cbl-independent manner [65] (Fig. 13.2).

The pharmacological inhibition of proteasome activity completely blocks c-Met degradation [31,35] while lysosomal inhibition also has been reported to inhibit c-Met degradation [10,27].

13.2.4 c-Met, c-Cbl and Cell Transformation

Studies using c-Met mutant Y1003F, to decouple the direct binding of c-Cbl to c-Met, showed that a defect in degradation of c-Met leads to increased signalling and cell transformation in vitro [51] and in vivo [2]. This mutant, although able to

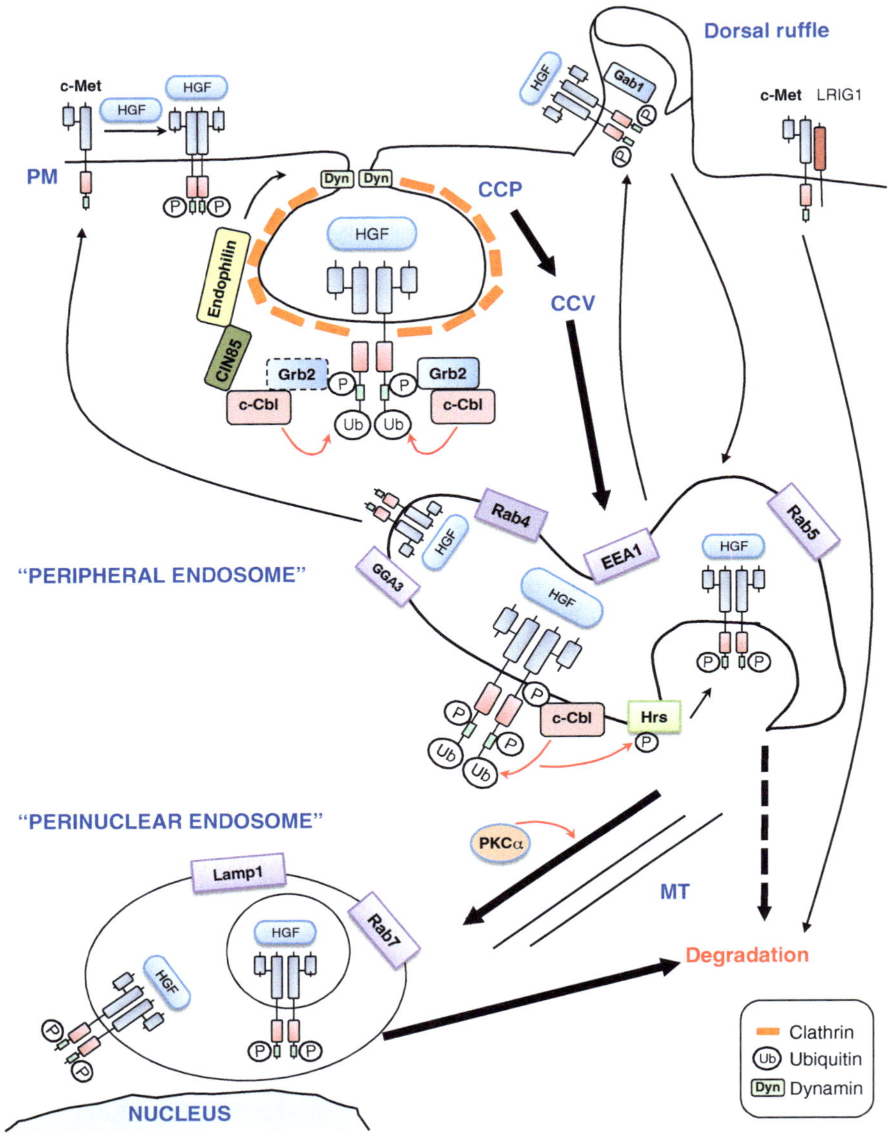

Fig. 13.2 c-Met trafficking. HGF binding triggers c-Met dimerisation and phosphorylation and recruitment into clathrin-coated pits (CCP). The mechanisms involved in c-Met internalisation include the binding of the ubiquitin ligase c-Cbl to c-Met *via* Grb2, leading to the ubiquitination of c-Met and the connection, through c-Cbl, between c-Met and the complex CIN85-endophilin leading to membrane curvature. The CCP is pinched off in a clathrin-coated vesicle (CCV), thanks to the activity of the GTPase, dynamin. c-Met is then recruited in the EEA1/Rab5/Rab4-positive peripheral endosome or sorting/early endosome where it remains phosphorylated. c-Cbl binds directly to Y1003 in the c-Met juxtamembrane domain, leading to c-Met ubiquitination and to the phosphorylation of Hrs; this then leads to c-Met sorting for ultimate degradation. However,

internalise and be recruited to Hrs-positive endosomes, is unable to phosphorylate Hrs [2]. In addition, c-Met activity recently has been shown to trigger a loss of c-Cbl in gastric cancer cell lines with c-Met amplification [37].

13.3 Influence of c-Met Endosomal Trafficking on Its Signalling

13.3.1 c-Met Endosomal Signalling

Although as a consequence of ligand-dependent internalisation, c-Met gets degraded; c-Met internalisation is not a mechanism for desensitisation only. A significant distribution of ligand-receptor complexes to intracellular compartments occurs during the first 30 min of stimulation, when c-Met receptor downregulation is not yet significant [35]. The process of c-Met degradation is slow such that even after 120 min of HGF stimulation, still 50% of the mature form of c-Met is present in the cell. Furthermore, endosomal c-Met remains bound to its ligand and activated [34,36].

Moreover endocytosis is required for an optimal c-Met signalling [34,36]. Indeed, inhibition of internalisation, using pharmacological means (such as the dynamin inhibitor dynasore), siRNA (clathrin or c-Cbl) or transfection of a dominant negative (K44A dynamin or AP180C), inhibits several HGF/c-Met-dependent signalling outputs such as the activation of ERK1/2 [36], STAT3 [34] and Rac1 [33,49].

13.3.2 c-Met Signalling from the Peripheral Endosome

From the peripheral endosome, c-Met activates ERK1/2, which further accumulates in focal complexes, to participate in regulation of cell migration. The atypical isoform of PKC, PKCε, promotes this ERK1/2 accumulation in focal complexes. Indeed PKCε knockdown, with siRNAs or inhibition of its activity with pharmacological approaches, inhibits the HGF-dependent ERK1/2 accumulation at the plasma membrane. The absence of PKCε (cells isolated from knock-out mice or transiently

Fig. 13.2 (continued) a proportion of c-Met may recycle either to the plasma membrane through interaction with Golgi-localised gamma ear-containing Arf-binding protein 3(GGA3) or to Gab1-positive dorsal ruffles, followed by the collapse of these ruffles and the subsequent delivery of c-Met to Rab5- and Hrs-positive endosomes. From the peripheral endosome, c-Met is trafficked along the microtubule network, under the control of PKCα activity, to accumulate progressively in the Lamp1/Rab7-positive perinuclear endosome [corresponding to the late endosome/multivesicular body (MVB)]. There, c-Met is still activated and is progressively targeted for degradation. Additionally, LRIG1 (leucine-rich repeats and immunoglobulin-like domains 1) may interact with c-Met at the plasma membrane, leading to c-Met degradation

knocked down with siRNA) prevents HGF-dependent cell migration. However, in the conditions where PKCε is absent or inactive, the total level of HGF-dependent ERK1/2 phosphorylation has increased in the cells (as assessed by Western blot). Thus it is believed that PKCε can dissociate activated ERK1/2 from the endosomal c-Met signalling complex, leading to ERK1/2 accumulation at the plasma membrane. Accordingly, colocalisations between PKCε and c-Met are observed on the peripheral endosomes. The consequence is stimulated cell migration on the one hand but a reduction of c-Met-dependent ERK1/2 phosphorylation on the other hand, as normally ERK1/2 activation is sustained through the increase in lifetime of its association to the endosomal c-Met complex. Thus although the signal has increased, since it is not at the correct location, the cell function cannot occur correctly. This study highlights the importance of signal localisation and indicates that signal intensity alone is insufficient to act as a unique determinant. Activation of molecules in incorrect locations will not provide an efficient cell response (Fig. 13.3).

13.3.3 c-Met Signalling from a Perinuclear Endosome

While the trafficking of c-Met to the perinuclear location is not required for its degradation [35], it plays an important role in its signalling as is shown by activation of the STAT3 pathway [34,43]. Under basal conditions, unphosphorylated STAT3 shuttles constitutively between the cytoplasm and the nucleus [41,55]. HGF stimulates STAT3 recruitment to c-Met, triggering STAT3 phosphorylation, dimerisation and nuclear translocation [8]. STAT3 appears to be able to associate with c-Met not only at the plasma membrane but also on peripheral and perinuclear endosomes and thus becomes activated at these locations. However, STAT3 is only weakly activated by HGF/c-Met (as compared, for example, to the stimulation with the cytokine oncostatin M). It is believed that the activated STAT3 detaches from c-Met and diffuses through the cytoplasm before reaching the nucleus to activate its target genes. When STAT3 is activated at the plasma membrane or on the peripheral endosome, its subsequent diffusion through the phosphatase-rich cytoplasm prevents its significant activation and nuclear accumulation. By contrast, when c-Met activates STAT3 on the perinuclear endosome, the proximity to the nucleus allows STAT3 to accumulate efficiently in the nucleus and STAT3 activation is now significant (as measured by Western blot). The process is very dynamic since an acute pharmacological inhibition of c-Met when located on the perinuclear compartment leads to an immediate significant reduction in STAT3 nuclear accumulation. The threshold of activation required to enable STAT3 nuclear accumulation is achieved because this locally activated STAT3 is protected from the cytoplasmic phosphatases.

The more efficient STAT3 signal from a perinuclear location might be due to the shuttling of STAT3 in and out of the nucleus, the dephosphorylated STAT3 exiting the nucleus but rapidly being re-phosphorylated by activated perinuclear c-Met. One alternative possible explanation would be that activated c-Met, at the plasma

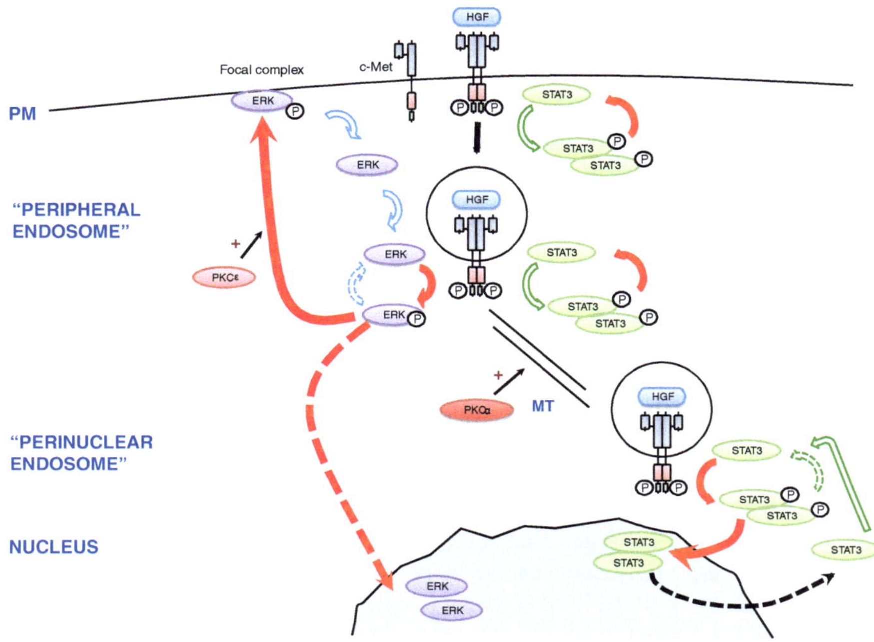

Fig. 13.3 c-Met endosomal signalling. HGF/c-Met efficient stimulation of ERK1/2 and STAT3 pathways requires c-Met endocytosis. Thus c-Met remains bound to HGF and activated in the peripheral endosome, from where it stimulates ERK1/2 activation. Phosphorylated ERK1/2 further accumulates in focal complexes. PKCε promotes activated ERK1/2 translocation to the focal complexes by dissociating activated ERK1/2 from the endosomal c-Met signalosome. This leads to an enhanced cell migration on the one hand but to a reduction of the cellular level of ERK1/2 phosphorylation on the other hand. Activated c-Met can bind and activate STAT3 at the plasma membrane as well as from the peripheral or perinuclear endosomes. However, the signal generated at the plasma membrane, or on the peripheral endosome, is weak, and the subsequent diffusion of STAT3 through the phosphatase-rich cytoplasm prevents its significant nuclear accumulation. By contrast, when c-Met activates STAT3 on the perinuclear endosome, the proximity to the nucleus allows STAT3 to accumulate efficiently in the nucleus. Interestingly, this perinuclear localisation of c-Met is not required for nuclear accumulation of activated ERK1/2 since c-Met strongly activates ERK1/2 at the peripheral endosome, and thus the threshold for nuclear accumulation is achieved despite diffusion through the phosphatase-rich cytoplasm

membrane and in the peripheral endosome membrane compartment, is selectively associated with a phosphatase that prevents a robust STAT3 phosphorylation. c-Met could be sorted away from the phosphatase in the juxtanuclear compartment. Nevertheless, these data provide strong evidence that the mechanism controlling STAT3 nuclear uptake upon c-Met activation is based primarily on where in the cell this signal is generated because of the weakness of the STAT3 signal. When the strength of the STAT3 signal is enhanced (using phosphatase inhibition or via a more robust agonist—e.g., oncostatin M), STAT3 nuclear accumulation is independent of the localisation of the trigger. This perinuclear localisation of c-Met is also not required for nuclear accumulation of activated ERK1/2, although c-Met endocytosis is required for the nuclear accumulation of both activated STAT3 and ERK1/2. In fact, unlike for STAT3, c-Met strongly activates ERK1/2 at the peripheral endosome, and thus the threshold for nuclear accumulation is achieved despite diffusion through the phosphatase-rich cytoplasm (Fig. 13.3). Interestingly, in this model, although c-Met trafficking to the perinuclear endosome is required for efficient STAT3 signalling, it is believed the two molecules do not need to traffic together along the microtubules, as was described for ERK5 and Trk which co-traffic from distal axons to the cell body, in order to induce ERK5 nuclear accumulation [73].

13.4 c-Met Endosomal Signalling and Tumorigenesis

The signalling pathways and cell functions (such as cell migration and survival) [3,32,34,36,39,48,64,66,72] controlled by endosomal signalling all are involved in tumorigenesis. This suggested that RTK endocytosis could play a role in cancer progression. Interestingly, some endocytic proteins, such as HIP1 (Huntingtin interacting protein 1) or Rab25, have been found overexpressed in human cancers [38,44], and this could affect RTK trafficking. Indeed HIP1 overexpression appears to increase EGFR internalisation on the one hand and to promote tumour formation in vivo on the other hand, suggesting a possible link between EGFR trafficking and tumorigenesis [57]. Such studies report on modifications of molecules, which impinge on RTK trafficking and signalling. Increase of RTK endocytosis directly leading to enhanced signalling and transformation, had not been reported until the demonstration of a direct link between c-Met endocytosis and tumorigenesis.

Thus certain point mutations in the kinase domain of c-Met, reported in cancer [62,63] and which lead to a constitutive activation of c-Met and subsequent cell transformation in vitro and in vivo [4,24,25,30], also trigger an enhanced endocytosis of c-Met. Furthermore, it is the combination of endosomal signalling and high activation that leads to cell transformation and in vivo tumorigenesis [33]. Classically, it had been assumed that the high activation of these c-Met mutants was sufficient to trigger their oncogenicity. The quantitative and/or qualitative modifications affecting these oncogenic c-Met mutants' signalling were thought to originate from the plasma membrane. This was considered as the whole requirement for tumorigenicity without taking into account the subcellular compartment where these signals were generated.

Thus the trafficking behaviours of the M1268T and D1246N c-Met mutants were compared to that of the Wt (wild type) form of the receptor, in NIH3T3 cells. It was observed initially that c-Met mutants were accumulated on endosomal compartments in a significantly higher proportion than c-Met Wt. This is, in fact, the result of an enhanced internalisation coupled to an increased stability of the mutants, as compared to c-Met Wt, under basal conditions (in absence of the ligand). While Wt c-Met is internalised weakly and then degraded after endocytosis, the two activated

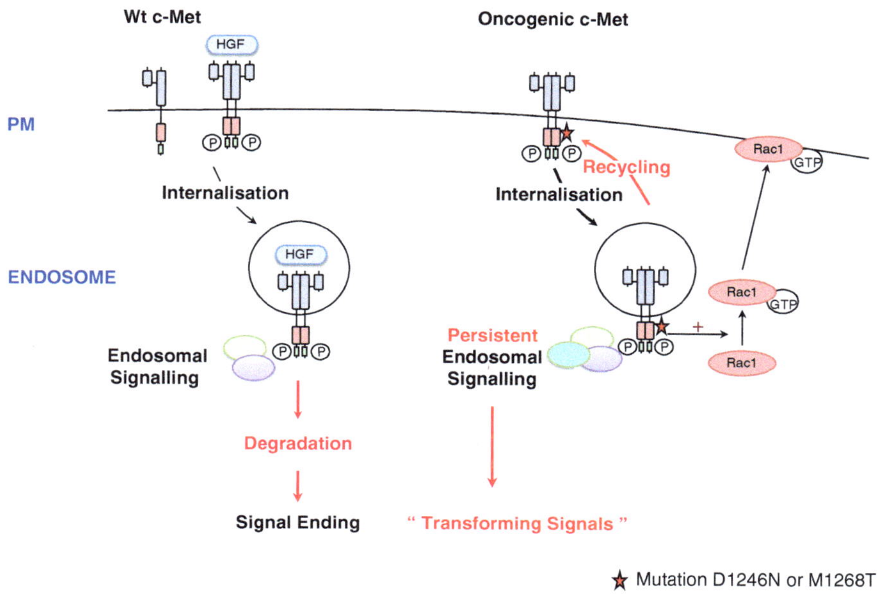

Fig. 13.4 Oncogenic c-Met endosomal signalling. While HGF bound c-Met Wt signals from endosomes, it gets degraded progressively, leading to the ending of signal generation. In contrast, constitutively activated c-Met mutants M1268T or D1246N constantly shuttle between the plasma membrane and endosomes, due to a constitutive endocytosis and recycling coupled to a defect in degradation. Consequently, the oncogenic c-Met accumulates on endosomes from where it generates a persistent endosomal signalling, leading to cell transformation and tumorigenesis. The GTPase Rac1 is activated by c-Met mutants on the endosome and is translocated to the plasma membrane where it is responsible for actin cytoskeleton remodelling and cell migration; thus it was identified as one of the "transforming signals"

mutants are internalised continuously into endosomes and, instead of being degraded, are recycled to the plasma membrane (Fig. 13.4). Thus, these mutants appear to shuttle perpetually between the plasma membrane and the endosomes.

The endocytosis of the c-Met mutants is dependent on clathrin, dynamin, Grb2 and c-Cbl as pharmacological inhibitors of dynamin (dynasore, dynol 34–2), siRNAs or shRNAs of clathrin, c-Cbl or Grb2 or the expression of dominant-negative forms of Grb2, all significantly reduced the mutant's endocytosis. Thus, it was concluded that the mutants appear to internalise through the same pathways as c-Met Wt.

The cells expressing c-Met mutants display a transformed phenotype as compared to the cells expressing Wt c-Met (which is poorly activated and which, in the absence of HGF, does not transform the cells). Thus mutant cells display a reduced spread morphology and actin stress fibres are lost. The mutants cells also are more migratory, grow in more numerous and bigger colonies in soft agar and trigger faster in vivo tumour growth and lung colonisation after tail vein injection, compared to the Wt cells. The transformed phenotype of the mutant cells is due to c-Met activity since c-Met knockdown reverts the transformed phenotype of the mutants expressing cells to the phenotype of the cells expressing Wt c-Met.

Interestingly, the inhibition of c-Met mutants' endocytosis also reverted the transformation of the c-Met mutants expressing cells both in vitro and in vivo. However, under these conditions, c-Met mutants still were activated to the same level as they were without blocking of their endocytosis. Conversely, and rather surprisingly, it was found that the level of internalisation of the mutants was not affected by pharmacological inhibition of their activation. This suggests that activation status and internalisation of the mutants are two separate events, although both are triggered by the mutation. Thus, upon endocytosis inhibition, the mutants were strongly reduced within the intracellular pool while they remained strongly activated at the plasma membrane. Despite their high activation, they lost their tumorigenic potential in vitro and in vivo because they no longer signalled from the correct subcellular location. This was the first time that the increased endocytosis of an RTK was shown to play a key role in tumorigenesis.

Interestingly, one of the two mutants, (D1246N), was resistant to characterised c-Met molecule inhibitors in vitro and in vivo. However, its oncogenic potential was impaired following endocytosis inhibition. This indicates that interfering with endocytic trafficking or endosomal signalling could constitute a new anti-tumour strategy which might be effective against tyrosine kinase inhibitor (TKI)-resistant cells [45].

Rac1 was identified as a major signalling molecule activated on endosomes by the c-Met mutants, and the activity of this molecule was shown to be responsible for the lost of actin stress fibres and acquisition of the increased cell migratory phenotype. Endocytosis indeed is required for c-Met mutants' dependent Rac1 activation, consistent with what has been shown previously in HeLa cells stimulated with HGF [48]. The model for the c-Met mutants' signalling is that the specific signals generated through endosomal signalling, coupled to their persistence due to the accumulation of the mutants on endosomes, may together constitute the "transforming signals" (Fig. 13.4).

In conclusion, c-Met activation on its own appears to be insufficient to confer oncogenicity. Additional spatial requirements seem necessary for the generation of transforming signals. c-Met mutants are oncogenic not only because they are constitutively activated but also because they signal from endosomes, thus resulting in a sustained signalling, which is qualitatively different from the signalling generated from the plasma membrane.

13.5 Conclusion and Perspectives

13.5.1 Conclusion

These studies on c-Met trafficking and signalling strongly suggest that c-Met exploits both its endocytosis and its subsequent intracellular trafficking to adapt its signalling capacity. For example, it is able to accumulate STAT3 efficiently in the nucleus when localised in a juxtanuclear compartment and endosomes constitute a dynamic modulating platform for c-Met signalling specificity. Furthermore, c-Met accumulation on endosomes leads to cell transformation.

13.5.2 Blocking c-Met Endocytosis and Trafficking Specifically

One important limitation in the study of the influence of c-Met trafficking on its signalling is that it has not hitherto been clear as to how to perturb c-Met endocytosis or trafficking specifically without interfering with the traffic of other cargoes, a way of modifying many cellular phenotypes independent of a role for c-Met. To date, the most specific way to block c-Met endocytosis was to interfere with Grb2 or c-Cbl expression levels or their binding to c-Met. A better understanding of which amino acid residues or protein domains bind to the endocytic or trafficking regulators/adaptors is required. Interestingly, one such study recently reported a dileucine motif located in the C-terminus of c-Met which regulates its endocytosis through interacting with the endocytic machinery, including adaptin β and caveolin-1 [13].

13.5.3 Understanding the Mechanisms of c-Met Endosomal Signalling

The signal relays between c-Met and its downstream pathways on the endosome have not been elucidated yet. There actually are a few examples reporting that, on endosomes, specific scaffold/adaptor proteins bring the signalling molecules into close proximity with each other and facilitate their activation. This is the case for p18-p14-MP1 complex which recruit MEK1 and ERK on late endosomes upon EGF stimulation [46,67], the Rab5 effector Appl1 which recruits AKT and its substrate GSK-3β in "Appl endosomes" upon IGF-1 stimulation and during zebrafish development [60], the scaffold β-arrestin2 which assembles Raf1-MEK1-ERK2 in the same complexes with angiotensin type 1 receptor antagonists (AT1aR) in endosomes [42], and the FYVE domain protein SARA which recruits TGFβ receptor to the endosome [28,69]. So far the distinction mainly has been made between signalling of RTKs on the endosome versus their signalling from the plasma membrane. Interestingly, some of the above studies indicate the probable existence of some "endosome-specific" signalling complexes. Thus, further studies should aim to determine the signalosome of c-Met on endosomes and whether c-Met triggers specific pathways from specific endosomal compartments.

13.5.4 Relevance of c-Met Endosomal Signalling in Cancer Diagnosis and Therapy

The recent study reporting a direct role for c-Met endocytosis in tumorigenesis represents a proof of principle that c-Met localisation on endosomes may be a determinant in c-Met-dependent cancer progression. It will be important now to establish the relevance of c-Met endosomal signalling in human cancer progression using human cancer cell lines as well as clinical tissues. Determining the endosomal localisation/activation of c-Met in clinical samples may establish the prognostic and therapeutic relevance of c-Met endosomal signalling. TKIs currently used in the clinic unfortunately lead to tumour resistance and this could be the case for c-Met TKI too, according to a recent study [11]. The mutant D1246N appears to be resistant to c-Met TKI. Since blocking endocytosis was able to overcome the resistance of this mutant, strategies which interfere with c-Met trafficking/endocytosis may prove beneficial for some patients. To achieve this, it is hoped that further studies will lead to approaches that inhibit c-Met endocytosis/trafficking specifically.

References

1. Abella JV, Parachoniak CA, Sangwan V, Park M (2010) Dorsal ruffle microdomains potentiate Met receptor tyrosine kinase signaling and down-regulation. J Biol Chem 285:24956–24967
2. Abella JV, Peschard P, Naujokas MA, Lin T, Saucier C, Urbe S, Park M (2005) Met/Hepatocyte growth factor receptor ubiquitination suppresses transformation and is required for Hrs phosphorylation. Mol Cell Biol 25:9632–9645
3. Assaker G, Ramel D, Wculek SK, Gonzalez-Gaitan M, Emery G (2010) Spatial restriction of receptor tyrosine kinase activity through a polarized endocytic cycle controls border cell migration. Proc Natl Acad Sci USA 107:22558–22563
4. Bardelli A, Longati P, Gramaglia D, Basilico C, Tamagnone L, Giordano S, Ballinari D, Michieli P, Comoglio PM (1998) Uncoupling signal transducers from oncogenic MET mutants abrogates cell transformation and inhibits invasive growth. Proc Natl Acad Sci USA 95:14379–14383
5. Bardelli A, Longati P, Gramaglia D, Stella MC, Comoglio PM (1997) Gab1 coupling to the HGF/Met receptor multifunctional docking site requires binding of Grb2 and correlates with the transforming potential. Oncogene 15:3103–3111
6. Birchmeier C, Birchmeier W, Gherardi E, Vande Woude GF (2003) Met, metastasis, motility and more. Nat Rev Mol Cell Biol 4:915–925
7. Bladt F, Riethmacher D, Isenmann S, Aguzzi A, Birchmeier C (1995) Essential role for the c-met receptor in the migration of myogenic precursor cells into the limb bud. Nature 376:768–771
8. Boccaccio C, Ando M, Tamagnone L, Bardelli A, Michieli P, Battistini C, Comoglio PM (1998) Induction of epithelial tubules by growth factor HGF depends on the STAT pathway. Nature 391:285–288
9. Brodin L, Low P, Shupliakov O (2000) Sequential steps in clathrin-mediated synaptic vesicle endocytosis. Curr Opin Neurobiol 10:312–320
10. Carter S, Urbe S, Clague MJ (2004) The met receptor degradation pathway: requirement for Lys48-linked polyubiquitin independent of proteasome activity. J Biol Chem 279:52835–52839
11. Cepero V, Sierra JR, Corso S, Ghiso E, Casorzo L, Perera T, Comoglio PM, Giordano S (2010) MET and KRAS gene amplification mediates acquired resistance to MET tyrosine kinase inhibitors. Cancer Res 70:7580–7590
12. Chmielowiec J, Borowiak M, Morkel M, Stradal T, Munz B, Werner S, Wehland J, Birchmeier C, Birchmeier W (2007) c-Met is essential for wound healing in the skin. J Cell Biol 177:151–162
13. Cho KW, Park JH, Park CW, Lee D, Lee E, Kim DJ, Kim KJ, Yoon SH, Park Y, Kim E, et al. (2012) Identification of a pivotal endocytosis motif in c-Met and selective modulation of HGF-dependent aggressiveness of cancer using the 16-mer endocytic peptide. Oncogene. 2012 Apr 23. [Epub ahead of print]

14. Comoglio PM, Boccaccio C (1996) The HGF receptor family: unconventional signal transducers for invasive cell growth. Genes Cells 1:347–354
15. Comoglio PM, Giordano S, Trusolino L (2008) Drug development of MET inhibitors: targeting oncogene addiction and expedience. Nat Rev Drug Discov 7:504–516
16. Cooper CS, Park M, Blair DG, Tainsky MA, Huebner K, Croce CM, Vande Woude GF (1984) Molecular cloning of a new transforming gene from a chemically transformed human cell line. Nature 311:29–33
17. Derksen PW, de Gorter DJ, Meijer HP, Bende RJ, van Dijk M, Lokhorst HM, Bloem AC, Spaargaren M, Pals ST (2003) The hepatocyte growth factor/Met pathway controls proliferation and apoptosis in multiple myeloma. Leukemia 17:764–774
18. Fan S, Ma YX, Gao M, Yuan RQ, Meng Q, Goldberg ID, Rosen EM (2001) The multisubstrate adapter Gab1 regulates hepatocyte growth factor (scatter factor)-c-Met signaling for cell survival and DNA repair. Mol Cell Biol 21:4968–4984
19. Furge KA, Zhang YW, Vande Woude GF (2000) Met receptor tyrosine kinase: enhanced signaling through adapter proteins. Oncogene 19:5582–5589
20. Gandino L, Longati P, Medico E, Prata M, Comoglio PM (1994) Phosphorylation of serine 985 negatively regulates the hepatocyte growth factor receptor kinase. J Biol Chem 269: 1815–1820
21. Garcia-Guzman M, Dolfi F, Zeh K, Vuori K (1999) Met-induced JNK activation is mediated by the adapter protein Crk and correlates with the Gab1 - Crk signaling complex formation. Oncogene 18:7775–7786
22. Gentile A, Trusolino L, Comoglio PM (2008) The Met tyrosine kinase receptor in development and cancer. Cancer Metastasis Rev 27:85–94
23. Gherardi E, Birchmeier W, Birchmeier C, Vande Woude G (2012) Targeting MET in cancer: rationale and progress. Nat Rev Cancer 12:89–103
24. Graveel C, Su Y, Koeman J, Wang LM, Tessarollo L, Fiscella M, Birchmeier C, Swiatek P, Bronson R, Vande Woude G (2004) Activating Met mutations produce unique tumor profiles in mice with selective duplication of the mutant allele. Proc Natl Acad Sci USA 101:17198–17203
25. Graveel CR, DeGroot JD, Su Y, Koeman J, Dykema K, Leung S, Snider J, Davies SR, Swiatek PJ, Cottingham S et al (2009) Met induces diverse mammary carcinomas in mice and is associated with human basal breast cancer. Proc Natl Acad Sci USA 106:12909–12914
26. Graziani A, Gramaglia D, dalla Zonca P, Comoglio PM (1993) Hepatocyte growth factor/scatter factor stimulates the Ras-guanine nucleotide exchanger. J Biol Chem 268:9165–9168
27. Hammond DE, Urbe S, Vande Woude GF, Clague MJ (2001) Down-regulation of MET, the receptor for hepatocyte growth factor. Oncogene 20:2761–2770
28. Hayes S, Chawla A, Corvera S (2002) TGF beta receptor internalization into EEA1-enriched early endosomes: role in signaling to Smad2. J Cell Biol 158:1239–1249
29. Ishibe S, Joly D, Liu ZX, Cantley LG (2004) Paxillin serves as an ERK-regulated scaffold for coordinating FAK and Rac activation in epithelial morphogenesis. Mol Cell 16:257–267
30. Jeffers M, Schmidt L, Nakaigawa N, Webb CP, Weirich G, Kishida T, Zbar B, Vande Woude GF (1997) Activating mutations for the met tyrosine kinase receptor in human cancer. Proc Natl Acad Sci USA 94:11445–11450
31. Jeffers M, Taylor GA, Weidner KM, Omura S, Vande Woude GF (1997) Degradation of the Met tyrosine kinase receptor by the ubiquitin-proteasome pathway. Mol Cell Biol 17: 799–808
32. Jekely G, Sung HH, Luque CM, Rorth P (2005) Regulators of endocytosis maintain localized receptor tyrosine kinase signaling in guided migration. Dev Cell 9:197–207
33. Joffre C, Barrow R, Menard L, Calleja V, Hart IR, Kermorgant S (2011) A direct role for Met endocytosis in tumorigenesis. Nat Cell Biol 13:827–837
34. Kermorgant S, Parker PJ (2008) Receptor trafficking controls weak signal delivery: a strategy used by c-Met for STAT3 nuclear accumulation. J Cell Biol 182:855–863
35. Kermorgant S, Zicha D, Parker PJ (2003) Protein kinase C controls microtubule-based traffic but not proteasomal degradation of c-Met. J Biol Chem 278:28921–28929

36. Kermorgant S, Zicha D, Parker PJ (2004) PKC controls HGF-dependent c-Met traffic, signalling and cell migration. EMBO J 23:3721–3734
37. Lai AZ, Durrant M, Zuo D, Ratcliffe CD, Park M (2012) Met kinase-dependent loss of the E3 ligase Cbl in gastric cancer. J Biol Chem 287:8048–8059
38. Lanzetti L, Di Fiore PP (2008) Endocytosis and cancer: an 'insider' network with dangerous liaisons. Traffic 9(12):2011–2021
39. Le Roy C, Wrana JL (2005) Signaling and endocytosis: a team effort for cell migration. Dev Cell 9:167–168
40. Li N, Lorinczi M, Ireton K, Elferink LA (2007) Specific Grb2-mediated interactions regulate clathrin-dependent endocytosis of the cMet-tyrosine kinase. J Biol Chem 282:16764–16775
41. Liu L, McBride KM, Reich NC (2005) STAT3 nuclear import is independent of tyrosine phosphorylation and mediated by importin-alpha3. Proc Natl Acad Sci USA 102:8150–8155
42. Luttrell LM, Roudabush FL, Choy EW, Miller WE, Field ME, Pierce KL, Lefkowitz RJ (2001) Activation and targeting of extracellular signal-regulated kinases by beta-arrestin scaffolds. Proc Natl Acad Sci USA 98:2449–2454
43. McShane MP, Zerial M (2008) Survival of the weakest: signaling aided by endosomes. J Cell Biol 182:823–825
44. Mosesson Y, Mills GB, Yarden Y (2008) Derailed endocytosis: an emerging feature of cancer. Nat Rev Cancer 8:835–850
45. Murphy JE, Padilla BE, Hasdemir B, Cottrell GS, Bunnett NW (2009) Endosomes: a legitimate platform for the signaling train. Proc Natl Acad Sci USA 106:17615–17622
46. Nada S, Hondo A, Kasai A, Koike M, Saito K, Uchiyama Y, Okada M (2009) The novel lipid raft adaptor p18 controls endosome dynamics by anchoring the MEK-ERK pathway to late endosomes. EMBO J 28:477–489
47. Nguyen L, Holgado-Madruga M, Maroun C, Fixman ED, Kamikura D, Fournier T, Charest A, Tremblay ML, Wong AJ, Park M (1997) Association of the multisubstrate docking protein Gab1 with the hepatocyte growth factor receptor requires a functional Grb2 binding site involving tyrosine 1356. J Biol Chem 272:20811–20819
48. Palamidessi A, Frittoli E, Garre M, Faretta M, Mione M, Testa I, Diaspro A, Lanzetti L, Scita G, Di Fiore PP (2008) Endocytic trafficking of Rac is required for the spatial restriction of signaling in cell migration. Cell 134:135–147
49. Palamidessi A, Testa I, Frittoli E, Barozzi S, Garre M, Mazza D, Di Fiore PP, Diaspro A, Scita G, Faretta M (2010) Understanding biological dynamics: following cells and molecules to track functions and mechanisms. Eur Biophys J 39:947–957
50. Parachoniak CA, Luo Y, Abella JV, Keen JH, Park M (2011) GGA3 functions as a switch to promote Met receptor recycling, essential for sustained ERK and cell migration. Dev Cell 20:751–763
51. Peschard P, Fournier TM, Lamorte L, Naujokas MA, Band H, Langdon WY, Park M (2001) Mutation of the c-Cbl TKB domain binding site on the Met receptor tyrosine kinase converts it into a transforming protein. Mol Cell 8:995–1004
52. Petrelli A, Gilestro GF, Lanzardo S, Comoglio PM, Migone N, Giordano S (2002) The endophilin-CIN85-Cbl complex mediates ligand-dependent downregulation of c-Met. Nature 416:187–190
53. Ponzetto C, Bardelli A, Zhen Z, Maina F, dalla Zonca P, Giordano S, Graziani A, Panayotou G, Comoglio PM (1994) A multifunctional docking site mediates signaling and transformation by the hepatocyte growth factor/scatter factor receptor family. Cell 77:261–271
54. Potempa S, Ridley AJ (1998) Activation of both MAP kinase and phosphatidylinositide 3-kinase by Ras is required for hepatocyte growth factor/scatter factor-induced adherens junction disassembly. Mol Biol Cell 9:2185–2200
55. Pranada AL, Metz S, Herrmann A, Heinrich PC, Muller-Newen G (2004) Real time analysis of STAT3 nucleocytoplasmic shuttling. J Biol Chem 279:15114–15123
56. Raiborg C, Stenmark H (2002) Hrs and endocytic sorting of ubiquitinated membrane proteins. Cell Struct Funct 27:403–408

57. Rao DS, Bradley SV, Kumar PD, Hyun TS, Saint-Dic D, Oravecz-Wilson K, Kleer CG, Ross TS (2003) Altered receptor trafficking in Huntingtin Interacting Protein 1-transformed cells. Cancer Cell 3:471–482
58. Rosario M, Birchmeier W (2003) How to make tubes: signaling by the Met receptor tyrosine kinase. Trends Cell Biol 13:328–335
59. Schaeper U, Gehring NH, Fuchs KP, Sachs M, Kempkes B, Birchmeier W (2000) Coupling of Gab1 to c-Met, Grb2, and Shp2 mediates biological responses. J Cell Biol 149:1419–1432
60. Schenck A, Goto-Silva L, Collinet C, Rhinn M, Giner A, Habermann B, Brand M, Zerial M (2008) The endosomal protein App 11 mediates Akt substrate specificity and cell survival in vertebrate development. Cell 133:486–497
61. Schmidt C, Bladt F, Goedecke S, Brinkmann V, Zschiesche W, Sharpe M, Gherardi E, Birchmeier C (1995) Scatter factor/hepatocyte growth factor is essential for liver development. Nature 373:699–702
62. Schmidt L, Duh FM, Chen F, Kishida T, Glenn G, Choyke P, Scherer SW, Zhuang Z, Lubensky I, Dean M et al (1997) Germline and somatic mutations in the tyrosine kinase domain of the MET proto-oncogene in papillary renal carcinomas. Nat Genet 16:68–73
63. Schmidt L, Junker K, Nakaigawa N, Kinjerski T, Weirich G, Miller M, Lubensky I, Neumann HP, Brauch H, Decker J et al (1999) Novel mutations of the MET proto-oncogene in papillary renal carcinomas. Oncogene 18:2343–2350
64. Scita G, Di Fiore PP (2010) The endocytic matrix. Nature 463:464–473
65. Shattuck DL, Miller JK, Laederich M, Funes M, Petersen H, Carraway KL 3rd, Sweeney C (2007) LRIG1 is a novel negative regulator of the Met receptor and opposes Met and Her2 synergy. Mol Cell Biol 27:1934–1946
66. Sorkin A, von Zastrow M (2009) Endocytosis and signalling: intertwining molecular networks. Nat Rev Mol Cell Biol 10:609–622
67. Teis D, Taub N, Kurzbauer R, Hilber D, de Araujo ME, Erlacher M, Offterdinger M, Villunger A, Geley S, Bohn G et al (2006) p14-MP1-MEK1 signaling regulates endosomal traffic and cellular proliferation during tissue homeostasis. J Cell Biol 175:861–868
68. Trusolino L, Bertotti A, Comoglio PM (2010) MET signalling: principles and functions in development, organ regeneration and cancer. Nat Rev Mol Cell Biol 11:834–848
69. Tsukazaki T, Chiang TA, Davison AF, Attisano L, Wrana JL (1998) SARA, a FYVE domain protein that recruits Smad2 to the TGFbeta receptor. Cell 95:779–791
70. Uehara Y, Minowa O, Mori C, Shiota K, Kuno J, Noda T, Kitamura N (1995) Placental defect and embryonic lethality in mice lacking hepatocyte growth factor/scatter factor. Nature 373:702–705
71. Veiga E, Cossart P (2005) Listeria hijacks the clathrin-dependent endocytic machinery to invade mammalian cells. Nat Cell Biol 7:894–900
72. Wang Y, Pennock S, Chen X, Wang Z (2002) Endosomal signaling of epidermal growth factor receptor stimulates signal transduction pathways leading to cell survival. Mol Cell Biol 22:7279–7290
73. Watson FL, Heerssen HM, Bhattacharyya A, Klesse L, Lin MZ, Segal RA (2001) Neurotrophins use the Erk5 pathway to mediate a retrograde survival response. Nat Neurosci 4:981–988
74. Zhang YW, Wang LM, Jove R, Vande Woude GF (2002) Requirement of Stat3 signaling for HGF/SF-Met mediated tumorigenesis. Oncogene 21:217–226

Chapter 14
Regulation of Epidermal Growth Factor Receptor Signaling by Endocytosis in Normal and Malignant Cells

Sergio Anastasi, Stefano Alemà, and Oreste Segatto

Abstract The epidermal growth factor receptor (EGFR) is a receptor tyrosine kinase whose physiological signaling activity regulates the morphogenesis and homeostasis of several tissues from worms to man. In contrast, aberrant signaling caused by overexpression or mutational activation of the EGFR plays a causal role in the pathogenesis of a number of human tumors. The fidelity of EGFR signals, which must be robust enough to convey instructive cues to the cell while also preventing the threat posed by excess receptor activity, is guaranteed by a complex regulatory circuitry. A pervasive role in EGFR regulation is played by endocytosis. Owing to its capacity to instruct degradation of activated EGFRs and reduce receptor expression at the cell surface, endocytosis has been regarded historically as the main cellular mechanism deputed to the attenuation of EGFR signaling. More recently, a great deal of attention has been focused on understanding endocytosis also as an element of spatial regulation of EGFR activity. Herein, we discuss molecular mechanisms controlling EGFR endocytosis, as they relate to the regulation of EGFR signal output and the implementation of EGFR-driven biological programs. We will then focus on reviewing the variegated mechanisms through which the EGFR escapes from downregulation in cancer cells. The emerging picture assigns to faulty endocytosis, in concert with constitutive catalytic activation, a prominent role in the ominous oncogenic conversion of EGFR.

S. Anastasi
Department of Experimental Oncology, Regina Elena Cancer Institute,
via delle Messi d'Oro, 156, 00158 Rome, Italy

Department of Physiology and Pharmacology, University of Rome La Sapienza,
00185 Rome, Italy

S. Alemà
Institute of Cell Biology and Neurobiology, CNR, Monterotondo, Italy

O. Segatto, MD (✉)
Department of Experimental Oncology, Regina Elena Cancer Institute,
via delle Messi d'Oro, 156, 00158 Rome, Italy
e-mail: segatto@ifo.it

Y. Yarden and G. Tarcic (eds.), *Vesicle Trafficking in Cancer*,
DOI 10.1007/978-1-4614-6528-7_14, © Springer Science+Business Media New York 2013

14.1 Introduction

The epidermal growth factor (EGF) receptor (EGFR), also known as ErbB1 or HER, is the founding member of the ErbB family of RTKs, which in vertebrates includes also ErbB2, ErbB3, and ErbB4 [1]. Signaling by the EGFR is involved in the execution of a number of cellular programs, ranging from cell survival and proliferation to cell locomotion [1]. The biological relevance of EGFR function is underscored by its evolutionary conservation and the phenotypes described in invertebrate model organisms, such as *C. elegans* and *D. melanogaster*, that carry mutations of either Egfr itself or components of the Egfr signaling axis [2, 3]. In the mouse, genetic ablation of *Egfr* has severe developmental consequences, whereas genetic manipulations that lead to excess receptor function disrupt tissue homeostasis by causing aberrant cell proliferation and igniting cell transformation [4]. Not unexpectedly, gain-of-function genetic lesions of *EGFR* are causally involved in the pathogenesis of several human tumors, including glioblastoma (GBM) and carcinomas of lung, colon, head, and neck [4, 5]. This notion has fueled intense drug discovery programs aimed at developing therapeutics capable of curbing EGFR expression and/or function in tumor cells reliant on EGFR oncogenic signaling, with several EGFR-targeted therapeutics being already licensed for clinical use [6, 7].

Owing to its relevance in cell physiology, tissue morphogenesis, and cancer pathogenesis and also as a result of being the first growth factor receptor ever to be characterized as RTK [8] and molecularly cloned [9, 10], the EGFR has become one of the most popular receptor models for dissecting the molecular composition, organizing principles and regulatory logic that underlie the architecture of RTK signaling circuits in normal and neoplastic cells. One of the topics that over the years has drawn the greatest attention is the role played by endocytosis in regulating the quality and quantity of EGFR signals in space and time [11–14]. The scope of this chapter is to provide an overview of how endocytic traffic impacts EGFR function and discuss how tumor cells may harness EGFR endocytosis to gain and/or maintain their oncogenic phenotype.

14.1.1 Essential Concepts in EGFR Regulation at the Cell Surface

The EGFR consists of an extracellular domain involved in ligand recognition and receptor dimerization, a single transmembrane region that anchors the receptor to the plasma membrane and an intracellular portion that contains the catalytic domain along with juxtamembrane and C-terminal regulatory regions. In its resting state, the receptor adopts an auto-inhibited conformation, which is held by intramolecular interactions precluding spurious catalytic activation [15, 16]. Ligand binding induces extensive structural rearrangements that lead to activation of the EGFR catalytic domain (Fig. 14.1). Key to this process is the stabilization of non-covalent dimers generated primarily by intermolecular contacts between the extracellular domains of ligand-bound receptors [15]. This is in turn conducive to allosteric activation of the EGFR kinase, a process driven by the formation of asymmetric dimers between juxtaposed catalytic domains [16] (Fig. 14.1). Ligand-activated EGFR dimers undergo in trans phosphorylation on specific tyrosine residues located in the receptor C-tail. Phosphorylated tyrosine residues create recognition codes for adaptors and enzymes containing phosphotyrosine binding motifs. Docking of these signaling proteins onto activated EGFR ignites downstream signal transduction [17].

Since the very inception of their signaling activity, ligand-bound EGFRs are targeted by inhibitory mechanisms that include the tonic activity of phosphotyrosine phosphatases (PTPs) [18] and negative feedback regulation of kinase activity imposed by phosphorylation of Thr 654 (mediated by protein kinase C [PKC]) and Thr 669 (mediated by extracellular regulated kinase [ERK]) [19]. EGF binding also induces rapid receptor endocytosis (Fig. 14.1). Endocytosed EGFR may follow different routes, which in turn are associated to different signaling outcomes. On one extreme, ligand-receptor complexes may be recycled to the cell surface, a condition permissive for reiteration of EGFR signaling; on the other, ligand-receptor complexes may be instructed to traffic to late endosomes, eventually undergoing degradation in lysosomes [12, 13]. Reduced expression of EGFR at the cell surface caused by endocytosis-dependent receptor degradation attenuates cellular responsiveness to incoming EGFR ligands, thus restraining EGFR signaling over time. This rather intuitive notion was initially derived from studies on cultured cells and was later solidified by developmental studies in worms [20] and flies [21], which showed that mutations inhibiting EGFR downregulation cause aberrant tissue morphogenesis as a consequence of excess EGFR activity.

14.2 Routes to EGFR Endocytosis and Fates of Internalized Receptors

In its inactive state, the EGFR undergoes endocytosis at very slow rates. In contrast, ligand binding and attendant triggering of EGFR tyrosine kinase activity are associated with rapid receptor internalization [22, 23]. EGFR catalytic competence is indeed a sine qua non for receptor endocytosis, as genetic [24, 25] or pharmacologic [26] manipulations that cause inhibition of EGFR kinase activity are associated with defective EGFR internalization.

The EGFR may use different routes for entering the cell, the major distinction being between clathrin-mediated and clathrin-independent endocytosis (Fig. 14.1). In clathrin-mediated endocytosis (CME), cargo is sorted into clathrin-coated pits (CCPs), specialized microdomains of the plasma membrane generated by polymerized clathrin. The clathrin coat, along with the longitudinal tension created by actin polymerization occurring at the membrane-proximal site of the bud, generates the tensile force necessary to progressively deform the plasma membrane from a

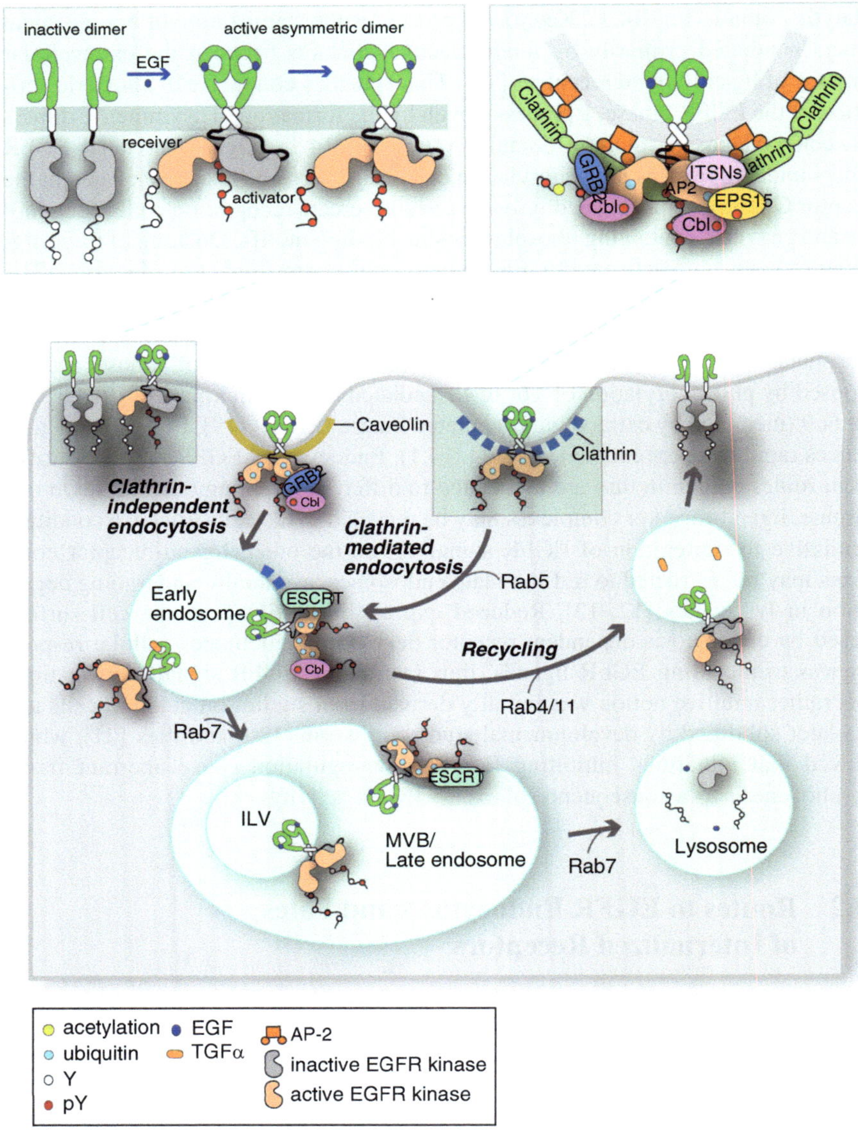

Fig. 14.1 Activation of EGFR and endocytic trafficking. Binding of EGF leads to the allosteric activation of the EGFR kinase (*upper left*) via formation of asymmetric kinase dimers. Accordingly, the large C lobe of one kinase (activator, *gray*) contacts the small N lobe of the adjacent partner kinase (receiver), causing activation of the latter (*orange*) and ensuing in trans phosphorylation of tyrosine residues located in the receptor C-tail (*red circles*). Activated EGFRs undergo internalization via clathrin-mediated (CME) and clathrin-independent endocytosis. CME is initiated at CCPs (*upper right*), into which EGFR is recruited via physical interactions with components of the CME machinery (*upper right*), such as the major clathrin adaptor AP-2 and accessory proteins, e.g., EPS15. CBL proteins participate in CME by acting as scaffold for accessory proteins such as ITSNs or ubiquitylating the EGFR itself and endocytic proteins. Ubiquitylated EGFR residues

shallow pit to a deep invagination. Eventually, the neck of the invagination undergoes constriction and fission, thanks to the activity of spiral-shaped dynamin oligomers. Clathrin-coated vesicles released in the cytoplasm undergo uncoating and subsequent fusion with the early endosome [27].

EGFR is sorted into nascent CCPs, thanks to its ability to interact with components of the molecular machinery engaged in CCP formation. These include AP-2, i.e., the major cargo adaptor complex involved in recruitment and polymerization of clathrin at the plasma membrane and accessory proteins (adaptors/scaffolds capable of interacting with AP-2 and/or participating in the mechanics of CCP formation), such as EPS15, epsin, and intersectins (ITSNs), among others [27, 28] (Fig. 14.1). Reviewing the large body of literature dealing with the molecular mechanisms responsible for sorting EGFR into CCPs is beyond the scope of this chapter (see refs. [12, 13]). Here, we will focus on a recent study that attempted to provide a unifying view of EGFR coupling to CME through an integrated mutational analysis of receptor determinants involved in molecular recognition of the CME machinery. Ubiquitylation of lysine residues located in the EGFR kinase domain (A), AP-2 interaction motifs located in the receptor C-tail (B), and Lys residues at positions 1155, 1158, and 1164, i.e., in the receptor C-tail, (C) were found to act redundantly to promote CME of EGFR [29]. Hence, simultaneous mutational inactivation of at least two of the above determinants (e.g., A + B) was required to impair CME of the EGFR to a significant extent, with simultaneous targeting of all three (A + B + C) being most effective. RNAi to GRB2 eliminated the residual CME of the triple mutant [29]. These data are fully consistent with EGFR activation being necessary for rapid receptor CME since (a) recruitment and activation of CBL proteins, i.e., the E3 ligases responsible for EGFR ubiquitylation, require EGFR kinase activity [30–32]; (b) conformational changes secondary to receptor activation are thought to unmask AP-2 recognition motifs in the receptor C-tail [29]; and (c) GRB-2 binds via its SH2 domain to pY1068 and pY1086 of EGFR [33]. Mechanistically, the interaction with AP-2 provides a straightforward explanation for EGFR recruitment into CCPs. Ubiquitylated Lys residues serve as recognition motifs for ubiquitin (Ub)-binding domains present in accessory proteins, such as EPS15, epsin, and EPS15R, which are redundantly involved in coupling activated EGFR to CCP assembly [34] via a web of molecular interactions [35, 36] (Fig. 14.1). As for GRB2, its role in mediating CME of EGFR had been ascribed to its ability to recruit CBL proteins onto the EGFR via SH3-mediated interactions

Fig. 14.1 (continued) (*blue circles*) are recognized by Ub-binding motifs present in endocytic proteins. A cluster of Lys residues between positions 1155 and 1164 of EGFR (*yellow circle*) is also required for CME of EGFR. A major route of clathrin-independent endocytosis is initiated in caveolae. This pathway has an absolute requirement for CBL-dependent EGFR ubiquitylation. Endocytic vesicles released into the cytosol undergo fusion with early endosomes. Receptors tagged by robust ubiquitylation are sorted into MVBs/late endosomes via an ESCRT-dependent process. EGFRs lacking robust ubiquitylation are recycled to the cell surface. For example, TGFα (*dark yellow ellipses*) dissociates from EGFR in early endosomes, which leads to receptor deubiquitylation and sorting to the recycling route. Different Rab proteins orchestrate the process of vesicle maturation/fusion

[30, 37, 38] (Fig. 14.1). However, the above data by Goh et al. suggest that GRB2 mediates CME of EGFR also via mechanism/s other than CBL-mediated ubiquitylation of the EGFR. CBL is thought to ubiquitylate and regulate key endocytic protein/s other than EGFR itself [39, 40] and also promote EGFR endocytosis via molecular interactions nucleated by its pro-rich C-terminal tail [41]. The mechanistic role of the K1155-1164 cluster in mediating CME of the EGFR is unclear. These lysine residues were found to be acetylated [29], but it is still unknown whether acetylation/deacetylation of the K1155-1164 cluster is in fact involved in regulating EGFR endocytosis.

While not strictly necessary for CME, CBL-driven EGFR ubiquitylation has been proposed to be indispensable for clathrin-independent endocytosis via caveolae [34] (Fig. 14.1). Caveolae are flask-shaped invaginations of the plasma membrane coated by oligomerized caveolin. Fission of caveolae releases vesicles (caveosomes) which eventually fuse with early endosomes [42, 43]. Ubiquitylated EGFRs are sorted into this endocytic pathway via physical interactions with the Ub-binding domain of EPS15, EPS15R, and epsin [34]. A third route of EGFR endocytosis exploits a clathrin- and caveolin-independent mechanism. EGFR activation may lead to the formation of circular dorsal ruffles (CDRs), which are ringlike actin-based protrusions of the plasma membrane [44]. CDRs undergo constriction and fusion, leading to the endocytic uptake of large areas of the cell membrane in form of elongated tubular structures, from which endocytic vesicles are eventually released. CDRs are transient, yet, within their 20–30 min lifetime, they are capable of promoting the endocytosis of a large fraction of ligand-engaged EGFRs.

What dictates the choice of the endocytic route through which the EGFR is cleared from the cell surface? While this decision could depend to some extent on the cell context, as suggested in the case of CDR-mediated endocytosis [45], it appears that CME and non-CME routes can be used differentially in the same cell depending on the sorting signals displayed by activated EGFR. An elegant model has been recently proposed whereby robust EGFR ubiquitylation occurs selectively under conditions of high receptor occupancy. Caveolae-dependent endocytosis is therefore observed only under these conditions, which, it should be noted, are also permissive for CME. Instead, at low receptor occupancy, EGFR lacks detectable ubiquitylation and undergoes endocytosis exclusively via the clathrin pathway (which, as discussed above, does not have an absolute requirement for EGFR ubiquitylation) [34, 46]. This model has important implications on EGFR post-endocytic sorting and signaling, as discussed below.

14.3 Post-endocytic Traffic of EGFR

Endocytic vesicles released from the plasma membrane fuse with early endosomes (EE), a process requiring the Rab5 GTPase. Importantly, EGFR activity drives GTP loading of Rab5, thus instigating vesicle fusion [47]. Early endosomes operate as key endocytic stations: the EGFR may be recycled to the cell surface from their tubular extensions or, in alternative, it may be retained onto the limiting surface of their vacuolar area and be eventually sorted into their lumen as EEs mature to become multivesicular bodies (MVBs)/late endosomes [12, 13] (Fig. 14.1).

A large and concordant body of literature supports the notion that EGFR ubiquitylation is the signal necessary to sort EGF-EGFR complexes into MVBs. Monomeric Ub and K63-linked polymeric Ub chains are appended to multiple Lys residues located in the receptor kinase domain [39, 40] and function as a recognition code for Ub-binding domains located in components of the evolutionarily conserved ESCRT (endosomal sorting complex required for transport) complex [48]. The CBL-associated E3 ligase activity must target the EGFR throughout its endocytic traffic to maintain the robust ubiquitylation necessary for the EGFR to be sequestered by the ESCRT machinery [49] (Fig. 14.1). High-level ubiquitylation likely ensures that the low-affinity binding reactions between Ub and Ub-binding domains present in ESCRT components are stabilized via the high avidity generated by multivalent interactions [48]. Sorting into MVBs is initiated by the physical interaction between the EGFR and the HRS/STAM complex (ESCRT-0) in clathrin-coated microdomains of early endosomes [50, 51]. ESCRT-I and ESCRT-II, whose components also contain Ub-binding domains, are then recruited sequentially onto ubiquitylated EGFR to promote invagination of the limiting membrane of MVBs. ESCRT-III eventually drives abscission of invaginated membrane domains, thus releasing EGFR into intraluminal vesicles (ILVs) [48] (Fig. 14.1). Of note, EGFR signaling drives formation of MVBs [52], again emphasizing the instructive role played by EGFR activity in the course of receptor endocytosis. Sorting into ILVs segregates ligand-receptor complexes from the cytoplasm, thus terminating the propagation of EGFR signals. Late endosomes eventually fuse with lysosomes, a process that seems to require Rab7 function [53] (Fig. 14.1). It is in the endolysosomes that both EGF and EGFR undergo proteolysis.

While a great deal of attention has been devoted to dissecting the mechanisms involved in EGFR sorting to MVBs, our understanding of how EGFR is recycled back to the cell surface is much less detailed. A K15R EGFR mutant (which is catalytically competent but unable to undergo ubiquitylation) interacts poorly with ESCRT-0, is not sorted into ILVs, and accumulates onto tubular extension of early endosomes [54]. Rather than accelerating receptor recycling, the K15R mutation enlarges the pool of EGFR molecules destined to be recycled [54]. Thus, recycling would appear to be the default fate of internalized EGFR, and escape from this route is ensured only by robust receptor ubiquitylation (Fig. 14.1). EPS15s, an isoform of EPS15 which lacks the ubiquitin interaction motifs, has been recently found to play a role in EGFR recycling [55]. Instead, EPS15b, an isoform lacking the three EH domains which couple EPS15 to the CME machinery, interacts with HRS, an ESCRT-0 component, and is implicated in sorting of EGFR to MVBs [56]. These data raise the intriguing possibility that molecular recognition of different EPS15 isoforms could instruct EGFR post-endocytic fate.

14.4 Ligand Identity Determines EGFR Endocytic Itinerary

The models discussed so far are mainly derived from studies employing EGF as EGFR agonist. However, there are six ligands known to be able of binding and activating the EGFR, namely, amphiregulin (AR), epigen (EPG), epiregulin (EPR), heparin-binding epidermal growth factor like (HB-EGF), transforming growth factor α (TGFα), and betacellulin (BTC), besides EGF itself [1]. Early studies led to the discovery that TGFα, at variance with EGF, dissociates from EGFR in the mildly acidic pH of early endosomes [57]. Ligand dissociation interrupts EGFR catalytic activity and is therefore followed by swift receptor dephosphorylation, disassembly of CBL-EGFR complexes, and receptor deubiquitylation [49, 58]. Consequently, EGFRs engaged by TGFα are recycled to the cell surface (Fig. 14.1).

Departing from the above dissimilarities between EGF and TGFα, a few studies have addressed the endocytic fate imposed to the EGFR by each of its ligands [59–62]. In a recent study employing the HepG2 cell line, TGFα, EPI, and AR were found to cause little, if any, EGFR degradation, consistent with their ability to drive receptor recycling. HB-EGF and BTC, instead, were found to drive persistent EGFR downregulation followed by receptor degradation; this correlated with protracted EGFR ubiquitylation and sorting into late endosomes. EGF-bound receptors had an intermediate behavior, in that both recycling to the plasma membrane and sorting to late endosomes could be observed [63]. While essentially confirmatory of previous publications [1], this study is nevertheless relevant since it was the first to carry out a comparative analysis using identical experimental methodologies in the same cellular background for all but one EGFR ligands.

Ligands eliciting prevalent EGFR recycling are imbued with higher mitogenic activity when compared to ligands capable of driving receptor downregulation [59, 60, 62]. Importantly, abrogation of CBL-dependent EGFR ubiquitylation increases the mitogenic potency of EGF, while being inconsequential in AR-stimulated cells [62]. Hence, it appears that distinct patterns of endocytic traffic are causally linked to differences in the biological potency of EGFR ligands.

14.5 Role of Inducible Feedback Inhibitors in EGFR Endocytosis

Cell surface receptors are engaged in the orchestration of cellular programs whose execution may span several hours. For instance, growth factor stimulation is required throughout the G1 phase of the cell cycle. In EGF-stimulated cells, a dense network of transcriptionally induced regulatory modules controls the execution of G1 traverse by targeting EGFR-activated signaling pathways at the transcriptional, translational, and posttranslational level [64]. Not surprisingly, the EGFR itself is also subject to transcriptionally induced feedback regulation during G1: positive loops reinforce EGFR signaling via autocrine production of EGFR ligands, whereas

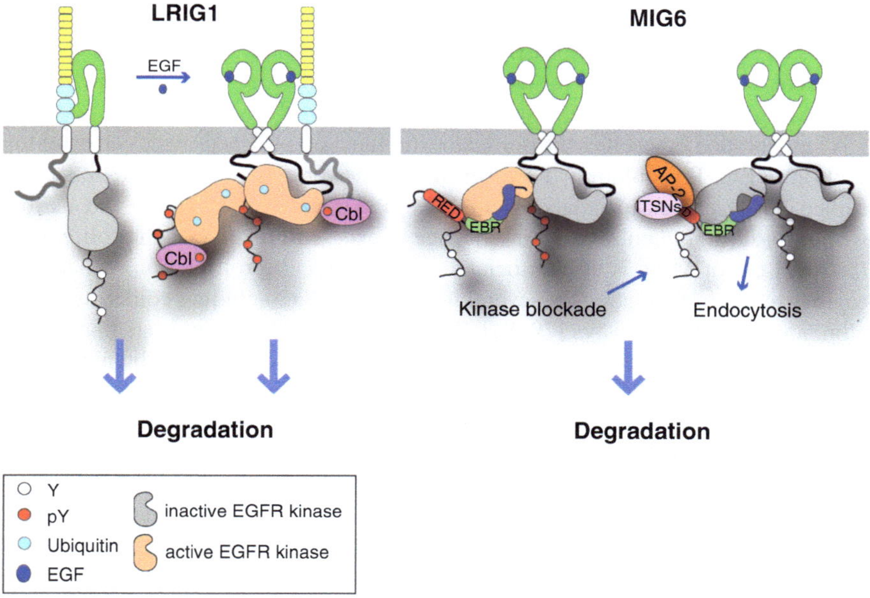

Fig. 14.2 Mechanisms of EGFR inhibition by LRIG1 and MIG6. *Left*: LRIG1 binds to ligand-free EGFRs via its extracellular portion comprising leucine-rich (*yellow*) and Ig-like (*light blue*) domains. Inactive EGFR molecules bound to LRIG1 are endocytosed and degraded via a poorly defined mechanism. A second mechanism involves activated EGFR inducing tyrosine phosphorylation of LRIG1-bound CBL, which is then capable of directing EGFR ubiquitylation. Under physiological conditions, it is probable that CBL is recruited to the EGFR both directly, as well as in an LRIG1-mediated fashion, to allow for maximal ubiquitylation to take place. Right: MIG6 binds to the kinase domain of ligand-activated EGFR through its EBR domain. Segment 2 of EBR (*purple*) binds to the kinase active site and interrupts catalytic activity; segment 1 (*green*) prevents the MIG6-bound receiver kinase from acting as activator upon reversal of the asymmetric dimer orientation [72]. Upon docking of the EBR onto the EGFR, the endocytic domain RED (*red*) interacts with endocytic proteins: AP-2 and intersectins (ITSNs) mediate MIG6-dependent endocytosis, whereas SNX8 directs early to late endosome sorting of MIG6-bound EGFRs (not shown)

negative feedback loops preside to EGFR downregulation [65, 66]. Four inducible feedback inhibitors (IFIs) of the EGFR have been identified so far, namely, LRIG1 (leucine-rich and immunoglobulin-like domains 1), MIG6 (mitogen-induced gene 6, also known as RALT or ERRFI1), SOCS4, and SOCS5 (members of the suppressor of cytokine signaling family). They are expressed in mid to late G1 in response to EGF stimulation and instruct endocytosis-dependent EGFR degradation [66]. LRIG1 [67, 68], SOCS4, and SOCS5 [69, 70] enhance EGFR ubiquitylation through mechanisms which may target inactive (i.e., ligand-free) and ligand-engaged EGFR molecules in both CBL-dependent and CBL-independent manner (Fig. 14.2). MIG6 inhibits allosteric activation of the EGFR by binding to the kinase domain of ligand-activated receptors [71, 72] (Fig. 14.2). Moreover, MIG6 is capable of driving also endocytosis and degradation of the kinase-inactive receptors to which it binds, owing to its capacity of recruiting components of the endocytic machinery such as AP-2, ITSNs, and sorting nexin 8 [73, 74] (Fig. 14.2). Interestingly, MIG6-dependent sorting of EGFR to late endosomes does not appear to require EGFR ubiquitylation [73].

A key question is whether the pro-endocytic function of IFIs provides an essential function in EGFR regulation. The affirmative answer comes from genetic studies in mice. Lrig1 expression marks stem cells in skin and gut epithelia, and its loss disrupts tissue architecture as a consequence of excess Egfr-driven cell proliferation [75–78]. Loss of Mig6 is also associated to aberrant cell proliferation, which may progress to tumor formation, in several mouse epithelia, including epidermis, liver, colon, and bronchial airways [79–81]. The available evidence suggests that loss of either Lrig1 or Mig6 maintains high Egfr expression, therefore heightening cells' responsiveness to limiting concentrations of Egfr ligands and propelling tissue hyperplasia [66, 78]. This model is congruent with (a) the bias of IFIs for enforcing degradation as the sole endocytic fate of EGFR; (b) the ability of LRIG1, SOCS4, and SOCS5 to target also ligand-free EGFRs, a scavenging function aimed at preventing EGFR activation; and (c) the dual function of MIG6 as kinase suppressor (acting synchronously with ligand binding to ablate receptor signaling) and inducer of EGFR degradation (acting diachronically to prevent subsequent rounds of receptor activation) [66].

14.6 Impact of Endocytic Traffic on EGFR Signaling

Whether destined to recycling or degradation, EGFRs present on the limiting membrane of endocytic vesicles abut with their cytoplasmic portion into the cytosol and are phosphorylated on tyrosine residues [82, 83], conditions indicative of signaling competence. Hence, an outstanding question concerns the quantitative and qualitative contribution of endocytic signaling (i.e., signals transmitted by EGFRs located on endocytic vesicles) to overall receptor output. For an in-depth discussion of the often conflicting literature related to this topic, the reader is referred to recent reviews [11, 14]. Here, we will focus on a few paradigms that illustrate different patterns of regulation impinging on EGFR signaling during endocytosis.

Removal of active EGFR from the cell surface terminates signaling events confined to the plasma membrane. This is the case of phospholipase C gamma (PLCγ), whose EGF-dependent activation entails a two-step process, namely, translocation to the plasma membrane via an SH2-dependent interaction with EGFR pTyr 992 and attendant Tyr phosphorylation by EGFR itself [84–87]. Activated PLCγ hydrolyzes PtdIns (4,5)P_2 (PIP_2) to generate the second messengers inositol-1,4,5 trisphosphate (IP_3) and diacylglycerol (DAG). PIP_2 can be detected at the plasma membrane but not on endosomes [88]. Thus, EGFR endocytosis leads to fast termination of EGF-induced PIP_2 hydrolysis [89] (Fig. 14.3). An opposite paradigm holds true for EGFR signaling via APPL1 and APPL2 (adaptor protein, phosphotyrosine interaction, PH domain, and leucine zipper containing), which are Rab5 effectors localized on immature early endosomes [90, 91]. EGFR activation and endocytosis causes

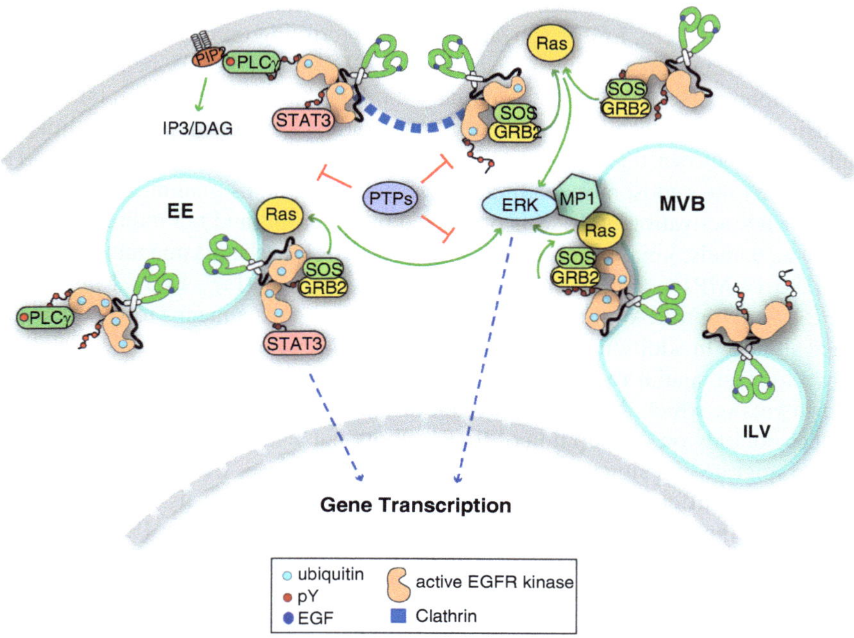

Fig. 14.3 Control of EGFR signaling by endocytosis. EGFR signaling is initiated at the plasma membrane upon receptor auto-phosphorylation and ensuing recruitment of signal transducers. Catalytically active receptors undergo endocytosis and are capable of transmitting signals until they are segregated into ILVs. Sustained EGFR catalytic activation is required to counteract phosphotyrosine phosphatases (PTPs). Recruitment and activation of PLCγ onto EGFR mediates hydrolysis of PIP_2, which is present only at the plasma membrane. Receptor endocytosis segregates EGFR-PLCγ complexes away from PIP_2 and terminates PIP_2 hydrolysis. EGFR recruits onto the plasma membrane of the guanine exchange factor SOS in complex with GRB2, thus activating RAS. In turn, RAS is recruited onto the MAPK scaffold KSR1 (not shown for simplicity) to activate ERK at the plasma membrane. Internalized EGFR still binds GRB2-SOS complexes, thus allowing for RAS activation to take place onto endosomes. On late endosomes, transmission of the RAS signal to the MAPK module is organized by the MP1 scaffold. EGFR activation causes translocation of STAT3 from cytoplasm to cell membranes. Endocytosis of the EGFR-STAT3 complex is required for STAT3 to migrate into the nucleus and regulate gene transcription

APPL proteins to be released from endosomes and migrate to the nucleus where they participate in chromatin remodeling [90]. In addition, increasing the residence time of EGFR in APPL endosomes strengthens ERK activation and AKT signaling in an APPL1-dependent manner in HeLa cells [91]. The PTB domain of APPL1 binds in vitro to EGFR [92] and APPL1 reportedly binds to AKT2 [93], consistent with EGFR and APPL proteins being assembled in a transient complex that regulates EGFR signaling on EEs. However, EGF-dependent AKT activation in mouse embryo fibroblasts lacking both APPL1 and APPL2 was shown to be normal [94], possibly indicating a subtle role of APPL proteins in regulating AKT signaling downstream to EGFR.

Endocytic regulation of EGFR signaling via the RAS-ERK pathway follows a more complex pattern. EGFR activates RAS by recruiting onto cell membranes the guanine exchange factor son of sevenless (SOS) via GRB2-SOS or SHC-GRB2-SOS complexes [95]. While it was initially thought that RAS activation takes place exclusively at the plasma membrane, a number of biochemical and cell imaging studies have shown that, upon EGF stimulation, EGFR, GRB2, SOS, and SHC co-localize onto endosomes, leading to RAS activation and signaling via ERK [96]. In fact, ERK activation by RAS is organized by different (and not interchangeable) scaffolds, namely, kinase suppressor of RAS 1 (KSR1) at the plasma membrane and the p18-p14-MP1 complex on late endosomes [96, 97] (Fig. 14.3). Maintaining EGFR signaling to RAS throughout the endocytic route is a means to generate sustained signals. In addition, however, activation of RAS at distinct endocytic stations allows for fast spatial resolution of RAS-dependent signals. This is illustrated by experiments in which the mislocalization of late endosomes had significant consequences not only on the duration of ERK activity but also on the ability of activated ERK to migrate and signal into the nucleus [98]. Likewise, ERK nuclear signaling was affected differentially depending on whether EGF-driven ERK activation occurred at the plasma membrane or on late endocytic stations [99]. Although K-RAS was shown to undergo CME upon EGF stimulation, there is also evidence that RAS proteins may reside stably on endosomes, with endocytosis being differentially required for activation of specific RAS isoforms [96]. Hence, endocytosis could also be a means to control EGFR coupling to different RAS isoforms.

Yet another scenario in which endocytic traffic imparts spatial control to EGFR signaling is that of receptor endocytosis being required for nuclear translocation of either EGFR itself [100–103] or EGFR-activated STAT3 molecules [104, 105] (Fig. 14.3).

The data discussed above provide cogent evidence that EGFRs residing on endosomes are coupled to downstream effectors. The actual impact of endocytic signaling on the execution of EGFR-driven biological programs, however, remains elusive. Gruenberg and colleagues set out to answer this question by addressing the impact of EGFR traffic on EGF-driven global gene transcription [99]. Prolonging the residence time of EGFR at the cell surface (by simultaneous RNAi to either clathrin heavy chain plus dynamin 2 or CBL plus CBLB) or on late endocytic stations (by RNAi to either ESCRT-0 or ESCRT-I components) caused a comparable gain of EGFR activity and attendant ERK signaling [99]. Inhibiting EGFR internalization increased expression of most genes in the EGF signature, recapitulating the effect of EGFR overexpression in the same experimental setting. In contrast, retention of EGFR on endosomes did not alter the global profile and architecture of the EGF-dependent transcriptome, although the expression of a small number of genes encoding cytokines and components of the NFkB pathway was found to be increased [99]. These data indicate that EGFR signaling at the cell surface is primarily responsible for triggering the EGF-driven transcriptional program and imply that endocytosis, in general, restrains this process. However, subtle qualitative aspects of the EGF-driven transcriptome appear to be regulated by signals originating from endosomes.

14.7 Dangerous Liaisons Between Oncogenic EGFR and Endocytosis

Gain-of-function genetic alterations of *EGFR*, in the form of amplification of wt alleles, single amino acid substitutions, or in frame deletions, play a causative role in the pathogenesis of several human tumors [106–108]. These genetic lesions, possibly in concurrence with increased availability of ErbB ligands aberrantly produced by autocrine or paracrine circuits [109], confer to EGFR the ability to signal constitutively. In turn, constitutive EGFR signaling pushes cells towards the acquisition of fundamental traits of the transformed phenotype [110]. We will now discuss how faulty receptor downregulation integrates constitutive catalytic activation of the EGFR to provide for unabated oncogenic signaling.

Gene amplification leads to massive accumulation of EGFR mRNA and protein. Increased EGFR density on the cell surface enhances spontaneous EGFR dimerization and thus favors receptor activation [111]. High-level activation of overexpressed EGFRs saturates the low-capacity CME compartment [112], increasing the residence time of activated EGFRs at the cell surface and precluding efficient receptor downregulation. Faulty endocytosis also propels lateral spreading of EGFR activation, thus igniting signaling by unoccupied EGFR molecules [113]. Moreover, EGFR overexpression in tumor cells is frequently accompanied by autocrine production of receptor ligands, such as TGFα, EPR, and AR [109], which, as discussed above, uncouple EGFR activation from endocytosis-dependent degradation. Finally, massive receptor overexpression is likely to represent also an escape route from IFIs, whose regulated expression is geared to handle physiological levels of EGFR.

Intrinsic refractoriness to downregulation characterizes the behavior of oncogenic EGFR mutants. EGFRvIII is found in 20–30 % of GBMs, most often in association with amplification of the wt*EGFR* allele [114]. EGFRvIII is expressed also in other type of tumors, most notably head and neck squamous cell carcinomas (HNSCC) [115]. EGFRvIII lacks the region of the extracellular domain encoded by exons 2 through 7 and, although unable to bind ligands, signals constitutively [116]. Importantly, EGFRvIII undergoes slow internalization, which is followed by recycling to the cell surface [117]. EGFRvIII binds poorly to CBL and CBL-associated proteins, is not ubiquitylated, and has very low turnover rates [117, 118] (Fig. 14.4). Defective coupling to the endocytic machinery has been proposed to be caused by the submaximal catalytic activity intrinsic to EGFRvIII. According to this model, refractoriness to downregulation is key to convert the low-level signaling activity of EGFRvIII into a persistent oncogenic stimulus [118].

Somatic mutations that target *EGFR* exons 18–21 occur in 10–20 % of non-small cell lung carcinomas (NSCLC). These mutations disrupt mechanisms responsible for maintaining the auto-inhibited conformation of the EGFR kinase domain, thus conferring variable degrees of constitutive catalytic activity [108, 119]. In addition, exon 18–21 mutations impair receptor downregulation (Fig. 14.4). Several reports converge on ascribing such deficit to reduced CBL-dependent ubiquitylation of mutant EGFRs [120–123]. Consequently, NSCLC EGFR mutants show an increased

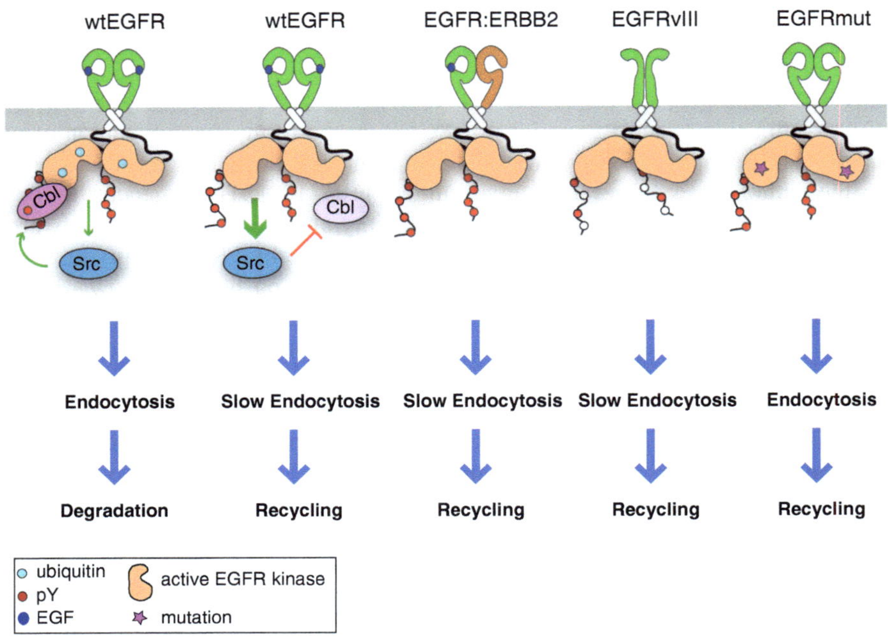

Fig. 14.4 Oncogenic EGFRs are degraded inefficiently as a consequence of poor coupling to CBL-dependent ubiquitylation. Under physiological conditions, regulated SRC activity facilitates CBL-dependent ubiquitylation of the EGFR, leading to receptor downregulation (*leftmost panel*). Constitutive SRC activation (as found, for example, in tumor cells overexpressing wt EGFR) inhibits CBL-dependent ubiquitylation and leads to poor receptor downregulation (second from *left*). Heterodimerization with ERBB2 also leads to defective EGFR endocytosis through different mechanisms, including poor coupling to CBL. Oncogenic conversion of EGFR by the EGFRvIII mutation or mutations in exons 18–21 (EGFRmut) also leads to defective receptor ubiquitylation

propensity to undergo recycling and co-localize extensively with SRC in recycling endosomes [124]. It must be noted that SRC cooperates with EGFR in cell transformation assays and signals downstream to mutated EGFR in NSCLC cell lines [125, 126]. Notably, high-level SRC activity is implied in negative regulation of CBL, as discussed below.

Receptors of the ErbB family are capable of forming combinatorial dimers, which confer great signaling versatility to the ErbB network. Heterodimerization is also essential to the workings of ErbB2 (a ligandless receptor prone to dimerize with other ligand-bound ErbB RTKs) and ErbB3 (a kinase-deficient receptor that acts as a trans-activator of catalytically proficient ErbB RTKs and signaling platform for PI-3K activation) [1]. Importantly, ErbB2 and ErbB3 couple inefficiently to the endocytic machinery [127]. Thus, EGFR molecules recruited into heterodimers with either ErbB2 or ErbB3 are prevalently recycled and hence degraded rather poorly [60, 128, 129]. Several mechanisms concur in altering the endocytic fate of the EGFR in the context of heterodimers with either ErbB2 or ErbB3.

The rates of endocytic uptake of EGFR-ErbB2 dimers are markedly lower when compared to those of EGFR homodimers [60, 129] (Fig. 14.4). In addition, heterodimerization with either ErbB2 or ErbB3 enhances ligand dissociation in the mildly acidic environment of early endosomes [60]. Lastly, both ErbB2 and ErbB3 do not couple to CBL [130, 131]. As a consequence, signals generated by EGFR-ErbB2 or EGFR-ErbB3 heterodimers are stronger and more durable than those elicited by EGFR homodimers and, predictably, have also higher oncogenic potency [1]. Dimerization with ErbB2 has been suggested to be a means through which exon 18–21 EGFR mutants escape downregulation in NSCLC cells [122]. Moreover, exon 18–21 EGFR mutants show a high propensity to dimerize also with ErbB3 [132], which therefore could contribute to divert NSCLC EGFR mutants from degradation.

Signaling pathways aberrantly activated by EGFR may also contribute to instigate refractoriness of EGFR to downregulation. A point in case is SRC, which, as alluded to above, lies downstream to EGFR and cooperates with EGFR in driving cell transformation. Besides contributing its own signaling potential, SRC also enhances EGFR oncogenic firing by inhibiting receptor degradation. SRC may quell CBL-dependent EGFR ubiquitylation by promoting either sequestration or degradation of CBL (Fig. 14.4). Sequestration of CBL is achieved via SRC-dependent phosphorylation of βPIX (Pak-interacting exchange factor beta). As a consequence, the guanine exchange factor activity of βPIX towards CDC42 is increased, along with the formation of a CDC42-βPIX complex which sequesters CBL away from the EGFR [133, 134]. βPIX phosphorylation and CBL sequestration are short-lived in EGF-stimulated normal cells, which have been proposed to provide for transient augmentation of EGFR signaling. Instead, in v-Src transformed cells, βPIX phosphorylation is constitutive and therefore provides for a robust enhancement of EGFR expression and signaling [133]. The latter mechanism is likely to be at work in tumors driven by oncogenic EGFR signaling and exhibiting high-level SRC activity. In addition, high SRC activity is associated with accelerated destruction of CBL [135]. It must be noted that functional SRC is required for CBL-mediated ubiquitylation of EGFR [136, 137], again suggesting that constitutive, rather than timely regulated, activation is key to subvert the role of SRC in EGFR regulation via CBL (Fig. 14.4). CBL-dependent EGFR ubiquitylation is also inhibited in HNSCC cell lines that overexpress cortactin. This is caused by an ill understood reduction of EGFR-CBL complex formation [138].

Owing to their role in regulating EGFR turnover, IFIs have come under scrutiny as potential tumor suppressors. Most of the studies published so far are essentially descriptive, but in a few instances, there is convincing evidence that loss of IFIs unleashes oncogenic EGFR signaling [66]. For example, homozygous deletions or epigenetic silencing of the *ERRFI1/MIG6* gene have been reported in GBM. These alterations occur with the highest frequency in GBMs that contain also *EGFR* amplification, consistent with a role of MIG6 as an EGFR suppressor [74, 139]. Indeed, restoration of MIG6 expression in GBM cell lines reduced their proliferation in in vitro assays. This was associated to enhanced sorting of EGFR from early to late endosomes and increased rates of EGFR degradation [74]. Interestingly, MIG6 was reported to be unable to promote downregulation of EGFRvIII [74]. However, loss of MIG6 is still likely to promote EGFRvIII-driven oncogenesis given that (a) EGFRvIII expression is detected almost invariably in GBM cells carrying concomitant amplification of the wt*EGFR* allele and (b) EGFRvIII oncogenic function has been ascribed also to its ability to foster EGFR activation [140].

At variance with Mig6, loss of Lrig1 in the mouse was not reported to increase the rate of spontaneous tumor formation [66]. However, a recent paper has reported that genetic ablation of Lrig1 in the mouse intestinal epithelium causes the development of highly penetrant duodenal adenomas marked by absence of *Apc* mutations and augmented expression and activation of Egfr and ErbB2 [76]. Interestingly, the phenotypes caused by genetic ablation of Lrig1 appear to be dependent upon the genetic background [78], suggesting that the tumor suppressor activity of Lrig1 in the mouse is currently underestimated. As a consequence, the investigation and modeling of LRIG1 loss in human tumors reliant on oncogenic EGFR signaling are bound to receive a great deal of attention in the near future.

In most of the above examples, we have discussed alterations of EGFR endocytosis that occur in tumor cells carrying amplified or mutated *EGFR* alleles. However, faulty endocytosis may buttress pathogenetically relevant EGFR signaling also in tumor cells in which EGFR does not act as a bona fide driver oncogene. For instance, gain-of-function p53 mutants instigate tumor cell invasiveness and proclivity to metastatic dissemination by promoting EGFR-driven random cell motility. This is due to formation of a molecular complex between EGFR and the $\alpha5\beta1$ integrin, which, by enhancing recycling of both EGFR and $\alpha5\beta1$, causes prolonged EGFR activity and stronger AKT signaling [141]. Another example is typified by the reduced turnover rates of EGFR observed in cells exposed to hypoxia and in cancer cells carrying loss of function mutations of the Von Hippel-Lindau (VHL) protein. The underlying mechanism involves reduced expression of rabaptin, a Rab5-binding protein. Loss of rabaptin attenuates Rab5-dependent vesicle fusion, thus prolonging the intracellular retention of active EGFR and attendant pro-survival signaling via AKT in the cancer hypoxic environment [142].

In summation, several mechanisms are capable of rendering the EGFR refractory to downregulation in tumor cells. Faulty endocytosis is an effective means of altering strength, duration, and compartmentalization of EGFR signaling, thus promoting cell survival, proliferation, and invasiveness.

14.8 Concluding Remarks

The role of endocytosis in controlling the quality and quantity of signals generated by the EGFR over timescales extending from minutes to hours is incontrovertible. There is growing appreciation that different endocytic stations serve as platforms from which the EGFR relays signals. However, the impact on cell physiology of compartmentalized EGFR signal output remains elusive since conclusions reached by different investigators are often conflicting. Advances in cell imaging methodologies and the ever increasing availability of chemical tools for acute inhibition of signaling proteins are expected to facilitate the analysis of EGFR signaling from endocytic stations. Nevertheless, major hurdles to the elucidation of compartmentalized EGFR signaling will still be represented by the biochemical and functional heterogeneity of endosomes as well as the numerous kinetic variables inherent to the analysis of receptor traffic. This complexity is likely to remain experimentally intractable, unless hypotheses are generated with the aid of mathematical models [143, 144].

We have also seen that endocytosis-dependent degradation of the EGFR is a key mechanism responsible for negative regulation of receptor signaling. Refractoriness to downregulation may be acquired via multiple and often concurrent molecular mechanisms and is in fact an integral component of EGFR oncogenic signaling. It will be important to understand whether altered EGFR endocytosis impacts on tumor responsiveness to anti-EGFR therapeutics.

Acknowledgements O.S. is supported by AIRC; S. Alemà is supported by MIUR (PRIN 2010) and CNR (FBQ). We apologize to our colleagues whose relevant work could not be cited due to space limitations.

References

1. Yarden Y, Sliwkowski MX (2001) Untangling the ErbB signalling network. Nat Rev Mol Cell Biol 2:127–137
2. Moghal N, Sternberg PW (2003) The epidermal growth factor system in Caenorhabditis elegans. Exp Cell Res 284:150–159
3. Shilo BZ (2003) Signaling by the Drosophila epidermal growth factor receptor pathway during development. Exp Cell Res 284:140–149
4. Sibilia M, Kroismayr R, Lichtenberger BM, Natarajan A, Hecking M, Holcmann M (2007) The epidermal growth factor receptor: from development to tumorigenesis. Differentiation 75:770–787
5. Pines G, Kostler WJ, Yarden Y (2010) Oncogenic mutant forms of EGFR: lessons in signal transduction and targets for cancer therapy. FEBS Lett 584:2699–2706
6. Ciardiello F, Tortora G (2008) EGFR antagonists in cancer treatment. N Engl J Med 358:1160–1174
7. Zhang J, Yang PL, Gray NS (2009) Targeting cancer with small molecule kinase inhibitors. Nat Rev Cancer 9:28–39
8. Ushiro H, Cohen S (1980) Identification of phosphotyrosine as a product of epidermal growth factor-activated protein kinase in A-431 cell membranes. J Biol Chem 255:8363–8365
9. Downward J, Yarden Y, Mayes E, Scrace G, Totty N et al (1984) Close similarity of epidermal growth factor receptor and v-erb-B oncogene protein sequences. Nature 307:521–527
10. Ullrich A, Coussens L, Hayflick JS, Dull TJ, Gray A, Tam AW et al (1984) Human epidermal growth factor receptor cDNA sequence and aberrant expression of the amplified gene in A431 epidermoid carcinoma cells. Nature 309:418–425
11. Disanza A, Frittoli E, Palamidessi A, Scita G (2009) Endocytosis and spatial restriction of cell signaling. Mol Oncol 3:280–296
12. Madshus IH, Stang E (2009) Internalization and intracellular sorting of the EGF receptor: a model for understanding the mechanisms of receptor trafficking. J Cell Sci 122:3433–3439
13. Sorkin A, Goh LK (2009) Endocytosis and intracellular trafficking of ErbBs. Exp Cell Res 315:683–696

14. Sorkin A, Von Zastrow M (2009) Endocytosis and signalling: intertwining molecular networks. Nat Rev Mol Cell Biol 10:609–622
15. Burgess AW, Cho HS, Eigenbrot C, Ferguson KM, Garrett TP et al (2003) An open-and-shut case? Recent insights into the activation of EGF/ErbB receptors. Mol Cell 12:541–552
16. Endres NF, Engel K, Das R, Kovacs E, Kuriyan J (2011) Regulation of the catalytic activity of the EGF receptor. Curr Opin Struct Biol 21:777–784
17. Lemmon MA, Schlessinger J (2010) Cell signaling by receptor tyrosine kinases. Cell 141:1117–1134
18. Kleiman LB, Maiwald T, Conzelmann H, Lauffenburger DA, Sorger PK (2011) Rapid phospho-turnover by receptor tyrosine kinases impacts downstream signaling and drug binding. Mol Cell 43:723–737
19. Red BM, Choi SH, Alvarado D, Moravcevic K, Pozzi A et al (2009) The juxtamembrane region of the EGF receptor functions as an activation domain. Mol Cell 34:641–651
20. Yoon CH, Lee J, Jongeward GD, Sternberg PW (1995) Similarity of sli-1, a regulator of vulval development in C. elegans, to the mammalian proto-oncogene c-cbl. Science 269:1102–1105
21. Pai LM, Barcelo G, Schupbach T (2000) D-cbl, a negative regulator of the Egfr pathway, is required for dorsoventral patterning in Drosophila oogenesis. Cell 103:51–61
22. Carpenter G, Cohen S (1976) 125I-labeled human epidermal growth factor. Binding, internalization, and degradation in human fibroblasts. J Cell Biol 71:159–171
23. Dunn WA, Connolly TP, Hubbard AL (1986) Receptor-mediated endocytosis of epidermal growth factor by rat hepatocytes: receptor pathway. J Cell Biol 102:24–36
24. Lamaze C, Schmid SL (1995) Recruitment of epidermal growth factor receptors into coated pits requires their activated tyrosine kinase. J Cell Biol 129:47–54
25. Wiley HS, Herbst JJ, Walsh BJ, Lauffenburger DA, Rosenfeld MG et al (1991) The role of tyrosine kinase activity in endocytosis, compartmentation, and down-regulation of the epidermal growth factor receptor. J Biol Chem 266:11083–11094
26. Sorkina T, Huang F, Beguinot L, Sorkin A (2002) Effect of tyrosine kinase inhibitors on clathrin-coated pit recruitment and internalization of epidermal growth factor receptor. J Biol Chem 277:27433–27441
27. Traub LM (2009) Tickets to ride: selecting cargo for clathrin-regulated internalization. Nat Rev Mol Cell Biol 10:583–596
28. McMahon HT, Boucrot E (2011) Molecular mechanism and physiological functions of clathrin-mediated endocytosis. Nat Rev Mol Cell Biol 12:517–533
29. Goh LK, Huang F, Kim W, Gygi S, Sorkin A (2010) Multiple mechanisms collectively regulate clathrin-mediated endocytosis of the epidermal growth factor receptor. J Cell Biol 189:871–883
30. Huang F, Sorkin A (2005) Growth factor receptor binding protein 2-mediated recruitment of the RING domain of Cbl to the epidermal growth factor receptor is essential and sufficient to support receptor endocytosis. Mol Biol Cell 16:1268–1281
31. Levkowitz G, Waterman H, Zamir E, Kam Z, Oved S et al (1998) c-Cbl/Sli-1 regulates endocytic sorting and ubiquitination of the epidermal growth factor receptor. Genes Dev 12:3663–3674
32. Levkowitz G, Waterman H, Ettenberg SA, Katz M, Tsygankov AY et al (1999) Ubiquitin ligase activity and tyrosine phosphorylation underlie suppression of growth factor signaling by c-Cbl/Sli-1. Mol Cell 4:1029–1040
33. Batzer AG, Rotin D, Urena JM, Skolnik EY, Schlessinger J (1994) Hierarchy of binding sites for Grb2 and Shc on the epidermal growth factor receptor. Mol Cell Biol 14:5192–5201
34. Sigismund S, Woelk T, Puri C, Maspero E, Tacchetti C et al (2005) Clathrin-independent endocytosis of ubiquitinated cargos. Proc Natl Acad Sci U S A 102:2760–2765
35. Praefcke GJ, Ford MG, Schmid EM, Olesen LE, Gallop JL et al (2004) Evolving nature of the AP2 alpha-appendage hub during clathrin-coated vesicle endocytosis. EMBO J 23:4371–4383
36. Schmid EM, Ford MG, Burtey A, Praefcke GJ, Peak-Chew SY et al (2006) Role of the AP2 beta-appendage hub in recruiting partners for clathrin-coated vesicle assembly. PLoS Biol 4:e262

37. Jiang X, Huang F, Marusyk A, Sorkin A (2003) Grb2 regulates internalization of EGF receptors through clathrin-coated pits. Mol Biol Cell 14:858–870
38. Stang E, Blystad FD, Kazazic M, Bertelsen V, Brodahl T et al (2004) Cbl-dependent ubiquitination is required for progression of EGF receptors into clathrin-coated pits. Mol Biol Cell 15:3591–3604
39. Huang F, Kirkpatrick D, Jiang X, Gygi S, Sorkin A (2006) Differential regulation of EGF receptor internalization and degradation by multiubiquitination within the kinase domain. Mol Cell 21:737–748
40. Huang F, Goh LK, Sorkin A (2007) EGF receptor ubiquitination is not necessary for its internalization. Proc Natl Acad Sci U S A 104:16904–16909
41. Soubeyran P, Kowanetz K, Szymkiewicz I, Langdon WY, Dikic I (2002) Cbl-CIN85-endophilin complex mediates ligand-induced downregulation of EGF receptors. Nature 416:183–187
42. Hansen CG, Nichols BJ (2009) Molecular mechanisms of clathrin-independent endocytosis. J Cell Sci 122:1713–1721
43. Sandvig K, Pust S, Skotland T, van Deurs B (2011) Clathrin-independent endocytosis: mechanisms and function. Curr Opin Cell Biol 23:413–420
44. Orth JD, Krueger EW, Weller SG, McNiven MA (2006) A novel endocytic mechanism of epidermal growth factor receptor sequestration and internalization. Cancer Res 66:3603–3610
45. Orth JD, McNiven MA (2006) Get off my back! Rapid receptor internalization through circular dorsal ruffles. Cancer Res 66:11094–11096
46. Sigismund S, Argenzio E, Tosoni D, Cavallaro E, Polo S et al (2008) Clathrin-mediated internalization is essential for sustained EGFR signaling but dispensable for degradation. Dev Cell 15:209–219
47. Barbieri MA, Roberts RL, Gumusboga A, Highfield H, Alvarez-Dominguez C et al (2000) Epidermal growth factor and membrane trafficking. EGF receptor activation of endocytosis requires Rab5a. J Cell Biol 151:539–550
48. Henne WM, Buchkovich NJ, Emr SD (2011) The ESCRT pathway. Dev Cell 21:77–91
49. Umebayashi K, Stenmark H, Yoshimori T (2008) Ubc4/5 and c-Cbl continue to ubiquitinate EGF receptor after internalization to facilitate polyubiquitination and degradation. Mol Biol Cell 19:3454–3462
50. Bache KG, Raiborg C, Mehlum A, Stenmark H (2003) STAM and Hrs are subunits of a multivalent ubiquitin-binding complex on early endosomes. J Biol Chem 278:12513–12521
51. Raiborg C, Bache KG, Gillooly DJ, Madshus IH, Stang E et al (2002) Hrs sorts ubiquitinated proteins into clathrin-coated microdomains of early endosomes. Nat Cell Biol 4:394–398
52. White IJ, Bailey LM, Aghakhani MR, Moss SE, Futter CE (2006) EGF stimulates annexin 1-dependent inward vesiculation in a multivesicular endosome subpopulation. EMBO J 25:1–12
53. Ceresa BP, Bahr SJ (2006) rab7 activity affects epidermal growth factor:epidermal growth factor receptor degradation by regulating endocytic trafficking from the late endosome. J Biol Chem 281:1099–1106
54. Eden ER, Huang F, Sorkin A, Futter CE (2012) The role of EGF receptor ubiquitination in regulating its intracellular traffic. Traffic 13:329–337
55. Chi S, Cao H, Wang Y, McNiven MA (2011) Recycling of the epidermal growth factor receptor is mediated by a novel form of the clathrin adaptor protein Eps15. J Biol Chem 286:35196–35208
56. Roxrud I, Raiborg C, Pedersen NM, Stang E, Stenmark H (2008) An endosomally localized isoform of Eps15 interacts with Hrs to mediate degradation of epidermal growth factor receptor. J Cell Biol 180:1205–1218
57. Derynck R (1992) The physiology of transforming growth factor-alpha. Adv Cancer Res 58:27–52
58. Longva KE, Blystad FD, Stang E, Larsen AM, Johannessen LE et al (2002) Ubiquitination and proteasomal activity is required for transport of the EGF receptor to inner membranes of multivesicular bodies. J Cell Biol 156:843–854

59. Kochupurakkal BS, Harari D, Di Segni A, Maik-Rachline G, Lyass L et al (2005) Epigen, the last ligand of ErbB receptors, reveals intricate relationships between affinity and mitogenicity. J Biol Chem 280:8503–8512
60. Lenferink AE, Pinkas-Kramarski R, van de Poll ML, van Vugt MJ, Klapper LN et al (1998) Differential endocytic routing of homo- and hetero-dimeric ErbB tyrosine kinases confers signaling superiority to receptor heterodimers. EMBO J 17:3385–3397
61. Shelly M, Pinkas-Kramarski R, Guarino BC, Waterman H, Wang LM et al (1998) Epiregulin is a potent pan-ErbB ligand that preferentially activates heterodimeric receptor complexes. J Biol Chem 273:10496–10505
62. Wilson KJ, Mill C, Lambert S, Buchman J, Wilson TR et al (2012) EGFR ligands exhibit functional differences in models of paracrine and autocrine signaling. Growth Factors 30:107–116
63. Roepstorff K, Grandal MV, Henriksen L, Knudsen SL, Lerdrup M et al (2009) Differential effects of EGFR ligands on endocytic sorting of the receptor. Traffic 10:1115–1127
64. Avraham R, Yarden Y (2011) Feedback regulation of EGFR signalling: decision making by early and delayed loops. Nat Rev Mol Cell Biol 12:104–117
65. Amit I, Citri A, Shay T, Lu Y, Katz M et al (2007) A module of negative feedback regulators defines growth factor signaling. Nat Genet 39:503–512
66. Segatto O, Anastasi S, Alema S (2011) Regulation of epidermal growth factor receptor signalling by inducible feedback inhibitors. J Cell Sci 124:1785–1793
67. Gur G, Rubin C, Katz M, Amit I, Citri A et al (2004) LRIG1 restricts growth factor signaling by enhancing receptor ubiquitylation and degradation. EMBO J 23:3270–3281
68. Laederich MB, Funes-Duran M, Yen L, Ingalla E, Wu X et al (2004) The leucine-rich repeat protein LRIG1 is a negative regulator of ErbB family receptor tyrosine kinases. J Biol Chem 279:47050–47056
69. Kario E, Marmor MD, Adamsky K, Citri A, Amit I et al (2005) Suppressors of cytokine signaling 4 and 5 regulate epidermal growth factor receptor signaling. J Biol Chem 280:7038–7048
70. Nicholson SE, Metcalf D, Sprigg NS, Columbus R, Walker F et al (2005) Suppressor of cytokine signaling (SOCS)-5 is a potential negative regulator of epidermal growth factor signaling. Proc Natl Acad Sci U S A 102:2328–2333
71. Anastasi S, Baietti MF, Frosi Y, Alema S, Segatto O (2007) The evolutionarily conserved EBR module of RALT/MIG6 mediates suppression of the EGFR catalytic activity. Oncogene 26:7833–7846
72. Zhang X, Pickin KA, Bose R, Jura N, Cole PA et al (2007) Inhibition of the EGF receptor by binding of MIG6 to an activating kinase domain interface. Nature 450:741–744
73. Frosi Y, Anastasi S, Ballaro C, Varsano G, Castellani L et al (2010) A two-tiered mechanism of EGFR inhibition by RALT/MIG6 via kinase suppression and receptor degradation. J Cell Biol 189:557–571
74. Ying H, Zheng H, Scott K, Wiedemeyer R, Yan H et al (2010) Mig-6 controls EGFR trafficking and suppresses gliomagenesis. Proc Natl Acad Sci U S A 107:6912–6917
75. Jensen KB, Watt FM (2006) Single-cell expression profiling of human epidermal stem and transit-amplifying cells: Lrig1 is a regulator of stem cell quiescence. Proc Natl Acad Sci U S A 103:11958–11963
76. Powell AE, Wang Y, Li Y, Poulin EJ, Means AL et al (2012) The Pan-ErbB negative regulator Lrig1 Is an intestinal stem cell marker that functions as a tumor suppressor. Cell 149:146–158
77. Suzuki Y, Miura H, Tanemura A, Kobayashi K, Kondoh G et al (2002) Targeted disruption of LIG-1 gene results in psoriasiform epidermal hyperplasia. FEBS Lett 521:67–71
78. Wong VW, Stange DE, Page ME, Buczacki S, Wabik A et al (2012) Lrig1 controls intestinal stem-cell homeostasis by negative regulation of ErbB signalling. Nat Cell Biol 14:401–408
79. Ferby I, Reschke M, Kudlacek O, Knyazev P, Pante G et al (2006) Mig6 is a negative regulator of EGF receptor-mediated skin morphogenesis and tumor formation. Nat Med 12:568–573

80. Jin N, Cho SN, Raso MG, Wistuba I, Smith Y et al (2009) Mig-6 is required for appropriate lung development and to ensure normal adult lung homeostasis. Development 136:3347–3356
81. Reschke M, Ferby I, Stepniak E, Seitzer N, Horst D et al (2009) Mitogen-inducible gene-6 is a negative regulator of epidermal growth factor receptor signaling in hepatocytes and human hepatocellular carcinoma. Hepatology 51:1383–1390
82. Burke P, Schooler K, Wiley HS (2001) Regulation of epidermal growth factor receptor signaling by endocytosis and intracellular trafficking. Mol Biol Cell 12:1897–1910
83. Di Guglielmo GM, Baass PC, Ou WJ, Posner BI, Bergeron JJ (1994) Compartmentalization of SHC, GRB2 and mSOS, and hyperphosphorylation of Raf-1 by EGF but not insulin in liver parenchyma. EMBO J 13:4269–4277
84. Nishibe S, Wahl MI, Hernandez-Sotomayor SM, Tonks NK, Rhee SG et al (1990) Increase of the catalytic activity of phospholipase C-gamma 1 by tyrosine phosphorylation. Science 250:1253–1256
85. Rotin D, Honegger AM, Margolis BL, Ullrich A, Schlessinger J (1992) Presence of SH2 domains of phospholipase C gamma 1 enhances substrate phosphorylation by increasing the affinity toward the epidermal growth factor receptor. J Biol Chem 267:9678–9683
86. Rotin D, Margolis B, Mohammadi M, Daly RJ, Daum G et al (1992) SH2 domains prevent tyrosine dephosphorylation of the EGF receptor: identification of Tyr992 as the high-affinity binding site for SH2 domains of phospholipase C gamma. EMBO J 11:559–567
87. Vega QC, Cochet C, Filhol O, Chang CP, Rhee SG et al (1992) A site of tyrosine phosphorylation in the C terminus of the epidermal growth factor receptor is required to activate phospholipase C. Mol Cell Biol 12:128–135
88. Haugh JM, Meyer T (2002) Active EGF receptors have limited access to PtdIns(4,5)P(2) in endosomes: implications for phospholipase C and PI 3-kinase signaling. J Cell Sci 115:303–310
89. Haugh JM, Schooler K, Wells A, Wiley HS, Lauffenburger DA (1999) Effect of epidermal growth factor receptor internalization on regulation of the phospholipase C-gamma1 signaling pathway. J Biol Chem 274:8958–8965
90. Miaczynska M, Christoforidis S, Giner A, Shevchenko A, Uttenweiler-Joseph S et al (2004) APPL proteins link Rab5 to nuclear signal transduction via an endosomal compartment. Cell 116:445–456
91. Zoncu R, Perera RM, Balkin DM, Pirruccello M, Toomre D et al (2009) A phosphoinositide switch controls the maturation and signaling properties of APPL endosomes. Cell 136:1110–1121
92. Jones RB, Gordus A, Krall JA, MacBeath G (2006) A quantitative protein interaction network for the ErbB receptors using protein microarrays. Nature 439:168–174
93. Mitsuuchi Y, Johnson SW, Sonoda G, Tanno S, Golemis EA et al (1999) Identification of a chromosome 3p14.3-21.1 gene, APPL, encoding an adaptor molecule that interacts with the oncoprotein-serine/threonine kinase AKT2. Oncogene 18:4891–4898
94. Tan Y, You H, Wu C, Altomare DA, Testa JR (2010) App l1 is dispensable for mouse development, and loss of Appl1 has growth factor-selective effects on Akt signaling in murine embryonic fibroblasts. J Biol Chem 285:6377–6389
95. Downward J (1994) The GRB2/Sem-5 adaptor protein. FEBS Lett 338:113–117
96. Fehrenbacher N, Bar-Sagi D, Philips M (2009) Ras/MAPK signaling from endomembranes. Mol Oncol 3:297–307
97. Nada S, Hondo A, Kasai A, Koike M, Saito K (2009) The novel lipid raft adaptor p18 controls endosome dynamics by anchoring the MEK-ERK pathway to late endosomes. EMBO J 28:477–489
98. Taub N, Teis D, Ebner HL, Hess MW, Huber LA (2007) Late endosomal traffic of the epidermal growth factor receptor ensures spatial and temporal fidelity of mitogen-activated protein kinase signaling. Mol Biol Cell 18:4698–4710
99. Brankatschk B, Wichert SP, Johnson SD, Schaad O, Rossner MJ et al (2012) Regulation of the EGF transcriptional response by endocytic sorting. Sci Signal 5:ra21

100. Carpenter G, Liao HJ (2009) Trafficking of receptor tyrosine kinases to the nucleus. Exp Cell Res 315:1556–1566
101. Wang YN, Yamaguchi H, Hsu JM, Hung MC (2010) Nuclear trafficking of the epidermal growth factor receptor family membrane proteins. Oncogene 29:3997–4006
102. Wang YN, Wang H, Yamaguchi H, Lee HJ, Lee HH et al (2010) COPI-mediated retrograde trafficking from the Golgi to the ER regulates EGFR nuclear transport. Biochem Biophys Res Commun 399:498–504
103. Wang YN, Yamaguchi H, Huo L, Du Y, Lee HJ et al (2010) The translocon Sec61beta localized in the inner nuclear membrane transports membrane-embedded EGF receptor to the nucleus. J Biol Chem 285:38720–38729
104. Xia L, Wang L, Chung AS, Ivanov SS, Ling MY et al (2002) Identification of both positive and negative domains within the epidermal growth factor receptor COOH-terminal region for signal transducer and activator of transcription (STAT) activation. J Biol Chem 277:30716–30723
105. Bild AH, Turkson J, Jove R (2002) Cytoplasmic transport of Stat3 by receptor-mediated endocytosis. EMBO J 21:3255–3263
106. Futreal PA, Coin L, Marshall M, Down T, Hubbard T et al (2004) A census of human cancer genes. Nat Rev Cancer 4:177–183
107. Santarius T, Shipley J, Brewer D, Stratton MR, Cooper CS (2010) A census of amplified and overexpressed human cancer genes. Nat Rev Cancer 10:59–64
108. Sharma SV, Settleman J (2009) ErbBs in lung cancer. Exp Cell Res 315:557–571
109. Wilson KJ, Gilmore JL, Foley J, Lemmon MA, Riese DJ (2009) Functional selectivity of EGF family peptide growth factors: implications for cancer. Pharmacol Ther 122:1–8
110. Hanahan D, Weinberg RA (2011) Hallmarks of cancer: the next generation. Cell 144:646–674
111. Chung I, Akita R, Vandlen R, Toomre D, Schlessinger J et al (2010) Spatial control of EGF receptor activation by reversible dimerization on living cells. Nature 464:783–787
112. Wiley HS (1988) Anomalous binding of epidermal growth factor to A431 cells is due to the effect of high receptor densities and a saturable endocytic system. J Cell Biol 107:801–810
113. Sawano A, Takayama S, Matsuda M, Miyawaki A (2002) Lateral propagation of EGF signaling after local stimulation is dependent on receptor density. Dev Cell 3:245–257
114. Furnari FB, Fenton T, Bachoo RM, Mukasa A, Stommel JM et al (2007) Malignant astrocytic glioma: genetics, biology, and paths to treatment. Genes Dev 21:2683–2710
115. Sok JC, Coppelli FM, Thomas SM, Lango MN, Xi S et al (2006) Mutant epidermal growth factor receptor (EGFRvIII) contributes to head and neck cancer growth and resistance to EGFR targeting. Clin Cancer Res 12:5064–5073
116. Huang HS, Nagane M, Klingbeil CK, Lin H, Nishikawa R et al (1997) The enhanced tumorigenic activity of a mutant epidermal growth factor receptor common in human cancers is mediated by threshold levels of constitutive tyrosine phosphorylation and unattenuated signaling. J Biol Chem 272:2927–2935
117. Grandal MV, Zandi R, Pedersen MW, Willumsen BM, van Deurs B et al (2007) EGFRvIII escapes down-regulation due to impaired internalization and sorting to lysosomes. Carcinogenesis 28:1408–1417
118. Schmidt MH, Furnari FB, Cavenee WK, Bogler O (2003) Epidermal growth factor receptor signaling intensity determines intracellular protein interactions, ubiquitination, and internalization. Proc Natl Acad Sci U S A 100:6505–6510
119. Pao W, Chmielecki J (2010) Rational, biologically based treatment of EGFR-mutant non-small-cell lung cancer. Nat Rev Cancer 10:760–774
120. Furukawa M, Nagatomo I, Kumagai T, Yamadori T, Takahashi R et al (2007) Gefitinib-sensitive EGFR lacking residues 746–750 exhibits hypophosphorylation at tyrosine residue 1045, hypoubiquitination, and impaired endocytosis. DNA Cell Biol 26:178–185
121. Padron D, Sato M, Shay JW, Gazdar AF, Minna JD et al (2007) Epidermal growth factor receptors with tyrosine kinase domain mutations exhibit reduced Cbl association, poor ubiquitylation, and down-regulation but are efficiently internalized. Cancer Res 67:7695–7702

122. Shtiegman K, Kochupurakkal BS, Zwang Y, Pines G, Starr A et al (2007) Defective ubiquitinylation of EGFR mutants of lung cancer confers prolonged signaling. Oncogene 26:6968–6978
123. Yang S, Qu S, Perez-Tores M, Sawai A, Rosen N et al (2006) Association with HSP90 inhibits Cbl-mediated down-regulation of mutant epidermal growth factor receptors. Cancer Res 66:6990–6997
124. Chung BM, Raja SM, Clubb RJ, Tu C, George M et al (2009) Aberrant trafficking of NSCLC-associated EGFR mutants through the endocytic recycling pathway promotes interaction with Src. BMC Cell Biol 10:84
125. Chung BM, Dimri M, George M, Reddi AL, Chen G et al (2009) The role of cooperativity with Src in oncogenic transformation mediated by non-small cell lung cancer-associated EGF receptor mutants. Oncogene 28:1821–1832
126. Leung EL, Tam IY, Tin VP, Chua DT, Sihoe ADL et al (2009) SRC promotes survival and invasion of lung cancers with epidermal growth factor receptor abnormalities and is a potential candidate for molecular-targeted therapy. Mol Cancer Res 7:923–932
127. Baulida J, Kraus MH, Alimandi M, Di Fiore PP, Carpenter G (1996) All ErbB receptors other than the epidermal growth factor receptor are endocytosis impaired. J Biol Chem 271:5251–5257
128. Haslekas C, Breen K, Pedersen KW, Johannessen LE, Stang E et al (2005) The inhibitory effect of ErbB2 on epidermal growth factor-induced formation of clathrin-coated pits correlates with retention of epidermal growth factor receptor-ErbB2 oligomeric complexes at the plasma membrane. Mol Biol Cell 16:5832–5842
129. Worthylake R, Opresko LK, Wiley HS (1999) ErbB-2 amplification inhibits down-regulation and induces constitutive activation of both ErbB-2 and epidermal growth factor receptors. J Biol Chem 274:8865–8874
130. Levkowitz G, Klapper LN, Tzahar E, Freywald A, Sela M et al (1996) Coupling of the c-Cbl protooncogene product to ErbB-1/EGF-receptor but not to other ErbB proteins. Oncogene 12:1117–1125
131. Waterman H, Alroy I, Strano S, Seger R, Yarden Y (1999) The C-terminus of the kinase-defective neuregulin receptor ErbB-3 confers mitogenic superiority and dictates endocytic routing. EMBO J 18:3348–3358
132. Rothenberg SM, Engelman JA, Le S, Riese DJ, Haber DA, Settleman J (2008) Modeling oncogene addiction using RNA interference. Proc Natl Acad Sci U S A 105:12480–12484
133. Feng Q, Baird D, Peng X, Wang J, Ly T et al (2006) Cool-1 functions as an essential regulatory node for EGF receptor- and Src-mediated cell growth. Nat Cell Biol 8:945–956
134. Wu WJ, Tu S, Cerione RA (2003) Activated Cdc42 sequesters c-Cbl and prevents EGF receptor degradation. Cell 114:715–725
135. Bao J, Gur G, Yarden Y (2003) Src promotes destruction of c-Cbl: implications for oncogenic synergy between Src and growth factor receptors. Proc Natl Acad Sci USA 100:2438–2443
136. Kassenbrock CK, Hunter S, Garl P, Johnson GL, Anderson SM (2002) Inhibition of Src family kinases blocks epidermal growth factor (EGF)-induced activation of Akt, phosphorylation of c-Cbl, and ubiquitination of the EGF receptor. J Biol Chem 277:24967–24975
137. Ravid T, Heidinger JM, Gee P, Khan EM, Goldkorn T (2004) c-Cbl-mediated ubiquitinylation is required for epidermal growth factor receptor exit from the early endosomes. J Biol Chem 279:37153–37162
138. Timpson P, Lynch DK, Schramek D, Walker F, Daly RJ (2005) Cortactin overexpression inhibits ligand-induced down-regulation of the epidermal growth factor receptor. Cancer Res 65:3273–3280
139. Duncan CG, Killela PJ, Payne CA, Lampson B, Chen WC et al (2010) Integrated genomic analyses identify ERRFI1 and TACC3 as glioblastoma-targeted genes. Oncotarget 1:265–277
140. Inda MM, Bonavia R, Mukasa A, Narita Y, Sah DW et al (2010) Tumor heterogeneity is an active process maintained by a mutant EGFR-induced cytokine circuit in glioblastoma. Genes Dev 24:1731–1745

141. Muller PA, Caswell PT, Doyle B, Iwanicki MP, Tan EH et al (2009) Mutant p53 drives invasion by promoting integrin recycling. Cell 139:1327–1341
142. Wang Y, Roche O, Yan MS, Finak G, Evans AJ et al (2009) Regulation of endocytosis via the oxygen-sensing pathway. Nat Med 15:319–324
143. Birtwistle MR, Kholodenko BN (2009) Endocytosis and signalling: a meeting with mathematics. Mol Oncol 3:308–320
144. Lazzara MJ, Lauffenburger DA (2009) Quantitative modeling perspectives on the ErbB system of cell regulatory processes. Exp Cell Res 315:717–725

Chapter 15
Stress-Driven Endocytosis of Tyrosine-Phosphorylated EGFR Leads to Tumorigenesis: The Critical Role of Oxidative Stress

Tzipora Goldkorn, Simone Filosto, and Samuel Chung

Abstract Epidermal growth factor receptor (EGFR) dysregulation drives several human pathological conditions, such as glioblastoma and non-small cell lung cancer, and as a result, EGFR activation, internalization, and downregulation are extensively studied for their contributions to tumorigenesis. However, while the mechanisms for ligand-dependent EGFR activation and internalization are quite well understood, current knowledge of ligand-independent mechanisms for EGFR activation and trafficking is obscured by somewhat conflicting data. Thus, unraveling the "unorthodox" machinery driving stress-induced EGFR activation and trafficking remains a very important task.

We present in this chapter the progression of the EGFR field, acknowledging mechanism(s) of stress-dependent activation, internalization, and trafficking of the EGFR. Emphasis is given within the context of cellular oxidative stress, which appears to be a common outcome of several types of stressors and pathological conditions, as observed in cancers and chemotherapy. In addition, c-Cbl, p38 MAPK, c-Src, and caveolin-1 are provided as examples of proteins contributing to the aberrant EGFR internalization and trafficking phenotype observed during

T. Goldkorn (✉)
Center for Comparative Respiratory Biology and Medicine (CCRBM), University of California School of Medicine, Davis, CA 95616, USA

Signal Transduction Internal Medicine, University of California School of Medicine, Room 6321, 451 E. Health Sciences Drive, Davis, CA 95616, USA
e-mail: ttgoldkorn@ucdavis.edu

S. Filosto • S. Chung
Center for Comparative Respiratory Biology and Medicine (CCRBM), University of California School of Medicine, Davis, CA 95616, USA

Y. Yarden and G. Tarcic (eds.), *Vesicle Trafficking in Cancer*,
DOI 10.1007/978-1-4614-6528-7_15,

cellular stress. Lastly, the role of the membrane itself in stress-induced EGFR internalization is discussed within the context of ceramide, a membrane sphingolipid generated during oxidative stress, and ceramide-enriched lipid rafts.

15.1 Introduction: Canonical EGFR Activation, Intracellular Trafficking, and Degradation

The EGFR (ErbB1) is a member of the ErbB family of receptor tyrosine kinases (RTKs), which also includes ErbB2, ErbB3, and ErbB4. These receptors have a vital role in normal cellular processes such as cell division, differentiation, and migration, and their overexpression or dysregulation has been linked to a variety of human cancers [1–3]. Hence, their activation, particularly that of the EGFR, has been a subject of intense studies.

The model of EGFR activation has been established wherein ligand binding induces receptor dimerization, leading to the activation of its intrinsic tyrosine kinase activity, autophosphorylation, and subsequent phosphorylation of downstream signaling molecules [3, 4] involved in cellular survival and proliferation. Therefore, to control cellular growth and tumorigenesis, the activation of the EGFR has to be tightly regulated in a process that includes degradation of the receptor. Upon EGF binding, the EGFRs are rapidly internalized from the cell surface through numerous pathways, including clathrin-coated pits [5]. Indeed, the inability of the EGFR to be downregulated via clathrin-mediated endocytosis and degradation has been linked to its oncogenicity [6].

Though the exact mechanism by which the EGFRs are recruited into the clathrin-coated pits is under investigation, the general paradigm of EGFR internalization is as follows. The EGF-activated EGFRs recruit Grb2 (growth factor receptor-bound protein 2) which plays a critical role in the binding of adaptor protein 2 complex (AP-2) [7, 8]. AP-2 subsequently recruits clathrin (cytosolic scaffold protein that mediates vesicle formation) to the plasma membrane which polymerizes to form a clathrin coat that provides the framework required for vesicular budding into the cell and eventual internalization of the activated EGFRs [9].

Internalized RTKs, such as the EGFR, undergo a general modulatory phase consisting of a two-step endocytosis process [5, 10–12]: (1) a fast ligand-dependent internalization step removes activated receptors from the cell membrane and segregates them in endosomes, followed by (2) sorting of internalized RTK molecules to lysosomes, where they go through degradation, or recycling back to the plasma membrane. Notably, EGFR classified for lysosomal degradation requires recruitment of c-Cbl (cellular Casitas B-lineage lymphoma), which ubiquitinates EGFR [13]. Expectedly, molecular studies impeding clathrin-mediated endocytosis resulted in no subsequent lysosomal degradation of the ligand-activated EGFR [4, 14, 15]. This could maintain prolonged downstream activation of pro-survival and proliferation molecules such as Akt and ERK1/2 [16], thereby enhancing cell proliferation and tumorigenesis [17].

15.2 Stress-Induced EGFR Activation and Intracellular Trafficking

ErbB proteins are not only activated by their associated ligands but may also be trans-modulated by other molecules and pathological states. For example, cellular stress conditions, such as exposure to ultraviolet (UV) irradiation, ionizing radiations, hypoxia, inflammatory cytokines, or H_2O_2, induce either tyrosine phosphorylation or serine/threonine phosphorylation of EGFR with direct consequences in EGFR signaling and intracellular trafficking [18–25]. In addition to direct effects on EGFR phosphorylation, such cellular stress may cause alterations in the cell membrane followed by accumulation and reorganization of surface-related receptor molecules [18, 19, 26]. However, the mechanisms and nature of the proteins involved are yet unknown.

15.3 Oxidative Stress, a Common Denominator of Cellular Stressors

Various cellular stressors, such as cytokines, chemotherapeutic drugs, UV irradiation [27], and CS exposure [28], may have a common denominator in their production of cellular oxidative (ox-) stress. Therefore, the purpose of this chapter is to provide perspective for the mechanistic studies of oxidative stress (ox-stress) modulation of EGFR and the subsequent effects on cell proliferation and tumorigenesis.

It was reported quite long ago that exposure to short wavelength UV increases EGFR tyrosine phosphorylation [29–31].

Huang and coworkers have suggested that UV-induced EGFR tyrosine phosphorylation involves formation of reactive oxygen species (ROS) [32]. In support of this hypothesis, they showed that the ROS scavenger N-acetylcysteine (NAC) inhibited UV-mediated receptor tyrosine phosphorylation. At the same time, it was also reported that ROS may disrupt specific cysteine-containing sequences of tyrosine phosphatases, thereby inhibiting their function [33].

These studies are in agreement with relatively new findings of Fisher et al. showing that UV light causes a ligand-independent EGFR Tyr autophosphorylation which could be recapitulated by H_2O_2-induced ox-stress. In that study, the inhibition of a receptor protein tyrosine phosphatase (RPTP-k) by ox-stress was emphasized as the potential mechanism, which contributes to EGFR Tyr phosphorylation following UV irradiation of keratinocytes [34].

However, the observation that UV irradiation causes EGFR kinase activation has been recently challenged by other groups that observed no tyrosine phosphorylation of EGFR following UV exposure of HeLa cells [23, 27].

Notably, these studies also investigated the intracellular trafficking of EGFR following UV treatment, and, within that context, Oksvold et al. [23] concluded that ox-stress does not have a role in the observed UV-dependent phenotype of EGFR trafficking. Therefore, it is still possible that for some unknown technical reasons, some studies that use UV irradiation report induction of ROS-dependent cellular ox-stress, which leads to EGFR kinase activation (and its Tyr autophosphorylation) [29, 31, 34], while in some other cases, no detectable ox-stress was generated by the UV exposure and therefore no EGFR kinase activation was observed [23, 27].

15.4 Stress-Dependent EGFR Internalization Is Ligand Independent but p38 MAPK Dependent

In contrast to the pronounced Tyr phosphorylation induced by EGF, several cellular stressors, including UV irradiation, cytokines, chemotherapy drugs, and lipid raft disruption (by MβCD), are unable to induce the specific phosphorylation of EGFR on tyrosine 1045. Therefore, as it has been shown under H_2O_2-induced ox-stress [21, 35], UV-irradiated EGFR could be internalized, but could not reach the lysosomes for degradation without having the c-Cbl binding site of phosphorylated Tyr-1045 [27, 36–39]. Furthermore, UV-mediated internalization of the EGFR has been shown to be dependent on specific residues of the cytoplasmic carboxy-terminal region of the receptor. However, different studies implicated different specific residues within that c-terminal region of the EGFR. For example, Oksvold et al. identified the serine residues 1046, 1047, 1057, and 1142 of EGFR as regulators of receptor internalization upon UV irradiation, while Zwang and Yarden implicated a short segment of the EGFR tail, which included residues 1002–1022 [27, 36].

Subsequent studies by Zwang and Yarden [27] clearly demonstrated that p38 MAPK (mitogen-activated protein kinase) was activated under such stressors as UV irradiation, cisplatin (chemotherapeutic agent), and cytokines such as TNF-alpha or interleukin-1 (IL-1). They found that p38 MAPK was essential for the EGFR internalization under these stress conditions in HeLa cells. Importantly, other groups also found that p38 MAPK is controlling EGFR internalization under cellular stress [39–41].

Yarden et al. further demonstrated that EGFR does undergo stress-induced phosphorylation dependent on p38 MAPK, which results in the phosphorylation of several serine and threonine residues within a short segment of EGFR (residues 1002–1022). These authors also reported that they observed some differences in the cellular localization of EGFR in response to cytokines or to UV exposure in per the different kinetics of p38 MAPK activation. In contrast to the EGF-induced endocytosis, these pathways involve no EGFR ubiquitination, thus leading to intracellular accumulation without EGFR degradation [27].

Importantly, the accumulation of EGFR tends to occur in different cellular sites. Stress-induced internalization due to cytokine stimulation results in receptor recycling to the plasma membrane. On the other hand, under UV irradiation, EGFR is blocked in early endosomes [27]. It was also shown that the EGFR phosphorylated by p38 MAPK was internalized via the clathrin-dependent pathway, which directed the endocytosed receptors to Rab5-containing early endosomes. As soon as p38

MAPK is inactivated, the internalized receptors may undergo dephosphorylation and cycle back to the cell surface [27].

Additional data generated by another group in HeLa cells [40] presented the same rationale: it was shown that it was enough to treat the cells with anisomycin, which activates p38 MAPK, to induce internalization of EGFR, and vice versa, specific inhibition of p38 MAPK stopped UV-induced EGFR endocytosis. Both Vergarajauregui et al. and Zwang and Yarden proposed that prompting the p38-mediated EGFR internalization represents a mechanism to preclude proliferative or survival signals under stress conditions and enhance the apoptotic mechanisms to get rid of the stress-defected cell.

In agreement with the above ideas, keratinocytes were treated with MβCD to disrupt lipid rafts and were found to subsequently display a ligand-independent internalization process that removes the EGFR from the cell surface in a p38 MAPK-dependent manner [39]. Other stress conditions such as keratinocyte monolayer scratch wounding could also induce this peculiar mechanism of EGFR internalization that is not ligand dependent but rather p38 dependent [41].

However, as already mentioned above, it seems that Zwang and Yarden did not observe any effects of ox-stress neither under UV irradiation of cells nor under cytokines or cisplatin treatments. Therefore, no activation of EGFR kinase and receptor autophosphorylation on tyrosines was reported [27]. This is in striking contradiction to reports by other groups [20, 34]. For example, Benhar et al. showed that cisplatin-induced p38 activation was blocked by treatments with the antioxidants glutathione (GSH) or N-acetylcysteine (NAC), as well as by more specific inhibitors of mitochondrial NAPDH oxidase [20]. Perhaps, some discrepancies between studies may evolve because of using different cells or simply different technical conditions.

The issue of induction of oxidative stress (ox-stress) could be very critical to the outcome of cell fate.

While the internalization of an EGFR that is not phosphorylated on tyrosines could indeed represent a conserved mechanism of preventing signals of growth in cells undergoing stress, it is equally possible that the internalization of an activated EGFR (aberrantly phosphorylated on tyrosines) without a negative feedback of receptor degradation could represent a survival mechanism under the same stress conditions and could also be one of the mechanisms adopted by cancerous cells to overcome the need for paracrine growth signaling (Figs. 15.1, 15.2 and 15.3).

Consistent with that possibility, several groups, including Levitzki et al. [20, 38], pointed out that tumor cells are known to contain elevated levels of ROS, as measured by H_2O_2 levels or by enhanced oxidative damage [42, 43]. Levitzki et al. found that cisplatin activates EGFR and Src in cancer cells, which could be explained by generation of ROS. This survival mechanism under stress, caused by cisplatin, could obviously counteract the drug's main mission of killing cancer cells via cross-linking their DNA [38]. Others have also considered the possible involvement of ROS as important regulators of stress responses in many cell types and implicated ROS in MAPK(s) activation [44–47].

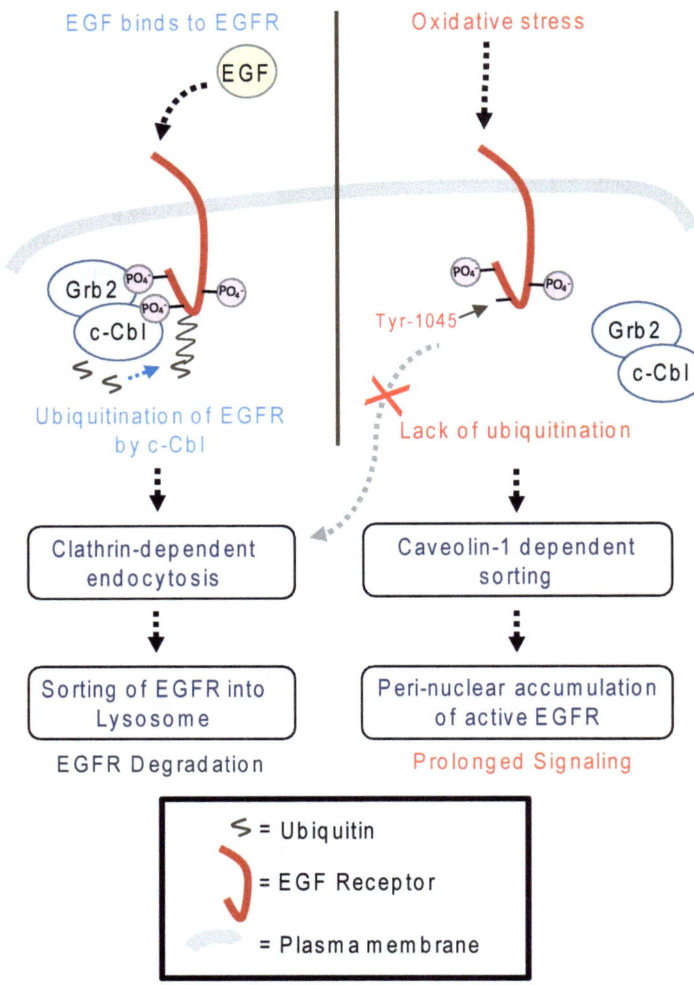

Fig. 15.1 EGFR is aberrantly phosphorylated under oxidative stress exposure, impairing (I) c-Cbl binding, (II) ubiquitination, and (III) subsequent clathrin-dependent endocytosis and sorting for lysosomal degradation

15.5 Cigarette Smoke and H_2O_2 Induce Oxidative Stress That Causes Aberrant EGFR Activation

(a) H_2O_2 is a ubiquitous molecule that is able to cross cell membranes freely. It is present in several air pollutants, including the vapor phase of CS. It is also detected in exhaled air of humans [48], and amounts of exhaled H_2O_2 appear greater in subjects with lung inflammation [49] and in smokers [50].

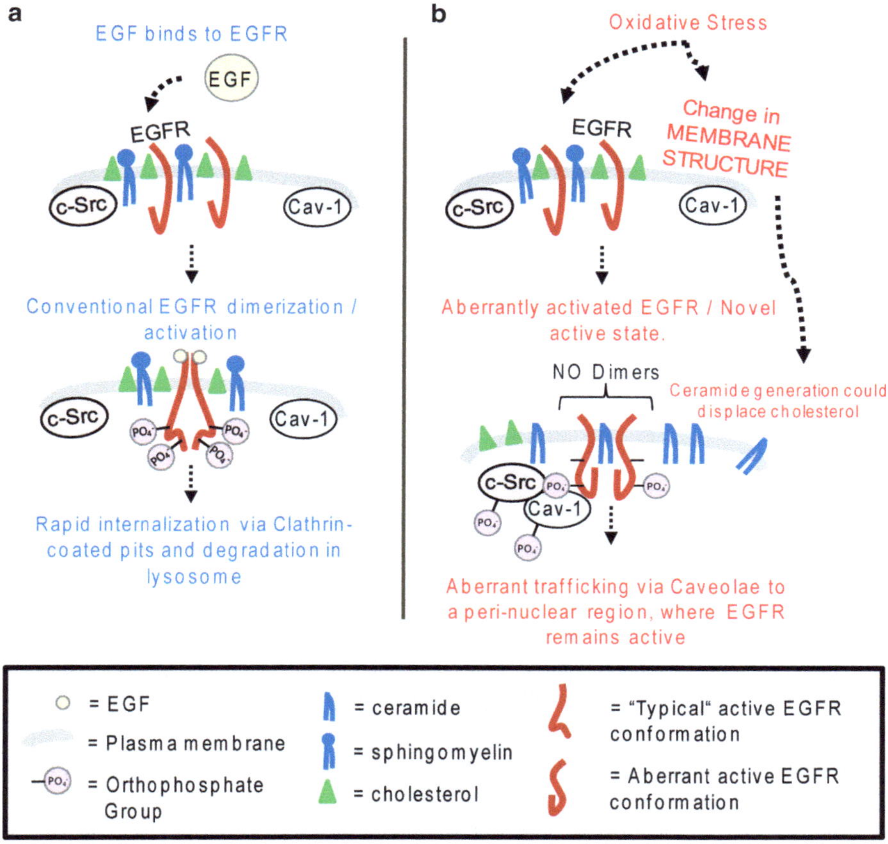

Fig. 15.2 EGFR exposed to oxidative stress acquires abnormal phosphorylation and aberrant activated conformation that impairs canonical dimerization. EGFR is bound to both active c-Src and phosphorylated Cav-1. This novel active state of EGFR could be supported by simultaneous changes in membrane structure/fluidity induced by oxidative stress-dependent increase in cellular ceramide levels

H_2O_2 belongs to the group of ROS that includes also superoxide anion (O^{2-}) and hydroxyl radical ($^{\cdot}OH$). Historically, the participation of ROS in diseases was explained with simplistic chemistry, in which critical cell proteins and lipids were randomly oxidized and rendered inactive for their roles in normal cell function [51]. The recognition that ROS function as signaling molecules has been more recent. It is well known that ROS regulate important steps in the signal transduction cascades and many critical cellular events [52–55]. As a result, ROS are involved in biologic processes ranging from normal tissue homeostasis to many human diseases. In this chapter, we present studies, which investigated the role of H_2O_2-induced ox-stress in cell proliferation and tumorigenesis through EGFR signaling [56, 57].

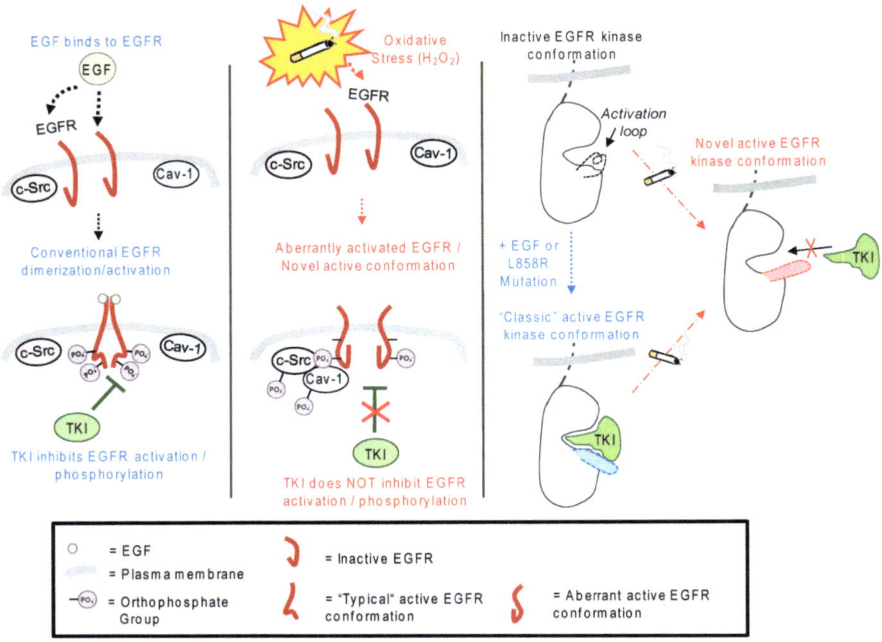

Fig. 15.3 Modeling EGFR structure/function alterations and change in kinase domain conformation that may lead to TKI resistance following oxidative stress exposure

(b) *Cigarette smoke produces H_2O_2-induced ox-stress*: Since H_2O_2 is known to be a major component of the gas phase of CS [58], a study was undertaken by the Goldkorn group to determine whether CS has a similar effect as H_2O_2 on EGFR activation in Human Airway Epithelial (HAE) cells, which included primary, immortalized, and transformed cells.

First, it was demonstrated that CS generates an amount of H_2O_2 that is comparable to the amount generated by the enzyme glucose oxidase (GO), which was used in earlier studies to artificially generate H_2O_2 [21, 35, 59]. Next, it was not only shown that CS can activate the EGFR in a dose- and time-dependent manner, but also that it aberrantly phosphorylated the receptor with a pattern of phosphorylation sites that was different from that induced by the ligand, EGF, but identical to the phosphorylation pattern induced by GO (namely, H_2O_2). Moreover, the EGFR was not activated and very little or no H_2O_2 could be detected in cell culture medium exposed to CS in the presence of glutathione (GSH), an antioxidant that specifically targets H_2O_2 and organic hydroperoxides, demonstrating that the H_2O_2-induced ox-stress was necessary for the activation of EGFR in HAE cells exposed to CS.

However, at the present time, there is no claim that H_2O_2 is the sole component in CS to cause the aberrant activation of the EGFR, although required. Rather, the previous knowledge of the effects of H_2O_2 on the EGFR was used to draw parallels to the effects of CS on the receptor and suggest that CS-generated H_2O_2 has a major role in the aberrant activation of the EGFR in the context of CS exposure [28].

(c) *The aberrant activation of EGFR by H_2O_2 and CS*: Several reports demonstrated the inactivation of protein tyrosine phosphatases (PTPs) by H_2O_2 and thus suggested that this was the cause responsible for EGFR-enhanced tyrosine phosphorylation under ox-stress [34, 56, 60–62]. For example, data published by Xu et al. indicated that H_2O_2 induced phosphorylation of EGFR on Y1068, while PTP activity was reduced to ~70 % of control. However, Goldkorn et al. as well as Reynolds et al. showed that H_2O_2 did not induce phosphorylation of a kinase-dead EGFR [59, 63], demonstrating that the kinase activity of the receptor is required for its activation by H_2O_2 and that the bulk inactivation of PTPs cannot explain the phenomenon of differential EGFR tyrosine phosphorylation when compared to the EGF-activated EGFR.

More specifically, it was reported that under both CS-induced H_2O_2 and direct H_2O_2 exposure, the pattern of EGFR phosphorylation is aberrant and different from the pattern of phosphorylation sites induced by EGF binding. Particularly, tyrosine (Tyr) 845 is robustly phosphorylated, and Tyr-1045 phosphorylation remains absent, which ultimately results in an active EGFR that is unable to undergo normal down-regulation, and demonstrates impaired trafficking and degradation.

Upon H_2O_2-induced ox-stress, the aberrantly activated EGFR is unable to bind c-Cbl and thus is neither ubiquitinated nor targeted to lysosomes for degradation. Instead, EGFR is strongly associated with phosphorylated Cav-1 in a Src-dependent manner and thus is recruited into caveolae and not into clathrin-coated pits [21, 28, 35, 56, 59]. Such ox-stress-dependent impairment of clathrin-mediated endocytosis and lysosomal degradation of the EGFR [4, 14, 15] result in prolonged downstream activation of pro-survival and proliferative molecules such as Akt and ERK1/2 [16]. This coincided with enhanced cell proliferation [64] and was shown to facilitate tumor promotion processes in the epithelial cell line T51B [65] as well as to mediate tumor promotion in other nonneoplastic rat liver epithelial cells [17].

To gain more insight into H_2O_2-induced EGFR signaling and hyperplasic responses, the role of the E3 ligase c-Cbl, as a possible link between ox-stress, EGFR signaling, and tumorigenesis was further examined.

15.6 c-Cbl and EGFR Under Oxidative Stress

(a) *c-Cbl does not bind EGFR under exposure to ox-stress*: It is well established that upon EGF stimulation of cells, c-Cbl binds directly to the EGFR via Tyr-1045 [35] and indirectly through the SH3 domain of Grb2 [66]. c-Cbl binding and its consequential phosphorylation result in the activation of the E3 ligase activity of c-Cbl, recruitment of the ubiquitin-conjugating enzyme Ubc-H7 [67], and EGFR ubiquitination.

On the other hand, upon mapping the EGFR phosphorylation sites, it was found that phosphorylation at Tyr-1045, the docking site for c-Cbl [13], was abrogated under ox-stress and thus could not be ubiquitinated and degraded. Therefore, it was suggested that this deficiency might have a key role in linking ox-stress, the EGFR, and tumorigenesis by conferring prolonged receptor signaling [21].

To gain a better understanding of how a receptor lacking c-Cbl binding may lead to tumorigenesis, the mutant EGFR (Tyr-1045 to Phe) was ectopically expressed in CHO cells to determine whether the lack of phosphorylation at this site is indeed the only cause for prolonged EGFR retention at the membrane under ox-stress. Additional findings suggested that the inability of the EGFR to bind c-Cbl under ox-stress is not solely due to its abrogated Tyr-1045 phosphorylation because the Y1045F mutant is still able to bind c-Cbl, probably indirectly via Grb2.

Indeed, other studies suggested that c-Cbl is recruited to the activated EGFR through both direct and indirect binding [68, 69]. Whereas direct c-Cbl-EGFR interaction is mediated through phosphorylated Tyr-1045 on EGFR [13], indirect c-Cbl-EGFR interaction is primarily mediated through Grb2. The SH3 domain of Grb2 binds to proline-rich sequences of c-Cbl, whereas the SH2 domain binds to autophosphorylated EGFR [8, 70]. Consistently, Huang and Sorkin [71] reported that knockdown of Grb2 by RNA interference inhibits clathrin-mediated endocytosis of the EGFR even following exposure of cells to EGF. This was linked to the impaired recruitment of the RING domain of c-Cbl to EGFR.

Interestingly, the two sites for Grb2 recruitment to the EGFR, Tyr-1068 and Tyr-1086, are phosphorylated under H_2O_2 exposure, but Grb2, which is still bound to c-Cbl, does not bind the EGFR. Furthermore, Shc, whose phosphorylation and binding to Grb2 had been shown to be responsible for Grb2 recruitment to the EGFR [8], is still phosphorylated under H_2O_2 exposure and is still able to bind to the EGFR. Therefore, it is possible that Grb2 fails to bind to the EGFR under ox-stress exposure due to conformational changes in the EGFR, and in fact, Goldkorn et al. [26, 72] have shown that ox-stress induces a novel active kinase conformer of the receptor, which is described in details at paragraph 8 of this chapter (below).

Undoubtedly, the studies with the Y1045F mutant (MT) EGFR [35] led to a better understanding of why c-Cbl fails to bind to the EGFR under ox-stress, which is due to both the abrogation of phosphorylation of Tyr-1045 and the lack of Grb2 binding, thereby resulting in a receptor unable to undergo normal internalization through the early endosomes and subsequent downregulation [59].

(b) *When is c-Cbl actually required for EGFR sorting?* Early on, several studies led to some confusion about when and where c-Cbl is required for EGFR internalization, as well as the role of EGFR Tyr-1045 phosphorylation in c-Cbl recruitment. For example, Jiang and Sorkin [68] showed data suggesting that the Y1045F EGFR MT was internalized despite its inability to undergo ubiquitination, whereas Mosesson et al. [73] demonstrated that the Y1045F MT is internalization resistant. At the same time, Duan et al. [74] showed in c-Cbl -/- mouse embryonic fibroblasts that the EGFR could still be internalized following EGF stimulation.

In order to gain a better understanding of the consequences of EGFR inability to recruit c-Cbl under ox-stress, the exact cellular compartment where c-Cbl-mediated ubiquitination is necessary for EGFR sorting was investigated. In particular, Goldkorn et al. showed [35] that the administration of PP1 (a Src family kinase inhibitor) to A549 cells blocked the EGF-induced phosphorylation of c-Cbl (but not of the EGFR) and the ubiquitination of the EGFR, without inhibiting the internalization of the receptor into early endosomes. Notably, an efficient binding of c-Cbl to the EGFR was observed upon EGF stimulation in the presence of PP1 [in both wild-type (WT) and Y1045F MT receptors], which implies that the tyrosine phosphorylation of c-Cbl is not required for its binding to the EGFR or for its role in EGFR entry into the early endosomes.

Subsequently, PP1 inhibitor was used to identify at which stage of EGFR intracellular sorting c-Cbl-mediated ubiquitination is required in EGFR trafficking. At extended time points, EGFR treated with PP1 and EGF remained associated with EEA1, an early endosomal marker. Only when PP1 was removed and c-Cbl phosphorylation and EGFR ubiquitination recovered, EGFR did migrate out of the early endosomes. Therefore, it turned out that the c-Cbl requirement and its mediation of ubiquitination may be very important for controlling EGFR exit from the early endosomes.

In summary, the major findings of that study [35] were that c-Cbl binding to the EGFR is sufficient to enhance receptor internalization, whereas the E3 ligase activity of c-Cbl and EGFR ubiquitination are required for EGFR trafficking out of the early endosomes and eventual transport to the lysosome for degradation of the receptor. In addition, that study also implied that the lack of Tyr-1045 phosphorylation during exposure to ox-stress is probably not the only factor in the inability of EGFR to enter the late endosomes. Rather, the role of Grb2 in the recruitment of c-Cbl as an adaptor added an additional level of complexity to the mechanism of EGFR sorting under ox-stress.

Importantly, the degree of intricacy was further expanded by other studies that demonstrated that several proteins with ubiquitin interaction domains are in fact required for EGFR transfer from the early endosomes to other vesicular bodies for degradation [75, 76]. For example, the proteins Hrs and Tsg 101 were shown to be involved in a large sorting complex that is "somehow" responsible for coupling EGFR transfer between early and late endosomes. Although the detailed mechanism has yet to be identified, one can imagine a situation where this large protein sorting complex binds to the ubiquitinated EGFR, thus allowing its transfer into the late endosomes.

15.7 EGFR Perinuclear Sorting Under Oxidative Stress

In light of the above studies, the next question would be how EGFR is specifically sorted under ox-stress. Findings by Goldkorn et al. [59] suggested that, under ox-stress, EGFR is able to undergo clathrin-independent endocytosis and is sorted to a perinuclear compartment, where it is not degraded and remains active. The mechanism of this trafficking involves activation of Src by H_2O_2, which subsequently phosphorylates Cav-1 at Tyr-14 and triggers the caveolar endocytosis of EGFR.

Cav-1 and Cav-2 hetero-oligomerize and form caveolae, which are integrated into the lipid raft [77–81]. It has been suggested that Cav-1 can function in caveolae in a manner analogous to the way clathrin adaptors draw membrane receptors to coated pits and/or drive membrane invagination and budding [77]. Cav-1 is known to interact directly with many signaling molecules through its caveolin-scaffolding domain at residues 82–101 [82, 83]. Indeed, EGFR has been reported to interact with the caveolin-scaffolding domain through a caveolin-binding sequence motif located in the intracellular kinase domain (residues 898–905) of the receptor [82, 84]. To elucidate the mechanism by which EGFR is being trafficked to the perinuclear compartment under ox-stress, the involvement of Cav-1 was investigated.

Expressing a wild-type (WT) Cav-1 or a Tyr-14 MT (Y14A) Cav-1 in a cell culture model, Goldkorn et al. demonstrated that EGFR constitutively associates with Cav-1. However, Cav-1 was phosphorylated on Y14 only in the presence of H_2O_2 and was subsequently accumulated together with EGFR in the perinuclear compartment. When either Src was inhibited or when the Y14A Cav-1 MT was overexpressed, EGFR accumulation was not observed at that unusual site [59].

Since caveolae-mediated endocytosis has been reported by others to be Src dependent, the data in this study are consistent with the notion that H_2O_2 induces caveolae-mediated endocytosis of EGFR and that Src-dependent Cav-1 Tyr-14 phosphorylation plays an essential role in the process.

Furthermore, although there have been reports that several Src family kinases (SFKs) are involved in ox-stress-induced phosphorylation of Cav-1 [85, 86], the observation that EGFR Tyr-845, a c-Src target [87], is robustly phosphorylated under ox-stress suggests that c-Src involvement is not coincidental. Consistently, Dittmann et al. have recently described a radiation-induced mode of EGFR trafficking to the nucleus, and the specific knockdown of c-Src (by siRNA) blocked EGFR phosphorylation at Y845 as well as the phosphorylation of Cav-1 at Y14, and resulted in the blockade of EGFR transport into the nucleus [24]. It should also be noted that Sanguinetti et al. [88] have reported that c-Abl (Abelson murine leukemia viral oncogene homolog) kinase expression is required for ox-stress-induced phosphorylation of Cav-1, although its eventual role in the perinuclear sorting of EGFR remains to be determined.

Even though the above study [59] demonstrated that Src-mediated phosphorylation of Cav-1 Tyr-14 was necessary for caveolar endocytosis of EGFR under ox-stress, the exact role of Cav-1 remains to be determined. Pelkmans et al. [89] suggested that the manner in which caveolar cargo is taken up and released depends on how the cargo interacts with Cav-1 as well as with other caveolar proteins and on how the cargo is influenced by compartment-specific signals such a pH changes. Goldkorn's model thus far indicates that H_2O_2 activation of EGFR results in aberrant receptor phosphorylation that precludes it from being sorted through clathrin-coated pits for eventual lysosomal degradation [21, 35]. Concomitantly, H_2O_2 (but not EGF) activates c-Src, which in turn phosphorylates Cav-1 at Y14 (a proposed trigger for caveolar endocytosis) and also dynamin-2 (at Y231/Y597), which is thought to localize at the neck of caveolae such that stimulation of its GTPase activity leads to vesicle fission [90–92]. Through this route of caveolar endocytosis, H_2O_2-activated EGFR is trafficked to a perinuclear region where continued receptor signaling is identified via Tyr-1173 phosphorylation, potentially contributing to prolonged proliferative signaling [26, 59] and tumorigenesis.

15.8 Activated Conformation of EGFR Under Oxidative Stress

Ox-stress activation of EGFR was found to be ligand independent, did not induce "classical" receptor dimerization, and was not inhibited by the TKIs AG1478, erlotinib (Tarceva) and gefitinib (Iressa) [26, 72]. Thus, an unprecedented, activated state of EGFR is generated under ox-stress. This activation mechanism was also described to be temperature dependent, suggesting the simultaneous involvement of membrane structure [26].

Recent data from Chung et al. [93], as well as the work of Bublil et al. [94], demonstrated that EGFR does not need an extracellular ligand to form dimers. EGFR continuously changes from a monomer to a dimer state, where the interactions between the intracellular domains are as important as that of the extracellular regions of the receptor in forming such dimers, thereby supporting the need of better understanding the mechanisms involved in the ligand-independent activation of EGFR and the physiological relevance of such mechanisms.

Very important indication that EGFR exposed to ox-stress undergoes a unique conformational change was supported by the finding that EGFR was strongly associated with c-Src. Moreover, the interaction between EGFR and c-Src was not dependent on the activation of c-Src because it persisted even in the presence of the c-Src kinase inhibitor PP1. Consistently, it was reported that under physiological conditions c-Src stably interacts with ErbB2, but not with EGFR/ErbB1, because of structural differences in the kinase domains of the two receptors [95]. Additionally, studies with the L858R EGFR MT also demonstrated that this MT could bind c-Src, whereas the WT EGFR under physiological conditions could not. The L858R EGFR MT was crystallized and shown to possess a protein conformation that differs from that of the WT EGFR at the level of the kinase domain, carrying a constitutively open "activating loop." Interestingly, the L858R EGFR MT was shown to have a similar functional phenotype to that of the WT EGFR exposed to ox-stress, previously described by Goldkorn et al. [21, 35, 59]: i.e., prolonged phosphorylation/activation, lack of Y1045 phosphorylation followed by lack of ubiquitination, impaired trafficking, and degradation and constitutive interaction with c-Src, without any ligand stimulation [96–99]. This further supported the idea that H_2O_2 induces a conformational change in the intracellular kinase domain of EGFR. Accordingly, dimerization of the extracellular domain could not be captured by the EDAC (1-Ethyl-3-(3-dimethylaminopropyl) carbodiimide) cross-linker neither for the WT EGFR under ox-stress [26, 72] nor for the L858R MT (unpublished observations).

Intriguingly, though WT EGFR under ox-stress appears to acquire a novel activated conformation, this conformation seems to be different from that of the L858R EGFR MT, which is known to be sensitive to TKIs [100]. Consistently, by employing a novel anti-EGFR antibody (α4-2 mAb) that is susceptible to the conformational changes induced by EGF binding to the receptor [101] or by the somatic mutation L858R, it was confirmed that CS/H_2O_2-generated ox-stress induces a unique active conformation of the receptor [26, 72], which is different from that of both the EGF-induced WT EGFR and the L858R EGFR MT. Furthermore, this unique conformation/activation mechanism appeared to be temperature dependent, suggesting the simultaneous involvement of membrane structure [26].

15.9 Is EGFR Aberrant Conformation Under Oxidative Stress Supported by Ceramide Generation?

Recently, Goldkorn et al. demonstrated that EGFR phosphorylation by ox-stress (CS/H_2O_2 exposure of HAE cells) is temperature dependent, suggesting membrane involvement [26]. Furthermore, the TKI AG1478 was ineffective in quenching EGFR phosphorylation by H_2O_2 in living cells, but capable of inhibiting EGFRs in a crude membrane fraction, where the normal membrane structure was destroyed. This supports the novel concept that the fluidity and structure of the membranes may be involved in either inducing or stabilizing the novel ox-stress-induced active conformation of EGFR [72], a theory encouraged by several interesting observations involving the EGFR and the membrane lipid components by other groups.

It was previously reported that disruption of cholesterol-enriched lipid rafts causes ligand-independent activation of EGFR [102]. One explanation was that a population of EGFRs is strongly associated with lipid rafts and their disruption by a cholesterol-sequestering agent, such as MβCD, would cause the relocalization of EGFR in non-raft portions of the plasma membrane where it could be activated due to its release from raft-associated inhibiting factors [103]. However, recent studies suggested the existence of at least two kinds of raft populations, the cholesterol-enriched and the ceramide-enriched rafts. The ceramide-enriched rafts are typically generated through cholesterol displacement by ceramide and are “less fluid” [104]. Therefore, Goldkorn et al. proposed a new role for ceramide generation under ox-stress exposure in stabilizing the EGFR aberrant activated conformation.

Developing this theory over the last 10 years, Goldkorn et al. have been investigating the mechanism of ceramide generation under ox-stress in the lungs, showing that ceramide levels are increased not only in the membranes of HAE cells but also in the lungs of mice exposed to H_2O_2 generated by cigarette smoke [105–110].

First, it was suggested that ceramide can induce the merging of lipid rafts in bigger signaling ceramide-enriched membrane platforms [104, 111–115].

Second, ceramide generation could support EGFR activation under ox-stress and may displace cholesterol in membrane rafts and thus support changes in the EGFR conformation, whereas cholesterol uptake in the plasma membrane could inhibit such ox-stress-induced activation of EGFR [26].

Third, because Goldkorn et al. were not able to demonstrate cholesterol depletion upon cellular ox-stress, it was suggested that the generation of cellular ceramide (as a result of exposure to ox-stress) may just displace cholesterol from the rafts, thereby disrupting cholesterol-enriched rafts and leading to EGFR relocalization to the more rigid ceramide-enriched rafts.

Indeed, it was recently shown for the first time that ox-stress-activated EGFR, as well as activated c-Src, colocalize within such ceramide-enriched regions of HAE cells. In particular, at early time points of ox-stress exposure (15 min), active EGFR and elevated ceramide colocalize primarily in the plasma membrane of the cells. Later on (30 min), such colocalization is observed mainly in a perinuclear region of the cells. Interestingly, as already mentioned above, Goldkorn et al. have previously shown that ox-stress-activated EGFR, unlike the EGF-stimulated receptor, is not internalized via clathrin-coated pits; while it is not degraded and remains active, EGFR can then traffic via caveolae to the perinucleus because of strong association with phosphorylated Cav-1 [35, 59]. Moreover, the ox-stress-activated EGFR colocalized with the early endosome marker EEA-1 [35, 59] and the recycling endosome marker Rab-11 (unpublished observation). Taken together, the generation of ceramide followed by cholesterol displacement may have a role in the aberrant trafficking of EGFR via caveolae to the perinuclear region under ox-stress exposure. Consistent with this hypothesis, it has been recently shown that ceramide generation increases the recruitment of Cav-1 into caveolae [116, 117].

Given the well-established pathophysiological function of EGFR and ceramide in carcinogenesis and severe lung injury, respectively, the data presented demonstrate that EGFR in airway epithelial cells exposed to ox-stress is not only aberrantly phosphorylated but also acquires a novel conformation that occurs in parallel with ceramide accumulation [110, 118–125]. Additionally, alterations in EGFR conformation are accompanied by c-Src binding to the receptor. Whether the change in EGFR conformation under ox-stress occurs as a result of simultaneous alterations in the membrane structure or happens independently and is only being stabilized by the simultaneous membrane changes requires additional studies.

15.10 Biological and Clinical Outcomes: Constitutively Active EGFR-Driven Non-small Cell Lung Cancer (NSCLC) Cells Are Sensitive to TKIs but Become Resistant Under (CS-Induced) Oxidative Stress

The EGFR and its downstream signaling are implicated in lung cancer development. Therefore, much effort was spent in developing specific tyrosine kinase inhibitors (TKIs) that bind to the EGFR ATP pocket, blocking EGFR phosphorylation/signaling. Clinical use of TKIs is effective in a subset of lung cancers with mutations in the EGFR kinase domain. However, these benefits are limited, and emergence of TKI resistance results in disease progression.

Canonically, resistance to TKIs is attributed to additional somatic mutations in EGFR, the T790M being the most common [126–128]. However, the first post-translational mechanism of EGFR resistance to TKIs (with no additional somatic mutations or kinase overexpression) was recently described in the context of ox-stress exposure of lung epithelial cells [26, 72]. Short (15–30 min) exposure of

HAE cells, both normal and transformed (NSCLC) cells, to CS or H_2O_2 causes aberrant phosphorylation/activation of EGFR, resulting in a conformation that is different from that induced by the ligand EGF or by some well-characterized activating somatic mutations of the EGFR kinase domain (such as the L858R). Unlike EGF-activated EGFR, CS/ox-stress-activated EGFR is not inhibited by TKIs (AG1478, erlotinib, gefitinib); in fact, the CS/ox-stress exposure induces TKI resistance even in the TKI-sensitive EGFR mutants (L858R and 746–750 deletion MT EGFR). These data demonstrated that CS/ox-stress exposure not only stimulates aberrant EGFR phosphorylation impairing receptor degradation but also induces a different EGFR conformation and signaling that are resistant to TKIs. Together, these findings offer new insights into CS-induced lung cancer development and TKI resistance.

Importantly, while the "classic" somatic mutations of EGFR and the CS-induced receptor alterations may result in similar phenotypes (in terms of EGFR downstream signaling and receptor trafficking), in all likelihood (as already mentioned above), they nonetheless represent distinct EGFR conformations and structure [129, 130]. This is most notably evidenced by the clinical findings that the sensitivity to TKIs is mainly observed in adenocarcinoma of nonsmokers, whereas CS exposure is primarily associated with squamous cell carcinoma and adenocarcinoma that are not sensitive to TKIs. Perhaps for this reason, nonsmoking adenocarcinoma patients who initially respond to TKIs develop resistance when they begin to smoke, while smoking patients (whose cancer developed in the setting of CS exposure) are resistant to TKIs in the first place. Therefore, we conclude that ox-stress-induced posttranslational changes in EGFR could provide an important mechanism of disease pathogenesis underlying TKI resistance in the context of cigarette smoking or any other source of ox-stress.

15.11 Summary

More than 3 decades of research have established a fairly defined mechanism of EGFR activation, internalization, and subsequent downregulation upon binding of its ligand EGF to the receptor's extracellular domain. However, much less is known about ligand-independent mechanisms of EGFR activation and trafficking. These "alternative" mechanisms substantiate the EGFR as a hallmark of several cancers, and thus defining them is extremely relevant to the understanding and treatment of EGFR-driven tumorigenesis.

In this chapter, we have discussed both known and postulated mechanism(s) of stress-dependent activation and subsequent internalization of the EGFR. We presented evidence demonstrating that the lack of c-Cbl recruitment to the EGFR plays a critical role in the stress-dependent EGFR aberrant endocytosis.

Although EGFR can be activated and/or trafficked within subcellular compartments following exposure to a variety of cell stressors, such as UV irradiation, cytokines, hydrogen peroxide (H_2O_2), and cigarette smoke, it appears that many, if not all, of these stressors share several similarities in the phenotype of EGFR activation and trafficking. Indeed, within the scope of this chapter, the role for cellular oxidative stress as a common denominator among cell stress events was presented. Both we and others found that following cell oxidative stress EGFR is aberrantly activated and eventually internalized via a path that differs from that induced by the its ligand EGF. This results in the EGFR either cycling back to the plasma membrane or accumulating in early endosomes, with no lysosomal degradation/downregulation.

Several molecular players of the aberrant EGFR trafficking have been proposed, such as c-Cbl, p38 MAPK (mitogen-activated protein kinase), c-Src, and caveolin-1 (Cav-1). However, the details of this complex machinery as well as the outcomes remain unresolved. On one hand, it was proposed that internalization of inactive EGFR following stress conditions can prevent its activation at the cell surface and thus block proliferative signaling in cells undergoing stress. On the other hand, an aberrantly tyrosine-phosphorylated/activated, and even internalized EGFR triggers survival signaling during stress conditions which could serve as a pathway usurped during tumorigenesis. Consistently, in the presence of cellular oxidative stress, the EGFR undergoes a robust and stable activation followed by activation of both survival and proliferative downstream signaling (such as Akt/protein kinase B and extracellular regulated protein kinase (ERK 1/2)).

Specifically, the abnormal processing of the EGFR under oxidative stress appears to stem from its (1) ligand-independent, (2) "monomeric," (3) tyrosine kinase inhibitor (TKI)-resistant, and (4) temperature-dependent activation, which is further characterized by c-Src binding to the receptor, an aberrant EGFR kinase conformation, and an aberrant pattern of EGFR phosphorylation sites. In addition, intriguing studies on EGFR activation and simultaneous membrane alteration suggested a role for ceramide-enriched lipid rafts in assisting the aberrant oxidative stress-generated activation of EGFR. However, whether the change in EGFR conformation under oxidative stress occurs as a result of simultaneous alterations in the membrane structure or happens independently and is subsequently stabilized by membrane alterations requires additional studies.

Ultimately, understanding the mechanism(s) of EGFR activation by oxidative stress will have important implications towards deciphering resistance to TKIs used clinically for treatment of NSCLC in addition to future drug development for EGFR-driven cancers.

References

1. Yarden Y (2001) The EGFR family and its ligands in human cancer. signalling mechanisms and therapeutic opportunities. Eur J Cancer 37(Suppl 4):S3–S8
2. Yarden Y, Sliwkowski MX (2001) Untangling the ErbB signalling network. Nat Rev Mol Cell Biol 2:127–137
3. Schlessinger J (2002) Ligand-induced, receptor-mediated dimerization and activation of EGF receptor. Cell 110:669–672
4. Schlessinger J (2000) Cell signaling by receptor tyrosine kinases. Cell 103:211–225

5. Waterman H, Yarden Y (2001) Molecular mechanisms underlying endocytosis and sorting of ErbB receptor tyrosine kinases. FEBS Lett 490:142–152
6. Franklin WA, Veve R, Hirsch FR, Helfrich BA, Bunn PA Jr (2002) Epidermal growth factor receptor family in lung cancer and premalignancy. Semin Oncol 29:3–14
7. Johannessen LE, Pedersen NM, Pedersen KW, Madshus IH, Stang E (2006) Activation of the epidermal growth factor (EGF) receptor induces formation of EGF receptor- and Grb2-containing clathrin-coated pits. Mol Cell Biol 26:389–401
8. Jiang X, Huang F, Marusyk A, Sorkin A (2003) Grb2 regulates internalization of EGF receptors through clathrin-coated pits. Mol Biol Cell 14:858–870
9. Rappoport JZ, Simon SM (2009) Endocytic trafficking of activated EGFR is AP-2 dependent and occurs through preformed clathrin spots. J Cell Sci 122:1301–1305
10. Haglund K, Di Fiore PP, Dikic I (2003) Distinct monoubiquitin signals in receptor endocytosis. Trends Biochem Sci 28:598–603
11. Wiley HS (2003) Trafficking of the ErbB receptors and its influence on signaling. Exp Cell Res 284:78–88
12. Marmor MD, Yarden Y (2004) Role of protein ubiquitylation in regulating endocytosis of receptor tyrosine kinases. Oncogene 23:2057–2070
13. Levkowitz G, Waterman H, Ettenberg SA, Katz M, Tsygankov AY et al (1999) Ubiquitin ligase activity and tyrosine phosphorylation underlie suppression of growth factor signaling by c-Cbl/Sli-1. Mol Cell 4:1029–1040
14. Hynes NE, Lane HA (2005) ERBB receptors and cancer: the complexity of targeted inhibitors. Nat Rev Cancer 5:341–354
15. Ogiso H, Ishitani R, Nureki O, Fukai S, Yamanaka M et al (2002) Crystal structure of the complex of human epidermal growth factor and receptor extracellular domains. Cell 110:775–787
16. Garrett TP, McKern NM, Lou M, Elleman TC, Adams TE et al (2002) Crystal structure of a truncated epidermal growth factor receptor extracellular domain bound to transforming growth factor alpha. Cell 110:763–773
17. Dawson JP, Berger MB, Lin CC, Schlessinger J, Lemmon MA et al (2005) Epidermal growth factor receptor dimerization and activation require ligand-induced conformational changes in the dimer interface. Mol Cell Biol 25:7734–7742
18. Rosette C, Karin M (1996) Ultraviolet light and osmotic stress: activation of the JNK cascade through multiple growth factor and cytokine receptors. Science 274:1194–1197
19. Hirota K, Murata M, Itoh T, Yodoi J, Fukuda K, Hirota K, Murata M, Itoh T, Yodoi J, Fukuda K (2001) Redox-sensitive transactivation of epidermal growth factor receptor by tumor necrosis factor confers the NF-kappa B activation. J Biol Chem 276:25953–25958
20. Benhar M, Engelberg D, Levitzki A (2002) Cisplatin-induced activation of the EGF receptor. Oncogene 21:8723–8731
21. Ravid T, Sweeney C, Gee P, Carraway KRK, Goldkorn T (2002) EGF receptor activation under oxidative stress fails to promote c-Cbl mediated down regulation. J Biol Chem 12:12
22. Bird TA, Saklatvala J (1990) Down-modulation of epidermal growth factor receptor affinity in fibroblasts treated with interleukin 1 or tumor necrosis factor is associated with phosphorylation at a site other than threonine 654. J Biol Chem 265:235–240
23. Oksvold MP, Huitfeldt HS, Ostvold AC, Skarpen E (2002) UV induces tyrosine kinase-independent internalisation and endosome arrest of the EGF receptor. J Cell Sci 115:793–803
24. Dittmann K, Mayer C, Kehlbach R, Rodemann HP (2008) Radiation-induced caveolin-1 associated EGFR internalization is linked with nuclear EGFR transport and activation of DNA-PK. Mol Cancer 7:69
25. Peng XH, Karna P, Cao Z, Jiang BH, Zhou M et al (2006) Cross-talk between epidermal growth factor receptor and hypoxia-inducible factor-1alpha signal pathways increases resistance to apoptosis by up-regulating survivin gene expression. J Biol Chem 281:25903–25914
26. Filosto S, Khan E, Tognon E, Becker C, Ashfaq M et al (2011) EGF receptor exposed to oxidative stress acquires abnormal phosphorylation and aberrant activated conformation that impairs canonical dimerization. PLoS One 6:e23240

27. Zwang Y, Yarden Y (2006) p38 MAP kinase mediates stress-induced internalization of EGFR: implications for cancer chemotherapy. EMBO J 25:4195–4206
28. Khan EM, Lanir R, Danielson AR, Goldkorn T (2008) EGF receptor exposed to cigarette smoke is aberrantly activated and undergoes perinuclear trafficking. FASEB J 22:910–917
29. Huang RP, Wu JX, Fan Y, Adamson ED (1996) UV activates growth factor receptors via reactive oxygen intermediates. J Cell Biol 133:211–220
30. Sachsenmaier C, Radler-Pohl A, Zinck R, Nordheim A, Herrlich P et al (1994) Involvement of growth factor receptors in the mammalian UVC response. Cell 78:963–972
31. Coffer PJ, Burgering BM, Peppelenbosch MP, Bos JL, Kruijer W (1995) UV activation of receptor tyrosine kinase activity. Oncogene 11:561–569
32. Huang C, Ma W, Bowden GT, Dong Z (1996) Ultraviolet B-induced activated protein-1 activation does not require epidermal growth factor receptor but is blocked by a dominant negative PKClambda/iota. J Biol Chem 271:31262–31268
33. Gross S, Knebel A, Tenev T, Neininger A, Gaestel M et al (1999) Inactivation of protein-tyrosine phosphatases as mechanism of UV-induced signal transduction. J Biol Chem 274:26378–26386
34. Xu Y, Tan LJ, Grachtchouk V, Voorhees JJ, Fisher GJ (2005) Receptor-type protein-tyrosine phosphatase-kappa regulates epidermal growth factor receptor function. J Biol Chem 280:42694–42700
35. Ravid T, Heidinger J, Gee P, Khan E, Goldkorn T (2004) c-Cbl-mediated ubiquitinylation is required for EGF receptor exit from the early endosomes. J Biol Chem 279:37153–37162
36. Oksvold MP, Thien CB, Widerberg J, Chantry A, Huitfeldt HS et al (2004) UV-radiation-induced internalization of the epidermal growth factor receptor requires distinct serine and tyrosine residues in the cytoplasmic carboxy-terminal domain. Radiat Res 161:685–691
37. Oksvold MP, Thien CB, Widerberg J, Chantry A, Huitfeldt HS et al (2003) Serine mutations that abrogate ligand-induced ubiquitination and internalization of the EGF receptor do not affect c-Cbl association with the receptor. Oncogene 22:8509–8518
38. Winograd-Katz SE, Levitzki A (2006) Cisplatin induces PKB/Akt activation and p38(MAPK) phosphorylation of the EGF receptor. Oncogene 25:7381–7390
39. Lambert S, Ameels H, Gniadecki R, Herin M, Poumay Y (2008) Internalization of EGF receptor following lipid rafts disruption in keratinocytes is delayed and dependent on p38 MAPK activation. J Cell Physiol 217:834–845
40. Vergarajauregui S, San Miguel A, Puertollano R (2006) Activation of p38 mitogen-activated protein kinase promotes epidermal growth factor receptor internalization. Traffic 7:686–698
41. Lambert S, Frankart A, Poumay Y (2010) p38 MAPK-regulated EGFR internalization takes place in keratinocyte monolayer during stress conditions. Arch Dermatol Res 302:229–233
42. Dorward A, Sweet S, Moorehead R, Singh G (1997) Mitochondrial contributions to cancer cell physiology: redox balance, cell cycle, and drug resistance. J Bioenerg Biomembr 29: 385–392
43. Toyokuni S, Okamoto K, Yodoi J, Hiai H, Toyokuni S, Okamoto K, Yodoi J, Hiai H (1995) Persistent oxidative stress in cancer. FEBS Lett 358:1–3
44. Adler V, Yin Z, Tew KD, Ronai Z (1999) Role of redox potential and reactive oxygen species in stress signaling. Oncogene 18:6104–6111
45. Kamata H, Hirata H (1999) Redox regulation of cellular signalling. Cell Signal 11:1–14
46. Karin M (1998) Mitogen-activated protein kinase cascades as regulators of stress responses. Ann N Y Acad Sci 851:139–146
47. Wang X, Martindale JL, Liu Y, Holbrook NJ (1998) The cellular response to oxidative stress: influences of mitogen-activated protein kinase signalling pathways on cell survival. Biochem J 333(Pt 2):291–300
48. Williams MD, Chance B (1983) Spontaneous chemiluminescence of human breath. Spectrum, lifetime, temporal distribution, and correlation with peroxide. J Biol Chem 258:3628–3631
49. Sznajder JI, Fraiman A, Hall JB, Sanders W, Schmidt G et al (1989) Increased H_2O_2 in the expired breath of patients with acute hypoxemic respiratory failure. Chest 96:606–612

50. Nowak D, Antczak A, Krol M, Pietras T, Shariati B et al (1996) Increased content of H_2O_2 in the expired breath of cigarette smokers. Eur Respir J 9:652–657
51. Halliwell B, Gutteridge JM (1990) Role of free radicals and catalytic metal ions in human disease: an overview. Methods Enzymol 186:1–85
52. Sen C, Packer L (1996) Antioxidant and redox regulation of gene transcription. FASEB J 10:709–720
53. Kamata H, Honda S, Maeda S, Chang L, Hirata H et al (2005) Reactive oxygen species promote TNFalpha-induced death and sustained JNK activation by inhibiting MAP kinase phosphatases. Cell 120:649–661
54. Li N, Karin M (1999) Is NF-kappaB the sensor of oxidative stress? FASEB J 13:1137–1143
55. Devary Y, Gottlieb RA, Smeal T, Karin M (1992) The mammalian ultraviolet response is triggered by activation of Src tyrosine kinases. Cell 71:1081–1091
56. Goldkorn T, Balaban N, Matsukuma K, Last J, Chan C et al (1998) EGF-Receptor phosphorylation and signaling are targeted by H_2O_2 redox stress. Am J Respir Cell Mol Biol 19:786–798
57. van der Vliet A, Hristova M, Cross CE, Eiserich JP, Goldkorn T (1998) Peroxynitrite induces covalent dimerization of epidermal growth factor receptors in A431 epidermoid carcinoma cells. J Biol Chem 273:31860–31866
58. Nakayama T, Kodama M, Nagata C (1984) Generation of hydrogen peroxide and superoxide anion radical from cigarette smoke. Gann 75:95–98
59. Khan E, Heidinger J, Levy M, Lisanti M, Ravid T et al (2006) EGF receptor exposed to oxidative stress undergoes Src- and caveolin-1-dependent perinuclear trafficking. J Biol Chem 281:14486–14493
60. Gamou S, Shimizu N (1995) Hydrogen peroxide preferentially enhances the tyrosine phosphorylation of epidermal growth factor receptor. FEBS Lett 357:161–164
61. Denu JM, Tanner KG (1998) Specific and reversible inactivation of protein tyrosine phosphatases by hydrogen peroxide: evidence for a sulfenic acid intermediate and implications for redox regulation. Biochemistry 37:5633–5642
62. Lee SR, Kwon KS, Kim SR, Rhee SG (1998) Reversible inactivation of protein-tyrosine phosphatase 1B in A431 cells stimulated with epidermal growth factor. J Biol Chem 273:15366–15372
63. Reynolds AR, Tischer C, Verveer PJ, Rocks O, Bastiaens PI (2003) EGFR activation coupled to inhibition of tyrosine phosphatases causes lateral signal propagation. Nat Cell Biol 5:447–453
64. Van Winkle LS, Isaac JM, Plopper CG (1997) Distribution of epidermal growth factor receptor and ligands during bronchiolar epithelial repair from naphthalene-induced Clara cell injury in the mouse. Am J Pathol 151:443–459
65. Huang RP, Peng A, Golard A, Hossain MZ, Huang R et al (2001) Hydrogen peroxide promotes transformation of rat liver non-neoplastic epithelial cells through activation of epidermal growth factor receptor. Mol Carcinog 30:209–217
66. Fujikawa K, de Aos Scherpenseel I, Jain SK, Presman E, Christensen RA et al (1999) Role of PI 3-kinase in angiopoietin-1-mediated migration and attachment-dependent survival of endothelial cells. Exp Cell Res 253:663–672
67. Yokouchi M, Kondo T, Houghton A, Bartkiewicz M, Horne WC et al (1999) Ligand-induced ubiquitination of the epidermal growth factor receptor involves the interaction of the c-Cbl RING finger and UbcH7. J Biol Chem 274:31707–31712
68. Jiang X, Sorkin A (2003) Epidermal growth factor receptor internalization through clathrin-coated pits requires Cbl RING finger and proline-rich domains but not receptor polyubiquitylation. Traffic 4:529–543
69. Waterman H, Katz M, Rubin C, Shtiegman K, Lavi S et al (2002) A mutant EGF-receptor defective in ubiquitylation and endocytosis unveils a role for Grb2 in negative signaling. EMBO J 21:303–313
70. Fukazawa T, Miyake S, Band V, Band H (1996) Tyrosine phosphorylation of Cbl upon epidermal growth factor (EGF) stimulation and its association with EGF receptor and downstream signaling proteins. J Biol Chem 271:14554–14559

71. Huang F, Kirkpatrick D, Jiang X, Gygi S, Sorkin A (2006) Differential regulation of EGF receptor internalization and degradation by multiubiquitination within the kinase domain. Mol Cell 21:737–748
72. Filosto S, Becker CR, Goldkorn T (2012) Cigarette smoke induces aberrant EGF receptor activation that mediates lung cancer development and resistance to tyrosine kinase inhibitors. Mol Cancer Ther 11:795–804
73. Mosesson Y, Shtiegman K, Katz M, Zwang Y, Vereb G et al (2003) Endocytosis of receptor tyrosine kinases is driven by mono-, not poly-ubiquitylation. J Biol Chem 278(24):21323–21326
74. Duan L, Miura Y, Dimri M, Majumder B, Dodge IL et al (2003) Cbl-mediated ubiquitinylation is required for lysosomal sorting of epidermal growth factor receptor but is dispensable for endocytosis. J Biol Chem 278:28950–28960
75. Bao J, Gur G, Yarden Y (2003) Src promotes destruction of c-Cbl: implications for oncogenic synergy between Src and growth factor receptors. Proc Natl Acad Sci U S A 100:2438–2443
76. Urbe S, Sachse M, Row PE, Preisinger C, Barr FA et al (2003) The UIM domain of Hrs couples receptor sorting to vesicle formation. J Cell Sci 116:4169–4179
77. Quest AF, Leyton L, Parraga M (2004) Caveolins, caveolae, and lipid rafts in cellular transport, signaling, and disease. Biochem Cell Biol 82:129–144
78. Nabi IR, Le PU (2003) Caveolae/raft-dependent endocytosis. J Cell Biol 161:673–677
79. Laude AJ, Prior IA (2004) Plasma membrane microdomains: organization, function and trafficking. Mol Membr Biol 21:193–205
80. Galbiati F, Razani B, Lisanti MP (2001) Emerging themes in lipid rafts and caveolae. Cell 106:403–411
81. Fra AM, Williamson E, Simons K, Parton RG (1995) De novo formation of caveolae in lymphocytes by expression of VIP21-caveolin. Proc Natl Acad Sci U S A 92:8655–8659
82. Ikezu T, Trapp BD, Song KS, Schlegel A, Lisanti MP et al (1998) Caveolae, plasma membrane microdomains for alpha-secretase-mediated processing of the amyloid precursor protein. J Biol Chem 273:10485–10495
83. Couet J, Li S, Okamoto T, Ikezu T, Lisanti MP (1997) Identification of peptide and protein ligands for the caveolin-scaffolding domain. Implications for the interaction of caveolin with caveolae-associated proteins. J Biol Chem 272:6525–6533
84. Couet J, Sargiacomo M, Lisanti MP (1997) Interaction of a receptor tyrosine kinase, EGF-R, with caveolins. Caveolin binding negatively regulates tyrosine and serine/threonine kinase activities. J Biol Chem 272:30429–30438
85. Cao H, Sanguinetti AR, Mastick CC (2004) Oxidative stress activates both Src-kinases and their negative regulator Csk and induces phosphorylation of two targeting proteins for Csk: caveolin-1 and paxillin. Exp Cell Res 294:159–171
86. Sanguinetti AR, Cao H, Corley Mastick C (2003) Fyn is required for oxidative- and hyperosmotic-stress-induced tyrosine phosphorylation of caveolin-1. Biochem J 376:159–168
87. Biscardi JS, Maa MC, Tice DA, Cox ME, Leu TH et al (1999) c-Src-mediated phosphorylation of the epidermal growth factor receptor on Tyr845 and Tyr1101 is associated with modulation of receptor function. J Biol Chem 274:8335–8343
88. Sanguinetti AR, Mastick CC (2003) c-Abl is required for oxidative stress-induced phosphorylation of caveolin-1 on tyrosine 14. Cell Signal 15:289–298
89. Pelkmans L, Burli T, Zerial M, Helenius A (2004) Caveolin-stabilized membrane domains as multifunctional transport and sorting devices in endocytic membrane traffic. Cell 118:767–780
90. Shajahan AN, Timblin BK, Sandoval R, Tiruppathi C, Malik AB et al (2004) Role of Src-induced dynamin-2 phosphorylation in caveolae-mediated endocytosis in endothelial cells. J Biol Chem 279:20392–20400
91. Shajahan AN, Tiruppathi C, Smrcka AV, Malik AB, Minshall RD (2004) Gbetagamma activation of Src induces caveolae-mediated endocytosis in endothelial cells. J Biol Chem 279:48055–48062
92. Oh P, McIntosh DP, Schnitzer JE (1998) Dynamin at the neck of caveolae mediates their budding to form transport vesicles by GTP-driven fission from the plasma membrane of endothelium. J Cell Biol 141:101–114

93. Chung I, Akita R, Vandlen R, Toomre D, Schlessinger J et al (2010) Spatial control of EGF receptor activation by reversible dimerization on living cells. Nature 464:783–787
94. Bublil EM, Pines G, Patel G, Fruhwirth G, Ng T et al (2010) Kinase-mediated quasi-dimers of EGFR. FASEB J 24(12):4744–4755
95. Kim H, Chan R, Dankort DL, Zuo D, Najoukas M et al (2005) The c-Src tyrosine kinase associates with the catalytic domain of ErbB-2: implications for ErbB-2 mediated signaling and transformation. Oncogene 24:7599–7607
96. Paez JG, Janne PA, Lee JC, Tracy S, Greulich H et al (2004) EGFR mutations in lung cancer: correlation with clinical response to gefitinib therapy. Science 304:1497–1500
97. Chung BM, Raja SM, Clubb RJ, Tu C, George M et al (2009) Aberrant trafficking of NSCLC-associated EGFR mutants through the endocytic recycling pathway promotes interaction with Src. BMC Cell Biol 10:84
98. Chung BM, Dimri M, George M, Reddi AL, Chen G et al (2009) The role of cooperativity with Src in oncogenic transformation mediated by non-small cell lung cancer-associated EGF receptor mutants. Oncogene 28:1821–1832
99. Shtiegman K, Kochupurakkal BS, Zwang Y, Pines G, Starr A et al (2007) Defective ubiquitinylation of EGFR mutants of lung cancer confers prolonged signaling. Oncogene 26:6968–6978
100. Yang CH (2008) EGFR tyrosine kinase inhibitors for the treatment of NSCLC in East Asia: present and future. Lung Cancer 60(Suppl 2):S23–S30
101. Ise N, Omi K, Miwa K, Honda H, Higashiyama S et al (2010) Novel monoclonal antibodies recognizing the active conformation of epidermal growth factor receptor. Biochem Biophys Res Commun 394:685–690
102. Lambert S, Vind-Kezunovic D, Karvinen S, Gniadecki R (2006) Ligand-independent activation of the EGFR by lipid raft disruption. J Invest Dermatol 126:954–962
103. Wang XQ, Yan Q, Sun P, Liu JW, Go L et al (2007) Suppression of epidermal growth factor receptor signaling by protein kinase C-alpha activation requires CD82, caveolin-1, and ganglioside. Cancer Res 67:9986–9995
104. Gulbins E, Kolesnick R (2003) Raft ceramide in molecular medicine. Oncogene 22:7070–7077
105. Goldkorn T, Balaban N, Shannon M, Chea V, Matsukuma K et al (1998) H2O2 acts on cellular membranes to generate ceramide signaling and initiate apoptosis in tracheobronchial epithelial cells. J Cell Sci 111:3209–3220
106. Levy M, Castillo SS, Goldkorn T (2006) nSMase2 activation and trafficking are modulated by oxidative stress to induce apoptosis. Biochem Biophys Res Commun 344:900–905
107. Filosto S, Fry W, Knowlton AA, Goldkorn T (2010) Neutral sphingomyelinase 2 (nSMase2) is a phosphoprotein regulated by calcineurin (PP2B). J Biol Chem 285:10213–10222
108. Chan C, Goldkorn T (2000) Ceramide path in human lung cell death. Am J Respir Cell Mol Biol 22:460–468
109. Lavrentiadou SN, Chan C, Ravid T, Tsaba A, van der Vliet A et al (2001) Ceramide-mediated apoptosis in lung epithelial cells is regulated by GSH. Am J Respir Cell Mol Biol 25:676–684
110. Filosto S, Castillo S, Danielson A, Franzi L, Khan E et al (2011) Neutral sphingomyelinase 2: a novel target in cigarette smoke-induced apoptosis and lung injury. Am J Respir Cell Mol Biol 44:350–360
111. Sot J, Ibarguren M, Busto JV, Montes LR, Goni FM et al (2008) Cholesterol displacement by ceramide in sphingomyelin-containing liquid-ordered domains, and generation of gel regions in giant lipidic vesicles. FEBS Lett 582:3230–3236
112. Megha LE (2004) Ceramide selectively displaces cholesterol from ordered lipid domains (rafts): implications for lipid raft structure and function. J Biol Chem 279:9997–10004
113. Megha SP, Kolter T, Bittman R, London E (2007) Effect of ceramide N-acyl chain and polar headgroup structure on the properties of ordered lipid domains (lipid rafts). Biochim Biophys Acta 1768:2205–2212

114. Goni FM, Alonso A (2009) Effects of ceramide and other simple sphingolipids on membrane lateral structure. Biochim Biophys Acta 1788:169–177
115. Contreras FX, Villar AV, Alonso A, Kolesnick RN, Goni FM (2003) Sphingomyelinase activity causes transbilayer lipid translocation in model and cell membranes. J Biol Chem 278:37169–37174
116. Yang Y, Yin J, Baumgartner W, Samapati R, Solymosi EA et al (2010) Platelet-activating factor reduces endothelial nitric oxide production: role of acid sphingomyelinase. Eur Respir J 36:417–427
117. Kuebler WM, Yang Y, Samapati R, Uhlig S (2010) Vascular barrier regulation by PAF, ceramide, caveolae, and NO - an intricate signaling network with discrepant effects in the pulmonary and systemic vasculature. Cell Physiol Biochem 26:29–40
118. Egloff AM, Grandis JR (2008) Targeting epidermal growth factor receptor and SRC pathways in head and neck cancer. Semin Oncol 35:286–297
119. Lieu C, Kopetz S (2010) The SRC family of protein tyrosine kinases: a new and promising target for colorectal cancer therapy. Clin Colorectal Cancer 9:89–94
120. Cai Z, Zhang H, Liu J, Berezov A, Murali R et al (2010) Targeting erbB receptors. Semin Cell Dev Biol 21(9):961–966
121. Lurje G, Lenz HJ (2009) EGFR signaling and drug discovery. Oncology 77:400–410
122. Goldkorn T, Filosto S (2010) Lung injury and cancer: mechanistic insights into ceramide and EGFR signaling under cigarette smoke. Am J Respir Cell Mol Biol 43:259–268
123. Petrache I, Natarajan V, Zhen L, Medler TR, Richter A et al (2006) Ceramide causes pulmonary cell apoptosis and emphysema: a role for sphingolipid homeostasis in the maintenance of alveolar cells. Proc Am Thorac Soc 3:510
124. Petrache I, Natarajan V, Zhen L, Medler TR, Richter AT et al (2005) Ceramide upregulation causes pulmonary cell apoptosis and emphysema-like disease in mice. Nat Med 11:491–498
125. Becker KA, Grassme H, Zhang Y, Gulbins E (2010) Ceramide in Pseudomonas aeruginosa infections and cystic fibrosis. Cell Physiol Biochem 26:57–66
126. Pao W, Miller VA, Politi KA, Riely GJ, Somwar R et al (2005) Acquired resistance of lung adenocarcinomas to gefitinib or erlotinib is associated with a second mutation in the EGFR kinase domain. PLoS Med 2:e73
127. Yun CH, Mengwasser KE, Toms AV, Woo MS, Greulich H et al (2008) The T790M mutation in EGFR kinase causes drug resistance by increasing the affinity for ATP. Proc Natl Acad Sci U S A 105:2070–2075
128. Kosaka T, Yatabe Y, Endoh H, Yoshida K, Hida T et al (2006) Analysis of epidermal growth factor receptor gene mutation in patients with non-small cell lung cancer and acquired resistance to gefitinib. Clin Cancer Res 12:5764–5769
129. Zhang X, Gureasko J, Shen K, Cole PA, Kuriyan J (2006) An allosteric mechanism for activation of the kinase domain of epidermal growth factor receptor. Cell 125:1137–1149
130. Jura N, Endres NF, Engel K, Deindl S, Das R et al (2009) Mechanism for activation of the EGF receptor catalytic domain by the juxtamembrane segment. Cell 137:1293–1307

Chapter 16
Internalisation, Endosomal Trafficking and Recycling of Integrins During Cell Migration and Cancer Invasion

Elena Rainero, Peter V.E. van den Berghe, and Jim C. Norman

Abstract There are now clear indications that endocytosis and recycling of integrins is important during cancer progression, invasion and metastasis. In this chapter, we will review the recent literature outlining the cellular mechanisms that control integrin endocytosis, endocytic trafficking and recycling and how these processes may be measured experimentally. The influence of endocytic processes on integrin function will then be considered, such as how the spatial dynamics of integrins are controlled in migrating cells. We will then focus on the various ways that integrin trafficking can influence cancer cell migration and invasion. In particular, we will highlight (a) how $\alpha5\beta1$ integrin and receptor tyrosine kinase (RTK) recycling is upregulated following $\alpha v\beta3$ inhibition as this provides a molecular explanation for the overall lack of efficacy of anti-αv integrin drugs in tumour angiogenesis; (b) how trafficking events can dictate whether certain GTPases, such as Rab25, function as tumour suppressors or promoters; and (c) how mutation of p53 drives invasion and metastasis by altering integrin and RTK recycling, and we will discuss whether components of these membrane trafficking pathways may be targeted therapeutically to reverse oncogenic drive in human cancers.

16.1 Introduction

A primary function of integrins is to mediate adhesion of cells to the extracellular matrix (ECM). Thus, integrins present at the plasma membrane bind extracellularly to ECM proteins, such as fibronectin, laminin and collagen, and associate intracellularly with the cytoskeleton and the cell's signalling machinery [1]. In adult organisms, integrin-mediated adhesion functions primarily to keep cells in their appropriate

E. Rainero • P.V.E. van den Berghe • J.C. Norman (✉)
Beatson Institute for Cancer Research, Garscube Estate, Glasgow G61 1BD, UK
e-mail: j.norman@beatson.gla.ac.uk

Y. Yarden and G. Tarcic (eds.), *Vesicle Trafficking in Cancer*,
DOI 10.1007/978-1-4614-6528-7_16,

place and to prevent their inappropriate movement around the body. However, during development and in situations such as wound healing, integrin-mediated adhesion becomes dynamic and acts to drive cell migration. Moreover, integral to a number of diseases is the acquisition of integrin-mediated migratory characteristics by cells that should rightly be stationary. Prominent amongst these pathologies is, of course, cancer metastasis [2, 3]. Indeed, one of the defining characteristics of metastatic cancer cells, and one that makes cancer so difficult to treat, is their ability to detach from the primary tumour and migrate away to form secondary tumours. The acquisition of migratory characteristics by cancer cells has been the focus of much attention by cell biologists over the last 15 years, and it is now clear that alterations in the behaviour of the cell's adhesion machinery and integrin signalling in particular are integral to the metastatic phenotype. For instance, cancer cells are established to invade the ECM by adopting either elongated (mesenchymal) or rounded (amoeboid) morphologies, and the switch between these two migratory modes is known to be dictated by the way that integrins signal to Rho subfamily GTPases [2, 3].

Many cell surface receptors, particularly those that are involved in the uptake of nutrients (such as the transferrin receptor (TfnR)), are continuously and rapidly internalised into endocytic compartments and then promptly returned (or recycled) from these endosomes to the plasma membrane; this is commonly described as an 'endo–exocytic cycle' [4]. Moreover, following internalisation to early endosomes (EEs), cycling receptors are returned to the plasma membrane via one of at least two temporally and spatially distinct recycling pathways. One pathway directly returns the receptors to the plasma membrane from EEs; this we term 'short-loop' recycling. Receptors not returned via the short-loop traffic from EEs to the perinuclear recycling compartment (PNRC) are recycled from here; this less direct route takes longer and so has been termed the 'long-loop' pathway.

Marc Bretscher was the first to report that some integrins are actively engaged in endo–exocytic cycling [5, 6]. These observations have been subsequently confirmed by many other laboratories, and it is now clear that, depending on the heterodimer in question, integrins can follow both the long- and short-loop pathways of endo–exocytic cycling and that this can be sufficient to route the entire plasma membrane pool of integrin through the endosomal system at least once every 30 min [7, 8]. It had been known for some time that inhibition of endocytosis profoundly disrupts the ability of cells to polarise their cytoskeleton and to migrate [9]. Therefore, the observations that integrins are internalised and recycled so quickly indicated a potentially important way for cells to control their migration and prompted a number of laboratories to study the regulation of integrin trafficking in more detail. As a result, it is now clear that the way that integrins are internalised and recycled does indeed control cell migration and has particular influence on the way that metastatic cancer cells move through and invade the ECM.

In this chapter, we aim to provide a comprehensive review of the literature which covers the role played by integrin internalisation and recycling in cancer cell migration and invasion. We will aim to emphasise not only the data and conclusions of studies reporting the consequences of disrupting integrin transport on cell migration and invasion but also the methodologies used to obtain these data. In view of this, we will devote a section of the chapter to looking at the various advantages and limitations of experimental approaches used to study integrin endocytosis and recycling and how the choice of a particular method may influence the interpretation of the results obtained. Following this, we will look sequentially at what is known about the pathways of integrin endocytosis and recycling and how these may be manipulated experimentally. Finally, we will cover the role played by integrin trafficking in cell migration and invasion and how this may occur mechanistically. Indeed, endo–exocytosis has been proposed to contribute to cell migration by:

(a) Mediating the bulk translocation of integrins from the back to the front of the cell (anterograde movement) [10]
(b) Transporting integrins from the cell front to the cell rear (retrograde movement) [11]
(c) Actively maintaining subpopulations of integrins within particular cellular regions (spatial restriction of cycling integrins) [12]
(d) Facilitating the recycling of other signalling receptors such as receptor tyrosine kinases (RTKs) [13]

Indeed, a view is now emerging that although there is little evidence that integrins are transported from the back to the front of migrating cells (as in point (a) above), it is now clear that retrograde transport and spatial restriction of cycling integrins does occur (points (b) and (c)). Moreover, recent work has shown that gain-of-faction mutants of the p53 tumour suppressor act to drive invasion by influencing integrin recycling which, in turn, affects the trafficking and signalling of RTKs.

16.2 Experimental Approaches for Measuring Integrin Transport

To draw meaningful conclusions concerning the role played by integrin trafficking during cell migration and invasion, it is necessary to be able to measure the rates at which integrins are internalised and recycled and ideally to visualise the sites at which these events occur in migrating cells. Moreover, as the processes of integrin internalisation and recycling are intrinsically linked (i.e. the cell must first internalise an integrin before it can recycle it), and also quite rapid, it is sometimes quite challenging to obtain metrics for internalisation and recycling that are determined as independently from each another as possible. There are a number of established ways to do this, and these fall primarily into two categories: 'biochemical' assays that provide information regarding the bulk rates at which integrins are internalised and recycled and 'imaging' assays that are primarily designed to visualise the spatial distribution of integrin trafficking but, as we will see, can also be used to determine rates of integrin recycling at different positions in the cell.

16.2.1 Assays to Determine Rates of Integrin Endocytosis and Recycling

In some cases, particularly when the endosomal pool of a cycling receptor is quite large, it is possible to infer alterations in a receptor's internalisation and recycling kinetics simply by measuring its expression at the cell surface. However, inhibition of internalisation or stimulation of recycling will both act to increase integrin surface expression. Furthermore, when a receptor's endosomal pool is quite small (as is the case for α5β1 [5, 14] and αvβ3 [14] integrins), even a large change in the rate of recycling will only generate limited alterations in receptor surface expression. Moreover, it is possible that, under certain circumstances, the internalisation and recycling rates for integrins are coordinately regulated (i.e. both rates may increase or decrease to the same extent) and this would not be expected to alter the steady-state levels of integrin at the plasma membrane. Thus, to study the influence that trafficking has on a cycling receptor's function, it is necessary to independently measure its internalisation and recycling as accurately as possible.

16.2.1.1 Internalisation

Assays have been developed in which integrins are labelled at the plasma membrane and allowed to internalise to intracellular compartments wherein they are protected from the action of reagents that remove or inactivate these tags [5, 14]. There are two main categories of internalisation assay depending on the method used to label the integrin:

(a) *Surface biotinylation/reduction.* In this approach, cell-surface proteins are biotinylated by incubating cells at 4°C with the membrane-impermeant labelling reagent, sulpho-NHS-SS-Biotin. Cells are then shifted to 37°C for various times to allow internalisation of surface proteins. Biotin remaining at the cell surface is removed by cleavage of the reducible disulphide bond within the labelling reagent by incubation at 4°C with a cell-impermeant reducing agent such as sodium mercaptoethane sulphonate (MesNa) or glutathione. The internalised fraction of receptors is protected from reduction and thus remains *biotin labelled.* Biotinylated integrins may then be selected for and quantified either by immunoprecipitation followed by Western blotting with streptavidin or by capture-ELISA (Fig. 16.1a) [14, 15].

Fig. 16.1 (continued) Cells are then shifted to 37°C for various times to allow internalisation of bound antibodies. Antibodies remaining at the cell surface are quenched with an antibody against this fluorophore, which results in reduction of the fluorescent signal. To measure recycling, the internalised antibodies are then 'chased' from the cell by incubation at 37°C for various times. Antibodies returning to the cell surface are then quenched by a second incubation with an antibody against the fluorophore. The remaining fluorescent signal within the cell may then be quantified by fluorescent multi-well plate reading, confocal microscopy or by flow cytometry

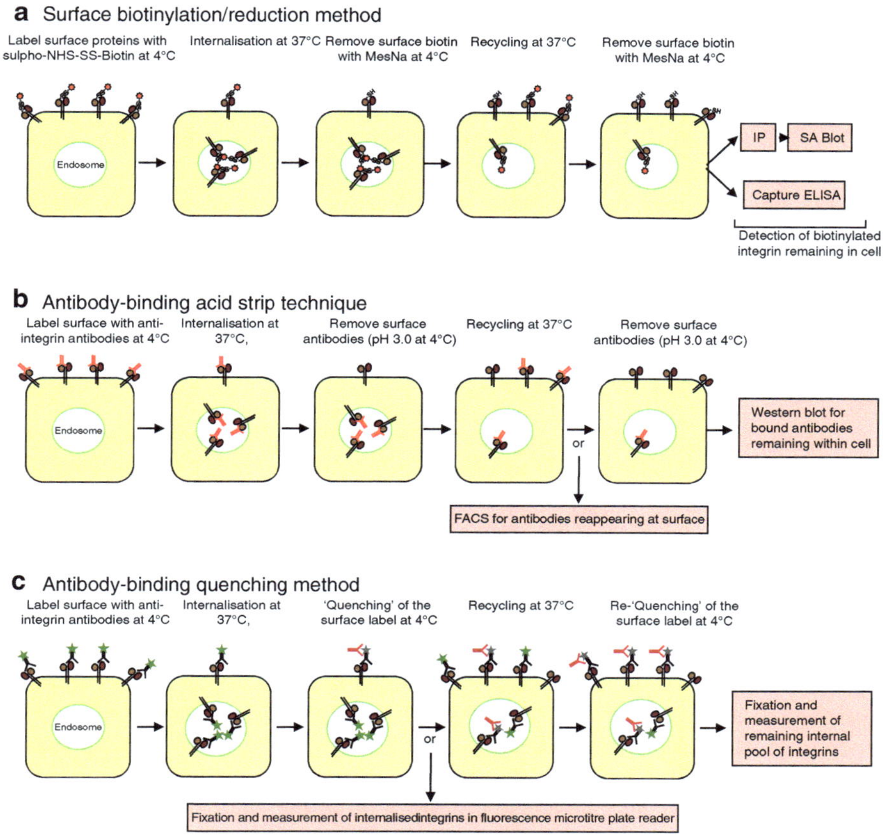

Fig. 16.1 'Biochemical' assays to determine integrin endocytosis and recycling. (**a**) Surface biotinylation/reduction method. To biotinylate receptors at the cell surface, cells are incubated at 4°C with the membrane impermeant labelling reagent, sulpho-NHS-SS-Biotin. Cells are then shifted to 37°C for various times to allow internalisation of surface proteins. Biotin remaining at the cell surface is removed by cleavage of the reducible disulphide bond within the labelling reagent by incubation at 4°C with a cell-impermeant reducing agent such as sodium mercaptoethane sulphonate (MesNa) or glutathione. The internalised fraction of receptors is protected from reduction and thus remains biotin labelled. To measure recycling, this internalised fraction of receptors may then be 'chased' from the cell by incubation at 37°C for various times. Biotin label that returns to the cell surface during this period is then removed by a second exposure to cell-impermeant reducing agent at 4°C. Biotinylated integrins may then be quantified by either immunoprecipitation followed by Western blotting with streptavidin or by capture-ELISA. (**b**) Antibody-binding acid strip technique. Integrins are labelled at the cell surface by incubation with anti-integrin antibodies at 4°C. Cells are then shifted to 37°C for various times to allow internalisation of bound antibodies. Antibodies remaining at the cell surface are removed by incubation at 4°C at low pH. To measure recycling, the internalised antibodies are then 'chased' from the cell by incubation at 37°C for various times. Antibodies returning to the cell surface are then removed by a second incubation at low pH. Antibodies remaining within the cell may then be quantified by Western blotting. Alternatively, if the second acid strip is omitted, antibodies reappearing at the surface may be detected by flow cytometry. (**c**) Antibody-'quenching' technique. Integrins are labelled at the cell surface by incubation with anti-integrin fluorophore-conjugated antibodies at 4°C.

(b) *Antibody-binding techniques*. To do this, integrins are labelled at the cell surface by incubation with anti-integrin antibodies at 4°C. Cells are then shifted to 37°C for various times to allow internalisation of bound antibodies. Antibodies remaining at the cell surface are removed by incubation at 4°C at low pH. The level of internalised antibody may then be quantified by Western blotting or flow cytometry (Fig. 16.1b) [16]. Alternatively, in a more recent adaptation of this assay [17], integrins may be labelled at the cell surface using an antibody which is covalently coupled to Alexa Fluor 488. Following internalisation, the fluorescence of Alexa Fluor 488-conjugated antibodies remaining at the cell surface is quenched with an antibody which alters the spectral properties of this fluorophore (Fig. 16.1c). The quantity of internalised antibody is then determined using fluorescence microtitre plate reader and/or its distribution assessed by fluorescence microscopy or flow cytometry.

The respective advantages and disadvantages of these two types of approaches are summarised in Table 16.1. Given the likelihood that antibody binding may alter the way that integrins are trafficked, an obvious advantage of the surface biotinylation/reduction method over the antibody-binding approaches is that the integrin is not bound to an antibody when it is internalised. The Ivaska laboratory has compared results obtained from the surface biotinylation/reduction and antibody-binding approaches and found them not to be significantly different [17]. However, Roberts et al. [14] report that antibody binding reduces recycling of $\alpha v\beta 3$ by approximately threefold, indicating that the influence of antibody binding in integrin internalisation and recycling may depend on the particular heterodimer and antibody in question. A principal advantage of antibody-binding approaches is that they may be easily combined with fluorescence microscopy to determine the post-endocytic trafficking and intracellular destination of the internalised integrin [18]. Moreover, conformation-specific antibodies may be used to report on the distribution of integrins in their various affinity states, and Arjonen et al. [17] have recently deployed this approach using the 12G10 and mAb13 antibodies to study trafficking of $\beta 1$ integrin in the active and inactive conformations, respectively. Conformation-specific antibodies (SNAKA51 and 9EG7) may also be incorporated into the surface biotinylation/reduction method to report on the differential kinetics of internalisation and recycling of particular active integrin conformations [19]. Furthermore, it should be noted that a disadvantage of both surface biotinylation/reduction and antibody-binding approaches is that they involve thermal cycling of the cells between 37 and 4°C and, given the effect that this may have on microtubule depolymerisation, this is a factor that needs to be carefully considered in the future.

Lastly, integrins begin to recycle very shortly following internalisation [7]. So, in order to obtain a measure of internalisation that is not underestimated due to early recycling, it is necessary to inhibit recycling whilst performing internalisation assays. To do this, the receptor recycling inhibitor, primaquine [20], may be included during the internalisation period. Although primaquine's mechanism of action is not closely defined, it is well established to block all known receptor recycling pathways in addition to opposing delivery of cargo from the trans-Golgi network to the plasma membrane [21].

Table 16.1 Comparison of key attributes and drawbacks of integrin trafficking methods

Method	Advantages	Caveats
Surface biotinylation/ reduction method	Interfaces well with capture ELISA to provide quantitative/linear detection of biotinylated receptors	Does not provide spatial information
	Suitable for both endogenous and overexpressed receptors	Temperature shifts are likely to distort microtubular cytoskeleton
	Small biotin tag is less disruptive than bound antibody	
	Use of conformation-specific antibodies for IP allows detection of active and inactive integrin conformations	
Antibody-binding acid strip technique and antibody-binding 'quenching' technique	Can be interfaced with immunofluorescence detection to track intracellular receptor journey	Antibody binding can alter receptor trafficking; $\alpha v\beta 3$ recycling is slowed two- to threefold by bound antibody [14]
	Use of conformation-specific antibodies allows tracking of active and inactive integrin conformations	Use of conformation-specific antibodies may alter the balance of 'active' vs. 'inactive' integrin
	Suitability for FACS allows simultaneous measurement from cell subpopulations	Many cell types cannot withstand the rigours of acid-stripping
Photoactivation in TIRF	Selective photoactivation at/near plasma membrane permits synchronous pulse labelling of endocytic cargo	Requires expression of fluorescently tagged integrin; does not report on endogenous receptors
	Yields spatial information re site of internalisation and destination of endocytic cargo	Not suitable for cells attached to 3D substrates such as cell-derived matrix
Photoactivation by confocal microscopy	Spatially defined photoactivation enables discrimination of recycling integrins from those following an endocytic route	Although ideally suited to the study of peripherally located recycling compartments (such as Rab25), photoactivation can occur above and below desired region, and this must be controlled for stop 2-photon laser may be more suited to photoactivation in the perinuclear region

The various advantages offered by biochemical and imaging approaches to the measurement of integrin trafficking are summarised in the table. Also considered are a number of drawbacks and caveats to each approach

16.2.1.2 Recycling

The assays to measure integrin recycling are formed by extension of the same approaches used to measure internalisation (Fig. 16.1) [14, 16, 17]. Thus, following labelling of internalised integrins by either the surface biotinylation/reduction or antibody-binding approaches, the internalised fraction may then be 'chased' back to the plasma membrane to determine the rate of recycling. Following this, the re-exposed label may be removed by a second round of MesNa reduction, quenching antibody or acid stripping (as appropriate), and the recycled integrin then expressed as a proportion of the amount of receptor that was endocytosed during the internalisation period.

16.2.2 Use of Photoactivatable Probes to Measure Integrin Traffic

It is becoming clear that much integrin endocytosis and recycling is restricted to defined cellular regions, such as the pseudopodial tips of invading cells and the leading edge of cells migrating on plastic [12, 22], indicating the pressing need to advance technologies that report on spatio-temporal aspects of integrin trafficking. One of the challenges with the imaging of trafficking is that, although it is relatively straightforward to determine how receptors are distributed spatially, the confusion of *endo- and exocytic vesicles* that are often travelling through the same cellular region in opposing directions makes it difficult to distinguish endocytic cargo from that which is being recycled to the cell surface. These problems are now being overcome by tagging proteins with photoactivatable GFP (paGFP) which enable the selective activation of fluorescence of a particular receptor subpopulation and the subsequent tracking of its movement within the cell both in time and space [23]. For instance, a combination of photoactivation and total internal reflection (TIRF) (to restrict the plane of photoactivation so that only paGFP-tagged receptors present at the cell surface become activated) has now been deployed to visualise $\alpha5\beta1$ endocytosis at fibrillar adhesions (Fig. 16.2a) [19]. Furthermore, when integrins are internalised, they are transported to recycling compartments that are variously positioned within migrating cells. Sometimes these are located next to the nucleus (as is the Rab11-positive PNRC) [14, 16, 24, 25], but often recycling compartments are positioned towards the cell front as is the case with the Rab25 compartment during migration on 3D matrices [12]. In this case, 'single point' photoactivation has been used to selectively activate the fluorescence of $\alpha5\beta1$ within the Rab25 compartment, thus allowing visualisation of spatially restricted integrin recycling to the plasma membrane at the tips of pseudopods as cells migrate on a 3D matrix (Fig. 16.2b) [12]. Furthermore, a recent study has reported the use of photoactivatable probes to study cadherin dynamics in tumours grown in mice, indicating that it may be feasible to image integrin trafficking as cells migrate through tissues [26]. However, although ideally suited to the study of peripherally located recycling compartments (such as Rab25), photoactivation can occur above and below desired target region, and this has compromised the application of photoactivatable probes to the study of integrin recycling from endosomes in the perinuclear region. Furthermore, activation of E-cadherin in tumours in situ was limited to a depth of 20 μm or less owing to light scattering by the tissue when using a standard laser for

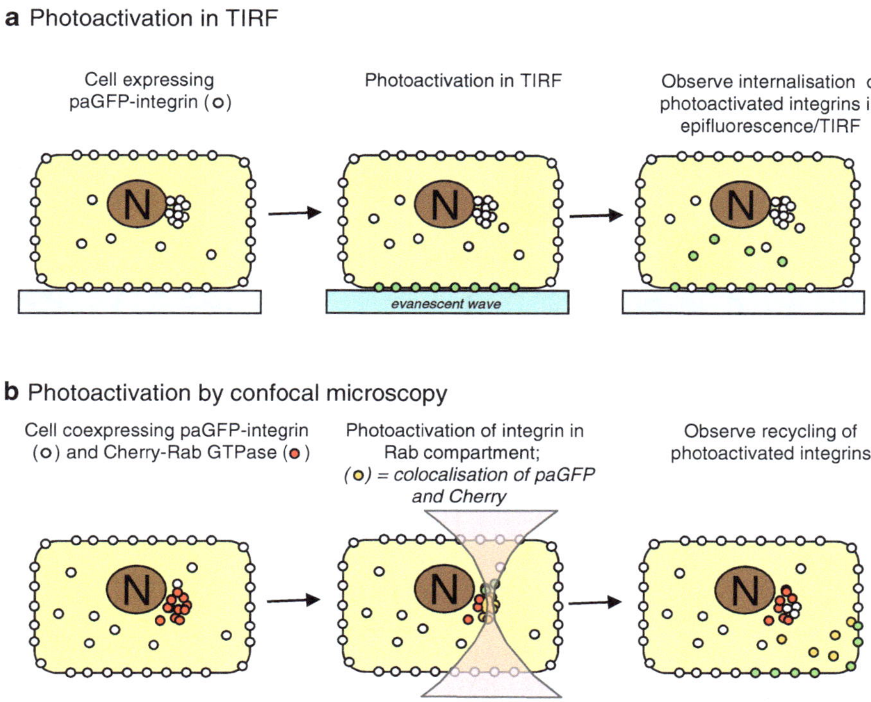

Fig. 16.2 Photoactivation methods for imaging integrin trafficking. (**a**) Photoactivation in total internal reflection (TIRF). Photoactivatable GFP-integrin (paGFP-integrin) is expressed in cells plated onto a glass coverslip. paGFP-integrin that is present at the ventral plasma membrane is selectively activated by shining a 405 nm laser laterally through the coverslip to generate a photoactivating evanescent field by TIRF. The subsequent trafficking of paGFP-integrin is then visualised by epifluorescence or TIRF microscopy. (**b**) Photoactivation by confocal microscopy. paGFP-integrin is expressed in cells and photoactivated using a 405 nm laser in confocal mode to tightly restrict the plane of activating fluorescence. Trafficking of photoactivated integrin is then visualised by confocal microscopy or wide-field epifluorescence

photoactivation [26]. Thus, it will be necessary to develop and refine the use of 2-photon lasers for the photoactivation of integrin probes to image integrin traffic from perinuclear compartments and to visualise vesicular dynamics in cancer cells as they invade through living tissue.

16.3 Pathways of Integrin Endocytosis

Integrins can be internalised via a number of known endocytic routes, including clathrin-dependent endocytosis [15, 18, 22, 27], clathrin-independent carriers (CLICs) [28], pathways that require caveolin [29] and macropinocytosis which is

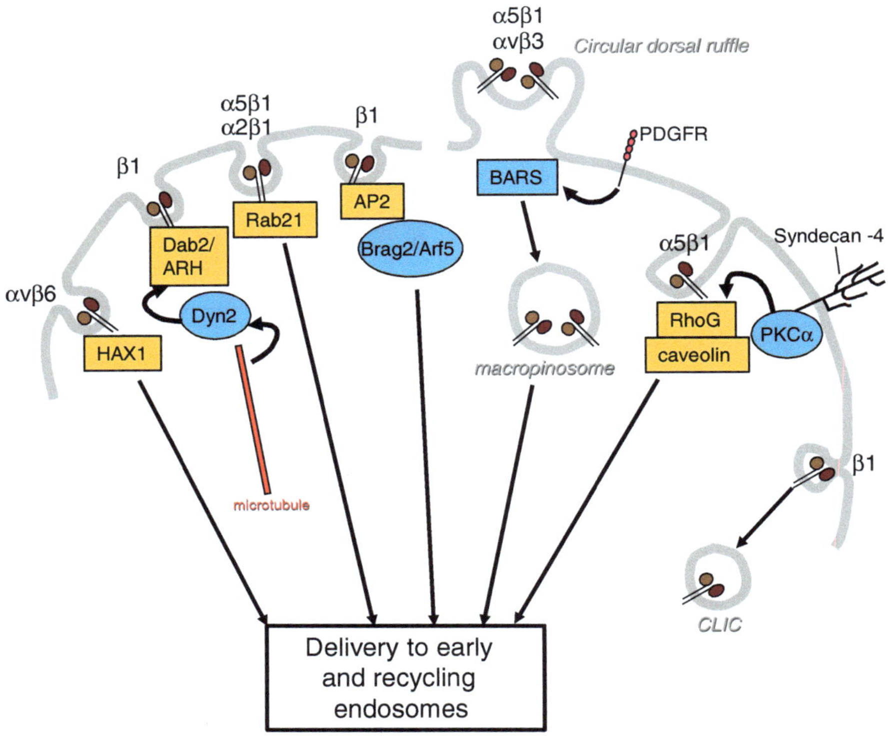

Fig. 16.3 Pathways to integrin endocytosis. A schematic representation of some of the known pathways to internalisation of integrins. Pathways that are clathrin dependent are depicted on the *left-hand side*, and those which do not require clathrin are on the *right*. Endocytic regulators and adaptors with a close physical association with integrins are in *yellow boxes*. Other regulatory elements are in *blue ovals/boxes*. The involvement of microtubules in triggering Dab2/ARH-dependent endocytosis is denoted by the *red microtubule*

associated with circular dorsal ruffles (CDRs) [30] (Fig. 16.3). Although it is clear that a given integrin heterodimer is not necessarily restricted to entering the cell via a particular route, for instance α5β1 is known to follow both clathrin-dependent and clathrin-independent routes into the cell, for the purposes of simplicity, we will deal with these internalisation pathways in separate subsections.

16.3.1 Clathrin-Dependent Endocytosis of Integrins

It is now well established that β1 integrins are internalised via clathrin-dependent mechanisms. The evidence for this comes from a number of different directions, but reports indicating a clear role for the unconventional clathrin adaptor, Dab2, in integrin endocytosis have provided the most persuasive arguments so far for the involvement of clathrin in integrin endocytosis [18, 22, 27]. In studying the mechanisms by which focal adhesions (FAs) are disassembled in response to microtubule regrowth following nocodazole-washout, the Gundersen lab found that a dynamin-dependent step was essential for this process [31] (Fig. 16.3). This implied that the endocytic process is associated with FA disassembly. Further work, involving a number of elegant TIRF imaging experiments, indicated that clathrin structures are indeed closely associated with disassembling FAs and that Dab2 and ARH (another unconventional clathrin adaptor) are required for integrin endocytosis and microtubule-induced FA assembly [22]. Using a similar nocodazole-washout system, the Kunz lab has also found that Dab2 is necessary for integrin endocytosis and FA disassembly following microtubule regrowth [18]. Coming from a different direction, the Cooper lab has deployed a novel mass spectrometry approach to look for proteins whose surface expression is particularly affected by Dab2 knockdown and in this way identified β1 integrin as prominent cargo of a Dab2- and clathrin-dependent endocytic step [27]. More recently, this lab has shown that Eps15 is required for Dab2-dependent β1 integrin internalisation, which further strengthens the role played by clathrin in integrin endocytosis [32]. Furthermore, *Numb*, which is thought to be a cargo-specific adaptor protein that links certain transmembrane receptors to elements of the clathrin machinery (such as Eps15 and α-adaptin), has been shown to influence integrin endocytosis [33]. In addition to Dab2 and ARH, other clathrin adaptors have been shown to control integrin endocytosis, and more recently Brag2 (a guanine nucleotide exchange factor [GEF] for Arf5 and Arf6) has been shown to drive β1 integrin endocytosis via association with the 'classic' plasma membrane clathrin adaptor, AP2 [34, 35] (Fig. 16.3).

The cytoplasmic portions of β subunits contain membrane proximal and membrane distal NxxY motifs, and these conserved sequences are known to recruit surface receptors, such as low-density lipoprotein (LDL) [36] or insulin receptors [37], to clathrin-coated structures, by allowing their interaction with adaptor proteins such as AP2. Even though initial studies, which utilised mutants either lacking the NxxY motifs or ones that contained mutated tyrosine residues, indicated no requirement for the conserved NxxY sequence in α5β1 internalisation [38], more recent approaches contradict this observation. Indeed, mutation of the membrane proximal or membrane distal NxxYs tyrosine to either alanine [39] or phenylalanine [40] markedly inhibits α5β1 integrin internalisation—but whether these motifs mediate physical association with the clathrin apparatus is not known. The integrin cytodomain NxxY motifs are certainly capable of binding to clathrin adaptors—the cytoplasmic domains of integrin β-subunits can associate with the PTB-binding domains of Dab2 and NUMB in vitro [41]—but the evidence that integrin endocytosis is driven via direct association with the clathrin machinery is not strong. Furthermore, although structural studies indicate that tyrosine to alanine substitutions in integrin NxxY motifs would be expected to disrupt association with PTB-binding domains, it is unlikely that the same is the case for tyrosine to phenylalanine substitution [42]. And, as mentioned above, mutation of NxxY tyrosines to either phenylalanine or alanine profoundly inhibits integrin endocytosis [39, 40]. Moreover, the Cooper *lab has stated clearly* that they are unable to detect physical association between β1 integrin and Dab2 [27]. A well-established role of the membrane proximal integrin NxxYs is to mediate association of integrins to actin-binding proteins such as talin and tensin [42, 43] which is necessary for recruiting integrins to focal and fibrillar adhesions, respectively, and integrin internalisation is known to occur from both these structures [19, 22]. Thus, a likely role of integrin NxxYs in endocytosis is not necessarily to mediate association with clathrin adaptors but to position integrins within the adhesive structures from which endocytosis occurs.

Despite lack of evidence for a functional physical association of integrins with clathrin adaptors, it is clear that integrins do directly recruit proteins to assist their endocytosis. The small GTPase Rab21 is now well established to interact with the conserved membrane-proximal GFFKR motif in the cytoplasmic tail of α-integrins [15, 40, 44] (Fig. 16.3). This is necessary for clathrin-dependent endocytosis of α5β1 integrin, and disruption of the interaction compromises integrin function in cell migration and cytokinesis. Furthermore, the cytoplasmic tail of β6-integrin interacts directly with *HS-associated protein X-1* (*HAX*-1) which is a component of a cortactin-containing complex involved in cell migration [45]. HAX-1 is required for the clathrin-dependent internalisation of αvβ6, and disruption of the β6-HAX-1 interaction impairs cancer invasion on ligands for αvβ6 integrin (Fig. 16.3).

16.3.2 Internalisation of Integrins by Clathrin-Independent Mechanisms

16.3.2.1 Caveolin-Dependent Mechanisms

It has been known for some time that cholesterol depletion can oppose integrin endocytosis, prompting the suggestion that lipid raft-like structures may play a role in internalising and trafficking integrins [46]. Indeed, a number of studies have shown that there is a possibility of physical association between integrins and caveolin-1 [47, 48]. Although it is still not clear how caveolin acts to drive endocytosis, there is good evidence that it is required for internalisation of α5β1 in myofibroblasts and that this pathway is responsible for integrin-mediated uptake and subsequent lysosomal degradation of the integrin ligand, fibronectin [29]. Caveolar endocytosis is normally thought to be dependent on dynamin. Indeed, a recent study reporting the role played by the heparin-binding protein, syndecan-4 in α5β1 endocytosis (see below) indicates that this integrin follows a caveolin- and dynamin-dependent pathway into the cell [49]. Taken together, an implication of these and other studies is that β1 integrins that are internalised via a caveolin-dependent mechanism are in the active conformation and, in some cases, likely to be ligand bound. Indeed, a likely scenario is that during ECM remodelling there is limited proteolysis of extracellular fibronectin fibrils and that the resulting fibronectin fragments which are associated with active-conformation β1 integrins are then endocytosed via a caveolin-dependent mechanism and degraded.

16.3.2.2 Clathrin-Independent Carriers and Macropinocytosis

Both clathrin and caveolar endocytosis are dynamin dependent, but recently it has become clear that integrins can enter the cell via mechanisms that do not require dynamin. Work conducted by the Parton, Mayor and McMahon labs has identified a category of morphologically distinct CLICs formed in a dynamin-independent fashion which are characterised by enrichment in GPI-anchored proteins and regulation by GTPases such as cdc42 and Arf1 and the BAR domain-containing rhoGAP protein, GRAF1 [28, 50, 51]. Recently, a mass spectrometry analysis of CLICs found them to contain β1 integrins [28] (Fig. 16.3). Moreover, uptake via CLICs is concentrated at the leading edge of migrating cells [52], indicating the possibility that these structures play a role in trafficking integrins during cell migration. Further work on the role played by key regulators of CLIC formation, such as cdc42 and GRAF1, in integrin endocytosis will help to establish the relevance of this clathrin- and dynamin-independent mechanism to integrin trafficking.

Many cell types, in particular fibroblasts, respond to treatment with growth factors (such as PDGF, EGF and VEGF) by assembling actin-rich ruffles on their dorsal surface [53]. These are termed CDRs. CDRs are produced in response to a burst of actin polymerisation which occurs as a result of Rac-activation downstream of growth factor signalling and, although the function of CDRs is largely unknown, they have been suggested to be part of an initial step leading to macropinocytosis [54]. A recent study has shown that β3 and β1 integrins are recruited to CDRs within minutes of growth factor addition and then subsequently internalised into macropinosomes [30] (Fig. 16.3). Moreover, internalisation of integrins into macropinosomes was opposed by knockdown of BARS, a protein which is required for the final fission stage of macropinocytosis. Given that oncogenes such as Ras are known to massively upregulate macropinocytosis [55], it will be interesting to determine the role played by macropinocytosis and BARS in integrin trafficking during cancer invasion and metastasis.

16.3.3 Regulation of Integrin Endocytosis

β1 integrin and PKCα have been shown to bind directly to one another in an interaction requiring both the integrin NxxY motifs and the V3 hinge domain of PKCα [56]. Although this interaction is required for chemotaxis of tumour cells toward gradients of growth factors, it is not clear whether this is owing to its influence on integrin endocytosis or recycling. Indeed, data on the role of PKCα in integrin trafficking is somewhat confusing. On the one hand, activation of PKCα with phorbol esters drives internalisation of β1 and its accumulation in recycling endosomes [57].

On the other hand, overexpression of PKCα has been shown to upregulate surface expression of β1 integrins, and this effect has been attributed to the PKC regulatory domain (RD) rather than the kinase domain [57]. Thus, it appears that PKCα may have both inhibitory and stimulatory influence on β1 integrin endocytosis and that some of these phenomena may involve both kinase-dependent and kinase-independent attributes of PKCα. Furthermore, as PKCα has a better established role in controlling caveolar endocytosis over clathrin-dependent pathways, and because PKC has been shown to drive α2β1 into caveolae [58], it would seem most likely that PKCα may control integrin internalisation via caveolin-dependent pathways.

More recently, PKCα has been shown to drive α5β1 endocytosis not by associating with the integrin itself, but with another fibronectin-binding receptor, syndecan-4 [49]. This study provides more weight to the argument that PKCα drives caveolin-dependent endocytosis of integrins. Since the pioneering work of Woods and Couchman in the 1980s and 1990s [59], it has been clear that integrin behaviour is affected by syndecan-4 which binds not to fibronectin's RGD sequences but to its heparin-binding domains. Bass et al. [49] have found that engagement of syndecan-4 with the heparin-binding domains of fibronectin triggers rapid endocytosis of α5β1 integrin through a caveolin-dependent pathway. Integral to this mechanism is that it is mediated by the recruitment of PKCα to syndecan-4's (not β1 integrin's) cytoplasmic domain and this leads to kinase activation. Activation of RhoG downstream of PKCα then mediates a signal which triggers α5β1 internalisation. Given that the nature of the physical association between PKCα and the syndecan-4 cytodomain is much better characterised than that formed between the PKCα and β1 integrin, it seems more likely that PKCα's control of integrin endocytosis proceeds via association with co-receptors, such as syndecan-4, rather than by direct association with the integrin (Fig. 16.3). However, owing to the lack of good reagents for looking at the syndecan-4's localisation, the physical relationship between the syndecan-4/PKCα complex and α5β1 integrin remains unclear. In addition to syndecan-4's control of α5β1 endocytosis, there are further examples of adhesion receptors other than integrins impacting on integrin internalisation. For instance, myelin-associated glycoprotein (MAG) induces de-adhesion and repulsion of growth cones by increasing local Ca^{2+} concentration to drive clathrin-dependent endocytosis of β1 integrins [60].

Another ECM-related parameter that may affect integrin endocytosis is the stiffness of the substrate to which cells are attached. In looking for a possible mechanism for how ECM stiffness alters cell-fate decisions, Du et al. [61] looked at endocytosis of β1 integrin in cells plated onto substrates of different stiffness. They found that integrins were internalised much more efficiently when cells were plated onto flexible substrates than on rigid ones and that this was via a caveolin-dependent and clathrin-independent mechanism. Given the importance of ECM stiffness to integrin signalling and the recent evidence that this influences cancer metastasis [62], it will be interesting to see how alterations in integrin endocytosis are responsible for increased aggression of tumours that reside within stiff collagen microenvironments.

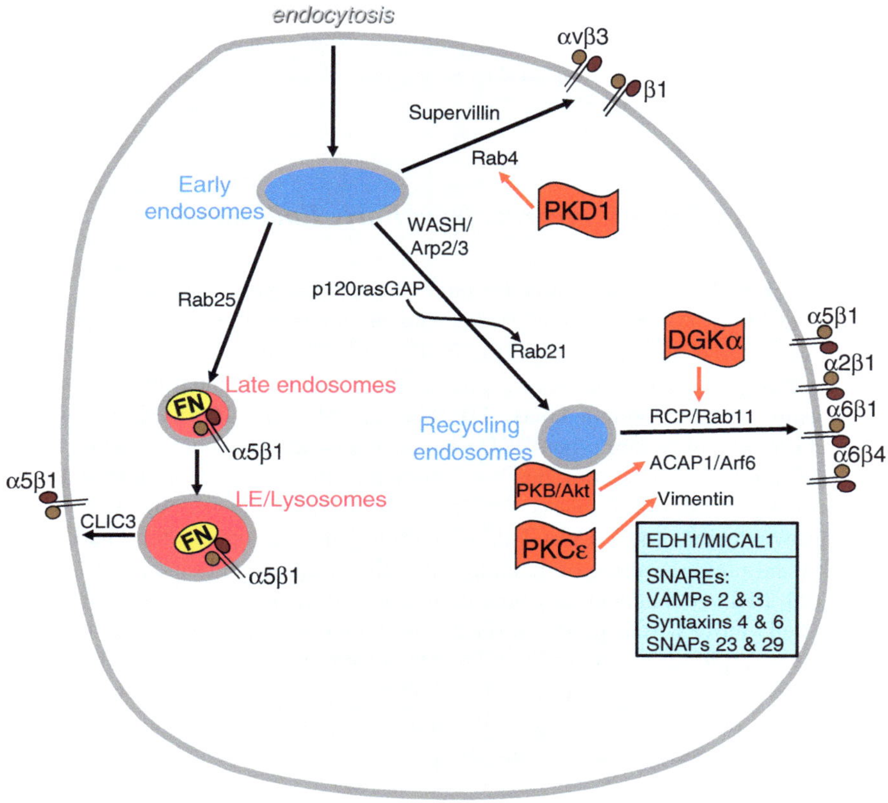

Fig. 16.4 Pathways that return internalised integrins to the plasma membrane. Integrins are internalised and delivered to early endosomes. From here, the heterodimers may follow three distinct routes back to the plasma membrane. (1) αvβ3 and β1 integrins may follow a Rab4-dependent 'short-loop' pathway that runs directly from early endosomes to the cell surface. (2) A number of heterodimers (including α5β1, α2β1, α6β1 and α6β4) may be delivered to recycling endosomes and return from this compartment. (3) α5β1 integrins which are engaged with fibronectin (FN) and in the active conformation are routed to late endosomes/lysosomes. Here they can either be degraded or, if CLIC3 is present, protected from degradation and returned to the plasma membrane

16.4 Post-endocytic Trafficking and Recycling of Integrins

Irrespective of whether they are endocytosed via clathrin-dependent or independent mechanisms, it appears that the first port of call for internalised receptors is the EEs [63] and it is here that they are sorted for degradation in late endosomes and lysosomes or for return to the plasma membrane (recycling). Return to the plasma membrane can occur via at least two distinct routes; in a direct and Rab4-dependent manner (short loop) or through a Rab11- and/or Arf6-dependent pathway (long loop) [7] (Fig. 16.4). The latter requires delivery of recycling cargo from sorting endosomes to the PNRC—which is located near to the microtubule organising centre and is often enriched in the Rab11 family GTPases—prior to trafficking back to the plasma membrane. These two recycling pathways differ with respect to the integrins they transport and the signalling cascades that modulate them, but are similar in that both are tightly controlled.

16.4.1 Recycling of Integrins from Early Endosomes

It is now clear that recycling of αvβ3 integrin occurs primarily through the Rab4-dependent, short-loop pathway which returns the integrin to the plasma membrane from EEs without passage through the PNRC [14]. Recruitment of αvβ3 into the Rab4 pathway relies on direct interaction between the C-terminal domain of the β3 cytoplasmic tail and protein kinase D1 (PKD1) [64] (Fig. 16.4). Moreover, this interaction can only occur when PKD1 is active and autophosphorylated on Ser^{916} [65]. Although this interaction has been studied primarily in the context of PKD1 activation by PDGF in fibroblasts, PKD1 can be activated and recruited to αvβ3 by other growth factors. Indeed VEGF drives recruitment of autophosphorylated PKD1 to αvβ3 and rapid recycling of this integrin to the plasma membrane in endothelial cells [66]. PKD1-dependent recycling contributes to maintaining persistent migration as cells move across plastic surfaces, and this is likely owing to its ability to facilitate the rapid recruitment of αvβ3 to newly forming focal complexes [65]. This has been reported to occur during cell spreading on vitronectin and to be required for assembly of αvβ3-containing focal complexes at the extreme front of migrating fibroblasts—which is consistent with the role played by PKD1 in maintaining persistent migration.

Given the role played by PKD1 in integrin recycling and cell migration, it is interesting to consider how the kinase may contribute to cancer progression and invasion. Interestingly, there is some disagreement in the literature as to the role played by PKD in cancer. Whilst some studies have shown that PKD promotes invasion [67], others have reported that it opposes invasive migration and is downregulated in aggressive metastatic disease [68–70]. Indeed, suppression of PKD1 can promote α5β1-integrin-dependent invasion of gastric cancer cells [69]. A likely explanation for this is that PKD1-regulated αvβ3 recycling acts to oppose recycling of α5β1 (see following sections). Thus, by promoting αvβ3 recycling, PKD1 negatively impacts α5β1-dependent invasion into fibronectin-rich matrices. Under these circumstances, reduction of invasion by PKD signalling is specifically due to the inhibition exerted by short-loop αvβ3 recycling over that of α5β1. However, in microenvironments that are fibronectin deficient, where invasion is dependent on αvβ3 and not α5β1, PKD would be expected to be a strong driver of tumour cell invasion.

More recently, it has become clear that Rab4- and PKD1-dependent routes are responsible for rapid recycling of integrins other than αvβ3. Following EGF treatment of HeLa cells, the raft-associated protein, supervillin, has been shown to drive β1 integrins into a Rab4-dependent rapid recycling pathway which is dependent on F-actin polymerisation [71] (Fig. 16.4). Interestingly, Arjonen et al. [17] have recently shown that β1 integrins can recycle via a Rab4-dependent short-loop pathway in breast cancer cells and that this is, like the supervillin pathway, inhibited by cytochalasin D indicating a requirement for actin polymerisation.

16.4.2 Rab11-Dependent Integrin Recycling

Integrins that do not recycle directly back to the plasma membrane via the short-loop pathway are normally then transferred to recycling endosomes (Fig. 16.4). In cells that are cultured on rigid substrates, a significant proportion of the cell's recycling endosomes are concentrated in the perinuclear region and form a structure which has been termed the PNRC. Over the last few years, there have been numerous reports indicating that a number of integrins, including α5β1, α2β1, α6β1, α6β4 and αLβ2, accumulate in the PNRC prior to their return to the cell surface [14, 16, 46, 57, 72–74]. Much of the machinery that controls β1 integrin recycling from the PNRC is shared with other cargoes that canonically follow this recycling route—such as the TfnR. For instance, EHD1, which is recruited to the PNRC by MICAL-L1, controls recycling of both integrins and TfnR [24, 75–77]. Moreover, the function of SNAREs (such as syntaxins 2, 3, 4 and 6, SNAP23, SNAP29, VAMP2 and VAMP3) that control the docking and fusion of vesicles carrying a range of recycling receptors has also been shown to be required for integrin recycling [72, 78–82] (Fig. 16.4).

A number of studies have identified elements that specifically control the recycling of integrins and which are not shared with the TfnR. We will discuss these in the following paragraphs:

(a) *PKB/Akt* acts via the phosphorylation of GSK-3β to enable delivery of α5β1 from the Rab11 compartment to the plasma membrane, but these kinases have no effect on TfnR recycling [25]. Microtubules have a role in Rab11 vesicle transport [83], and GSK-3β could phosphorylate microtubule-binding proteins, such as tau or APC [84], to influence integrin recycling. Consistent with this, PKB/GSK-3 has been shown to control Rab11-dependent recycling of α6β4 integrin and to drive cancer cell invasion that is dependent on this integrin via a mechanism that likely involves alterations to the microtubular cytoskeleton [74]. Another way in which PKB/Akt can afford selective regulation of β1 integrin recycling is via the phosphorylation of the ArfGAP protein, ACAP1. ACAP1 functions as a transport effector to recruit cargoes, such as TfnR and the Glut4 glucose transporter into an Arf6-regulated recycling pathway [85]. But, when ACAP1 is phosphorylated at Ser^{554} by PKB/Akt, it can bind not only to TfnR but also to the β1 integrin cytoplasmic tail [86]. Thus, in contrast to TfnR recycling, the role of ACAP1 in β1 integrin recycling requires its phosphorylation by Akt, thus providing a critical link between growth factor signalling and cell migration and invasion (Fig. 16.4).
(b) *PKCε* has a well-supported role in integrin-dependent cell migration, and this novel PKC isoform likely influences these processes by selectively controlling β1

integrin trafficking [87]. PKCε regulates β1 integrin recycling by phosphorylating vimentin to release integrin transport vesicles from cytoskeletal elements in the perinuclear region, but, like PKB/Akt, this kinase does not affect TfnR transport [88] (Fig. 16.4).

(c) *Rab21 and p120RasGAP* both bind to similar regions in the α integrin cytoplasmic domain, and recent evidence indicates that these two proteins act sequentially to specifically control progression of integrins through the recycling pathway [44]. Rab21 binds to a sequence which is found in membrane proximal regions of both the α2 and α5 integrin cytodomains, and this association promotes integrin internalisation and trafficking through EEs to the PNRC. Once the integrin has reached the recycling compartment, p120RasGAP competitively displaces Rab21 from the membrane proximal region of the α-cytodomain and takes its place. p120RasGAP then promotes delivery of integrin-containing vesicles from the PNRC to the cell surface (Fig. 16.4). Consistently, disruption of p120RasGAP expression leads to suppression of a nonpersistent mode of migration, most likely as a consequence of decreased recycling of α5β1 from the PNRC. This competitive binding mechanism may provide a conceptually straightforward and rational explanation for how an integrin may process through the endocytic pathway and is the first paper to report such a competitive hand-on-hand relay in the progression of cargo from one compartment to another. A recent report has shown that p120RasGAP may drive tumour cell invasion by increasing the activation of Src [89], and it will be interesting to determine whether this is owing to effects on recycling of α5β1 which would be likely to increase activation of Src family kinases.

(d) *Rab-coupling protein (RCP) and diacylglycerol kinase-*α. RCP, a member of the FIP family of Rab11-interacting proteins, plays an important role in regulating integrin trafficking, but does not influence Tfn recycling [13, 90]. In situations when αvβ3 integrin is inhibited by drugs such as Cilengitide, or when cells express mutant p53 (see below), RCP is recruited to α5β1 integrin at recycling vesicles and integrin recycling then becomes RCP dependent. Interestingly, recruitment of RCP to α5β1 enables RCP-dependent recycling of RTKs, such as EGFR and cMET, which drives cancer cell invasion and scattering [13, 90, 91]. RCP possesses a C2 domain in its N-terminal region which has been shown to bind particularly strongly to phosphatidic acid [92]. Phosphatidic acid may be produced in the cell via two main routes: via phospholipase D-mediated hydrolysis of phosphatidyl lipids or by phosphorylation of diacylglycerol by diacylglycerol kinases [93, 94]. Diacylglycerol kinase-α (DGKα) is required for RCP-mediated recycling of α5β1, but does not seem to play a role in recycling of internalised transferrin [95] (Fig. 16.4). Moreover, DGKα and RCP's C2-domain are both required for RCP to be tethered to the tips of invading pseudopods as cancer cells migrate through 3D matrices (Fig. 16.5a). Thus, deletion of the C2 domain or inhibition of DGKα signalling reduces the number of RCP-positive structures that are tethered at pseudopod tips and correspondingly increases the vesicle population in the perinuclear region without apparently restricting the overall mobility of RCP endosomes and their ability to move up

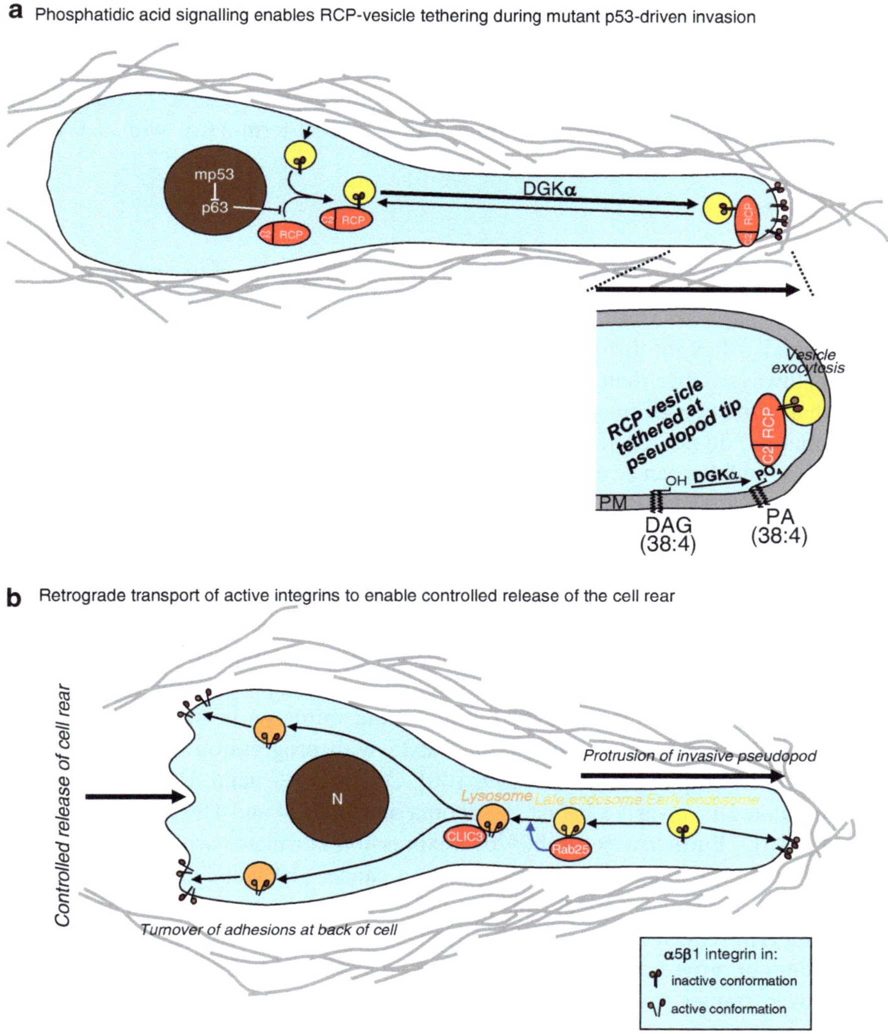

Fig. 16.5 Different integrin recycling pathways are directed either to the cell front or the cell rear. (**a**) Expression of mutant p53 inhibits p63 which promotes recruitment of RCP to endosomal α5β1. Association of α5β1 with RCP is not a DGK-dependent event and does not require RCP's PA-binding C2 domain. RCP/integrin recycling vesicles can then move up and down the pseudopod shaft, and the role of DGKα is to generate phosphatidic acid (PA) species which enable the tethering of RCP at pseudopod tips, an event that requires RCP's C2 domain. Inhibition or silencing of DGKα most significantly affects the interconvertion of 38:4 DAG to 38:4 PA. The 38:4 species of PA is therefore depicted as the most likely to be involved in tethering RCP at pseudopod tips. (**b**) Active-conformation α5β1 integrin is transported from early to late endosomes under control of Rab25. Upon arrival at the late endosomes/lysosomal compartment, active α5β1 is trafficking in lysosomes to the back of the invading cell, and this is dependent on CLIC3. Meanwhile, α5β1 that remains in the inactive conformation is recycled at the cell front. CLIC3 and the retrograde trafficking of α5β1 is required for turnover of adhesions which allows the controlled and coordinated release of the cell rear as the tumour cells move forward through the ECM

and down the pseudopod shaft [95]. Established vesicle tethers commonly possess a Rab GTPase/cargo-binding portion that is linked by an extended coiled-coil region to a domain that senses membrane curvature [96]. These observations suggest that RCP may function as a vesicle tether which interacts C-terminally with Rab11 and the integrin cargo vesicle and N-terminally with a PA-rich region of the plasma membrane.

16.4.3 Integrin Recycling from Late Endosomes/Lysosomes

It is generally thought that the majority of internalised integrins are routed to recycling pathways rather than sent to lysosomes for degradation. However, it is now clear that a fraction of integrins which are ligand engaged (and thus in the active conformation) do not reach recycling endosomes but are sent to late endosomes and lysosomes, and this can be associated with increased integrin degradation [97]. More recently the Rab11 GTPase, Rab25 has been found to localise to a late endosomal-type compartment and contribute to the delivery of active integrin to lysosomes [11]. In pancreatic and ovarian cancers, Rab25 expression is associated with upregulation of a lysosomal protein called CLIC3, which is structurally related to glutathione transferase, and CLIC3 prevents degradation of $\alpha5\beta1$ within lysosomes and allows integrin recycling from this compartment to the plasma membrane (Fig. 16.4).

At present, the role of Rab25 in development and progression of cancer is unclear. Loss of Rab25 is associated with tumour initiation in the colon [98], and several studies have connected loss of Rab25 with progression of breast cancer, particularly ER-negative subtypes [99, 100]. On the other hand, Rab25 is found to be upregulated in invasive breast carcinoma and in ER- and Her2-positive breast cancers [101]. Furthermore, Rab25 overexpression correlates with decreased survival and increased aggressiveness of ovarian cancer [100] and enhanced invasive migration of ovarian cancer cells in vitro [12]. The discovery of CLIC3 as a regulator of integrin recycling from lysosomes may resolve conflicts regarding Rab25's role in tumour progression. Indeed, it is likely that when CLIC3 is present and cells can recycle (rather than degrade) lysosomally routed integrins, Rab25 promotes $\alpha5\beta1$-dependent invasion. However, in tumours where the CLIC3 pathway is absent, integrins trafficked to lysosomes will likely be degraded, thus suppressing tumour aggressiveness.

16.5 Integrin Trafficking During Cell Migration

Integrin endocytosis and recycling has been proposed to influence cell migration in various ways. It is perhaps unsurprising that there is evidence that the endo–exocytic cycle contributes to adhesion assembly and disassembly by respectively delivering and removing integrins from the plasma membrane [22, 34]. But there are also indications that integrin trafficking modulates cell migration in more complex ways.

For instance, endocytic trafficking has been shown to promote both the spatial restriction of integrins within defined cellular locales [12] and to redistribute integrins en masse from one end of the cell to the other [11, 40]. Furthermore, the way that integrins signal to pathways that are key to cell migration, such as the activation of Rho GTPases, is influenced by the way that integrins are trafficked [65, 102].

16.5.1 Integrin Endocytosis Controls Focal Adhesion Turnover

Integrin endocytosis can regulate turnover and stability of FAs, thereby directly affecting cell adhesion and migration. In order to specifically address the mechanisms involved in the regulation of FA turnover, Ezratty and co-workers developed a model system in which the regrowth of microtubules after nocodazole washout induces the disassembly of FAs [31]. Indeed, microtubules can contact β1 integrin-containing FAs and induce their disassembly. This process is independent of Rac and Rho small GTPases, but requires focal adhesion kinase (FAK) and dynamin-2, a large GTPase necessary for endocytosis. Moreover, dynamin-2 itself partially co-localises with FAs, and the binding to FAK is required for FA targeting and integrin internalisation [18] indicating that an endocytic event mediates FA disassembly. As previously discussed, the clathrin adaptors Dab2 and ARH mediate the targeting of clathrin to FAs, and this drives internalisation of the integrin and FA disassembly [22]. Using TIRF microscopy, clathrin positive structures have been shown to co-localise with focal adhesion components, such as FAK and vinculin, and to leave the TIRF field together with β1 integrin. After internalisation, β1 co-localises with Rab5, marker of early endosomes, and Rab11, a marker of recycling endosomes, indicating the integrins are then delivered to the PNRC. Interestingly, the downregulation of any component of this integrin internalisation system (Dab2, ARH, dynamin-2) impairs cell motility [22]. Furthermore, the activation state of β1 integrin may play an important role in endocytosis via this pathway; indeed, the ligand-activated β1 conformation, recognised by the 12G10 antibody, has been shown to be preferentially internalised in a Dab2 and clathrin-dependent fashion from FAs [18]. The endocytic machinery needs to be specifically targeted to adhesive sites prior to their disassembly, and it has been shown that the local production of phosphatidyl inositol-4,5-bisphosphate (PI4,5P2) by type I phosphatidylinositol phosphate kinase beta (PIPKIβ) orchestrates the recruitment of components of the endocytic machinery to adhesion sites, thus leading to β1 integrin internalisation and FA disassembly [103].

16.5.2 Spatial Restriction Versus En Masse Redistribution of Integrins

By constantly pumping endocytosed receptors back to the very domains from which they were internalised, endo–exocytic cycling can function to spatially restrict receptors within defined regions of the plasma membrane. Indeed, RTK signalling is localised to the leading edge of migrating Drosophila border cells by an endo–exocytic cycle that involves dynamin, Cbl and Sprint [104]. More recently, endo–exocytic cycling has been shown to restrict the localisation of active Rac to CDRs in a process involving dynamin-dependent endocytosis which transfers Rac from the ruffles to endosomes containing a GEF for Rac [105]. This leads to Rac activation and active Rac then returns to the plasma membrane by an ARF6-dependent recycling route, and this maintains the concentration of Rac within the CDR and thus the polymerisation of actin within this structure.

When integrin endo–exocytic cycling was originally described by Marc Bretscher, it was proposed that this may contribute to cell migration by internalising integrins at the back of the cell and then transporting them forwards within vesicles for re-exocytosis at the cell front [10]. However, although such en masse redistribution of integrin occurs during cytokinesis and post-mitotic spreading [40], existing evidence does not support a model in which vesicles take integrins from the back to the front of cells. There is currently no evidence that endocytic rates are augmented at the cell rear. Conversely, most studies indicate increased rates of internalisation and localisation of clathrin-dependent endocytosis toward the leading and not the trailing edge of migrating cells [106]. Consistent with this, it appears likely that integrins are internalised at FAs near the leading edge [18] and then transported backwards to the PNRC [107]. Traffic emerging from the perinuclear region, including recycling cargos, travels forwards in migrating cells to sites of focal exocytosis near the lamellipodium [108–110], indicating the likelihood that integrins are both internalised and recycled near the lamellipodium, thus spatially restricting a population of integrins to the front portion of the cell.

When cancer cells migrate in a 3 dimensional (3D) environment, endo–exocytic cycling of integrins in the inactive conformation is restricted at the tip of invasive protrusion by the action of Rab25 [11, 12] (Fig. 16.5b). Experiments with photoactivatable paGFP-$\alpha5$ clearly show that $\alpha5\beta1$ integrin leaves Rab25 positive vesicles and reaches the plasma membrane at the tip of the pseudopod. Most of the photoactivated integrin is then maintained in dynamic equilibrium between the Rab25 vesicle and the plasma membrane in the front portion of the cell without diffusing backwards. This localised targeting of $\alpha5\beta1$ appears to promote extension of the invasive pseudopod. However, in the same cells, a smaller pool of $\alpha5\beta1$ in the active conformation is not spatially restricted to the pseudopod tip, but redistributed from the cell front to the back of the cell [11]. Indeed, integrins that remain in the active conformation are sorted into Rab25-positive late endosomes and therein transferred to lysosomes further back in the cell. This population of lysosomes are enriched in CLIC3, which prevents degradation of the active integrins and facilitates their return to the plasma membrane in the rearward portion of the cell. CLIC3 is required for efficient retraction of the cell rear, indicating that retrograde integrin trafficking may be associated with the release of lysosomal components, such as proteases that promote the detachment of adhesions to allow the cell to move forward (Fig. 16.5b). Consistent with this, it has recently been shown that exocytosis of Rab27 vesicles (which are largely thought to be late endosomal/lysosomal in nature) occurs at the back of migrating neutrophils and that this contributes to release of adhesions at the cell rear [111].

The endocytic pathway has been shown to contribute to the en masse redistribution of integrins during cytokinesis and post-mitotic spreading. During mid telophase, Rab21-positive endosomes containing β1 integrins move en masse toward the mid-body whereupon they concentrate the integrin at the cleavage furrow and anchor this structure to the substratum [40]. This process appears to be reversed following cytokinesis as the integrins are transported away from the cleavage furrow toward opposing poles of the daughter cells as they migrate away from one another.

More recently, it has become clear that Rab21's capacity to traffic integrin to and from the cytokinetic furrow is likely to be important in preventing genetic instability. Rab21 levels are often downregulated in cancer, and the Ivaska lab tested the hypothesis that this may lead to aberrant cell division and thus contribute to tumour progression [112]. They found that suppression of Rab21 levels, or disruption of the association between Rab21 and integrin α-cytotails, decreases the fidelity of chromosome segregation events during mitosis. The resulting chromosomal instability is associated with increased oncogenic transformation and is of the type that is commonly observed in cancers and thought to drive cancer progression in humans.

16.5.3 Integrin Trafficking Dictates Rho GTPase Signalling

Rho family GTPases are central to the regulation of cell migration, and one of the ways that integrins are known to influence cell movement is by modulating the activities of these GTPases. The relative expression levels of αvβ3 and α5β1 integrins determine how cells migrate—when αvβ3 expression is high (and α5β1 is not expressed), cells migrate persistently due to downstream Rac activation that allows the formation of a flat lamellipodium at the cell front [113]. However, when α5β1 is expressed in the absence of αvβ3, cells migrate with low persistence owing to activation of RhoA downstream of α5β1 which (via ROCK phosphorylation) inactivates the actin severing protein cofilin leading to the lamellipodial collapse and the formation of less well-organised protrusions [113].

Rho GTPase signalling is known to be influenced by endosomal transport. As mentioned previously, Tiam-1 catalyses the exchange of GTP onto Rac when they meet on endosomal membranes, and this promotes spatial restriction of active Rac in CDRs [105]. Endocytic dynamics are also known to coordinate RhoA signalling. Indeed the collagen internalisation receptor, Endo180, acts from endosomes near the back of the cell to promote detachment of the cell rear [114]. And, even when the relative expression levels αvβ3 and α5β1 remain the same, the way that these integrins are trafficked can influence the balance of Rho and Rac signalling [65]. When αvβ3 can recruit PKD1 and recycle via the Rab4-dependent short loop, signalling via this integrin to Rac tends to be dominant (over α5β1 signalling to Rho-ROCK), and cells assemble a broad flat lamellipodium and migrate persistently into scratch wounds. Conversely, when this is impaired by inhibiting either short-loop trafficking (by disrupting association with PKD1, or by opposing Rab4 function) or by suppressing αvβ3 levels or its ligand-binding capacity (with Cilengitide), cells move less persistently [13, 65]. Interestingly, this is not so much a direct consequence of reduced αvβ3 function, but due to a compensatory increase in α5β1 trafficking and signalling that occurs following inhibition of αvβ3 trafficking or signalling. Indeed, αvβ3 exerts tonic inhibition over α5β1 recycling through the RCP- and Rab11-dependent pathway. Thus, when αvβ3 (or its trafficking) is blocked, this inhibition is relieved and α5β1 recycling is increased leading to RhoA-ROCK activation, cofilin phosphorylation and the acquisition of a less persistent mode of migration into scratch wounds.

16.5.4 Integrin Trafficking and Signalling Influences Cancer Progression by Regulating the Endocytosis and Recycling of Other Cargoes

Integrin engagement influences the recycling of other receptors, such as RTKs, in a way that has relevance to cancer progression, and this is apparent both in terms of tumour angiogenesis and invasive migration of tumour cells. Experiments in some in vivo and ex vivo models indicated that inhibition of αvβ3 can oppose cell migration and angiogenesis, and these observations prompted the development of RGD-mimetic integrin inhibitors which target αvβ3 [115]. The highly potent cyclic peptide αv integrin inhibitor Cilengitide (cyclo-RGDfNmeV; which inhibits the ligand-binding capacity of αvβ3 and αvβ5 integrins) has been evaluated as a potential anticancer agent in clinical trials, but with disappointing results [116, 117]. The lack of efficacy of Cilengitide is likely explained by recent studies demonstrating that αvβ3 integrin can play inhibitory roles both during tumour angiogenesis and in invasive migration of tumour cells and that it does this by influencing receptor trafficking [118].

Indeed, when αvβ3 is present and able to engage ligand, sorting of VEGFR2 into the recycling pathway is slow and, correspondingly, VEGFR2 is degraded rapidly following addition of VEGF [118] (Fig. 16.6a). Following blockade of αvβ3 with Cilengitide, however, VEGFR2 is rapidly recycled via a Rab4-regulated pathway, and this protects VEGFR2 from degradation. The resulting increase in cell-surface levels of VEGFR2 promotes endothelial cell migration, angiogenic sprouting and tumour vascularisation. Given the previous literature on the anti-angiogenic effects of αvβ3 integrin-blocking agents, some workers expressed surprise that blockade of αvβ3 can promote tumour angiogenesis [119]. But to oppose angiogenesis, Cilengitide must be used at a much higher concentrations (μM) than is necessary to block the binding of vitronectin and fibrinogen to isolated immobilised αvβ3. Indeed the IC_{50} for the inhibition of vitronectin-αvβ3 binding by Cilengitide is 0.58 nM [120], a concentration not dissimilar to that found by Reynolds et al. [118] to promote tumour angiogenesis and VEGFR2 trafficking. Furthermore, results obtained with low nM concentrations of Cilengitide are consistent with the β3 integrin knockout phenotype; $\beta 3^{-/-}$ mice display markedly enhanced tumour angiogenesis, and this, again, is attributable to increased levels of VEGFR2 [121].

As well as promoting angiogenesis, inhibition of αvβ3 activates the trafficking of other receptors to drive invasion and migration of cancer cells. RCP can associate with both αvβ3 and α5β1 integrins, but treatment with Cilengitide displaces RCP from αvβ3 and promotes its recruitment to α5β1 to increase recycling of this integrin [13]. Moreover, inhibition of αvβ3 enables the α5β1-RCP complex to also recruit EGFR. This leads to increased EGFR recycling and enhanced EGFR signalling (Fig. 16.6b). Indeed, inhibition of αvβ3 potentiates EGFR autophosphorylation and modulates its downstream signalling to PKB/Akt in a way that is dependent on RCP. Given that PKB/Akt promotes cancer cell invasion [122], it is likely that much of the contribution made by RCP-dependent α5β1/EGFR to invasive migration is attributable to its ability to potentiate PKB/Akt signalling. These studies indicate that a key function of αvβ3 integrin is to suppress trafficking of promigratory receptors. Thus, when αvβ3 is inhibited, this increases recycling of other integrins and their associated RTKs to drive downstream signalling leading to enhanced cancer metastasis and tumour angiogenesis. Clearly, these studies indicate that inhibition of αvβ3 is not a good strategy to pursue as a potential cancer therapy. Moreover, it is clear that before pressing forward with strategies to inhibit adhesion receptors, it is necessary to consider the effects that anti-integrins might have on the trafficking and signalling of other integrins and adhesion receptors.

The p53 tumour suppressor is frequently lost in cancer. However, in many human tumours, p53 expression is not lost, but the gene acquires mutations that disrupt its ability to bind to DNA, and a growing body of evidence now supports additional gain-of-function roles for mutant p53s in cancer invasion and metastasis. Indeed, introduction of mutant p53s into p53 null mice increases the incidence of highly metastatic carcinomas [123], and this is associated with enhanced cancer cell migration and invasion [90, 91, 124]. One way that mutant p53s can achieve prometastatic gain of function is by binding to and inhibiting the p53-family member, p63. And p63, in particular the TAp63 isoform, is now known to play an important role in tumour suppression [125, 126]. Reduction of p63 levels in cancer cell lines increases invasive migration, and expression of mutant p53 recapitulates the effects of TAp63 loss, indicating that mutant p53's gain of function may operate by specifically inhibiting the TA isoform of p63 [90, 124].

Two recent studies indicate that mutant p53's achieves its gain of function to drive cancer cell invasion and migration by inhibiting TAp63 which leads to recruitment of RCP to α5β1 [90, 91]. This then channels α5β1 integrin and RTKs, such as EGFR and cMET, into an RCP-dependent recycling pathway. The consequences of this are increased downstream signalling—which in the case of EGFR is to PKB/Akt [90], and cMET's ability to activate the MEK-ERK cascade is also enhanced in an RCP-dependent fashion [91]—which, in turn, leads to disruption of cell–cell junctions, increased cell migration on 2D surfaces and enhancement of cancer cell invasion into 3D microenvironments (Fig. 16.6b).

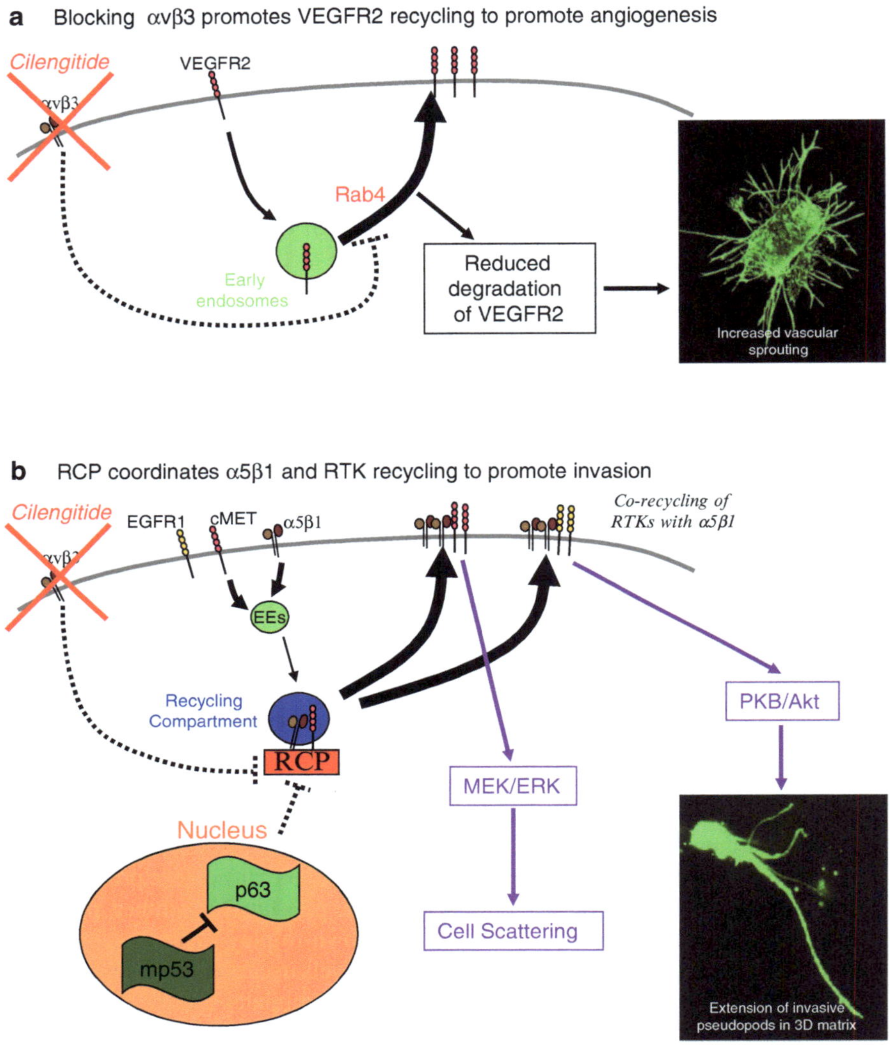

Fig. 16.6 Role of integrins in controlling RTK trafficking. (**a**) αvβ3 inhibitors promote activation of a Rab4-regulated pathway that diverts VEGFR2 from the degradative route to carry it back to the plasma membrane. This boosts the levels of VEGFR2 at the cell surface to promote the endothelial cell migration, sprouting and tubulation that drives tumour angiogenesis. (**b**) Inhibition of αvβ3 or expression of mutant p53 promotes assembly of an endosomal complex containing RCP, α5β1 and EGFR or cMET. α5β1, EGFR and cMET are then coordinately recycled to the plasma membrane. This potentiates signalling which contributes to cell scattering and to extension of invasive pseudopods and migration of tumour cells into 3D matrices

16.6 Concluding Remarks

We have seen over the course of this chapter that there are now clear indications that integrin trafficking is important during cancer progression, and we assert that this must be considered whilst assessing the efficacy of emerging potential anticancer agents. Indeed, observations that α5β1 and RTK recycling is upregulated following αvβ3 inhibition not only reveal the likely mechanism and molecular explanation for the overall lack of efficacy of anti-αv integrin drugs, but also highlight the need to consider any potential compensatory changes in recycling of one receptor following the inhibition of another. We have also seen that certain GTPases, such as Rab25, can be considered to be either tumour suppressors or promoters depending on the cancer type and cellular context. And the discovery of CLIC3 as a protein that protects lysosomally routed integrins from degradation and allows their recycling may dictate whether Rab25 acts to promote or oppose tumour progression. Furthermore, the recently described requirement for Rab21-mediated integrin trafficking in maintaining genetic stability highlights the need to consider the possibility that targeting integrin function may promote chromosomal aberrations that would restrict the efficacy of any potential anticancer agents. Lastly, we have seen that mutation of p53 (which is considered to be possibly one of the most significant events in cancer progression) drives invasion and metastasis by altering integrin and RTK recycling, and further work will determine which components of the RCP-dependent recycling pathway may be targeted therapeutically to reverse mutant p53's oncogenic drive in human cancers.

Acknowledgements Work in the Norman lab is funded by Cancer Research UK. Elena Rainero is supported by a fellowship from the West of Scotland Women's Bowling Association.

References

1. Hynes RO (2002) Integrins: bidirectional, allosteric signaling machines. Cell 110(6):673–687
2. Sahai E (2005) Mechanisms of cancer cell invasion. Curr Opin Genet Dev 15(1):87–96
3. Sahai E, Marshall CJ (2002) RHO-GTPases and cancer. Nat Rev Cancer 2(2):133–142
4. Grant BD, Donaldson JG (2009) Pathways and mechanisms of endocytic recycling. Nat Rev Mol Cell Biol 10(9):597–608
5. Bretscher MS (1989) Endocytosis and recycling of the fibronectin receptor in CHO cells. EMBO J 8(5):1341–1348
6. Bretscher MS (1992) Circulating integrins: alpha 5 beta 1, alpha 6 beta 4 and Mac-1, but not alpha 3 beta 1, alpha 4 beta 1 or LFA-1. EMBO J 11(2):405–410
7. Caswell PT, Norman JC (2006) Integrin trafficking and the control of cell migration. Traffic 7(1):14–21
8. Pellinen T, Ivaska J (2006) Integrin traffic. J Cell Sci 119(pt 18):3723–3731
9. Altankov G, Grinnell F (1995) Fibronectin receptor internalization and AP-2 complex reorganization in potassium-depleted fibroblasts. Exp Cell Res 216(2):299–309
10. Bretscher MS (1996) Moving membrane up to the front of migrating cells. Cell 85(4):465–467

11. Dozynkiewicz MA, Jamieson NB, Macpherson I, Grindlay J, van den Berghe PV, von Thun A et al (2012) Rab25 and CLIC3 collaborate to promote integrin recycling from late endosomes/lysosomes and drive cancer progression. Dev Cell 22(1):131–145
12. Caswell PT, Spence HJ, Parsons M, White DP, Clark K, Cheng KW et al (2007) Rab25 associates with alpha5beta1 integrin to promote invasive migration in 3D microenvironments. Dev Cell 13(4):496–510
13. Caswell PT, Chan M, Lindsay AJ, McCaffrey MW, Boettiger D, Norman JC (2008) Rab-coupling protein coordinates recycling of alpha5beta1 integrin and EGFR1 to promote cell migration in 3D microenvironments. J Cell Biol 183(1):143–155
14. Roberts M, Barry S, Woods A, van der Sluijs P, Norman J (2001) PDGF-regulated rab4-dependent recycling of alphavbeta3 integrin from early endosomes is necessary for cell adhesion and spreading. Curr Biol 11(18):1392–1402
15. Pellinen T, Arjonen A, Vuoriluoto K, Kallio K, Fransen JA, Ivaska J (2006) Small GTPase Rab21 regulates cell adhesion and controls endosomal traffic of {beta}1-integrins. J Cell Biol 173(5):767–780
16. Powelka AM, Sun J, Li J, Gao M, Shaw LM, Sonnenberg A et al (2004) Stimulation-dependent recycling of integrin beta1 regulated by ARF6 and Rab11. Traffic 5(1):20–36
17. Arjonen A, Alanko J, Veltel S, Ivaska J (2012) Distinct recycling of active and inactive beta1 integrins. Traffic. 13(4):610–625
18. Chao WT, Kunz J (2009) Focal adhesion disassembly requires clathrin-dependent endocytosis of integrins. FEBS Lett 583(8):1337–1343
19. Valdembri D, Caswell PT, Anderson KI, Schwarz JP, Konig I, Astanina E et al (2009) Neuropilin-1/GIPC1 signaling regulates alpha5beta1 integrin traffic and function in endothelial cells. PLoS Biol 7(1):e25
20. Reid PA, Watts C (1990) Cycling of cell-surface MHC glycoproteins through primaquine-sensitive intracellular compartments. Nature 346(6285):655–657
21. Strous GJ, Du Maine A, Zijderhand-Bleekemolen JE, Slot JW, Schwartz AL (1985) Effect of lysosomotropic amines on the secretory pathway and on the recycling of the asialoglycoprotein receptor in human hepatoma cells. J Cell Biol 101(2):531–539
22. Ezratty EJ, Bertaux C, Marcantonio EE, Gundersen GG (2009) Clathrin mediates integrin endocytosis for focal adhesion disassembly in migrating cells. J Cell Biol 187(5):733–747
23. Lippincott-Schwartz J, Patterson GH (2008) Fluorescent proteins for photoactivation experiments. Methods Cell Biol 85:45–61
24. Jovic M, Naslavsky N, Rapaport D, Horowitz M, Caplan S (2007) EHD1 regulates beta1 integrin endosomal transport: effects on focal adhesions, cell spreading and migration. J Cell Sci 120(pt 5):802–814
25. Roberts MS, Woods AJ, Dale TC, Van Der Sluijs P, Norman JC (2004) Protein kinase B/Akt acts via glycogen synthase kinase 3 to regulate recycling of alpha v beta 3 and alpha 5 beta 1 integrins. Mol Cell Biol 24(4):1505–1515
26. Serrels A, Timpson P, Canel M, Schwarz JP, Carragher NO, Frame MC et al (2009) Real-time study of E-cadherin and membrane dynamics in living animals: implications for disease modeling and drug development. Cancer Res 69(7):2714–2719
27. Teckchandani A, Toida N, Goodchild J, Henderson C, Watts J, Wollscheid B et al (2009) Quantitative proteomics identifies a Dab2/integrin module regulating cell migration. J Cell Biol 186(1):99–111
28. Howes MT, Kirkham M, Riches J, Cortese K, Walser PJ, Simpson F et al (2010) Clathrin-independent carriers form a high capacity endocytic sorting system at the leading edge of migrating cells. J Cell Biol 190(4):675–691
29. Shi F, Sottile J (2008) Caveolin-1-dependent beta1 integrin endocytosis is a critical regulator of fibronectin turnover. J Cell Sci 121(pt 14):2360–2371
30. Gu Z, Noss EH, Hsu VW, Brenner MB (2011) Integrins traffic rapidly via circular dorsal ruffles and macropinocytosis during stimulated cell migration. J Cell Biol 193(1):61–70
31. Ezratty EJ, Partridge MA, Gundersen GG (2005) Microtubule-induced focal adhesion disassembly is mediated by dynamin and focal adhesion kinase. Nat Cell Biol 7(6):581–590

32. Teckchandani A, Mulkearns EE, Randolph TW, Toida N, Cooper JA (2012) The clathrin adaptor Dab2 recruits EH domain scaffold proteins to regulate integrin beta1 endocytosis. Mol Cell Biol 23(15):2905–2916
33. Nishimura T, Kaibuchi K (2007) Numb controls integrin endocytosis for directional cell migration with aPKC and PAR-3. Dev Cell 13(1):15–28
34. Dunphy JL, Moravec R, Ly K, Lasell TK, Melancon P, Casanova JE (2006) The Arf6 GEF GEP100/BRAG2 regulates cell adhesion by controlling endocytosis of beta1 integrins. Curr Biol 16(3):315–320
35. Moravec R, Conger KK, D'Souza R, Allison AB, Casanova JE (2012) BRAG2/GEP100/IQSec1 interacts with clathrin and regulates alpha5beta1 integrin endocytosis through activation of Arf5. J Biol Chem 287(37):31138–31147
36. Chen WJ, Goldstein JL, Brown MS (1990) NPXY, a sequence often found in cytoplasmic tails, is required for coated pit-mediated internalization of the low density lipoprotein receptor. J Biol Chem 265(6):3116–3123
37. Rajagopalan M, Neidigh JL, McClain DA (1991) Amino acid sequences Gly-Pro-Leu-Tyr and Asn-Pro-Glu-Tyr in the submembranous domain of the insulin receptor are required for normal endocytosis. J Biol Chem 266(34):23068–23073
38. Vignoud L, Usson Y, Balzac F, Tarone G, Block MR (1994) Internalization of the alpha 5 beta 1 integrin does not depend on "NPXY" signals. Biochem Biophys Res Commun 199(2):603–611
39. Margadant C, Kreft M, de Groot DJ, Norman JC, Sonnenberg A (2012) Distinct roles of Talin and Kindlin in regulating integrin alpha5beta1 function and trafficking. Curr Biol 22(17):1554–1563
40. Pellinen T, Tuomi S, Arjonen A, Wolf M, Edgren H, Meyer H et al (2008) Integrin trafficking regulated by Rab21 is necessary for cytokinesis. Dev Cell 15(3):371–385
41. Calderwood DA, Fujioka Y, de Pereda JM, Garcia-Alvarez B, Nakamoto T, Margolis B et al (2003) Integrin beta cytoplasmic domain interactions with phosphotyrosine-binding domains: a structural prototype for diversity in integrin signaling. Proc Natl Acad Sci U S A 100(5):2272–2277
42. Garcia-Alvarez B, de Pereda JM, Calderwood DA, Ulmer TS, Critchley D, Campbell ID et al (2003) Structural determinants of integrin recognition by talin. Mol Cell 11(1):49–58
43. McCleverty CJ, Lin DC, Liddington RC (2007) Structure of the PTB domain of tensin1 and a model for its recruitment to fibrillar adhesions. Protein Sci 16(6):1223–1229
44. Mai A, Veltel S, Pellinen T, Padzik A, Coffey E, Marjomaki V et al (2011) Competitive binding of Rab21 and p120RasGAP to integrins regulates receptor traffic and migration. J Cell Biol 194(2):291–306
45. Ramsay AG, Keppler MD, Jazayeri M, Thomas GJ, Parsons M, Violette S et al (2007) HS1-associated protein X-1 regulates carcinoma cell migration and invasion via clathrin-mediated endocytosis of integrin alphavbeta6. Cancer Res 67(11):5275–5284
46. Fabbri M, Di Meglio S, Gagliani MC, Consonni E, Molteni R, Bender JR et al (2005) Dynamic partitioning into lipid rafts controls the endo-exocytic cycle of the alphaL/beta2 integrin, LFA-1, during leukocyte chemotaxis. Mol Biol Cell 16(12):5793–5803
47. Galvez BG, Matias-Roman S, Yanez-Mo M, Vicente-Manzanares M, Sanchez-Madrid F, Arroyo AG (2004) Caveolae are a novel pathway for membrane-type 1 matrix metalloproteinase traffic in human endothelial cells. Mol Biol Cell 15(2):678–687
48. Wickstrom SA, Lange A, Hess MW, Polleux J, Spatz JP, Kruger M et al (2010) Integrin-linked kinase controls microtubule dynamics required for plasma membrane targeting of caveolae. Dev Cell 19(4):574–588
49. Bass MD, Williamson RC, Nunan RD, Humphries JD, Byron A, Morgan MR et al (2011) A syndecan-4 hair trigger initiates wound healing through caveolin- and RhoG-regulated integrin endocytosis. Dev Cell 21(4):681–693
50. Hansen CG, Nichols BJ (2009) Molecular mechanisms of clathrin-independent endocytosis. J Cell Sci 122(pt 11):1713–1721
51. Lundmark R, Doherty GJ, Howes MT, Cortese K, Vallis Y, Parton RG et al (2008) The GTPase-activating protein GRAF1 regulates the CLIC/GEEC endocytic pathway. Curr Biol 18(22):1802–1808

52. Doherty GJ, Ahlund MK, Howes MT, Moren B, Parton RG, McMahon HT et al (2011) The endocytic protein GRAF1 is directed to cell-matrix adhesion sites and regulates cell spreading. Mol Biol Cell 22(22):4380–4389
53. Orth JD, McNiven MA (2006) Get off my back! Rapid receptor internalization through circular dorsal ruffles. Cancer Res 66(23):11094–11096
54. Orth JD, Krueger EW, Weller SG, McNiven MA (2006) A novel endocytic mechanism of epidermal growth factor receptor sequestration and internalization. Cancer Res 66(7):3603–3610
55. Bar-Sagi D, Hall A (2000) Ras and Rho GTPases: a family reunion. Cell 103(2):227–238
56. Parsons M, Keppler MD, Kline A, Messent A, Humphries MJ, Gilchrist R et al (2002) Site-directed perturbation of protein kinase C-integrin interaction blocks carcinoma cell chemotaxis. Mol Cell Biol 22(16):5897–5911
57. Ng T, Shima D, Squire A, Bastiaens PI, Gschmeissner S, Humphries MJ et al (1999) PKCalpha regulates beta1 integrin-dependent cell motility through association and control of integrin traffic. EMBO J 18(14):3909–3923
58. Upla P, Marjomaki V, Kankaanpaa P, Ivaska J, Hyypia T, Van Der Goot FG et al (2004) Clustering induces a lateral redistribution of alpha 2 beta 1 integrin from membrane rafts to caveolae and subsequent protein kinase C-dependent internalization. Mol Biol Cell 15(2):625–636
59. Couchman JR, Woods A (1999) Syndecan-4 and integrins: combinatorial signaling in cell adhesion. J Cell Sci 112(pt 20):3415–3420
60. Hines JH, Abu-Rub M, Henley JR (2010) Asymmetric endocytosis and remodeling of beta1-integrin adhesions during growth cone chemorepulsion by MAG. Nat Neurosci 13(7):829–837
61. Du J, Chen X, Liang X, Zhang G, Xu J, He L et al (2011) Integrin activation and internalization on soft ECM as a mechanism of induction of stem cell differentiation by ECM elasticity. Proc Natl Acad Sci U S A 108(23):9466–9471
62. Levental KR, Yu H, Kass L, Lakins JN, Egeblad M, Erler JT et al (2009) Matrix crosslinking forces tumor progression by enhancing integrin signaling. Cell 139(5):891–906
63. Naslavsky N, Weigert R, Donaldson JG (2003) Convergence of non-clathrin- and clathrin-derived endosomes involves Arf6 inactivation and changes in phosphoinositides. Mol Biol Cell 14(2):417–431
64. Woods AJ, White DP, Caswell PT, Norman JC (2004) PKD1/PKCmu promotes alphavbeta3 integrin recycling and delivery to nascent focal adhesions. EMBO J 23(13):2531–2543
65. White DP, Caswell PT, Norman JC (2007) alpha v beta3 and alpha5beta1 integrin recycling pathways dictate downstream Rho kinase signaling to regulate persistent cell migration. J Cell Biol 177(3):515–525
66. di Blasio L, Droetto S, Norman J, Bussolino F, Primo L (2010) Protein kinase D1 regulates VEGF-A-induced alphavbeta3 integrin trafficking and endothelial cell migration. Traffic 11(8):1107–1118
67. Ochi N, Tanasanvimon S, Matsuo Y, Tong Z, Sung B, Aggarwal BB et al (2011) Protein kinase D1 promotes anchorage-independent growth, invasion, and angiogenesis by human pancreatic cancer cells. J Cell Physiol 226(4):1074–1081
68. Eiseler T, Doppler H, Yan IK, Goodison S, Storz P (2009) Protein kinase D1 regulates matrix metalloproteinase expression and inhibits breast cancer cell invasion. Breast Cancer Res 11(1):R13
69. Kim M, Jang HR, Kim JH, Noh SM, Song KS, Cho JS et al (2008) Epigenetic inactivation of protein kinase D1 in gastric cancer and its role in gastric cancer cell migration and invasion. Carcinogenesis 29(3):629–637
70. Zhang K, Ye C, Zhou Q, Zheng R, Lv X, Chen Y et al (2007) PKD1 inhibits cancer cells migration and invasion via Wnt signaling pathway in vitro. Cell Biochem Funct 25(6):767–774
71. Fang Z, Takizawa N, Wilson KA, Smith TC, Delprato A, Davidson MW et al (2010) The membrane-associated protein, supervillin, accelerates F-actin-dependent rapid integrin recycling and cell motility. Traffic 11(6):782–799

72. Skalski M, Coppolino MG (2005) SNARE-mediated trafficking of alpha5beta1 integrin is required for spreading in CHO cells. Biochem Biophys Res Commun 335(4):1199–1210
73. Strachan LR, Condic ML (2004) Cranial neural crest recycle surface integrins in a substratum-dependent manner to promote rapid motility. J Cell Biol 167(3):545–554
74. Yoon SO, Shin S, Mercurio AM (2005) Hypoxia stimulates carcinoma invasion by stabilizing microtubules and promoting the Rab11 trafficking of the alpha6beta4 integrin. Cancer Res 65(7):2761–2769
75. Jovic M, Kieken F, Naslavsky N, Sorgen PL, Caplan S (2009) Eps15 homology domain 1-associated tubules contain phosphatidylinositol-4-phosphate and phosphatidylinositol-(4,5)-bisphosphate and are required for efficient recycling. Mol Biol Cell 20(11):2731–2743
76. Naslavsky N, Caplan S (2011) EHD proteins: key conductors of endocytic transport. Trends Cell Biol 21(2):122–131
77. Sharma M, Giridharan SS, Rahajeng J, Naslavsky N, Caplan S (2009) MICAL-L1 links EHD1 to tubular recycling endosomes and regulates receptor recycling. Mol Biol Cell 20(24):5181–5194
78. Day P, Riggs KA, Hasan N, Corbin D, Humphrey D, Hu C (2011) Syntaxins 3 and 4 mediate vesicular trafficking of alpha5beta1 and alpha3beta1 integrins and cancer cell migration. Int J Oncol 39(4):863–871
79. Hasan N, Hu C (2010) Vesicle-associated membrane protein 2 mediates trafficking of alpha-5beta1 integrin to the plasma membrane. Exp Cell Res 316(1):12–23
80. Luftman K, Hasan N, Day P, Hardee D, Hu C (2009) Silencing of VAMP3 inhibits cell migration and integrin-mediated adhesion. Biochem Biophys Res Commun 380(1):65–70
81. Rapaport D, Lugassy Y, Sprecher E, Horowitz M (2010) Loss of SNAP29 impairs endocytic recycling and cell motility. PLoS One 5(3):e9759
82. Veale KJ, Offenhauser C, Whittaker SP, Estrella RP, Murray RZ (2010) Recycling endosome membrane incorporation into the leading edge regulates lamellipodia formation and macrophage migration. Traffic 11(10):1370–1379
83. Casanova JE, Wang X, Kumar R, Bhartur SG, Navarre J, Woodrum JE et al (1999) Association of Rab25 and Rab11a with the apical recycling system of polarized Madin-Darby canine kidney cells. Mol Biol Cell 10(1):47–61
84. Cohen P, Frame S (2001) The renaissance of GSK3. Nat Rev Mol Cell Biol 2(10):769–776
85. Dai J, Li J, Bos E, Porcionatto M, Premont RT, Bourgoin S et al (2004) ACAP1 promotes endocytic recycling by recognizing recycling sorting signals. Dev Cell 7(5):771–776
86. Li J, Ballif BA, Powelka AM, Dai J, Gygi SP, Hsu VW (2005) Phosphorylation of ACAP1 by Akt regulates the stimulation-dependent recycling of integrin beta1 to control cell migration. Dev Cell 9(5):663–673
87. Ivaska J, Whelan RD, Watson R, Parker PJ (2002) PKC epsilon controls the traffic of beta1 integrins in motile cells. EMBO J 21(14):3608–3619
88. Ivaska J, Vuoriluoto K, Huovinen T, Izawa I, Inagaki M, Parker PJ (2005) PKCepsilon-mediated phosphorylation of vimentin controls integrin recycling and motility. EMBO J 24(22):3834–3845
89. Chan PC, Chen HC (2012) p120RasGAP-mediated activation of c-Src is critical for oncogenic Ras to induce tumor invasion. Cancer Res 72(9):2405–2415
90. Muller PA, Caswell PT, Doyle B, Iwanicki MP, Tan EH, Karim S et al (2009) Mutant p53 drives invasion by promoting integrin recycling. Cell 139(7):1327–1341
91. Muller PA, Trinidad AG, Timpson P, Morton JP, Zanivan S, van den Berghe PV et al (2012) Mutant p53 enhances MET trafficking and signalling to drive cell scattering and invasion. Oncogene. Epub 2012/05/15.
92. Lindsay AJ, McCaffrey MW (2004) The C2 domains of the class I Rab11 family of interacting proteins target recycling vesicles to the plasma membrane. J Cell Sci 117(pt 19):4365–4375
93. Jenkins GM, Frohman MA (2005) Phospholipase D: a lipid centric review. Cell Mol Life Sci 62(19–20):2305–2316

94. Merida I, Avila-Flores A, Merino E (2008) Diacylglycerol kinases: at the hub of cell signalling. Biochem J 409(1):1–18
95. Rainero E, Caswell PT, Muller PA, Grindlay J, McCaffrey MW, Zhang Q et al (2012) Diacylglycerol kinase alpha controls RCP-dependent integrin trafficking to promote invasive migration. J Cell Biol 196(2):277–295
96. Brown FC, Pfeffer SR (2010) An update on transport vesicle tethering. Mol Membr Biol 27(8):457–461
97. Lobert VH, Brech A, Pedersen NM, Wesche J, Oppelt A, Malerod L et al (2010) Ubiquitination of alpha 5 beta 1 integrin controls fibroblast migration through lysosomal degradation of fibronectin-integrin complexes. Dev Cell 19(1):148–159
98. Nam KT, Lee HJ, Smith JJ, Lapierre LA, Kamath VP, Chen X et al (2010) Loss of Rab25 promotes the development of intestinal neoplasia in mice and is associated with human colorectal adenocarcinomas. J Clin Invest 120(3):840–849
99. Cheng JM, Volk L, Janaki DK, Vyakaranam S, Ran S, Rao KA (2010) Tumor suppressor function of Rab25 in triple-negative breast cancer. Int J Cancer 126(12):2799–2812
100. Cheng KW, Lahad JP, Kuo WL, Lapuk A, Yamada K, Auersperg N et al (2004) The RAB25 small GTPase determines aggressiveness of ovarian and breast cancers. Nat Med 10(11):1251–1256
101. Agarwal R, Jurisica I, Mills GB, Cheng KW (2009) The emerging role of the RAB25 small GTPase in cancer. Traffic 10(11):1561–1568
102. Sadok A, Pierres A, Dahan L, Prevot C, Lehmann M, Kovacic H (2009) NADPH oxidase 1 controls the persistence of directed cell migration by a Rho-dependent switch of alpha2/alpha3 integrins. Mol Cell Biol 29(14):3915–3928
103. Chao WT, Ashcroft F, Daquinag AC, Vadakkan T, Wei Z, Zhang P et al (2010) Type I phosphatidylinositol phosphate kinase beta regulates focal adhesion disassembly by promoting beta1 integrin endocytosis. Mol Cell Biol 30(18):4463–4479
104. Jekely G, Sung HH, Luque CM, Rorth P (2005) Regulators of endocytosis maintain loc.lized receptor tyrosine kinase signaling in guided migration. Dev Cell 9(2):197–207
105. Palamidessi A, Frittoli E, Garre M, Faretta M, Mione M, Testa I et al (2008) Endocytic trafficking of Rac is required for the spatial restriction of signaling in cell migration. Cell 134(1):135–147
106. Rappoport JZ, Simon SM (2003) Real-time analysis of clathrin-mediated endocytosis during cell migration. J Cell Sci 116(pt 5):847–855
107. Laukaitis CM, Webb DJ, Donais K, Horwitz AF (2001) Differential dynamics of alpha 5 integrin, paxillin, and alpha-actinin during formation and disassembly of adhesions in migrating cells. J Cell Biol 153(7):1427–1440
108. Hopkins CR, Gibson A, Shipman M, Strickland DK, Trowbridge IS (1994) In migrating fibroblasts, recycling receptors are concentrated in narrow tubules in the pericentriolar area, and then routed to the plasma membrane of the leading lamella. J Cell Biol 125(6):1265–1274
109. Prigozhina NL, Waterman-Storer CM (2004) Protein kinase D-mediated anterograde membrane trafficking is required for fibroblast motility. Curr Biol 14(2):88–98
110. Schmoranzer J, Kreitzer G, Simon SM (2003) Migrating fibroblasts perform polarized, microtubule-dependent exocytosis towards the leading edge. J Cell Sci 116(pt 22):4513–4519
111. Singh RK, Liao W, Tracey-White D, Recchi C, Tolmachova T, Rankin SM et al (2012) Rab27a-mediated protease release regulates neutrophil recruitment by allowing uropod detachment. J Cell Sci 125(pt 7):1652–1656
112. Hognas G, Tuomi S, Veltel S, Mattila E, Murumagi A, Edgren H et al (2012) Cytokinesis failure due to derailed integrin traffic induces aneuploidy and oncogenic transformation in vitro and in vivo. Oncogene 31(31):3597–3606
113. Danen EH, van Rheenen J, Franken W, Huveneers S, Sonneveld P, Jalink K et al (2005) Integrins control motile strategy through a Rho-cofilin pathway. J Cell Biol 169(3):515–526
114. Sturge J, Wienke D, Isacke CM (2006) Endosomes generate localized Rho-ROCK-MLC2-based contractile signals via Endo180 to promote adhesion disassembly. J Cell Biol 175(2):337–347

115. Nisato RE, Tille JC, Jonczyk A, Goodman SL, Pepper MS (2003) alphav beta 3 and alphav beta 5 integrin antagonists inhibit angiogenesis in vitro. Angiogenesis 6(2):105–119
116. Stupp R, Ruegg C (2007) Integrin inhibitors reaching the clinic. J Clin Oncol 25(13):1637–1638
117. Tucker GC (2006) Integrins: molecular targets in cancer therapy. Curr Oncol Rep 8(2):96–103
118. Reynolds AR, Hart IR, Watson AR, Welti JC, Silva RG, Robinson SD et al (2009) Stimulation of tumor growth and angiogenesis by low concentrations of RGD-mimetic integrin inhibitors. Nat Med 15(4):392–400
119. Weller M, Reardon D, Nabors B, Stupp R (2009) Will integrin inhibitors have proangiogenic effects in the clinic? Nat Med 15(7):726; author reply 7
120. Dechantsreiter MA, Planker E, Matha B, Lohof E, Holzemann G, Jonczyk A et al (1999) N-methylated cyclic RGD peptides as highly active and selective alpha(V)beta(3) integrin antagonists. J Med Chem 42(16):3033–3040
121. Reynolds LE, Wyder L, Lively JC, Taverna D, Robinson SD, Huang X et al (2002) Enhanced pathological angiogenesis in mice lacking beta3 integrin or beta3 and beta5 integrins. Nat Med 8(1):27–34
122. Irie HY, Pearline RV, Grueneberg D, Hsia M, Ravichandran P, Kothari N et al (2005) Distinct roles of Akt1 and Akt2 in regulating cell migration and epithelial-mesenchymal transition. J Cell Biol 171(6):1023–1034
123. Oren M, Rotter V (2010) Mutant p53 gain-of-function in cancer. Cold Spring Harb Perspect Biol 2(2):a001107
124. Adorno M, Cordenonsi M, Montagner M, Dupont S, Wong C, Hann B et al (2009) A mutant-p53/Smad complex opposes p63 to empower TGFbeta-induced metastasis. Cell 137(1):87–98
125. Flores ER, Sengupta S, Miller JB, Newman JJ, Bronson R, Crowley D et al (2005) Tumor predisposition in mice mutant for p63 and p73: evidence for broader tumor suppressor functions for the p53 family. Cancer Cell 7(4):363–373
126. Guo X, Keyes WM, Papazoglu C, Zuber J, Li W, Lowe SW et al (2009) TAp63 induces senescence and suppresses tumorigenesis in vivo. Nat Cell Biol 11(12):1451–1457

Chapter 17
Antibody-Mediated Receptor Endocytosis: Harnessing the Cellular Machinery to Combat Cancer

Gabi Tarcic and Yosef Yarden

Abstract Monoclonal antibodies targeting specific surface antigens of cancer cells are rapidly becoming the main stay drugs in specific diseases, such as lymphoma and breast cancer. Therapeutic antibodies almost invariably induce endocytosis of their antigens, and this attribute is already harnessed as a strategy to deliver cytotoxic payloads into cancer cells. The therapeutic potential, however, extends to direct antitumor activity of naked (unconjugated) antibodies, but the contribution of antibody-induced endocytosis to antitumor effects is variable and remains largely unclear. Interestingly, mixtures of monoclonal antibodies, each engaging a distinct epitope of the same antigen, synergistically induce receptor degradation and correspondingly collaborate in tumor inhibition. Here we describe several examples of therapeutic and experimental antibodies, with an emphasis on growth factor receptors and the possibility that future immunotherapy will employ specific antibody combinations, which robustly strip tumors of their most essential receptors.

17.1 Introduction

When Georges Köhler and César Milstein first generated monoclonal antibodies (mAbs) [1], it was believed that the new method would provide a platform for the delivery of "magic bullets," able to specifically eradicate malignant and other lesions. Accordingly, through the extremely high target specificity of mAbs, a toxic payload such as a bacterial toxin or a radioisotope might be delivered to specific target tissues, with minimal toxicity and adverse effects. Remarkably, however, the prerequisite of a toxin or an isotope proved unnecessary with specific cell surface antigens, like the cluster of differentiation 20 (CD20) and the rodent form

---

G. Tarcic • Y. Yarden (✉)
Department of Biological Regulation, Weizmann Institute of Science, Rehovot 76100, Israel
e-mail: yosef.yarden@weizmann.ac.il

Y. Yarden and G. Tarcic (eds.), *Vesicle Trafficking in Cancer*,
DOI 10.1007/978-1-4614-6528-7_17,

of NEU/HER2, an oncogenic receptor harboring a single transforming mutation within the transmembrane segment [2]. Because murine antibodies to CD20 or HER2, from either rodent or human origin [3], exhibited anticancer actions by their own, but administrating murine antibodies to patients inevitably evokes human anti-mouse antibodies (HAMA), the original murine mAbs were genetically manipulated to minimize HAMA. This was initially achieved by the generation of mouse-human chimeric mAbs [4] and later by antibody humanization [5] and the development of completely human mAbs in transgenic mice [6, 7].

Since the approval in 1997 of an anti-CD20 antibody (rituximab) for lymphoma treatment and trastuzumab, in 1998, for metastatic breast cancer, several more chimeric, humanized, or fully human antibodies were approved for clinical application in cancer and in other diseases (see Table 17.1). As reviewed below, mechanisms underlying the therapeutic effects of antibodies may be divided into immune-mediated cell killing and direct effects on tumor cells. This chapter will cover primarily one direct mechanism of action of naked antibodies, namely, antibody-mediated endocytosis and degradation of surface antigens, such as growth factor receptors. Readers are referred to other recent reviews that cover clinical aspects of anticancer antibodies [8, 9] and their immunomodulatory actions [10].

17.2 Mechanisms of Anticancer Effects Involving the Immune System

Figure 17.1 schematically presents potential mechanisms of tumor cell killing (or growth arresting) by antibodies. When dealing with naked antibodies, relevant direct mechanisms entail blocking the function of the antigen, for example, curtailing the angiogenic action of the vascular endothelial growth factor (VEGF) by an antibody called bevacizumab or cetuximab-mediated blocking of the binding of the epidermal growth factor (EGF) to the EGF receptor. Immune-mediated cell killing mechanisms include antibody-dependent cellular cytotoxicity (ADCC) and complement-dependent cytotoxicity (CDC). So far, most of the antibodies that have successfully been applied in clinical wards are intact immunoglobulin G (IgG) molecules. The Fc domain of such antibodies is particularly important for the recruitment of immune cells, whereas antigen binding and antibody bivalence are essential for the majority of the direct mechanisms.

17.2.1 Antibody-Dependent Cellular Cytotoxicity

ADCC involves recruitment of Fc receptor expressing immune effector cells, such as natural killer lymphocytes, to antibody-decorated tumor cells, followed by the killing of the antigen-expressing cancer cells [11]. Several lines of evidence indicate that ADCC plays critical roles in cancer cell targeting by certain antibodies.

Table 17.1 Monoclonal antibodies currently approved for therapy

Brand name	Antibody	Target	Indication	Company	Approval date
ReoPro	Abciximab	PIIb/IIIa	Cardiovascular	Johnson & Johnson	1984
OKT3	Muronomab-CD3	CD3	Autoimmune	Johnson & Johnson	1986
Rituxan	Rituximab	CD20	Cancer	Genentech	1997
Simulect	Basilixumab	CD25	Autoimmune	Novartis	1998
Synagis	Palivizumab	RSV	Infection	MedImmune	1998
Remicade	Infliximab	TNFα	Autoimmune	Johnson & Johnson	1998
Herceptin	Trastuzumab	HER2	Cancer	Genentech/Roche	1998
Campath	Alemtuzumab	CD52	Cancer	Genzyme	2001
Zevalin	Ibritumomab tuxan	CD20	Cancer	Biogen Idec	2002
Humira	Adalimumab	TNFα	Autoimmune	Abbot	2002
Xolair	Omalizumab	IgE	Autoimmune	Genentech/Roche	2003
Bexxar		CD20	Cancer	Corixa/GSK	2003
Erbitux	Cetuximab	EGFR	Cancer	ImClone/Lilly	2004
Avastin	Bevacizumab	VEGF	Cancer	Genentech/Roche	2004
Tysabri	Natalizumab	α4 integrin	Autoimmune	Biogen Idec	2004
Actemra	Tocilizumab	IL-6R	Autoimmune	Chugai/Roche	2005
Vectibix	Panitumumab	EGFR	Cancer	Amgen	2006
Lucentis	Ranibizumab	VEGF	Macular degeneration	Genentech/Roche	2006
Soliris	Eculizumab	C5	Hematology	Alexion	2007
Cimzia	Certolizumab pegol	TNFα	Autoimmune	UCB	2008
Simponi	Golimumab	TNFα	Autoimmune	Johnson & Johnson	2009
Ilaris	Canakinumab	IL1B	Inflammatory	Novartis	2009
Stelara	Ustekinumab	IL12/23	Autoimmune	Johnson & Johnson	2009
Arzerra	Ofatumumab	CD20	Cancer	Genmab	2009
Prolia	Denosumab	RANKL	Bone loss	Amgen	2010
Benlysta	Belimumab	BLyS	Autoimmune	Human Genome Science/GSK	2011
Yervoy	Ipilimumab	CTLA-4	Cancer	BMS	2011
Adcetris	Brentuximab vedotin	CD30	Cancer	Seattle Genetics	2011

Listed are the generic names and the respective clinical indications of mAbs which were approved over the last 28 years for treatment of patients. Note that anticancer antibodies are being used either in combination with cytotoxic regimens (chemo- or radiotherapy) or as single agents

Studies involving Fc-gamma receptor-deficient mice demonstrated that trastuzumab and rituximab engaged both activation (Fc-gamma RIIIa) and inhibitory (Fc-gamma RIIB) antibody receptors on myeloid cells. Mice deficient in the inhibitory receptor showed enhanced ADCC. In contrast, mice deficient in activating Fc receptors as well as antibodies engineered to disrupt Fc binding to those receptors were unable to arrest tumor growth in vivo [12]. Consistent with the prediction that

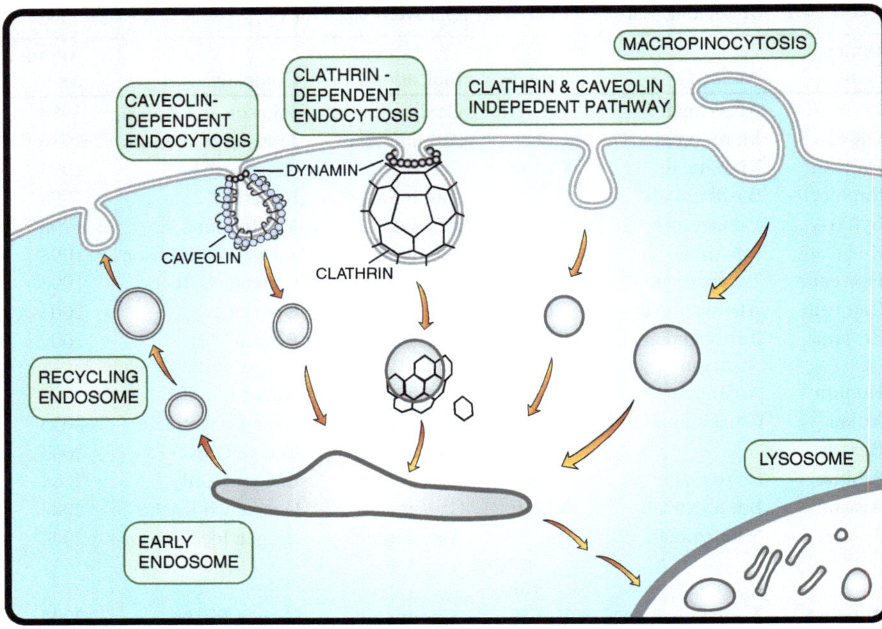

Fig. 17.1 Mechanisms underlying the antitumor actions of therapeutic antibodies. Both immunological (ADCC and CDC) and non-immunological mechanisms are schematically depicted, and examples of specific mAbs are indicated. Note that the immunological mechanisms depend on an intact Fc portion of the therapeutic antibody, which recruits natural killer (NK) lymphocytes and components of the complement. Antibody-mediated blockade of the vascular endothelial growth factor (VEGF) can inhibit angiogenesis. Likewise, anti-receptor antibodies either interfering with growth factor binding or accelerating receptor downregulation may promote growth arrest or induce apoptosis

an optimal antibody against tumors would bind preferentially to activation Fc receptors and minimally to the inhibitory partner (Fc-gamma RIIB), the presence of the 158V allotype of Fc-gamma RIIIa, which displays a higher affinity for human IgG1 and increased ADCC, associated with a greater objective response rate in a group of follicular non-Hodgkin lymphoma patients treated with rituximab (as compared with the more frequent 158F allotype) [13]. Similarly, Fc-gamma receptor polymorphisms were found to be associated with clinical outcome of patients with breast cancer who received trastuzumab: the Fc-gamma RIIIa-158V/V genotype was significantly correlated with objective response rate and progression-free survival in a group of 54 patients with HER2-amplified breast cancer receiving trastuzumab plus taxane [14]. According to another line of evidence, combining rituximab with chemotherapy and GM-CSF (granulocyte-macrophage colony-stimulating factor), a cytokine that causes granulocyte and monocyte expansion and increases CD20 expression, is beneficial due to enhanced ADCC [15].

17.2.2 Complement-Dependent Cytotoxicity

The complement system comprises a group of proteins, which are circulating as inactive precursors. The precursors represent a series of zymogen activators (C1–C9) or inhibitory proteases, able to initiate an amplifying cascade of cleavage events leading to activation of the membrane attack complex (MAC), which exerts cytolytic effects on targets cells. Although there is some evidence for the contribution of CDC to antitumor activity of mAbs in carcinoma, its involvement in hematologic malignancies is more pronounced, probably due the accessibility of components of the complement. Alemtuzumab, a humanized anti-CD52 monoclonal antibody, which is clinically effective on B-cell chronic lymphocytic leukemia (B-CLL), cannot induce ADCC, but it is able to initiate a CDC response [16]. Likewise, two anti-CD20 antibodies, rituximab and ofatumumab, can cause CDC-mediated cell killing in vitro [17]. Using a non-immunodeficient mice model of murine lymphoma ectopically expressing human CD20, it was shown that rituximab fully protected animals in an NK cell-independent manner, in line with the finding that knockout animals lacking C1q, the first component of the classical complement pathway, derived no benefit from rituximab [18]. Consistent with a critical role for CDC in rituximab action, it was demonstrated that C1qA polymorphisms affects the clinical response to rituximab therapy of follicular lymphoma [19]. Likewise, it was shown that complement depletion (using cobra venom factor) markedly reduced the efficacy of rituximab in lymphoma xenograft models, but such depletion had no effect on the therapeutic activity of another mAb that does not cause CDC [20]. Importantly, equivalent immunotherapy occurred in the presence or absence of NK cells, and $F(ab')_2$ fragments of the other mAb were able to provide substantial immunotherapy, indicating that both CDC and non-Fc-dependent mechanisms are involved in immunotherapy in this lymphoma model.

17.3 Target Cell-Dependent Mechanisms

Diverse target cell-dependent mechanisms have been shown to mediate the cytotoxic effect of antitumor mAbs, including inhibiting cellular proliferation, decreasing angiogenesis, and accelerating receptor internalization. Presumably, the majority of mAbs act through several direct pathways, in addition to their ability to recruit the immune system. An interesting example is provided by mAbs to EGFR. EGFR can bind seven different growth factors, but both cetuximab and panitumumab are able to inhibit binding of all ligands. Both antibodies are approved for colorectal cancer, and it seems that neither is effective on tumors expressing a mutant form of KRAS [21]. Because RAS acts downstream of EGFR, it is likely that cetuximab and panitumumab inhibit tumor growth by displacing the natural ligands of EGFR, thereby preventing cell survival, especially when the mAb is combined with chemotherapy. In line with ADCC-independent mechanisms, panitumumab, a fully human IgG2 molecule, cannot mount a robust ADCC response; hence, other mechanisms might be involved in the therapeutic effects of panitumumab. Trastuzumab, an antibody approved for breast and gastric cancer, presents a very different case. This antibody targets HER2, a ligand-less kin of EGFR, which is overexpressed in 10–25 % of breast, gastric, and other tumor types. Because no known ligand can bind with HER2, other mechanisms of trastuzumab action have been proposed. For example, the mAb might inhibit cleavage of HER2, which normally generates an intracellular, catalytically active fragment called $p95^{HER2}$ [22, 23]. Alternatively, antibody treatment might disrupt ligand-independent HER2/HER3 interactions and uncouples the heterodimeric receptor complex from PI3K activity, leading to downregulation of AKT signaling [24].

Enhancing apoptosis and interfering with mechanisms controlling the cell cycle in tumors are considered major routes of mAb action. Apoptosis is the process of programmed cell death in which dying cells undergo disassembly into apoptotic bodies engulfed by phagocytes. There are two main routes to apoptosis, the extrinsic (triggered by cytokines) and intrinsic pathways, and both involve activation of the caspase cascade, a series of proteases that cleave effector proteins essential for the apoptotic process. Several therapeutic mAbs have been shown to activate the intrinsic apoptotic pathway: rituximab can activate apoptosis either by accelerating calcium fluxes or by direct induction of the intrinsic pathway [25]. Trastuzumab has been shown to induce apoptosis by both inhibiting signaling pathways downstream to HER2/ErbB-2, as well as by causing TRAIL-induced apoptosis [26]. Apparently, by stimulating the AKT pathway, HER2 overexpression inhibits apoptosis induction by the tumor necrosis factor [27], but downregulation of HER2 reduces AKT activity and sensitizes tumor cells to both TNF and TRAIL. In analogy, in vitro studies that used head and neck cancer cells showed that another mAb, cetuximab (anti-EGFR), can induce accumulation of cells in G1, and this is accompanied by a decrease in the percentage of cells in S phase. In parallel, the antibody increased expression of the pro-apoptotic protein Bax and decreased abundance of the anti-apoptotic regulator Bcl-2 [28].

Inhibition of the cell cycle is another prominent mode of action by which mAbs are able to affect tumor cells. The cell cycle is regulated by cyclin-dependent kinases (CDKs), which are activated by cyclins and inhibited by specific proteins, such as $p27^{Kip}$, which arrests cells in the G1/S phase. Rituximab can cause cell cycle arrest by upregulating the expression of $p27^{Kip}$ [29]. Similarly, the effect of cetuximab on cell cycle progression has been attributed to $p27^{Kip}$-mediated G1 arrest [30], and trastuzumab-treated mammary tumor cells undergo G1 arrest by a mechanism that leads to release of $p27^{Kip}$ from inhibitory proteins [31]. While excessive proliferation of cancer cells allows expansion of tumor mass, the ability of tumors to induce angiogenesis, thereby receive nutrients and oxygen, is a major limiting factor. Consequently, inhibition of the angiogenic process holds great benefits in restricting tumor growth [32]. One such mAb, bevacizumab, which targets all isoforms of the vascular endothelial growth factor A (VEGF-A), has been approved in combination with chemotherapy for treatment of colorectal cancer. Transcription of VEGF mRNA is induced by low oxygen conditions (hypoxia), as well as by different growth factors and cytokines, including EGF family members. Accordingly, when tested in mice, trastuzumab induced normalization of the vasculature in an experimental human breast tumor [33]. The underlying mechanism might involve mAb-induced downregulation of VEGF production by tumor cells [34].

Another target cell-dependent mechanism, by which certain mAbs might exert their therapeutic effects, is antibody-mediated receptor internalization. While ligand-induced internalization of activated receptors represent a relatively well understood physiological process, which is commonly regarded as a major desensitization step [35], the process of antibody-mediated internalization is less characterized. Because the clinical implications of this process are increasingly recognized, we devote the rest of the chapter to the current knowledge pertaining to this potential mode of antibody action.

17.4 A Primer to Receptor Endocytosis

While the plasma membrane provides an efficient barrier, it also needs to act as a sensor of extracellular signals. Several mechanisms have evolved to allow such sensing, the most notable of which is membrane-bound proteins that specifically interact with their cognate ligands. These receptors fall into several families, such as G protein-coupled receptors (GPCR), receptor tyrosine kinases (RTK), integrins, and ion channels. Receptor activation takes place following binding of specific ligands, thus allowing transfer of extracellular signals to the cellular interior. In addition to structural changes, receptor activation involves posttranslational modifications, such as phosphorylation and ubiquitination. This transfer of information is commonly followed by an adaptation to the new activated state, and this entails primarily receptor internalization and degradation in lysosomes ("downregulation"). Alternatively, the internalized receptor can be destined to recycling through the endocytic machinery (reviewed in [36, 37]).

Although the majority of endocytic pathways converge into a common vesicular compartment, called the late endosome or the multi-vesicular body, they differ in their mechanisms and cargo specificities (see Fig. 17.2). There are five main endocytic routes that follow receptor activation: clathrin-mediated endocytosis (CME), caveolae-mediated endocytosis (CavME), clathrin- and caveolin-independent endocytosis, macropinocytosis, and phagocytosis. While the first three modes correspond to relative small membrane patches, the latter two pathways involve internalization of large patches (>1 μM) and will not be discussed in the context of antibody-induced endocytosis. Originally identified by studying the uptake of low-density lipoprotein, CME is initiated by receptor clustering in clathrin-coated pits [38]. This is followed by membrane invagination, vesicle scission, and fusion with early endosomes (EE). Two groups of GTPases act in this process: dynamins (which promote membrane constriction and fission) and RAB family members, which promote vesicle budding, trafficking, and fusion [39]. The second mode of endocytosis is instigated in specialized forms of lipid rafts enriched in caveolin. Dynamins, as

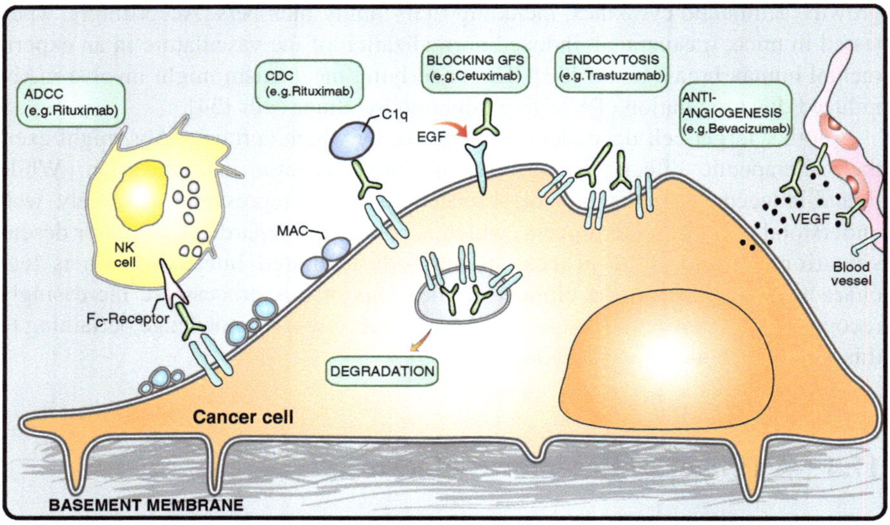

Fig. 17.2 Pathways of receptor endocytosis. A schematic diagram of the major routes of receptor endocytosis and their dependencies on dynamins or on coat proteins (clathrin and caveolin). Note the convergence of pathways at the level of early endosomes. Two versions of the clathrin- and caveolin-independent pathway exist, either a dynamin-dependent or a dynamin-independent route, but only the latter is shown

well as protein kinase C and Src family kinases drive membrane budding into larger caveosomes (caveolin-enriched organelles), which eventually recycle or fuse with EEs, thereby converge with the CME pathway [40]. The third route of endocytosis involves neither clathrin nor caveolin, but in some instances it utilizes dynamin and the small GTPase RHOA [41]. The formed intracellular vesicles are termed clathrin- and caveolin-independent carriers, which also converge into EEs, albeit in some cases this process involves an intermediate termed GPI-AP-enriched early endosomal compartments (GEECs) [42]. Hence, virtually all internalized vesicles converge at the EE, an organelle characterized by both RAB5 and the early endosomal antigen-1 (EEA1).

17.5 Antibody-Mediated Receptor Internalization

Antibodies directed at various surface antigens variably initiate receptor endocytosis, and this may modulate growth of cancer cells (see a summary list in Table 17.2). The nature of the antigen, the specific epitope engaged by the antibody, as well as antibody valence critically determine the rate and route of receptor endocytosis, as well as the fraction of internalized receptors that undergo recycling. For example, when engaged by mAbs, RTKs are commonly internalized and rapidly recycled, but antibody-mediated endocytosis of CD20 is variable and

Table 17.2 Monoclonal antibodies capable of inducing antitumor effects in model systems

Target	Antibody	Type	Effect	Mechanism	References
Alpha3/beta1 integrin	SW1	scFv	Increased cell adhesion and migration		[1]
Transferrin receptor	42/6	Mouse mAb	In vitro cell growth inhibition	Ligand mimetic, causes iron deprivation	[2]
	OKT9	Mouse mAb	In vitro cell growth inhibition	Ligand mimetic, causes iron deprivation	[3]
	RI7 208	Mouse mAb	In vitro cell growth inhibition	Ligand mimetic, causes iron deprivation	[4]
	Anti Rat-TfR IgG3-AV	Mouse mAb conjugated to avidin	Apoptosis in vitro	Ligand mimetic, causes iron deprivation	[5]
	H7	scFv	In vitro cell growth inhibition	Ligand mimetic, causes iron deprivation	[6]
	F12CH	scFv	Apoptosis and autophagic cell death in vitro. Decreased growth of erythroleukemia xenografts	Ligand mimetic, causes iron deprivation	[7]
	A24	Mouse mAb	Apoptosis in ATL cells ex vivo	Ligand mimetic, routing of TfR to degradation. Causes iron deprivation	[8]
	A24	Mouse mAb	Decreased tumor growth and tumor establishment is MCL xenografts	Ligand mimetic, routing of TfR to lysosomes. Causes iron deprivation	[9]

(continued)

Table 17.2 (continued)

Target	Antibody	Type	Effect	Mechanism	References
IGF-IR	MAB 391	Mouse mAb	Reduced anchorage-independent growth in vitro	Lysosomal degradation	[10]
	EM164	Mouse mAb	Cell cycle arrest and apoptosis in vitro. Suppression of growth in pancreatic xenografts	Antagonist	[11]
	A12 (cixutumumab)	Human mAb	Cell cycle arrest and apoptosis in vitro. Suppression of growth in pancreatic, breast and colon xenografts	Antagonist	[12]
		scFv-Fc	Decreased proliferation in vitro. Decreased IGF-IR expression in breast tumor xenografts	Agonist, induces degradation	[13]
	CP-751871 (figitumumab)	Human mAb	Inhibits breast and colorectal cancer tumor cell growth in vivo	Antagonistic	[14]
	19D12	Human mAb	In vitro cell growth inhibition and decreased growth in ovarian tumor cell xenografts	Antagonistic, induces ADCC	[15]
	H7C10	Humanized mAb	In vitro cell growth inhibition and decreased growth in breast and NSCLC tumor cell xenografts	Antagonistic, induces ADCC	[16]
	h10H5	Humanized mAb	In vitro cell growth inhibition and decreased growth in neuroblastoma tumor cell xenografts	Antagonist, decreases glucose uptake	[17]
	11A4, 9A2, 8A1	Human mAb	In vitro and in vivo cell growth inhibition	Antagonist	[18]
IR	EM164	Mouse mAb	Decreased growth in breast tumor cell xenografts	IGF-IR targeting mAb casing degradation of IR residing in lipid rafts	[19]

EGFR	C225 (Cetuximab)	Chimeric mAb	Inhibition of prostate tumor cell growth in vitro and in vivo	Antagonist	[20, 21]
	Sym004 (992+1024)	Combination of two human mAbs	In vitro cell growth inhibition and decreased growth in head and neck, NSCLC, and carcinoma tumor cell xenografts	Degradation of EGFR	[22]
ErbB2	4D5 (Trastuzumab)	Humanized mAb	Inhibition of proliferation in vitro and in vivo	Agonist	[23]
	L26	Mouse mAb	Inhibition of gastric cell growth in vivo	In trans-inhibition of ligand binding	[24]
	2C4 (Pertuzumab)	Humanized mAb	Inhibition of tumor cell growth in vivo	Inhibition of heterodimerization with EGFR	[25]
	L26+4D5, L26+N12, L431+N12	Mouse mAbs+humanized mAb	In vitro cell growth inhibition and decreased tumor growth in gastric cancer in vivo	Internalization of receptor by cross-linking at the cell membrane	[26, 27]
ErbB3	MP-RM1	Mouse mAb	In vitro and in vivo cell growth inhibition of melanoma and prostate cancer cells	Antagonist, inhibition of heterodimerization	[28]

A list of selected experimental cancer therapeutic mAbs. Antibody's name, effect in vitro and/or in animals, proposed mechanisms of action and the respective cell surface antigen are indicated

seems uncoupled from recycling [43]. The internalizing capacities of mAbs are relevant to cancer therapy for several reasons. For one, covalent conjugates of cytotoxic molecules and mAbs, such as an antibody to CD22 [44] or an antibody to HER2/ERBB-2 [45], can be delivered into cancer cells by means of receptor-mediated endocytosis. Below we focus on another therapeutic scenario, namely, the variable ability of naked mAbs to directly modulate tumor cell growth while undergoing endocytosis.

17.5.1 Antibodies to Integrins

Integrins comprise a family of transmembrane receptors present as heterodimers of alpha and beta subunits. There are 19 alpha subunits and 8 beta subunits, together yielding 25 or more different heterodimers. Integrins are expressed in almost all cell types, and they act mainly as adhesion molecules mediating cell-cell and cell-extracellular matrix (ECM) interactions [46]. Upon ligand binding, integrins are able to transduce signals into the cell by recruiting adaptor and signaling proteins. Additionally, integrins transmit "inside-out" signals by means of conformational switches. Acting as ECM attachment sites, integrins are involved in diverse processes, including morphogenesis and tumor progression. For example, the receptor for laminin, $\alpha3\beta1$ integrin, is involved in kidney and lung development [47]. This integrin regulates cell migration and wound healing [48], as well as angiogenesis [49]. Once activated, $\alpha3\beta1$ can promote pro-tumorigenic gene expression in breast cancer cells, which identifies the respective integrin as a potential therapeutic target [50]. Accordingly, a phage display screen aimed at identifying tumor-specific single-chain variable fragments (scFv), isolated three clones, two of them recognized $\alpha3\beta1$ integrin and caused receptor internalization [51]. Interestingly, one of the clones was studied in details and found to induce functional effects, such as ligand mimetic cell adhesion and migration. It is worth noting that enhanced integrin trafficking, which depends on the RAB-coupling protein (RCP), is implicated in p53-mediated progression of carcinomas [52], probably because persistent migration necessitates rapid turnover of integrin-based adhesion sites [53]. Whether or not scFv constructs, which target specific integrins and undergo internalization, could have significant impact on cancer therapy remains to be seen.

17.5.2 Antibodies to the Transferrin Receptor (TfR)

The importance of iron for cell growth and metabolism has long been appreciated; iron is bound by circulating transferrin molecules and its association with cells is mediated by the TfR. This type II transmembrane protein is abundantly expressed on the surface of proliferating cells, which underscores potential utilization for cancer therapy. On binding of iron-loaded transferrin, the receptor internalizes via

CME into endosomes, but upon acidification the iron is released intracellularly. Afterwards, the iron-free transferrin and TfR recycle back to the cell surface [54]. A mAb able to internalize TfR has been reported as early as in 1982 [55]. The mAb was able to block binding of transferrin and consequently inhibited cell growth. Following this study, additional mAbs were generated, including an antibody able to bind at a site distinct from the ligand-binding site, but still capable of inducing rapid receptor internalization [56]. Using phage display and screening for scFv fragments internalizing into the HER2-overexpressing breast tumor cell line SKBR3, three internalizing scFv were identified, two specific to HER2 and one to TfR [57]. The mechanism of TfR internalization involved activation of the receptor by acting as ligand mimetics. More recent studies identified additional scFvs that bind and internalize TfR by acting as ligand mimetics and competing with transferrin [58]. When these monovalent scFvs were converted to a bivalent antibody format, their growth inhibitory effects in vitro and in a xenograft model of erythroleukemia were markedly increased.

Another approach to exploit TfR entails the identification of receptor overexpressing malignancies. For instance, Oliver Hermine and his colleagues focused on adult T-cell leukemia/lymphoma (ATL), an aggressive disease associated with the human T-cell lymphotropic virus type-1 (HTLV-1). Due to TfR overexpression, they employed a mAb able to displace iron-bound transferrin and observed apoptosis of ATL tumor cells [59]. Apparently, the cytotoxic effect was achieved by inducing TfR internalization and rerouting from the recycling pathway, thereby causing iron deprivation. Another malignancy they targeted is mantle cell lymphoma (MCL) [60], an aggressive non-Hodgkin's lymphoma subtype, highly expressing TfR. Injection of the mAb to mice with preestablished MCL xenografts decreased growth rate and also prevented tumor establishment when the mAb was injected together with the tumor cells. The antitumorigenic effects were attributed to apoptosis of the TfR-expressing cells and mAb-induced rapid endocytosis of the receptor. Interestingly, the mAb-bound TfR was routed to lysosomes in a CME-dependent process, thus halting recycling and eventually causing detrimental iron deficiency. In conclusion, TfR has emerged as a potential drug target, especially for cancer types highly expressing TfR; inhibition of this receptor halts proliferation due to the dependence of proliferating cells on iron uptake.

17.5.3 Antibodies to Type I Insulin-Like Growth Factor Receptor

The receptor for the insulin-like growth factor 1 (IGF-IR) belongs to the RTK family and shows high homology to the insulin receptor. While initially thought to be functionally redundant with the insulin receptor, it is clear that IGF-IR has many unique roles in regulating cellular proliferation and apoptosis. For example, IGF-IR plays an important role in cancer progression by promoting mitogenesis and inducing anti-apoptotic effects [61]. Upon ligand binding, the intracellular kinase is activated via autophosphorylation, causing conformational changes and recruitment of various substrates, which the IGF-IR later phosphorylates, thereby the signal undergoes amplification and executes transcriptional responses. Considerable evidence has been accumulated attributing pivotal roles for IGF-IR in lung, breast, prostate, gastric, and colon tumors, to name a few. In these cases overexpression of either IGF-IR itself or one of its ligands, IGF-I or IGF-II, has been observed, in line with autocrine loops [62]. Relative to other RTKs, the activated IGF-IR is considered a stable protein, with slow CME internalization rates that require both receptor phosphorylation and ubiquitination.

Antibodies targeting IGF-IR have already been described in the early 1980s. However, the observation that antibodies can accelerate receptor degradation, in addition to blocking receptor functions, originated from experiments that employed breast cancer cells treated with a neutralizing mAb [63]. While short-term treatments (~1 h) inhibited autophosphorylation and signaling, longer incubations (>4 h) resulted in receptor degradation, which was mediated in part by lysosomes. Functionally, treatment with the mAb reduced anchorage-independent growth, underpinning its therapeutic potential. Similarly, another mAb, EM164, was found to antagonize IGF-IR signaling by inhibiting ligand binding [64]. Unlike treatment with the ligand, upon binding to IGF-IR, the mAb caused receptor downregulation, as well as inhibited cell proliferation due to cell cycle arrest and induction of apoptosis. When tested in vivo, EM164 suppressed the growth of pancreatic cancer xenografts, both as single agents and more potently in combination with chemotherapy. Yet another antibody, A12 (cixutumumab), was identified in a phage display screen [65]. An interesting cross talk between the IGF-IR and insulin receptor (IR) was uncovered when cells treated with EM164 were checked for IR levels [66]: EM164 induced internalization not only of IGF-IR but also of IR in both cell culture models and in xenografts. Moreover, the co-internalization process was localized to lipid rafts, in which IGR-IR and IR normally reside. Disruption of lipid rafts partially rescued the downregulation of IR but did not affect IGF-IR, indicating that the internalization mechanism was not dependent on lipid rafts. In analogy to EM164, mAb A12 also inhibits receptor activation and downstream signaling, concomitant with internalization and degradation. Along with in vitro growth inhibition, the authors assessed the in vivo antitumor effects of this antibody in breast, pancreatic, and colon cancer models. In all cases, A12 treatment exerted over 70 % inhibition of tumor growth.

Although both studies showed that antibody-mediated IGF-IR internalization does not require receptor activation, an scFv fused to a human Fc domain (scFv-Fc) that acts as an agonist, was also shown to induce downregulation of IGF-IR [67]. This scFv-Fc activated IGF-IR in a fashion identical to IGF-I and induced downstream signaling but also partially inhibited xenograft growth of mammary cancer cells. This discrepancy might be explained by a refractory period exerted by the antibody, due to receptor internalization and degradation. Even though the antibody acts as an agonist similar to IGF-I, unlike this natural ligand, the scFv-Fc causes more effective internalization and degradation of the receptor, thereby inhibiting further activation and decreasing cell proliferation.

Interestingly, unlike the relatively slow IGF1-induced downregulation of IGF-IR and weak ubiquitination, antibody-induced receptor degradation and ubiquitination are more robust [68]. Two lysine residues in the IGF-IR activation loop (Lys-1138 and Lys-1141) were mapped, which nucleate polyubiquitin chains through both Lys-48 and Lys-29 linkages. Mutation of these ubiquitinated lysine residues resulted in decreased mAb-induced IGF-IR internalization and downregulation, as well as a reduced cellular response to mAb treatment. Interestingly, these sites were weakly ubiquitinated upon IGF1 stimulation. Moreover, efficient receptor endocytosis and delivery to lysosomes necessitated ubiquitination of these sites. Importantly, cancer cell lines that do not undergo ubiquitination and internalization upon mAb treatment were identified, which likely identifies a mechanism that confers resistance to anti-IGF-IR antibodies. In summary, it appears that antibody-mediated internalization of the IGF-IR is a major mode of action enabling antitumor effects by the aforementioned mAbs, as well as by additional anti-IGF-IR antibodies [69–71], including fully human antibodies, which are currently being tested on cancer patients.

17.5.4 Antibodies to ErbB-Family Receptors

The founder of the RTK family, EGFR (also called ERBB-1), belongs to a subfamily, which also includes HER2/ERBB-2, a ligand-less receptor [72], ERBB-4, and ERBB-3, a receptor characterized by an extremely weak kinase activity [73]. Because the four receptors can differentially bind 11 different growth factors and they can form up to 10 different homo- and heterodimeric complexes, each binds a unique combination of cytoplasmic signaling proteins, it has been proposed that the ERBB group functions within a framework of a layered signaling network [74]. Importantly, the ERBB network is frequently involved in human malignancies, such as various carcinomas and brain tumors. For example, EGFR has been found to be overexpressed, amplified, or mutated in multiple human tumors, including cancers of the breast, head and neck, lung, colon, ovary, and brain tumors of glial origin [75]. Likewise, amplification of the gene encoding HER2/ERBB-2 was originally observed in both breast and ovarian tumors [76], and this observation was later extended to other types of solid tumors. Consistent with a driving role of the ERBB network in cancer progression [77], kinase inhibitors directed at EGFR/ERBB-1 have been approved already in 2004 to treat lung tumors expressing mutant forms of EGFR [78]. Likewise, mAbs directed at EGFR/ERBB-1 are routinely used in the treatment of colorectal and head and neck cancer [79]. In analogy, lapatinib, a drug able to inhibit the kinase activity of HER2/ERBB-2, is approved for treatment of breast cancer with amplification of the HER2 gene, while trastuzumab, a humanized mAb specific to HER2/ERBB-2, is approved for the same clinical indication. Unfortunately, resistance to these molecular targeted therapies inevitably evolves in patients, which has motivated a major current effort to understand the underlying mechanisms and accordingly apply drug combinations able to delay the onset of resistance. One recent example is a combination of trastuzumab and pertuzumab, another anti-HER2 antibody, which prolonged progression-free survival of HER2-positive breast cancer patients [80].

17.5.5 Antibodies to HER2/ERBB-2

Preventing posttranslational insertion of the rodent form of HER2 within the plasma membrane, by using a fusion protein approach, abolished its transforming potential [81]. Likewise, intracellular antibodies able to arrest HER2 in the endoplasmic reticulum prevented cellular transformation [82], which established the notion that antibody-mediated removal of HER2/ERBB-2 from the plasma membrane might reduce oncogenicity. The first mAb to HER2/ERBB-2 was generated in mice immunized with fibroblasts ectopically overexpressing the oncogenic rat homolog of HER2 (called NEU) [83]. Later studies showed that mAb treatment reversed the transformed phenotype of NEU-transformed fibroblasts, as determined by anchorage-independent growth, and this was associated with the rapid and reversible loss of both cell surface and total cellular NEU protein [83]. While each mAb of a set of anti-NEU antibodies was able to cause partial eradication of tumors, the administration of mixtures of antibodies reactive with two distinct regions on the p185-NEU molecule resulted in synergistic antitumor effects and eradication of tumors [84]. Synergistic effects of mAbs to the human HER2/ERBB-2 protein on xenografts of human breast cancer cells were later confirmed [85]. In vitro, the more effective mAb mixture was also more effective than the single mAbs in inducing ADCC and CDC, inhibiting cell growth, inducing apoptosis, and inhibiting the secretion of the VEGF. Likewise, synergistic antitumor effects were confirmed using another set of mAbs to HER2 [86]. Moreover, it was concluded that pairs comprising an antibody reactive with the dimerization site of HER2/ERBB-2 (similar to pertuzumab) and an antibody recognizing another distinct epitope better than other pairs inhibit HER2-overexpressing tumors. Because the superiority of antibody combinations extended to tumor cell cultures, the authors assumed that non-immunological mechanisms contribute to mAb synergy. For example, they demonstrated an ability of mAb combinations to remarkably enhance endocytosis, ubiquitination, and intracellular degradation of HER2/ERBB-2. Yet another potential mechanism of synergy emerged from a study that combined trastuzumab and pertuzumab in an animal model [87]. The authors proposed that the strongly enhanced antitumor activity was mainly due to the differing but complementary mechanisms of action of pertuzumab and trastuzumab, namely, inhibition of HER2 dimerization (by pertuzumab) and prevention of HER2 cleavage (by trastuzumab), which generates p95HER2, a disregulated intracellular kinase fragment.

Presumably, depending on their specific epitope on HER2, mAbs act by either intercepting signaling pathways or by weakly inducing receptor endocytosis when singly applied. This notion might explain early reports demonstrating the ability of specific mAbs, including 4D5, the father of trastuzumab, to activate NEU/HER2 and consequently downregulate receptor abundance [88–90]. Nevertheless, several reports concluded that trastuzumab cannot downregulate HER2/ERBB-2. For example, morphological analyses showed that trastuzumab does not influence HER2/ERBB-2 distribution but instead recycles passively with the internalized protein [91]. Similarly, another study observed no antibody-mediated depletion of

HER2 but did observe a reduction in phosphorylation of AKT, probably due to recruitment of PTEN to the plasma membrane [92]. In line with this model, it was shown that trastuzumab treatment disrupts ligand-independent HER2/HER3 interactions, leading to downregulation of AKT signaling [24]. Yet another study that questioned an endocytosis-related mechanism employed two anti-ERBB-2 scFv molecules that induced receptor internalization [93]. The authors found no correlation between the extent of HER2/ERBB-2 internalization and inhibition of cell growth.

In line with an epitope-specific mode of action, screening of a battery of mAbs to HER2/ERBB-2 for their ability to elicit receptor internalization, phosphorylation, and growth inhibition uncovered two potential mechanisms: one class of mAbs accelerated endocytosis whereas another class blocked heterodimerization with other ERBBs [94]. Interestingly, endocytosis and degradation of HER2/ERBB-2 might depend on the ability of specific mAbs to induce ubiquitination of HER2 by recruiting the CBL ubiquitin ligase [95]. Unlike the endocytosis-inducing mAbs, the action of which might be limited to HER2-overexpressing cancer cells, mAbs able to block formation of ligand-induced heterodimers can block proliferation of cancer cells, such as prostate cancer cells, which express moderate levels of HER2 [96]. Interestingly, because HER2 decelerates the rate of EGFR endocytosis in the context of EGFR-HER2 heterodimers [97], by blocking heterodimer formation, the latter class of mAbs might act also by removing EGFR and other growth factor receptors from the cell surface [98]. In conclusion, mAb-induced downregulation of HER2 represents only one out of a list of potential mechanisms of HER2-targeted immunotherapy. Nevertheless, this mechanism can be augmented by applying appropriate mixtures of mAbs, especially those able to block heterodimer formation by HER2 and synergistically sort the oncoprotein for lysosomal degradation.

17.5.5.1 Antibodies to EGFR/ERBB-1

The clinically approved mAbs directed against EGFR, cetuximab, and panitumumab are ligand competitive, meaning that they can displace EGF-like growth factors and hence inhibit receptor activation and downstream signaling [99]. In-depth analyses of the biological effects of cetuximab uncovered multiple in vitro mechanisms, including cell cycle arrest, inhibition of angiogenesis, invasion and metastasis, as well as induction of apoptosis. It is notable that cetuximab induces dimerization of EGFR, without activation of the intrinsic tyrosine kinase, resulting in receptor downregulation, and this effect appears to be important for its growth inhibitory capacity [100]. Experiments that employed a radio-labeled form of cetuximab confirmed endocytosis of the mAb, albeit with a slower rate than that of EGF, but the internalized mAb recycled more effectively than internalized EGF [101]. In addition, internalization of the mAb, in contrast to that of EGF, was independent of the receptor's tyrosine kinase activity, in line with distinct endocytic processing of EGF and cetuximab.

In similarity to the synergistic internalizing effects of combinations of mAbs directed at HER2/ERBB-2, certain pairs of anti-EGFR antibodies accelerate receptor endocytosis and degradation [102]. To enhance endocytosis, the mAbs must engage non-overlapping antigenic epitopes of EGFR. Interestingly, the mechanism of endocytosis appears distinct from the one underlying EGF-induced receptor downregulation: no kinase activity or cytoplasmic domains of EGFR were necessary. The authors proposed a "lattice model," attributing robust endocytosis to the size of the cluster of receptors engaged by the combination of mAbs. Another study showed that highly potent mAb combinations reduced surface receptor levels by up to 80 % with a halftime of 0.5–5 h, through a mechanism consistent with mAb-mediated inhibition of EGFR recycling [103]. Like anti-EGFR mAbs in current clinical use, Sym004, a mixture of two anti-EGFR mAbs inhibited cancer cell growth and survival by blocking ligand binding and receptor activation [104]. However, unlike the other antibodies, Sym004 induces rapid and efficient EGFR downregulation and superior antitumor efficacy in vivo. In conclusion, clinical development of Sym004 and other mAb combinations may herald a departure from a monoclonal to oligo-clonal mixtures of mAbs able to effectively control tumor growth in patients.

17.6 Concluding Remarks

Antibody-induced endocytosis of oncogenic (or survival-mediating) antigens emerges as a common process, which might contribute to cancer therapy, along with other immunological and non-immunological mechanisms. The relative contribution of each mechanism to therapy seems to depend on the nature of the antigen; antibody-induced downregulation of RTK (e.g., IGF-IR and HER2) is especially effective. Currently, the identity of the endocytic route(s) and the respective molecular drivers are largely unknown. Likewise, the necessity of specific cytoplasmic domains of the internalizing receptor is not understood, let alone potential post-translational modifications (e.g., ubiquitination and phosphorylation). Nevertheless, the reported ability of antibody mixtures, mimicking a polyclonal antiserum, to enhance receptor endocytosis not only provides hints as to the nature of the endocytic pathway but also focuses the attention on the ill-defined recycling of internalized antigens back to the plasma membrane.

Importantly, resolving molecular mechanisms underlying antibody-based endocytosis bears clinical implications. Since the approval of the first monoclonal antibody for cancer therapy, in 1997, many additional antibodies have entered routine application in oncology institutions, but response rates and durations are far from being satisfactory. Both primary resistance of patients as well as resistance that evolves following several months of patient treatment with a monoclonal antibody severely limit pharmacological efficacy. In-depth understanding of antibody-induced endocytosis is expected to uncover ways to enhance cancer immunotherapy, strategies to better combine immunotherapy with chemotherapy, as well as means to delay the onset of patient resistance to specific antibodies.

References

1. Kohler G, Milstein C (1975) Continuous cultures of fused cells secreting antibody of predefined specificity. Nature 256:495–497
2. Drebin JA, Link VC, Stern DF, Weinberg RA, Greene MI (1985) Down-modulation of an oncogene protein product and reversion of the transformed phenotype by monoclonal antibodies. Cell 41:697–706
3. Hudziak RM, Lewis GD, Winget M, Fendly BM, Shepard HM et al (1989) p185HER2 monoclonal antibody has antiproliferative effects in vitro and sensitizes human breast tumor cells to tumor necrosis factor. Mol Cell Biol 9:1165–1172
4. Boulianne GL, Hozumi N, Shulman MJ (1984) Production of functional chimaeric mouse/human antibody. Nature 312:643–646
5. Riechmann L, Clark M, Waldmann H, Winter G (1988) Reshaping human antibodies for therapy. Nature 332:323–327
6. Fishwild DM, O'Donnell SL, Bengoechea T, Hudson DV, Harding F et al (1996) High-avidity human IgG kappa monoclonal antibodies from a novel strain of minilocus transgenic mice. Nat Biotechnol 14:845–851
7. Mendez MJ, Green LL, Corvalan JR, Jia XC, Maynard-Currie CE et al (1997) Functional transplant of megabase human immunoglobulin loc. recapitulates human antibody response in mice. Nat Genet 15:146–156
8. Scott AM, Wolchok JD, Old LJ (2012) Antibody therapy of cancer. Nat Rev Cancer 12:278–287
9. Ben-Kasus T, Schechter B, Sela M, Yarden Y (2007) Cancer therapeutic antibodies come of age: targeting minimal residual disease. Mol Oncol 1:42–54
10. Weiner LM, Surana R, Wang S (2010) Monoclonal antibodies: versatile platforms for cancer immunotherapy. Nat Rev Immunol 10:317–327
11. Iannello A, Ahmad A (2005) Role of antibody-dependent cell-mediated cytotoxicity in the efficacy of therapeutic anti-cancer monoclonal antibodies. Cancer Metastasis Rev 24:487–499
12. Clynes RA, Towers TL, Presta LG, Ravetch JV (2000) Inhibitory Fc receptors modulate in vivo cytotoxicity against tumor targets. Nat Med 6:443–446
13. Cartron G, Dacheux L, Salles G, Solal-Celigny P, Bardos P et al (2002) Therapeutic activity of humanized anti-CD20 monoclonal antibody and polymorphism in IgG Fc receptor FcgammaRIIIa gene. Blood 99:754–758
14. Musolino A, Naldi N, Bortesi B, Pezzuolo D, Capelletti M et al (2008) Immunoglobulin G fragment C receptor polymorphisms and clinical efficacy of trastuzumab-based therapy in patients with HER-2/neu-positive metastatic breast cancer. J Clin Oncol 26:1789–1796
15. Olivieri A, Lucesole M, Capelli D, Gini G, Montanari M et al (2005) A new schedule of CHOP/rituximab plus granulocyte-macrophage colony-stimulating factor is an effective rescue for patients with aggressive lymphoma failing autologous stem cell transplantation. Biol Blood Marrow Transplant 11:627–636
16. Lundin J, Kimby E, Bjorkholm M, Broliden PA, Celsing F et al (2002) Phase II trial of subcutaneous anti-CD52 monoclonal antibody alemtuzumab (Campath-1H) as first-line treatment for patients with B-cell chronic lymphocytic leukemia (B-CLL). Blood 100:768–773
17. Pawluczkowycz AW, Beurskens FJ, Beum PV, Lindorfer MA, van de Winkel JG et al (2009) Binding of submaximal C1q promotes complement-dependent cytotoxicity (CDC) of B cells opsonized with anti-CD20 mAbs ofatumumab (OFA) or rituximab (RTX): considerably higher levels of CDC are induced by OFA than by RTX. J Immunol 183:749–758
18. Di Gaetano N, Cittera E, Nota R, Vecchi A, Grieco V et al (2003) Complement activation determines the therapeutic activity of rituximab in vivo. J Immunol 171:1581–1587
19. Racila E, Link BK, Weng WK, Witzig TE, Ansell S et al (2008) A polymorphism in the complement component C1qA correlates with prolonged response following rituximab therapy of follicular lymphoma. Clin Cancer Res 14:6697–6703
20. Cragg MS, Glennie MJ (2004) Antibody specificity controls in vivo effector mechanisms of anti-CD20 reagents. Blood 103:2738–2743

21. Amado RG, Wolf M, Peeters M, Van Cutsem E, Siena S et al (2008) Wild-type KRAS is required for panitumumab efficacy in patients with metastatic colorectal cancer. J Clin Oncol 26:1626–1634
22. Xia W, Liu LH, Ho P, Spector NL (2004) Truncated ErbB2 receptor (p95ErbB2) is regulated by heregulin through heterodimer formation with ErbB3 yet remains sensitive to the dual EGFR/ErbB2 kinase inhibitor GW572016. Oncogene 23:646–653
23. Molina MA, Codony-Servat J, Albanell J, Rojo F, Arribas J et al (2001) Trastuzumab (herceptin), a humanized anti-Her2 receptor monoclonal antibody, inhibits basal and activated Her2 ectodomain cleavage in breast cancer cells. Cancer Res 61:4744–4749
24. Junttila TT, Akita RW, Parsons K, Fields C, Lewis Phillips GD et al (2009) Ligand-independent HER2/HER3/PI3K complex is disrupted by trastuzumab and is effectively inhibited by the PI3K inhibitor GDC-0941. Cancer Cell 15:429–440
25. Bubien JK, Zhou LJ, Bell PD, Frizzell RA, Tedder TF (1993) Transfection of the CD20 cell surface molecule into ectopic cell types generates a Ca2+ conductance found constitutively in B lymphocytes. J Cell Biol 121:1121–1132
26. Cuello M, Ettenberg SA, Clark AS, Keane MM, Posner RH et al (2001) Down-regulation of the erbB-2 receptor by trastuzumab (herceptin) enhances tumor necrosis factor-related apoptosis-inducing ligand-mediated apoptosis in breast and ovarian cancer cell lines that overexpress erbB-2. Cancer Res 61:4892–4900
27. Hudziak RM, Lewis GD, Shalaby MR, Eessalu TE, Aggarwal BB et al (1988) Amplified expression of the HER2/ERBB2 oncogene induces resistance to tumor necrosis factor alpha in NIH 3T3 cells. Proc Natl Acad Sci USA85:5102–5106
28. Huang SM, Bock JM, Harari PM (1999) Epidermal growth factor receptor blockade with C225 modulates proliferation, apoptosis, and radiosensitivity in squamous cell carcinomas of the head and neck. Cancer Res 59:1935–1940
29. Bezombes C, Grazide S, Garret C, Fabre C, Quillet-Mary A et al (2004) Rituximab antiproliferative effect in B-lymphoma cells is associated with acid-sphingomyelinase activation in raft microdomains. Blood 104:1166–1173
30. Peng D, Fan Z, Lu Y, DeBlasio T, Scher H et al (1996) Anti-epidermal growth factor receptor monoclonal antibody 225 up-regulates p27KIP1 and induces G1 arrest in prostatic cancer cell line DU145. Cancer Res 56:3666–3669
31. Lane HA, Motoyama AB, Beuvink I, Hynes NE (2001) Modulation of p27/Cdk2 complex formation through 4D5-mediated inhibition of HER2 receptor signaling. Ann Oncol 12(suppl 1):S21–S22
32. Ferrara N, Kerbel RS (2005) Angiogenesis as a therapeutic target. Nature 438:967–974
33. Izumi Y, Xu L, di Tomaso E, Fukumura D, Jain RK (2002) Tumour biology: herceptin acts as an anti-angiogenic cocktail. Nature 416:279–280
34. Petit AM, Rak J, Hung MC, Rockwell P, Goldstein N et al (1997) Neutralizing antibodies against epidermal growth factor and ErbB-2/neu receptor tyrosine kinases down-regulate vascular endothelial growth factor production by tumor cells in vitro and in vivo: angiogenic implications for signal transduction therapy of solid tumors. Am J Pathol 151:1523–1530
35. Zwang Y, Yarden Y (2009) Systems biology of growth factor-induced receptor endocytosis. Traffic 10:349–363
36. Platta HW, Stenmark H (2011) Endocytosis and signaling. Curr Opin Cell Biol 23:393–403
37. Sorkin A, von Zastrow M (2009) Endocytosis and signalling: intertwining molecular networks. Nat Rev Mol Cell Biol 10:609–622
38. Goldstein JL, Anderson RG, Brown MS (1979) Coated pits, coated vesicles, and receptor-mediated endocytosis. Nature 279:679–685
39. Zerial M, McBride H (2001) Rab proteins as membrane organizers. Nat Rev Mol Cell Biol 2:107–117
40. Parton RG, Simons K (2007) The multiple faces of caveolae. Nat Rev Mol Cell Biol 8:185–194
41. Lamaze C, Dujeancourt A, Baba T, Lo CG, Benmerah A et al (2001) Interleukin 2 receptors and detergent-resistant membrane domains define a clathrin-independent endocytic pathway. Mol Cell 7:661–671

42. Mayor S, Pagano RE (2007) Pathways of clathrin-independent endocytosis. Nat Rev Mol Cell Biol 8:603–612
43. Pulczynski S, Boesen AM, Jensen OM (1994) Modulation and intracellular transport of CD20 and CD21 antigens induced by B1 and B2 monoclonal antibodies in RAJI and JOK-1 cells—an immunofluorescence and immunoelectron microscopy study. Leuk Res 18: 541–552
44. DiJoseph JF, Dougher MM, Kalyandrug LB, Armellino DC, Boghaert ER et al (2006) Antitumor efficacy of a combination of CMC-544 (inotuzumab ozogamicin), a CD22-targeted cytotoxic immunoconjugate of calicheamicin, and rituximab against non-Hodgkin's B-cell lymphoma. Clin Cancer Res 12:242–249
45. Mathew J, Perez EA (2011) Trastuzumab emtansine in human epidermal growth factor receptor 2-positive breast cancer: a review. Curr Opin Oncol 23:594–600
46. Hynes RO (2002) Integrins: bidirectional, allosteric signaling machines. Cell 110:673–687
47. Kreidberg JA, Donovan MJ, Goldstein SL, Rennke H, Shepherd K et al (1996) Alpha 3 beta 1 integrin has a crucial role in kidney and lung organogenesis. Development 122:3537–3547
48. Margadant C, Raymond K, Kreft M, Sachs N, Janssen H et al (2009) Integrin alpha3beta1 inhibits directional migration and wound re-epithelialization in the skin. J Cell Sci 122:278–288
49. da Silva RG, Tavora B, Robinson SD, Reynolds LE, Szekeres C et al (2010) Endothelial alpha3beta1-integrin represses pathological angiogenesis and sustains endothelial-VEGF. Am J Pathol 177:1534–1548
50. Subbaram S, Dipersio CM (2011) Integrin alpha3beta1 as a breast cancer target. Expert Opin Ther Targets 15:1197–1210
51. Gao C, Mao S, Ronca F, Zhuang S, Quaranta V et al (2003) De novo identification of tumor-specific internalizing human antibody-receptor pairs by phage-display methods. J Immunol Methods 274:185–197
52. Muller PAJ, Caswell PT, Doyle B, Iwanicki MP, Tan EH et al (2010) Mutant p53 drives invasion by promoting integrin recycling. Cell 139:1327–1341
53. Dozynkiewicz MA, Jamieson NB, Macpherson I, Grindlay J, van den Berghe PV et al (2012) Rab25 and CLIC3 collaborate to promote integrin recycling from late endosomes/lysosomes and drive cancer progression. Dev Cell 22:131–145
54. Wang J, Pantopoulos K (2011) Regulation of cellular iron metabolism. Biochem J 434:365–381
55. Trowbridge IS, Lopez F (1982) Monoclonal antibody to transferrin receptor blocks transferrin binding and inhibits human tumor cell growth in vitro. Proc Natl Acad Sci USA 79:1175–1179
56. Weissman AM, Klausner RD, Rao K, Harford JB (1986) Exposure of K562 cells to anti-receptor monoclonal antibody OKT9 results in rapid redistribution and enhanced degradation of the transferrin receptor. J Cell Biol 102:951–958
57. Poul MA, Becerril B, Nielsen UB, Morisson P, Marks JD (2000) Selection of tumor-specific internalizing human antibodies from phage libraries. J Mol Biol 301:1149–1161
58. Crepin R, Goenaga AL, Jullienne B, Bougherara H, Legay C et al (2010) Development of human single-chain antibodies to the transferrin receptor that effectively antagonize the growth of leukemias and lymphomas. Cancer Res 70:5497–5506
59. Moura IC, Lepelletier Y, Arnulf B, England P, Baude C et al (2004) A neutralizing monoclonal antibody (mAb A24) directed against the transferrin receptor induces apoptosis of tumor T lymphocytes from ATL patients. Blood 103:1838–1845
60. Lepelletier Y, Camara-Clayette V, Jin H, Hermant A, Coulon S et al (2007) Prevention of mantle lymphoma tumor establishment by routing transferrin receptor toward lysosomal compartments. Cancer Res 67:1145–1154
61. Baserga R, Peruzzi F, Reiss K (2003) The IGF-1 receptor in cancer biology. Int J Cancer 107:873–877
62. Pollak MN, Schernhammer ES, Hankinson SE (2004) Insulin-like growth factors and neoplasia. Nat Rev Cancer 4:505–518

63. Hailey J, Maxwell E, Koukouras K, Bishop WR, Pachter JA et al (2002) Neutralizing anti-insulin-like growth factor receptor 1 antibodies inhibit receptor function and induce receptor degradation in tumor cells. Mol Cancer Ther 1:1349–1353
64. Maloney EK, McLaughlin JL, Dagdigian NE, Garrett LM, Connors KM et al (2003) An anti-insulin-like growth factor I receptor antibody that is a potent inhibitor of cancer cell proliferation. Cancer Res 63:5073–5083
65. Burtrum D, Zhu Z, Lu D, Anderson DM, Prewett M et al (2003) A fully human monoclonal antibody to the insulin-like growth factor I receptor blocks ligand-dependent signaling and inhibits human tumor growth in vivo. Cancer Res 63:8912–8921
66. Sachdev D, Singh R, Fujita-Yamaguchi Y, Yee D (2006) Down-regulation of insulin receptor by antibodies against the type I insulin-like growth factor receptor: implications for anti-insulin-like growth factor therapy in breast cancer. Cancer Res 66:2391–2402
67. Sachdev D, Li SL, Hartell JS, Fujita-Yamaguchi Y, Miller JS et al (2003) A chimeric humanized single-chain antibody against the type I insulin-like growth factor (IGF) receptor renders breast cancer cells refractory to the mitogenic effects of IGF-I. Cancer Res 63:627–635
68. Mao Y, Shang Y, Pham VC, Ernst JA, Lill JR et al (2011) Polyubiquitination of insulin-like growth factor I receptor (IGF-IR) activation loop promotes antibody-induced receptor internalization and down-regulation. J Biol Chem 286:41852–41861
69. Shang Y, Mao Y, Batson J, Scales SJ, Phillips G et al (2008) Antixenograft tumor activity of a humanized anti-insulin-like growth factor-I receptor monoclonal antibody is associated with decreased AKT activation and glucose uptake. Mol Cancer Ther 7:2599–2608
70. Runnels HA, Arbuckle JA, Bailey KS, Nicastro PJ, Sun D et al (2010) Human monoclonal antibodies to the insulin-like growth factor 1 receptor inhibit receptor activation and tumor growth in preclinical studies. Adv Ther 27:458–475
71. Cohen BD, Baker DA, Soderstrom C, Tkalcevic G, Rossi AM et al (2005) Combination therapy enhances the inhibition of tumor growth with the fully human anti-type 1 insulin-like growth factor receptor monoclonal antibody CP-751,871. Clin Cancer Res 11:2063–2073
72. Klapper LN, Glathe S, Vaisman N, Hynes NE, Andrews GC et al (1999) The ErbB-2/HER2 oncoprotein of human carcinomas may function solely as a shared coreceptor for multiple stroma-derived growth factors. Proc Natl Acad Sci USA 96:4995–5000
73. Fumin S, Telesco SE, Liu Y, Radhakrishnan R, Lemmon MA (2010) ErbB3/HER3 intracellular domain is competent to bind ATP and catalyze autophosphorylation. Proc Natl Acad Sci USA 107:7692–7697
74. Yarden Y, Sliwkowski MX (2001) Untangling the ErbB signalling network. Nat Rev Mol Cell Biol 2:127–137
75. Hynes NE, MacDonald G (2009) ErbB receptors and signaling pathways in cancer. Curr Opin Cell Biol 21:177–184
76. Slamon DJ, Godolphin W, Jones LA, Holt JA, Wong SG et al (1989) Studies of the HER-2/neu proto-oncogene in human breast and ovarian cancer. Science 244:707–712
77. Witsch E, Sela M, Yarden Y (2010) Roles for growth factors in cancer progression. Physiology (Bethesda) 25:85–101
78. Pao W, Chmielecki J (2010) Rational, biologically based treatment of EGFR-mutant non-small-cell lung cancer. Nat Rev Cancer 10:760–774
79. Baselga J (2006) Targeting tyrosine kinases in cancer: the second wave. Science 312:1175–1178
80. Baselga J, Cortes J, Kim SB, Im SA, Hegg R et al (2012) Pertuzumab plus trastuzumab plus docetaxel for metastatic breast cancer. N Engl J Med 366:109–119
81. Flanagan JG, Leder P (1988) Neu protooncogene fused to an immunoglobulin heavy chain gene requires immunoglobulin light chain for cell surface expression and oncogenic transformation. Proc Natl Acad Sci USA 85:8057–8061
82. Beerli RR, Wels W, Hynes NE (1994) Intracellular expression of single chain antibodies reverts ErbB-2 transformation. J Biol Chem 269:23931–23936

83. Drebin JA, Stern DF, Link VC, Weinberg RA, Greene MI (1984) Monoclonal antibodies identify a cell-surface antigen associated with an activated cellular oncogene. Nature 312:545–548
84. Drebin JA, Link VC, Greene MI (1988) Monoclonal antibodies reactive with distinct domains of the neu oncogene-encoded p185 molecule exert synergistic anti-tumor effects in vivo. Oncogene 2:273–277
85. Spiridon CI, Ghetie MA, Uhr J, Marches R, Li JL et al (2002) Targeting multiple Her-2 epitopes with monoclonal antibodies results in improved antigrowth activity of a human breast cancer cell line in vitro and in vivo. Clin Cancer Res 8:1720–1730
86. Ben-Kasus T, Schechter B, Lavi S, Yarden Y, Sela M (2009) Persistent elimination of ErbB-2/HER2-overexpressing tumors using combinations of monoclonal antibodies: relevance of receptor endocytosis. Proc Natl Acad Sci USA 106:3294–3299
87. Scheuer W, Friess T, Burtscher H, Bossenmaier B, Endl J et al (2009) Strongly enhanced antitumor activity of trastuzumab and pertuzumab combination treatment on HER2-positive human xenograft tumor models. Cancer Res 69:9330–9336
88. Sarup JC, Johnson RM, King KL, Fendly BM, Lipari MT et al (1991) Characterization of an anti-p185HER2 monoclonal antibody that stimulates receptor function and inhibits tumor cell growth. Growth Regul 1:72–82
89. Yarden Y (1990) Agonistic antibodies stimulate the kinase encoded by the neu protooncogene in living cells but the oncogenic mutant is constitutively active. Proc Natl Acad Sci USA 87:2569–2573
90. Hurwitz E, Stancovski I, Sela M, Yarden Y (1995) Suppression and promotion of tumor growth by monoclonal antibodies to ErbB-2 differentially correlate with cellular uptake. Proc Natl Acad Sci USA 92:3353–3357
91. Austin CD, De Maziere AM, Pisacane PI, van Dijk SM, Eigenbrot C et al (2004) Endocytosis and sorting of ErbB2 and the site of action of cancer therapeutics trastuzumab and geldanamycin. Mol Biol Cell 15:5268–5282
92. Longva KE, Pedersen NM, Haslekas C, Stang E, Madshus IH (2005) Herceptin-induced inhibition of ErbB2 signaling involves reduced phosphorylation of Akt but not endocytic down-regulation of ErbB2. Int J Cancer 116:359–367
93. Neve RM, Nielsen UB, Kirpotin DB, Poul MA, Marks JD et al (2001) Biological effects of anti-ErbB2 single chain antibodies selected for internalizing function. Biochem Biophys Res Commun 280:274–279
94. Klapper LN, Vaisman N, Hurwitz E, Pinkas-Kramarski R, Yarden Y et al (1997) A subclass of tumor-inhibitory monoclonal antibodies to ErbB-2/HER2 blocks crosstalk with growth factor receptors. Oncogene 14:2099–2109
95. Klapper LN, Waterman H, Sela M, Yarden Y (2000) Tumor-inhibitory antibodies to HER-2/ErbB-2 may act by recruiting c-Cbl and enhancing ubiquitination of HER-2. Cancer Res 60:3384–3388
96. Agus DB, Akita RW, Fox WD, Lewis GD, Higgins B et al (2002) Targeting ligand-activated ErbB2 signaling inhibits breast and prostate tumor growth. Cancer Cell 2:127–137
97. Lenferink AE, Pinkas-Kramarski R, van de Poll ML, van Vugt MJ, Klapper LN et al (1998) Differential endocytic routing of homo- and hetero-dimeric ErbB tyrosine kinases confers signaling superiority to receptor heterodimers. EMBO J 17:3385–3397
98. Hughes JB, Berger C, Rodland MS, Hasmann M, Stang E et al (2009) Pertuzumab increases epidermal growth factor receptor down-regulation by counteracting epidermal growth factor receptor-ErbB2 heterodimerization. Mol Cancer Ther 8:1885–1892
99. Mendelsohn J, Baselga J (2000) The EGF receptor family as targets for cancer therapy. Oncogene 19:6550–6565
100. Fan Z, Masui H, Altas I, Mendelsohn J (1993) Blockade of epidermal growth factor receptor function by bivalent and monovalent fragments of 225 anti-epidermal growth factor receptor monoclonal antibodies. Cancer Res 53:4322–4328
101. Jaramillo ML, Leon Z, Grothe S, Paul-Roc B, Abulrob A et al (2006) Effect of the anti-receptor ligand-blocking 225 monoclonal antibody on EGF receptor endocytosis and sorting. Exp Cell Res 312:2778–2790

102. Friedman LM, Rinon A, Schechter B, Lyass L, Lavi S et al (2005) Synergistic down-regulation of receptor tyrosine kinases by combinations of mAbs: implications for cancer immunotherapy. Proc Natl Acad Sci USA 102:1915–1920
103. Spangler JB, Neil JR, Abramovitch S, Yarden Y, White FM et al (2010) Combination antibody treatment down-regulates epidermal growth factor receptor by inhibiting endosomal recycling. Proc Natl Acad Sci USA 107:13252–13257
104. Pedersen MW, Jacobsen HJ, Koefoed K, Hey A, Pyke C et al (2010) Sym004: a novel synergistic anti-epidermal growth factor receptor antibody mixture with superior anticancer efficacy. Cancer Res 70:588–597

Index

Y. Yarden and G. Tarcic (eds.), *Vesicle Trafficking in Cancer*,
DOI 10.1007/978-1-4614-6528-7, © Springer Science+Business Media New York 2013

D

MIX
Papier aus verantwortungsvollen Quellen
Paper from responsible sources
FSC® C105338

If you have any concerns about our products,
you can contact us on
ProductSafety@springernature.com

In case Publisher is established outside the EU,
the EU authorized representative is:
Springer Nature Customer Service Center GmbH
Europaplatz 3, 69115 Heidelberg, Germany

Printed by Libri Plureos GmbH
in Hamburg, Germany